ESO ASTROPHYSICS SYMPOSIA
European Southern Observatory

Series Editor: Philippe Crane

Springer-Verlag Berlin Heidelberg GmbH

Hans Ulrich Käufl
Ralf Siebenmorgen (Eds.)

The Role of Dust in the Formation of Stars

Proceedings of the ESO Workshop
Held at Garching, Germany,
11–14 September 1995

 Springer

Volume Editors

Hans Ulrich Käufl
European Southern Observatory
Karl-Schwarzschild-Strasse 2
D-85748 Garching, Germany

Ralf Siebenmorgen
ESA
Villafranca del Castillo
Satellite Tracking Station
P.O. Box 50727
E-28080 Madrid, Spain

Series Editor

Philippe Crane
European Southern Observatory
Karl-Schwarzschild-Strasse 2
D-85748 Garching, Germany

Cataloging-in-Publication Data applied for

Die Deutsche Bibliothek – CIP-Einheitsaufnahme

The role of dust in the formation of stars : Proceedings of the ESO workshop, held at
Garching, Germany, 11–14 September 1995 / Hans Ulrich Käufl ; Ralf Siebenmorgen (ed.). –
Berlin ; Heidelberg ; New York ; Barcelona ; Budapest ; Hong Kong ; London ; Milan ;
Paris ; Santa Clara ; Singapore ; Tokyo : Springer, 1996
(ESO astrophysics symposia)

NE: Käufl, Hans Ulrich [Hrsg.]; European Southern Observatory

ISBN 978-3-662-22515-8   ISBN 978-3-540-68594-4 (eBook)
DOI 10.100/978-3-540-68594-4

Typesetting: Camera ready by authors/editors
SPIN: 10517774   55/3142-543210 - Printed on acid-free paper

Preface

This ESO workshop, which took place in September 1995 on a topic that at a first glance could be considered rather specialized, attracted an unpredictably large number of scientists. This certainly reflects the importance of this field, which has lost its seemingly esoteric character, in a wider astrophysical context.

To give as much room as possible in these proceedings to the targeted talks, no presentation of the Very Large Telescope Observatory has been included. All readers missing such a presentation are reminded that up-to-date in-depth information about the VLT status is available electronically.[1]

Papers were given concerning observations in the entire electromagnetic spectrum from x-rays to mm-waves, i.e., exceeding 22 octaves in frequency. The VLT as any ground-based optical observatory can only address at best 7 octaves. Nevertheless the VLT, most likely the only ground-based observatory specifically designed to access all these 7 octaves of the electromagnetic spectrum practically in parallel, will undoubtedly be a tool of extreme value to this field.

This workshop took place only days before *ISO*, Europe's *Infrared Space Observatory*, was successfully launched. In its lifetime, which apparently will substantially exceed all expectations, ISO will undoubtedly make important observations (and discoveries!) in this field. Combining data from ISO and the VLT, two extremely complementary observatories, will greatly enhance the understanding of star formation. It is of particular advantage that the VLT is located in the southern hemisphere, because this is the part in the sky where most galactic star forming regions are located. Moreover the VLT will allow for in-depth studies of star formation in the Magellanic Clouds. The influence of metallicity (i.e., dust abundance and composition) on star formation can hence be tested. This workshop demonstrated that the improved observational capabilities are well complemented by progress in the fields of numerical simulation and laboratory experiments. Talks were given on how magnetic fields reign the protostellar collaps and on grain–grain interaction, and there was also a demonstration of a model–model collision.

At this point the editors wish to thank the scientific organizing committee and all participants and authors. Our thanks go also to all ESO staff contributing to this conference in general, but especially to the conference secretary Christina Stoffer.

Garching/Vilspa, June 1996

Hans Ulrich Käufl
Ralf Siebenmorgen

---

[1] http://http.hq.eso.org/vlt/

Contents

Part II   Observations of Dust Factories

Part III   Observational Results Based on IRAS, COBE or Balloon Borne Platforms

Part IV   Vega-Type, T-Tauri, λ Bootis and Herbig Ae/Be Stars

Part V   Properties of Dust
           Around Young Stellar Objects

Part VI   Ices and Laboratory Studies

Part VII   Radiative Transfer

Part VIII   Dust as a Catalytic Agent
for Star Formation

Part IX   Miscellaneous

Part X   Concluding Remarks

List of Participants

Fred ADAMS adams@pablo.physics.lsa.umich.edu
University of Michigan, Physics Department

Nancy AGEORGES nageorge@eso.org
European Southern Observatory, Garching

Yuri AIKAWA aikawa@yso.mtk.nao.ac.jp
National Astronomical Observatory of Japan (NAOJ), Tokyo

Philippe ANDRÈ andre@sapvxg.saclay.cea.fr
CEA - Centre d'Etudes de Saclay, Service d'Astrophysique

Pawel ARTYMOWICZ pawel@astro.su.se
Stockholm Observatory

Guido BARBARO Barbaro@astrpd.pd.astro.it
Universita di Padova, Dip. di Astronomia

Mary BARSONY fun@nusun.ucr.edu
University of California, Riverside

Paolo BERNASCONI berna@scsun.unige.ch
Observatoire de Geneve

Nicole BERRUYER nicole@obs-nice.fr
Observatoire de la Côte d'Azur, Nice

Alain BLANCHARD blanchard@astro.u-strasbg.fr
Observatoire Astronomique de Strasbourg

Armando BLANCO blanco@lecce.infn.it
University of Lecce, Dept. of Physics

Patrice BOUCHET pbouchet@eso.org
ESO Chile

Harold BUTNER butner@dtm.ciw.edu
Carnegie Institution of Washington, DTM

Murray CAMPBELL mfcampbe@colby.edu
Colby College, Dept. of Physics & Astronomy

Ivo ČERMÁK cermak@kosmo.mpi-hd.mpg.de
MPI für Kernphysik, Heidelberg

Rolf CHINI rchini@mpifr-bonn.mpg.de
MPI für Radioastronomie, Bonn

Glenn CIOLEK ciolek@sirius.astro.uiuc.edu
Univ, of Chicago, Dept. of Astron. & Astrophys.

Frank CLARK fclark@plh.af.mil
Phillips Laboratory (AFGL)

Martin COHEN cohen@bkyast.berkeley.edu

University of California, Berkeley, Radio Astronomy Laboratory

Luigi COLANGELI   colangeli@astrna.na.astro.it
Osservatorio Astronomico di Capodimonte

José Carlos CORREIA   j.c.correia@qmw.ac.uk
Queen Mary and Westfield College, London

Pierre COX   p427cox@mpifr-bonn.mpg.de and
Obs. de Marseille   cox@obmara.cnrs-mrs.fr

Douglas CURRIE   currie@khaos.umd.edu
University of Maryland, Dept. of Physics

Louis d'HENDECOURT   ldh@iaslab.ias.fr
Institut d'Astrophysique Spatiale - CNRS, Orsay

John DANZIGER   jdanzige@eso.org
ESO, Garching

Emmanuel DARTOIS   dartois@ias.fr
Institut d'Astrophysique Spatiale - CNRS, Orsay

William DENT   dent@jach.hawaii.edu
Joint Astronomy Centre, Hilo

Carsten DOMINIK   dominik@strw.leidenuniv.nl
Leiden University Observatory

Michael EGAN   egan@pldac.plh.af.mil
Phillips Laboratory, Geophysics Directorate

Pascale EHRENFREUND   pascale@strwchem.strw.LeidenUniv.nl
Leiden University Observatory

James EMERSON   j.p.emerson@qmw.ac.uk
Queen Mary and Westfield College, London

Nye EVANS   ae@astro.keele.ac.uk
Keele University, Department of Physics

Michèle GERBALDI   gerbaldi@iap.fr
Institut d'Astrophysique, Paris

Teresa GIANNINI   teresa@sunir.ifsi.fra.cnr.it
IFSI - CNR, Frascati

Philippe GIOVANETTI   giova@cesr.cnes.fr
C.E.S.R., Toulouse

Tim GLEDHILL   tmg@star.herts.ac.uk
University of Hertfordshire, Dept. of Physical Sciences

Mayo GREENBERG   mayo@rulhl1.leidenuniv.nl
Leiden University, Astrophysics Lab.

Eike GUENTHER   e.guenther@qmw.ac.uk
Queen Mary and Westfield College, London

Thomas HAYWARD   hayward@astrosun.tn.cornell.edu
  Cornell University, Dep. of Astronomy

Thomas HENNING   henning@fred.astro.uni-jena.de
  Max-Planck-Gesellschaft AG, Jena

Melvin HOARE   mgh@mpia-hd.mpg.de
  MPI für Astronomie, Heidelberg

Hartmut HOLWEGER   pas59@rz.uni-kiel.d400.de
  Universität Kiel

Željko IVEZIĆ   ivezic@asta.pa.uky.edu
  University of Kentucky, Dep. of Physics & Astronomy

Ant JONES   ant@dusty.arc.nasa.gov
  NASA Ames Research Center, Space Science Division

Hans Ulrich KÄUFL   hukaufl@eso.org
  ESO, Garching

Wilhelm KEGEL   kegel@astro.uni-frankfurt.de
  Univ. Frankfurt, Institut für theoretische Physik

Carsten KOEMPE   koempe@betty.astro.uni-jena.de
  Universitäts-Sternwarte, Jena

Jacek KRELOWSKI   jacek@astri.uni.torun.pl
  Nicolaus Copernicus Univ., Inst. of Astronomy, Toruna, Poland

Monika KRESS   kress@dusty.arc.nasa.gov
  NASA Ames Research Center, Space Science Div.

Natalia KRIVOVA   krivov@aispbu.spb.su
  St. Petersburg University, Astronomical Institute

Endrik KRÜGEL   p309ekr@mpifr-bonn.mpg.de
  MPI für Radioastronomie, Bonn

Maria KUN   kun@ogyalla.konkoly.hu
  Konkoly Observatory, Hungarian Academy of Sciences, Budapest

Jean-Pierre LAFON   jpj.lafon@obspm.fr
  Obs. de Paris, DASGAL, Meudon

Pierre-Olivier LAGAGE   lagage@sapvxa.saclay.cea.fr
  CEA Saclay, Service d'Astrophysique, DAPNIA

Ralf LAUNHARDT   launh@sol.astro.uni-jena.de
  Max-Planck-Gesellschaft AG, Jena

David LEISAWITZ   leisawitz@stars.gsfc.nasa.gov
  NASA Goddard Space Flight Center

Aigen LI   agli@strw.leidenuniv.nl
  Sterrewacht Leiden

Dario LORENZETTI   dloren@hp.ifsi.fra.cnr.it

IFSI - CNR, Frascati

Sarah MADDISON   `maddison@hypatia.maths.monash.edu.au`
Monash University, Dep. of Mathematics

Alexander MEN'SHCHIKOV   `sascha@georg.astro.uni-jena.de`
Max-Planck-Gesellschaft AG, Jena

Vito MENNELLA   `mennella@astrna.na.astro.it`
Osservatorio Astronomico di Capodimonte

Dante MINNITI   `dminniti@eso.org`
ESO, Garching

Joseph MONAGHAN   `joe.monaghan@sci.monash.edu.au`
Monash University, Dep. of Mathematics

Guy MONNET   `gmonnet@eso.org`
ESO, Garching

Frédérique MOTTE   `motte@gag.observ-gr.fr`
Observatoire de Grenoble

David MOUILLET   `mouillet@gag.observ-gr.fr`
Observatoire de Grenoble

Telemachos MOUSCHOVIAS   `tchm@astro.uiuc.edu`
University of Illinois, Dept. of Astronomy, Urbana

Takenori NAKANO   `nakano@nro.nao.ac.jp`
Nobeyama Radio Observatory, National Astronomical Observatory

Thorsten NECKEL   `neckel@mpia-hd.mpg.de`
MPI für Astronomie, Heidelberg

Dieter NÜRNBERGER   `nurnberg@astro.uni-wuerzburg.de`
Universitat Wurzburg, Astronomisches Institut

René OUDMAIJER   `roud@ic.ac.uk`
Imperial College of Science, Blackett Laboratory, London

Gernot PAATZ   `gpaatz@hp2.lsw.uni-heidelberg.de`
Landessternwarte Heidelberg

Maria Elisabetta PALUMBO   `mepalumbo@astrct.ct.astro.it`
Universita di Catania, Istituto di Astronomia

Eric PANTIN   `pantin@sapvxg.saclay.cea.fr`
CEA - Centre d'Etudes de Saclay, Service d'Astrophysique

Beate PATZER   `patzer@physik.TU-Berlin.de`
TU Berlin, Institut für Astronomie und Astrophysik

Stefano PEZZUTO   `pezzuto@le.infn.it`
University of Lecce, Dept. of Physics

Jonathan RAWLINGS   `jcr@star.ucl.ac.uk`
University College London, Dept. of Physics & Astronomy

Andrea RICHICHI   `richichi@arcetri.astro.it`
Osservatorio Astrofisico di Arcetri

Isabelle RISTORCELLI   `ristorce@cesr.cnes.fr`
C.E.S.R., Toulouse

Michael ROSA   `mrosa@eso.org`
ST-ECF, Garching

Ralf SABLOTNY   `sablotny@fred.astro.uni-jena.de`
Max-Planck-Gesellschaft AG, Jena

Paolo SARACENO   `saraceno@sunir2.ifsi.fra.cnr.it`
IFSI - CNR, Frascati

Wolfgang SCHMITT   `schmitt@georg.astro.uni-jena.de`
Max-Planck-Gesellschaft AG, Jena

Willem SCHUTTE   `schutte@strwchem.strw.LeidenUniv.nl`
Leiden University Observatory

Russell SHIPMAN   `shipman@plh.af.mil`
Phillips Laboratory, Geophysics Directorate

Ralf SIEBENMORGEN   `rsiebenm@iso.vilspa.esa.es`
ESA-Villafranca del Castillo Satellite Tracking Station

Gregory SLOAN   `sloan@ssa1.arc.nasa.gov`
NASA Ames Research Center

Craig SMITH   `c-smith@adfa.oz.au`
University College, ADFA, Dept. of Physics, Canberra

Marco SPAANS   `spaans@strw.leidenuniv.nl`
Leiden University Observatory

Thomas STANKE   `stanke@astro.uni-wuerzburg.de`
Universität Würzburg, Astronomisches Institut

Jakob STAUDE   `staude@mpia-hd.mpg.de`
MPI für Astronomie, Heidelberg

Jürgen STEINACKER   `stein@georg.astro.uni-jena.de`
Max-Planck-Gesellschaft AG, Jena

Francesco STRAFELLA   `strafella@le.infn.it`
University of Lecce, Dept. of Physics

Ryszard SZCZERBA   `szczerba@iras.ucalgary.ca`
The University of Calgary, Dept. of Physics & Astronomy

Mauricio TAPIA   `tapia@bufadora.astrosen.unam.mx`
U.N.A.M., Instituto de Astronomia

Stephen TAYLOR   `sdt@star.ucl.ac.uk`
University College London, Dept. of Physics & Astronomy

Teresa TEIXEIRA   `T.C.Teixeira@qmw.ac.uk`

Queen Mary and Westfield College, London

Alexander TIELENS   `tielens@dusty.arc.nasa.gov`
NASA Ames Research Center

Elisabetta TOMMASI   `elisa@sunir2.ifsi.fra.cnr.it`
IFSI - CNR, Frascati

Viktor TÓTH   `lvtoth@innin.elte.hu`
Eötvös University, Budapest

Mario van den ANCKER   `mario@astro.uva.nl`
Astronomical Institute, Amsterdam

Nikolai VOSHCHINNIKOV   `nvv@aispbu.spb.su`
St. Petersburg University, Astronomical Institute

Christoffel WAELKENS   `christoffel@ster.kuleuven.ac.be`
Katholieke Universiteit Leuven, Instituut voor Sterrenkunde

Helen WALKER   `h.walker@rl.ac.uk`
CCLRC - Rutherford Appleton Laboratory

Derek WARD-THOMPSON   `dwt@roe.ac.uk`
Royal Observatory Edinburgh

Mark WARDLE   `mark@holly.pas.rochester.edu`
University of Rochester, Dept. of Physics and Astronomy

Walter WEGNER   `walter@astri.uni.torun.pl`
Pedagogical University, Dept. of Mathematics

Helmut WIESEMEYER   `p713hwi@mpifr-bonn.mpg.de`
MPI für Radioastronomie, Bonn

Bogdan WSZOLEK   `bogdan@oa.uj.edu.pl`
Jagiellonian University, Astronomical Observatory, Cracow, Poland

Günther WUCHTERL   `wuchterl@amok.ast.univie.ac.at`
Universität Wien, Institut für Astronomie

Harold YORKE   `yorke@astro.uni-wuerzburg.de`
Universität Würzburg, Astronomisches Institut

Annie ZAVAGNO   `annie@orion.ifsi.fra.cnr.it`
IFSI - CNR, Frascati

Hans ZINNECKER   `hzinnecker@aip.de`
Astrophysikalisches Institut Potsdam

Victor ZUBKO   `maouas@gluk.apc.org`
                          `zubko@astri.uni.torun.pl`
Main Astronomical Obs., NAS, Kiev

Robert ZYLKA   `rzylka@mpifr-bonn.mpg.de`
MPI für Radioastronomie, Bonn

Part I

Ground-Based Observations of Young Stellar Objects

High-Angular Resolution Near-Infrared Observations of the Circumstellar Environment of Young Stellar Objects

N. Ageorges[1,2] and A. Eckart[1]

[1] Max Planck Institut für Extraterrestrische Physik, Giessenbachstrasse
D-85748 Garching b. München, Germany
[2] Laboratoire d'Astrophysique, BP 53 X, F-38041 Grenoble Cedex, France

Abstract. Using the MPE SHARP camera at the ESO NTT, we conducted a study of several young stars selected on the basis of their high degree of linear polarization at optical wavelength. Observing at high angular resolution in the near-infrared, we can observe closer to the star. We can thus study the direct environment of the sources. Some results of our speckle survey are presented and discussed. To complete our study of these objects, we made two-dimensional speckle polarimetry measurments. It is briefly described here and results are illustrated on the InfraRed Nebula of Chamaeleon.

1 Results of the Speckle Survey

The speckle survey, of young stellar objects (YSOs) selected to have a high degree of linear polarization at optical wavelength, has been completed in two observing runs (July-August 1993 and April-May 1994). 28 YSOs have been observed mostly in Chamaeleon and Ophiuchus. We found nine binaries (5 new ones), 3 multiple systems (all newly discovered) and 4 extended sources. From the two extended sources discovered by this study, one is also a binary: V536 Aql (see Ageorges et al., 1994). A good example of our achievement is given by one of the multiple source discovered, where the intensity ratio between the main component and the second one (2.45^o away) is of the order 550! (Ageorges et al., 1995, in preparation)

For sources found punctual, our high angular resolution (0.2" in K band) allows to give an upper limit (30 AU at 150 pc) for the size of an eventual disk or the separation between multiple components. This is illustrated for example by ROX 14, which we found punctual but is a binary of 20 mas separation (lunar occultation, Simon et al., 1995). For these sources, the VLT (and much more the VLTI) project will allow resolutions high enough (60 mas for the VLT and less than 10 mas for the VLTI) to resolve these sources (in the closest molecular clouds: Taurus, Chamaeleon, ...). It will then be very interesting to know the exact degree of multiplicity of YSOs; since there is growing evidence that the usual product of star formation is a multiple system (Mathieu 1994).

We present hereafter an example of our results, obtained on the multiple source (result of this study) AS 310.

1.1   AS 310

The brightest component of this known binary has been identified as an Herbig Ae/Be star (Finkenzeller & Mundt, 1984). This source is embedded in a complex brigth nebula, which is associated with an HII region. The coincidence of IRAS sources and CO emission peaks in this region suggests that parts of this molecular cloud have sites of star formation other than the optically visible HII region (Hunter et al. 1990). An other argument in accordance with the idea that star formation is still undergoing there comes from Altenhoff et al. (1994). They found a substantial excess of 250GHz radiation over that found at cm-wavelength and suggest that this extra flux density originates from thermal radiation of dust. The presence of this gas is stressed by the observations of Henning et al. (1994) at 1.3mm. From their measurments, with the assumption that, at this wavelength, the medium is optically thin, they derive a gas mass of 30 $M_{\odot}$.

In our speckle observations, we found (Ageorges et al., 1995, in preparation) a compact stellar cluster with $\approx$ 10 sources in the central 20" $\times$ 20". Combining our photometric and polarimetric speckle data with data obtained at different wavelengths, we now want to investigate the nature of this probably young star forming region.

2   Two-Dimensional Speckle Polarimetry

2.1   Instrumentation and Observing Mode

To complete our study of the previously mentioned sources, we performed speckle polarimetric measurments using SHARP-I at the ESO NTT. A wire-grid on a BaF_2 substrate has been installed on a computer controlled rotating table fixed in front of the entrance window of the SHARP camera. We made measurments for each source at 6 different position angle separed by 30 degrees. For each source of interest, we also observed an unpolarised calibrator, to measure the instrumental polarization. For each position angle of the polarizer, we got images that we reduced with the simple shift-and-add algorithm (Christou 1991); we then fitted a cosine function through fluxes obtained for a source from the six images obtained at different position angle of the polarizer. The calibration of Stokes'parameters is done following Goodrich (1986).

2.2   Example of Results: IRN Cha

The infrared nebula of Chamaeleon has first been observed by Schwartz & Henize (1983). Its total dimensions are 125" $\times$ 35", with a principal axis at 75°. Its western lobe is much smaller than the eastern one. Cohen & Schwartz (1984; CS84 hereafter) did photometric observations, from which they derive a bolometric luminosity of $14.4L_{\odot}$ and an A_V of 10. They also notice a small north-south elongation at 52 and 100μm. Linear polarization measurments have been done by Scarrott et al. (1987) at optical wavelength and Mc Gregor et al. (1994) in H band (1.65μm).

We did linear polarization measurements, on this source, in J(1.25µm) and H(1.65µm) in April 1994. We found in corresponding apertures a comparable degree of polarization as the above mentionned authors (32% at 160° in a 8" aperture in H band, and 22% in a 1.4" aperture centered on the brightest knot in the nebula). As in Scarrott et al. (1987), in our maps, the polarization vectors are asymmetrically distributed around the source, with high degree of polarization (40-60%) in the lobes. Such a distribution is typical of a reflection nebula illuminated from the inside. We nevertheless notice a deviation from this centro-symmetric pattern in form of a parallel band of polarization between the two lobes of the nebula. Such a band is interpreted by polarization models (see e.g. Bastien & Ménard 1990, BM90 hereafter) as due to the presence of an optically thick circumstellar disk.

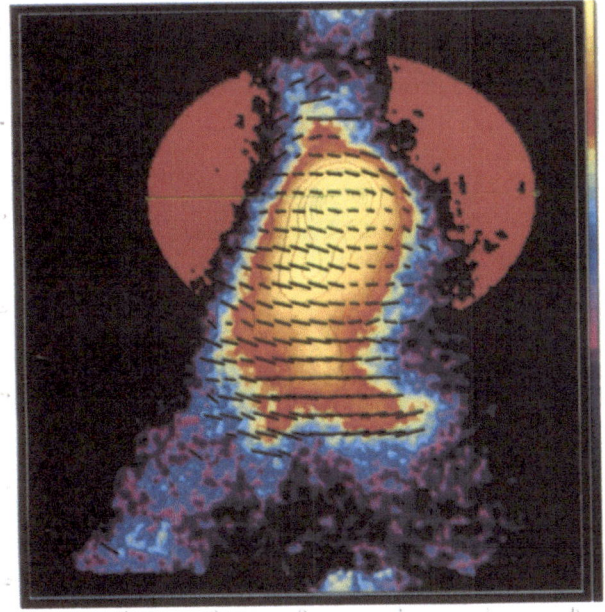

Fig. 1. Calibrated H band polarization map of IRN Cha overplotted on an intensity color map. The red ellipse represents our disk model.

A comparison of our maps of polarization with disk models indicates that the disk is tilted between 50 and 90 degrees, with a most probable angle of 70°, that we assumed in the following discussion. This angle plus the gap, between the two lobes of the nebula, that we assume to be due to high extinction, leads us to a minimum disk radius of the order of 800AU. An other argument going in the sense of this disk not beeing seen edge-on comes from polarization models (BM90): this nebula is highly asymmetric and the parallel band of polarization is not centered between the two lobes of the nebula but close (within 0.3") to the brightest source component.

From the 100μm flux measurments of CS84, we derive a dust mass of the order of $1.56.10^{-4}M_\odot$, assuming a dust temperature of 65K (CS84). This mass estimation is also in agreement with $C^{18}O$ data from Haikala et al. (private communication). Combined with our estimate of the disk size, we then get a surface density of 0.068g.cm^2. From our observations, we estimate the extinction, A_K to 2.8, and thus $A_V \approx 18.67$, in better accordance with the fact that we do not see the source of illumination of this nebula. From this, we derive N_{H_2} of $1.69.10^{22}$cm^{-2}, that leads us to a surface density of the order of 0.056g.cm^2. The good agreement between the two estimates is considered as an argument in favor of our disk model.

3   Comments

This work is the result of a close collaboration between the Observatoire de Grenoble and the MPE.; it also involves F. Ménard & J.L. Monin.

References

Ageorges N., Ménard F., Monin J.L., Eckart A., 1994, A&A 283, L5-L8

Altenhoff W.J., Thum C., Wendkler H.J., 1994, A&A 281, 161-183

Bastien P., Ménard F., 1990, Ap. J. 364, 232-241

Christou J.C., 1991, Experimental Astr. 2, 27

Cohen M., Schwartz R.D., 1984, Ap.J. 89, 277-279

Finkenzeller U., Mundt R., 1984, A&A Supp. Series 55, 109-141

Goodrich R.W., 1986, Ap.J. 311, 882-894

Henning T., Launhardt R., Steinacker J., Thamm E., 1994, A&A 291, 546-556

Hunter D.A., Thronson H.A. Jr., Wilton C., 1990, A.J. 100, 1915-1926

McGregor P.J., Harrison T.E., Hough J.H., Bailey J.A., 1994, M.N.R.A.S. 267, 755-765

Mathieu R.D., 1994, ARAA 32, 465

Scarrott S.M. et al., 1987, M.N.R.A.S. 228, 827-831

Schwartz R.D., Henize K.G., 1983, A.J. 88, 1665-1669

Submillimeter Dust Continuum Emission as a Probe of Protostellar Evolution

Philippe André and Sylvain Bontemps

Service d'Astrophysique, CEA Saclay, F-91191 Gif-sur-Yvette Cedex, France

Abstract. The strength and spatial distribution of the (sub)millimeter continuum emission from young stellar objects (YSOs) are powerful diagnostics of the progressive decrease and condensation of circumstellar mass during protostellar evolution. In particular, the ratio of submm to bolometric luminosity may be used as a practical evolutionary indicator for embedded YSOs. We illustrate the usefulness of this age-ordering indicator by studying outflow evolution through the embedded protostellar phase: There is good evidence that outflow activity declines dramatically with time from the youngest Class 0 objects to the most evolved Class I sources.

1 Class 0 or Submillimeter Protostars

(Sub)millimeter continuum observations conducted in the last five years with large radiotelescopes such as the IRAM 30 m and the JCMT have proved very successful in supplementing infrared data to define a complete empirical sequence (Class 0 → Class I → Class II → Class III) for the early evolution of low-mass YSOs (e.g. André & Montmerle 1994 – AM). Class 0 sources, which are the youngest objects in this sequence, are characterized by strong dust continuum emission at $\lambda > 300$ μm but very little or no emission shortward of 10 μm. More specifically, Class 0 objects are defined by the following observational properties (see André, Ward-Thompson, & Barsony 1993 – AWB):

- (1) Indirect evidence for a central YSO, as indicated for instance by the detection of compact centimeter continuum VLA emission or the presence of a collimated CO outflow.

- (2) Centrally peaked but extended submillimeter continuum emission tracing the presence of a spheroidal circumstellar dust envelope.

- (3) High ratio of submillimeter to bolometric luminosity: $L_{submm}/L_{bol} >> 5 \times 10^{-3}$, where L_{submm} is measured longward of 350 μm. (In practice, this often means a spectral energy distribution resembling a single temperature blackbody at $T \sim 15$-30 K.)

Property (1) distinguishes Class 0 objects from pre-protostellar condensations (e.g., Ward-Thompson et al., this volume), while properties (2) and (3) distinguish them from Class I sources and pre-main sequence stars (Class II and Class III sources).

In contrast to Class I sources which show up in surveys with near-IR arrays, most known Class 0 objects (e.g., VLA1623, L1448-C, NGC2264G-VLA2, HH24MMS) were originally discovered through (continuum or line) observations at submil-

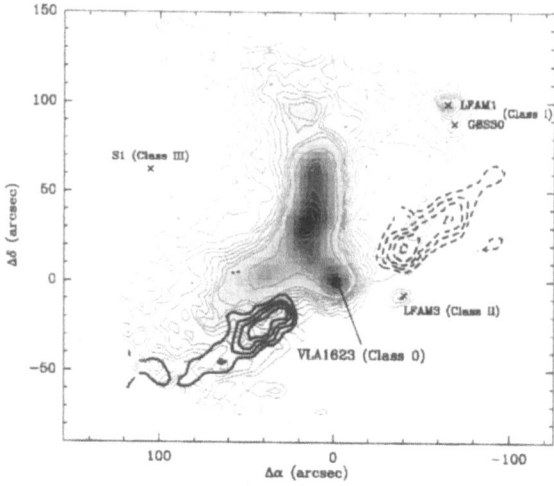

Fig. 1. Grey-scale 1.3 mm dust continuum map of the ρ Oph A cloud core region obtained with the IRAM 30 m telescope and the MPIfR 19-channel bolometer array. A CSO map of the jet-like CO(3-2) outflow emanating from VLA 1623 is superposed (blue-shifted lobe shown by thick solid contours; red-shifted lobe by dashed contours).

limeter or longer radio wavelengths (see Bontemps, André, & Ward-Thompson 1995a for a recent example).

1.1   Age Ordering and Interpretation of Class 0 Sources

Millimeter dust continuum maps of optically thin dust emission may be used to make quantitative estimates of the circumstellar masses surrounding YSOs of all types, and are therefore a useful tool to track their evolution (see AM and Fig. 1). In a statistical sense at least, larger amounts of circumstellar material are expected to surround younger stellar objects. In this spirit, AWB proposed an age ordering of embedded YSOs based on the L_{submm}/L_{bol} ratio. While the (integrated) submillimeter luminosity L_{submm} of a protostellar source provides a relative measure of its (total) circumstellar mass M_{env}, the bolometric luminosity L_{bol} may be used to infer the central stellar mass $M_{\star}$ on the basis of existing mass–luminosity relations for protostars (see AM for details). In the youngest sources, dominated by an envelope rather than a disk, L_{submm}/L_{bol} should thus tend to reproduce the variations of the mass ratio $M_{env}/M_{\star}$, which is expected to decrease with protostellar age. The formal boundary between Class 0 and Class I sources ($L_{submm}/L_{bol} \sim 5 \times 10^{-3}$) was set so as to correspond to a mass ratio $M_{env}/M_{\star} = 1$, assuming the most plausible relations between L_{bol} and $M_{\star}$ on the one hand and between L_{submm} and M_{env} on the other hand (see AWB for details). Class 0 sources are thus *conceptually* different from the other classes of YSOs: *they are protostars which have yet to accrete the bulk of their final stellar mass* ($M_{env} \gg M_\star$). For this reason, they can potentially tell us a lot about the physics of protostellar collapse (e.g., Motte et al., this volume).

The scarcity of Class 0 YSOs in young stellar populations such as the ρ Ophiuchi embedded cluster provides additional evidence that they are significantly younger (probable age $\lesssim 10^4$ yr) than Class I sources (which have an estimated lifetime $\sim 10^5$ yr).

2   The Decline of Outflow Activity with Age

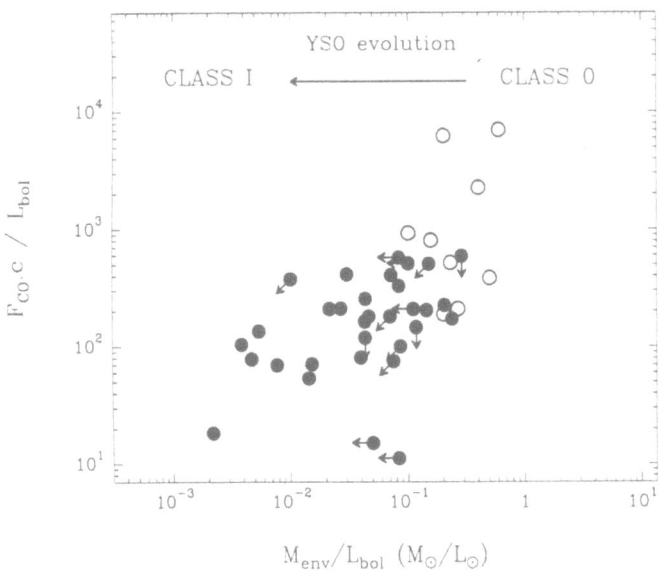

Fig. 2. Outflow efficiency $F_{CO}\,c/L_{bol}$ versus M_{env}/L_{bol} for a sample of nearby Class I (filled circles) and Class 0 (open circles) YSOs (Bontemps et al. 1995b). This diagram, which is essentially free of any luminosity effect, shows a clear decline of outflow strength from Class 0s to Class Is.

Class 0 protostars tend to drive highly collimated or "jet–like" CO molecular outflows (see, e.g., Fig. 1 and Bachiller et al. 1995). The mechanical luminosities of these outflows are often of the same order as the bolometric luminosities of the central sources (e.g., AWB). In contrast, while there is growing evidence that some outflow activity exists throughout the embedded phase, the CO outflows from Class I sources tend to be poorly collimated and much less powerful than those from Class 0 sources.

In an effort to quantify this evolution of molecular outflows during the protostellar phase, we have recently obtained and analyzed a homogeneous set of

CO(2–1) data around a large sample of low-luminosity ($L_{bol} < 50 \, L_\odot$), nearby ($d < 450$ pc) embedded YSOs, including 36 Class I sources and 9 Class 0 sources (Bontemps et al. 1995b). Our results show that essentially *all* embedded YSOs have some degree of outflow activity, suggesting the outflow phase and the infall/accretion phase coincide. This is consistent with the idea that accretion cannot proceed without ejection and that outflows are directly powered by accretion (e.g., Ferreira & Pelletier 1995).

In our study, Class 0 objects are found to lie an order of magnitude above the well-known correlation between outflow momentum flux and bolometric luminosity holding for Class I sources. This confirms that Class 0 objects differ qualitatively from Class I sources, independently of inclination effects. On the other hand, outflow momentum flux is roughly proportional to circumstellar envelope mass in our *entire* sample (i.e., including both Class I and Class 0 sources). As illustrated in Figure 2, this new correlation is independent of the F_{CO}–L_{bol} correlation and most likely results from a progressive decrease of outflow power with time during the accretion phase.

Since many theoretical models of bipolar outflows (e.g., Ferreira & Pelletier 1995) predict a direct proportionality between accretion and ejection, Bontemps et al. (1995b) further suggest that the observed decline in outflow energetics reflects a corresponding decrease in the mass accretion/infall rate: In this view, $\dot{M}_{acc}$ would decline from $\sim 10^{-5} \, M_\odot \mathrm{yr}^{-1}$ for the youngest Class 0 protostars to $\sim 10^{-7} \, M_\odot \mathrm{yr}^{-1}$ for the most evolved Class I sources and most active T Tauri stars.

Because of their prominence and their youth, Class 0 outflows offer a unique opportunity to learn more about the ejection mechanism and entrainment process. Recent submm polarimetry and interferometric CO observations of Class 0 sources suggest that molecular outflows are not collimated by large-scale magnetic fields (Holland et al. 1995) and are primarily entrained in jet bow shocks (Gueth et al. 1995).

References

André, P. (1995) Ap&SS **224**, 29

André, P., & Montmerle, T. (1994) ApJ **420**, 837 (AM)

André, P., Ward-Thompson, D., Barsony, M. (1993) ApJ **406**, 122 (AWB)

Bachiller, R., Guilloteau, S., Dutrey, A. et al. (1995) A&A **299**, 857

Bontemps, S., André, P., Terebey, S., & Cabrit, S. (1995b) A&A, submitted

Bontemps, S., André, P., & Ward-Thompson, D. (1995a) A&A, **297**, 98

Ferreira, J., & Pelletier, G. (1995) A&A **295**, 807

Gueth, F., Guilloteau S., & Bachiller, R. (1995) A&A, in press

Holland, W.S., Greaves, J.S., Ward-Thompson, D., & André, P. (1995) A&A, in press

Compact Molecular Outflows from Young Stellar Objects in L1641

J.C. Correia[1], M. Griffin[1], P. Saraceno[2], and A. Zavagno[2]

[1] Physics Dept., Queen Mary & Westfield College, Mile End Road, London E1 4NS, England
[2] CNR - Istituto di Fisica dello Spazio Interplanetario, Casella Postale 27, I-00044 Frascati, Rome, Italy

Abstract. We report ^{12}CO J = 3–2 observations of compact molecular outflows around two young stellar objects in L1641. We have discovered a new outflow associated with VLA 1, the powering source of the well known optical flow HH 1-2. Preliminary calculations have shown that this outflow is one of the weakest, youngest and most compact ever seen. We have also mapped the more extended outflow from VLA 3 previously detected by Chernin and Masson (1995), confirming the existence of two young and very compact molecular outflows in this region.

1 Introduction

L1641 is a dark molecular cloud in the southern part of the Ori A complex. It is at a distance of 470 pc and is a region of low- and intermediate-mass star formation. This region contains the well known Herbig-Haro objects, HH 1 and HH 2. These objects, separated by 3′, are moving in opposite directions constituting a bipolar optical outflow, emanating from the powering source VLA 1, lying midway between then (Pravdo et al. 1985; Rodriguez et al. 1990).

Chernin and Masson (1995) mapped the region around VLA 1 in CO J = 2–1 with 30″ resolution and found an outflow with a red-shifted lobe which peaks at (−0.8′, 0.5′) from VLA 1. They claim that this outflow is powered not by VLA 1 but by VLA 3, a radio continuum source 1.3′ to the northwest of the center of HH 1-2 flow. They did not detect any outflow associated with VLA 1 but suggested that observations with a beam less than 0.25′ might separate out a possible weak HH 1-2 outflow from the stronger VLA 3 outflow.

We have mapped the same region in CO J = 3–2 with 14″ resolution and we have found a very compact and young molecular outflow centered on VLA 1, with an axis well aligned with the HH 1-2 flow axis. We have also detected and mapped the more extended outflow associated with VLA 3 reported by Chernin and Masson, confirming the existence of two young and very compact molecular outflows in this region.

2 Observations

The CO J = 3–2 (ν = 345.796 GHz) observations presented in this work were carried out in March and October 1994 at the James Clerk Maxwell Telescope

(JCMT) on Mauna Kea, Hawaii. At 345 GHz, the beamwidth was 14″ and the main beam efficiency was 0.58. The 345 GHz receiver was used in conjunction with the "Dutch" Autocorrelation Spectrometer (DAS). We integrated for several minutes per point, achieving typically rms noise levels of 0.1 K.

3   Results

A map of the high-velocity CO emission in the direction of VLA 1 is presented in Fig. 1. The map is centered on VLA 1 (R.A. $(1950) = 05^h 33^m 57^s$, DEC. $(1950) = -06° 47' 55''$), covering an area of approximately $1' \times 1'$. A half beam spacing (7.5″) was used to cover the area. From the map we can see that the peaks of red- and blue-shifted emission are separated by only 30″. However the two lobes are cleary distinct from each other which sugests that the axis of the outflow is close to the plane of the sky. This axis is very well aligned with the HH 1–2 axis and VLA 1 is in the center of the two lobes.

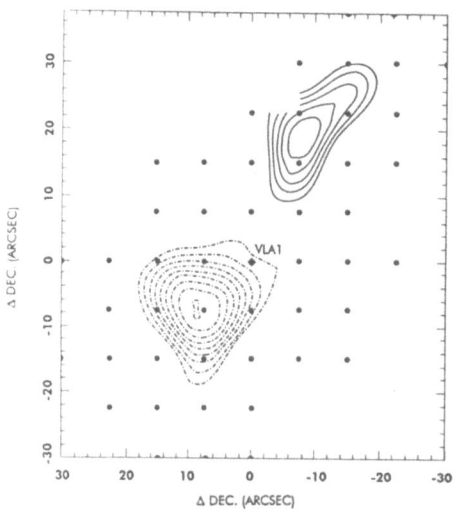

Fig. 1. CO J = 3–2 map in the direction of VLA 1. Red-shifted emission is shown by dashed lines and corresponds to an integrated intensity between 14-20 km s^{-1}. Blue-shifted emission is shown by solid lines and corresponds to an integration between 0-5 km s^{-1}. The contour interval is 0.3 K km s^{-1} and the lowest contour is 2 K km s^{-1}.

The mass, momentum and the energy associated with each lobe have been calculated using the same approach taken by Scoville et al. (1986). We have considered an excitation temperature (T_{EX}) of 30 K and an optical depth (τ) of 3, since, according to Parker et al. (1991), most outflows in dark clouds have optically thick emission. The results are summarized in Table 1. The momentum and energy values were calculated assuming that θ, the angle between the line of sight and the outflow axis, is 45°. If $\theta > 45°$ then these values are lower limits. We estimate a dynamical timescale of 4.2×10^3 yr for the outflow. If $\theta > 45°$ then this is an upper limit making this one of the youngest outflows ever detected.

Table 1. VLA 1 – Outflow Physical Parameters

	V_{min} (km s^{-1})	V_{max} (km s^{-1})	Mass (M$_\odot$)	Momentum (M$_\odot$ km s^{-1})	Energy ($\times 10^{42}$ ergs)
Blue lobe	0	5	0.001	−0.005	0.328
Red lobe	14	20	0.003	0.020	1.57

Submillimetre continuum observations of VLA 1 (see Zavagno et al., *this conference*) indicate that VLA 1 may be a Class 0 source.

Fig. 2 shows the red and the blue lobes of the VLA 3 outflow. The map covers an area of roughly $1' \times 2'$ and it is clear that this outflow is much more extended than the previous one. The blue peak is very close to VLA 3 and the two peaks are separated by only $20''$. The large overlap suggests that θ must be smaller than $45°$. The outflow axis is well aligned with that of VLA 1.

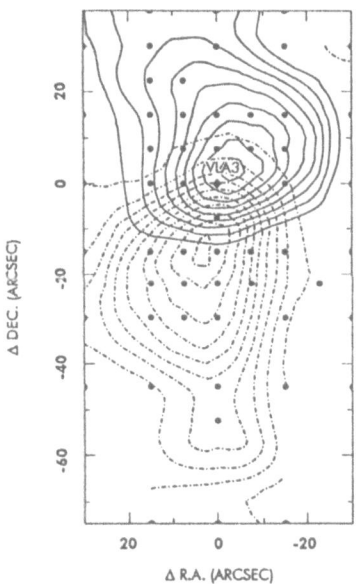

Fig. 2. CO J = 3–2 map in the direction of VLA 3. The red-shifted emission is shown by dashed lines and corresponds to an integrated intensity between 12-30 km s^{-1}. The blue-shifted emission is shown by solid lines and corresponds to an integration between -10-5 km s^{-1}. In the blue lobe, the lowest contour is 1 K km s^{-1} and the contour interval is 1 K km s^{-1} and in the red lobe the lowest contour is 2 K km s^{-1} and the contour interval is 2.5 K km s^{-1}.

The calculated physical parameters are summarized in Table 2. We have again used $T_{EX} = 30$ K and $\tau = 3$. The momentum and energy values were calculated assuming $\theta = 0°$. If $\theta > 0°$ then these values are underestimated. We estimate a dynamical timescale of 2.1×10^3 yr but this is a lower limit since we have assumed $\theta = 0°$. The outflow could be an order of magnitude older than this, and a timescale of 10^4 yr is probably more likely.

Table 2. VLA 3 – Outflow Physical Parameters

	V_{min} (km s^{-1})	V_{max} (km s^{-1})	Mass (M$_\odot$)	Momentum (M$_\odot$ km s^{-1})	Energy ($\times 10^{42}$ ergs)
Blue lobe	-10	5	0.061	-0.522	53.8
Red lobe	12	30	0.251	2.164	226

4   Conclusions

1. VLA 1 does indeed possess an outflow. This outflow is extremely weak, compact and young.
2. The red and blue lobes of the molecular emission are closer to the central source than the HH objects, consistent with the idea that the outflow dynamical timescales measured from high velocity molecular line emission can underestimate the actual age of the outflow. The present scenario seems to sugest that we might be observing a particular phase of an episodic outflow.
3. Submillimetre continuum observations of VLA 1 indicate that VLA 1 may be a Class 0 source.
4. VLA 3 is confirmed as a source with a more extended and powerful outflow.

5   Acknowledgements

JCC is supported by a grant from JNICT – Programa PRAXIS XXI: BD/2961/94.

References

Chernin, L., & Masson, C. (1995), ApJ, **443**, 181
Parker, N. D., Padman, R., & Scott, P. F. (1991), MNRAS, **252**, 442
Pravdo, S. H., Rodriguez, L. F., Curiel, S., Cantó, J., Torrelles, J. M., Becker, R. H., & Sellgren, K. (1985), ApJ, **293**, L35
Rodriguez, L. F., Ho, P. T. P., Torrelles, J. M., Curiel, S., & Cantó, J. (1990), ApJ, **352**, 645
Scoville, N. Z., Sargent, A. I., Sanders, D. B., Claussen, M. J., Masson, C. R., Lo, K. Y., & Phillips, T. G. (1986), ApJ, **303**, 416
Zavagno, A., Molinari, S., Tommasi, E., Griffin, M. & Saraceno, P. (1995), *this conference*

Anatomy of a Spatially Resolved Dust Disc Around a B-Type YSO

W.R.F. Dent[1], C. Racela[2], and F. Rosengarten[2]

(1) Joint Astronomy Centre, 660 N. Aohoku Pl., Hilo, Hawaii 96720, USA
(2) SSTP Student, Hawaii Community College, Hilo, Hawaii 96720, USA

Abstract. Deep maps of the sub-millimetre dust emission around the high-mass star G35.2N have clearly resolved the edge-on circumstellar disc of outer diameter $\sim 10^5$au. The data have been compared with the results from a 2-D numerical radiative transfer program in order to determine the underlying disc density and temperature structure. We find that a flared disc with height $\propto r^{1.5}$ and midplane density $\propto r^{-2}$ can fit the data at radii $\geq 20,000au$. However, at smaller radii, the midplane density decreases as r^{-1}. The results are compared with estimates of the disc temperature structure using the molecular tracers H_2CO and NH_3; we find that H_2CO gives much higher temperatures than either the dust model or NH_3.

1 Observational Data

1.1 Why choose G35.2N?

Embedded *high*-mass YSO discs have been generally neglected when compared to T-Tauri stars. But in the closest cases, it may more easy to spatially resolve their surrounding dust and gas. G35.2N is one of the closest, relatively isolated, high mass YSOs (eg Dent et al., 1985). All the previous data suggests that it is very close to edge-on, making interpretation somewhat easier. Furthermore, there is strong evidence from molecular line velocity shifts of a flattened cloud in Keplerian motion about the central star, indicating the presence of a dynamically stable, large rotating disc.This is important, as it implies that we are not simply observing a large interstellar cloud near the star, but that the material is *gravitationally bound* and directly associated with the YSO. Other parameters of the object are:

- Distance = 2000 pc
- Bolometric luminosity = $2 \times 10^4 L_\odot$ (B0.5 ZAMS)
- No associated optical emission
- Known CO outflow close to the plane of the sky, at $PA \approx 45^\circ$
- Prominent near-ir bipolar nebula

1.2 Images

We have made maps with high dynamic range at $\lambda 800$ and $450 \mu m$ using the JCMT with the continuum bolometer UKT14; fig. 1 shows the map at 450 μm.

We also have K-band ($\lambda 2.2\mu m$) continuum images of the reflection nebula taken with UKIRT (see also Walther et al., 1990). A spectral energy distribution (SED) of the total integrated emission from the cloud over the wavelength range 2.2 to $1100\mu m$ has been compiled from published data.

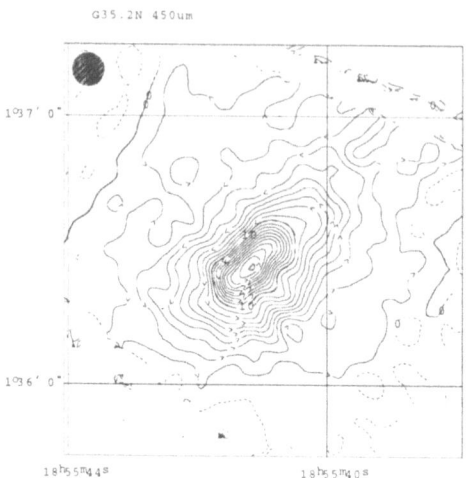

Fig. 1. Map of G35.2N at $450\mu m$ taken with the JCMT. Beamsize is 7.5"

2   Model

A 2-D radiative transfer program with a reprocessing disc was used to derive the temperature structure; from this we can form simulated images at each wavelength as well as the SED (eg Lefévre et al., 1983; Dent, 1988). To emulate the density distribution of typical cloud collapse models, we describe the dust density (n_d) as a function of the radial distance r, and distance above the plane z, using the following:

$$n_d = n_{pl} \times exp(-(z/h)^2) \tag{1}$$

where the density in the central plane, n_{pl} decreases with radius:

$$n_{pl} = n_{po} \times r^{-\alpha} \tag{2}$$

and the disc scale height, h increases with radius:

$$h = h_o \times r^\beta \tag{3}$$

The height and density scaling factors h_o and n_{po}, the power law parameters α and β, and the disc inclination angle i are optimised to give the best agreement with cross sections through the sub-mm images, the structure of the near-infrared scattering nebula, the disc mass and the SED.

3   Derived Disc Structure

We assume the value of i to be $\sim 10°$ (ie the disc is viewed close to edge-on), mainly from the near-symmetry of the reflection nebula. One characteristic of the sub-mm data that is important for the derived model is that the central emission peak is *resolved* into an extended plateau along the plane direction. No single power-law for midplane density α could be found that would fit this extended central emission "plateau" as well as the more rapid decrease at larger radii. However, a change in α from 1.0 at $r \leq 0.1pc$ to 2.0 at larger radii could fit the data reasonably well. This "flat" core is similar to the shallow density gradient recently found around some class 0 YSO's (eg Barsony & Chandler, 1993). The best value of β required to match the bipolar nebula was 1.5; values of 0.0 (flat disc) or 1.0 (constant opening angle disc) did not give large-enough scattering surfaces to produce a bright nebula, and values greater than 2.0 gave near-ir nebulae that were too small. For images of the model disc see fig. 2.

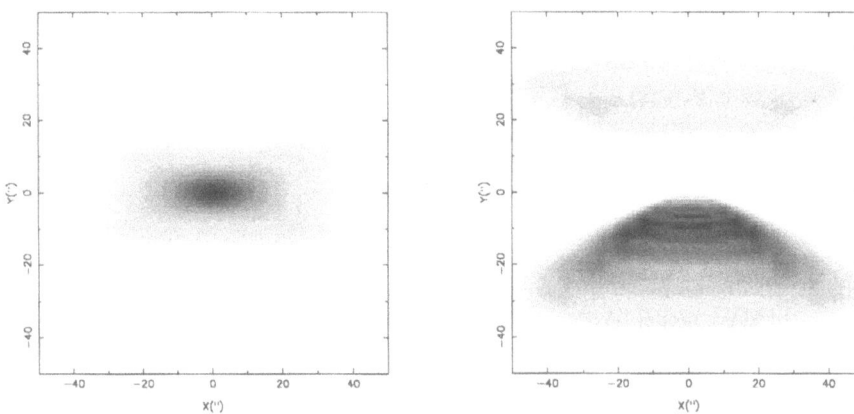

Fig. 2. Simulated images of model at wavelengths of 450 and $2.2\mu m$. The scale is shown in arcsec, assuming the parameters pertaining to G35.2N (see text). Note the faint cross-shaped low-intensity structure at $450\mu m$ arising from the heated disc surface.

4   Comparison of Dust and Molecular Tracers

VLA observations of NH_3 (Little et al., 1985) indicate a factor of 2 density drop within 0.1pc; however, this did not fit our sub-mm data, and may be caused by a change in relative abundance. Spectra of multiple transitions of H_2CO were taken across the disc, although the resolution was insufficient to obtain more than 3-6 spatially independent measurements. Both VLA and single-dish

NH$_3$ observations have been published, from which kinetic temperatures can be derived. Figure 3 compares the model dust temperature distribution in the plane of the disc (z=0) with the molecular data. The beam sizes are indicated by the horizontal bars. The agreement with NH$_3$ is reasonably accurate; however, H$_2$CO gives abnormally high values in the outer disc. The reason for this is unclear, although it may be caused by heating from the high-velocity CO outflow. The results suggest that temperatures derived from H$_2$CO should be treated with some caution.

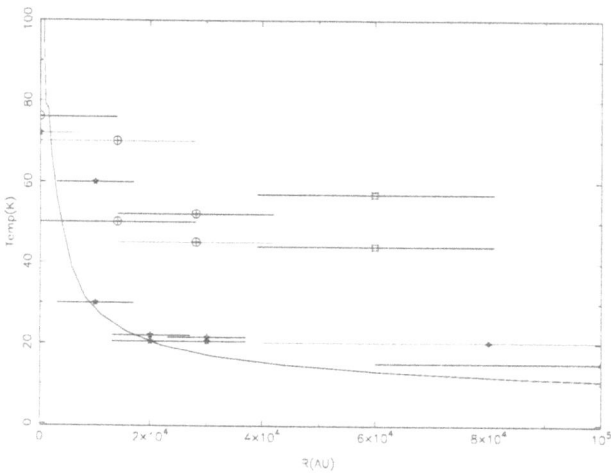

Fig. 3. Plot of temperature vs. radial distance along the disc plane. The solid line shows the dust model, crosses represent values derived from NH$_3$, circles are those from H$_2$CO transitions near 360GHz and squares are from transitions near 218GHz.

5   References

Barsony & Chandler, 1993, ApJ, 406, L71
Butner et al., 1991, ApJ, 376, 636
Chandler et al., 1990, MNRAS, 243, 330
Dent et al., 1985, A & A, 146, 375
Dent, 1988, ApJ 325, 252
Lefévre, Daniel & Bergeat, 1983, A & A, 121, 51
Little et al., 1985, MNRAS 217, 227
Kenyon & Hartman, 1987, ApJ, 323, 714
Lin & Pringle, 1990, ApJ, 358, 575
Walther et al., 1990, ApJ, 356, 544

Mid-IR PAH Emission in YSOs WL 16 & WL 22

J.P. Emerson

Dept of Physics, Queen Mary & Westfield College, Mile End Road, London E1 4NS

Abstract. We present 8–24μm spectra of the YSOs WL 16 & WL 22 which lie in the Ophiuchus star forming region. Both were modelled by Adams Lada & Shu 1987 as Class I objects with deep 9.7μm silicate features. Our spectroscopy however shows PAH features which are not seen in other Ophiuchus YSOs studied. The significance of the PAHs for our understanding of the nature of these objects is discussed. We also present a 8–24μm spectrum of the Class I YSO El 29. A spectrum of YLW13b shows that although it shares a lack of X-ray emission with WL 16 & WL 22 it has no mid-IR PAH emission.

1 Introduction

WL 16 & WL 22 are well-studied young stellar objects (YSOs) deeply embedded in a molecular cloud core and were first identified by the near-infrared survey of the ρ Ophiuchus star-forming region by Wilking & Lada (1983).

They have usually been classified as Class I (low mass protostellar objects) but more recently it has been suggested they are more evolved Class II objects of intermediate mass nearer the main sequence, seen through very high interstellar extinction (eg Comeron et al. (1993)). Together with El 29 (an undisputed Class I) WL 16 & WL 22 were the three Ophiuchus cloud core objects modelled by Adams Lada & Shu (1987), henceforth ALS, as protostellar sources. Their model spectra fitted the known spectral-energy distributions including supposed deep 9.7μm silicate absorption features indicated by the photometric data of Lada & Wilking (1984), and indeed were selected for modelling mainly because of this strong silicate absorption feature. In Fig. 1 we show the WL 16 photometric points in the 10μm region and the ALS model.

2 PAH Features

During an attempt to more precisely determine the strength of the silicate absorption feature in these objects we used CGS3 on UKIRT to obtain spectra with a resolution $\Delta\lambda/\lambda \sim 55$ in the 8-14μm region and ~ 75 in the 16-24μm region. To our surprise the 8-14μm spectrum we obtained for WL 16, shown in Fig. 1, did not show a strong silicate feature as we had expected but a spectrum dominated by the mid-IR PAH emission bands. (The same 8-14μm spectrum was independently obtained by Hanner et al. (1992) and is identical with ours within the uncertainties). In fact PAH emission in the 3.29μm feature in WL 16 had been reported by Tanaka et al. (1990), who found that this was the only one of a sample of 31 Ophiuchus sources to show the feature.

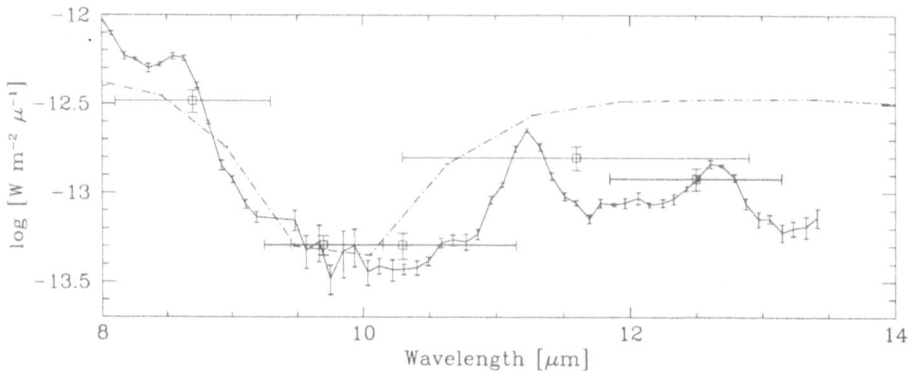

Fig. 1. 8–14μm spectrum of WL 16: square symbols (photometric points, bandwidths indicated), dot-dash curve (ALS model), curve with error bars (observed spectrum).

It can be seen from the spectrum in Fig. 1 that the strong hydrocarbon emission features occuring between 7μm & 13μm completely dominate the luminosity in this wavelength range. In fact there is approximately 0.7 $L_\odot$ in the observed 8.2μm band (assuming a distance to ρ Oph of 160 pc), and from images of the spatially resolved emission (Emerson et al 1996) we estimate that the total hydrocarbon luminosity between 7μm & 13μm is around 2$L_\odot$, somewhat higher than the Hanner et al. (1992) estimate of $\sim 1 L_\odot$ obtained from a single aperture spectrum. Estimates of the bolometric luminosity of WL 16 range up to 57 L$_\odot$ (Comeron et al. 1993).

PAH emission is thought to be due to fluorescence following absorption of UV photons and 3.29μm PAH emission from YSOs appears to have been detected only towards Herbig Ae/Be stars, and even then in only 20% of those observed (Brooke et al 1993). The presence of mid-IR PAH emission thus supports the idea that WL 16 & WL 22 are not low mass YSOs but intermediate mass ones. They are thus probably deeply embedded Herbig Ae stars. The ALS models for WL 16 & WL 22 should be considered in the light of this new information.

The strength of the PAH features, which bracket any 9.7μm absorption, appears to have caused a considerable overestimate of the silicate optical depth based on the earlier broad-band photometry, as may be understood by studying Fig. 1. However there could be an underlying silicate absorption feature and Hanner et al.(1992) concluded that, although there is no direct evidence for a silicate feature in their spectrum, silicate absorption arising in the extended foreground cloud is probably present with an optical depth of $\sim$ 0.7. In Fig. 2 our 8–24μm spectra of WL 16, WL 22 & El 29 are shown together with 8–14μm spectra of the protoplanetary nebula IRAS21282+5050, a well studied PAH emitter, and the Ophiuchus YSO YLW 13b (which is discussed below). The dips in the 20μm spectrum of El 29 reflect imperfect cancellation of atmospheric features.

The PAH features in WL 16 are notable for their strength relative to the continuum, and the prominence of the plateau emission and the strength of the

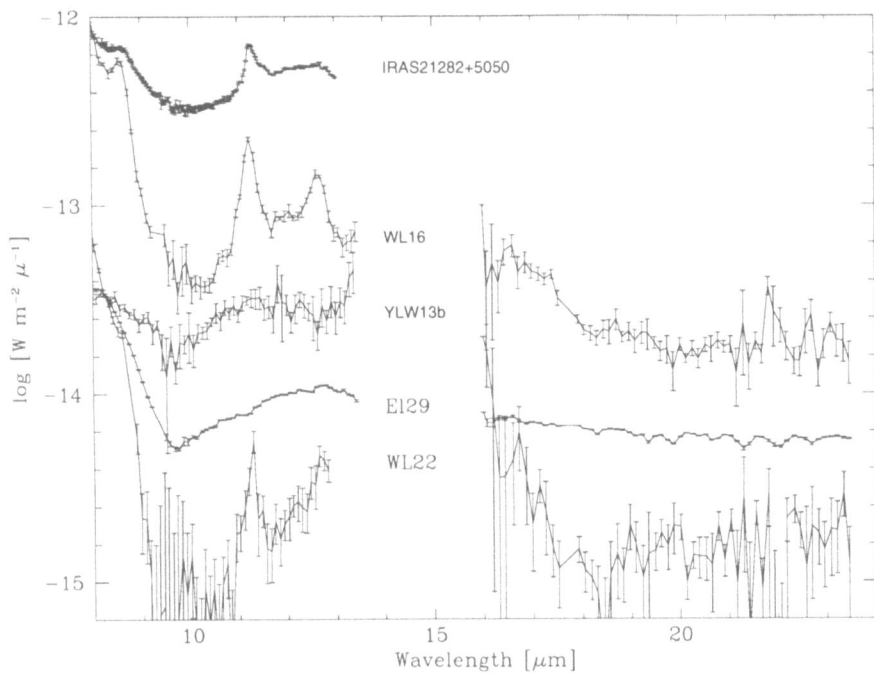

Fig. 2. 8-24μm spectra. To avoid overlaps spectra have been shifted vertically as follows. WL 16 (+0.0), WL 22 (-0.7), El29 (-1.9), YLW13b (-0.3), IRAS21282+5050 (-0.2)

12.7 μm feature. Comparing with the spectrum of El 29, which shows a clear silicate absorption feature with an optical depth of 1.5 (Hanner et al 1995) it can be seen that there is indeed probably some silicate absorption present in WL 16. WL 22 is fainter than WL 16 but, within the uncertainties, its 8–14μm spectrum is similar to that of WL 16 and it is likely another example of a deeply embedded Ae star exciting PAH emission. Relative to their 8-14μm emission, WL 22 is ~ 3 times, & El 29 is ~ 2 times brighter at 20μm than WL 16. WL 16 thus has a relative lack of 20μm emission. This is probably related to the well known absence of sub-mm emission from WL 16 (André & Montmerle, 1994).

Clearly considerable caution has to be exercised in interpreting photometric points in the 10μm region in terms of silicate features, and it is desirable to obtain spectra to properly characterise the spectrum. PAH emission can be very important in the mid-IR, even in highly embedded objects. Elsewhere in this volume we show that the mid-IR PAH emission in WL 16 is extended and flattened (Emerson et al 1996)

3   X-rays from YSOs in Ophiuchus

Casanova et al (1995) in a ROSAT based study of YSOs in Ophiuchus, found that most Class II/IIIs in Ophiuchus are associated with X-ray emission, and suggested that one of the effects of X-rays might be to cause PAHs to be released from dust grains. However they noted that there were 11 Class II/III YSOs in Ophiuchus that did not show X-ray emission, and were unable to assign a reason to this. Two of these 11 are WL 16 & WL 22, the same two YSOs that stand out as having mid-IR PAH emission (although no extensive survey of the 8–14μm spectra in Ophiuchus has yet been made). It seemed interesting that WL 16 & WL 22 appeared in two lists of peculiar objects and this raised the prospect that there might be some connection between the lack of detectable X-ray emission and the presence of mid-IR PAH emission.

A possible mechanism might be that there may be so much dust around WL 16 & WL 22 that the X-rays are completely absorbed by it, and these absorbed X-rays cause ejection of PAHs from larger dust particles, the X-rays thus being responsible for the presence of the PAH material. The PAH excitation mechanism would involve UV photons as usual. WL 16 & WL 22 are the two brightest of the 11 objects noted by Casanova et al (1995), so to investigate this speculation we obtained an 8–14μm spectrum of the next brightest object in this sample YLW13b. The spectrum is shown in Fig. 2 but shows silicate absorption with no evidence for PAH emission. Although it is intriguing that WL 16 & WL 22 are weak in X-ray emission and strong in mid-IR features the lack of PAHs in YLW13b does not encourage attribution of a physical connection between the mid-IR emission and absoption of X-rays.

4   Conclusion

Spectra, rather than photometry, should be used to define the presence of silicate features. Modelling of WL 16 & WL 22, and no doubt other objects, should include PAH emission and attempts to match the 8–24μm spectra. PAH emission may be more widely present in YSOs than is widely appreciated, and the source of the PAHs needs to be understood.

References

Adams, F.C., Lada, C.J., Shu, F. (1987): ApJ, **312**, 788
André, P., Montmerle, T. (1994): ApJ, **420**, 837
Brooke, T.Y., Tokunaga A.T., Strom, S.E. (1993): AJ, **106**, 656
Casanova, S., Montmerle, T., Feigelson, E.D., André, P. (1995): ApJ, **439**, 752
Comeron, F., Rieke, G.H., Burrows, A., Rieke, M.J., (1993): ApJ, **416**, 185
Emerson, J.P., Moore, T.J., Skinner C.J., Meixner, M.M. (1996): this volume
Hanner, M.S., Brooke, T.Y., Tokunaga, A.T. (1995): ApJ **438**, 250
Hanner, M.S., Tokunaga, A.T., Geballe, T.R., (1992): ApJ, **395**, L111
Lada, C.J., Wilking, B.A. (1984): ApJ **287**, 610
Tanaka M., Sato, S., Nagata, T., Yamamoto, T. (1990): ApJ, **352**, 724
Wilking, B.A., Lada, C.J. (1983): ApJ, **274**, 698

Mid-IR Imaging of YSOs:
the Hydrocarbon Emission Features in WL 16

J.P. Emerson[1], T.J.T. Moore[2], C.J. Skinner[3], and M.M. Meixner[4]

[1] Dept of Physics, Queen Mary & Westfield College, Mile End Road, London E1 4NS
[2] Physics Dept, University of New South Wales, ADFA, Canberra ACT 2600, Australia
[3] Space Telescope Science Institute, 3700 San Martin Drive, Baltimore, MD 21218
[4] Dept of Astronomy, MC 221, University of Illinois, Urbana- Champaign, IL 61801

Abstract. Arcsecond resolution images of the unusual young object WL 16 in Ophiuchus, taken at wavelengths of 8.2, 10 and 11.3μm, show an extended flattened envelope of fluorescing hydrocarbon molecules. The extended hydrocarbon emission probably traces a flattened, equatorial distribution of circumstellar material, or may arise in bipolar lobes. To the limit of achieved sensitivity, the faint 10μm continuum has a surface-brightness distribution that is not distinguishable from those at 8.2μm and 11.3μm, where the luminosity is known to be dominated by the PAH emission features. We conclude that the 10μm continuum arises either from non-equilibrium heating of small dust grains that are well mixed with the hydrocarbons or is quasi-continuous emission from the PAH particles themselves, rather than thermal equilibrium emission from macroscopic dust grains, and that there is no significant extinction or silicate absorption variation across the source. An origin of the extended emission in a disk rather than bipolar lobes gives less problems with currently available data and implies that WL 16 is a relatively evolved, highly obscured, Herbig Ae star whose equatorial plane has been almost cleared of normal dust, leaving only fluorescing hydrocarbons and larger coagulated particles as a possibly transient fossil of the original circumstellar disc.

1 Introduction, Observations, & Results

We present high-resolution mid-infrared imaging observations of the optically invisible YSO WL 16 in three 9.2% resolution wavebands at 8.2μm (covering the 7.7μm C–C stretch and 8.6μm C–H in-plane bend hydrocarbon emission features), at 11.3μm (covering the 11.3μm C–H out-of-plane bend and the 12μm plateau features), and at 10μm (for the thermal continuum). The data were obtained in order to investigate the relative spatial structure of the PAH and continuum emission, with a view to understanding the excitation and distribution of the hydrocarbon emission (Hanner et al 1992, Emerson 1996).

The observations were made at UKIRT using the 10μm camera developed at the Space Sciences Laboratory at the University of California at Berkeley (Keto et al. 1992). The imaging element was a rectangular 10×64 pixel, gallium-doped silicon hybrid detector array and the plate scale was 0.39'' per pixel.

Figures 1 – 3 show the surface-brightness distributions at 8.2μm, 10μm and 11.3μm, respectively. The 10μm contours (Fig. 2) have been smoothed by convolution with a 2-dimensional, $\sigma = 0.5''$ Gaussian function. WL 16 is extended at all three wavelengths having an elliptical projected structure with major axis orientated at p.a. ∼ 60° (east of north), and an extent along the major axis of 8″ to 10% of the peak surface brightness. Within the uncertainties the surface-brightness distribution is very similar at all three wavelengths and also consistent with the 12.5μm image presented by Hoffman et al. (1994) and the 8.6 & 11.2μm images presented by Deutsch et al (1995). 3.8μm speckle observations by Zinnecker et al. (1988) showed an unresolved (< 0.06″) core surrounded by a roughly symmetrical 3.5″halo, a smaller extent and different shape than found in our mid-IR data.

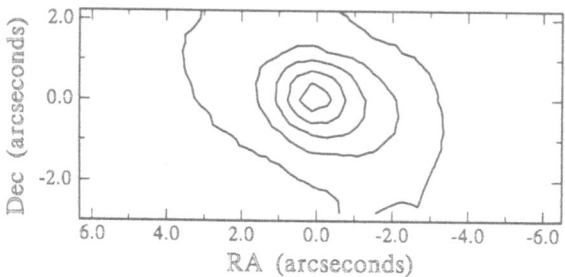

Fig. 1. WL 16 at 8.2μm. Contours are plotted at intervals of 600 mJy arcsec^{-2}(18σ) from 300 mJy arcsec^{-2}(9σ). Offsets are relative to RA 16^h 24^m 00.3^s, Dec -24° 30′ 44″ (1950.0).

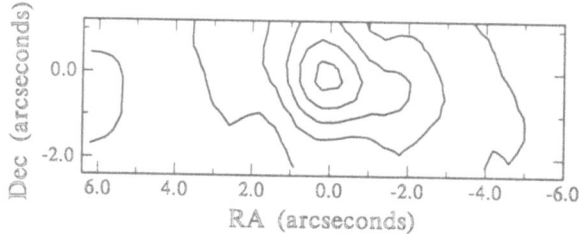

Fig. 2. WL 16 at 10μm, smoothed. Contours are plotted at intervals of 30 mJy arcsec^{-2}(6σ in the smoothed image) from 25 mJy arcsec^{-2}(8σ).

2   Discussion

The Full Width of the source to 10% corresponds to 1280×640 AU at the distance of the Ophiuchus star forming region. These extents are broadly consistent with models of PAH emission around Herbig Ae/Be stars (Natta & Krügel 1995).

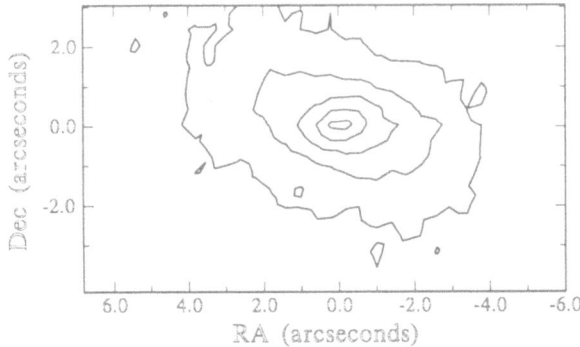

Fig. 3. WL 16 at 11.3μm. Contours are plotted at intervals of 150 mJy arcsec^{-2} (6σ) from 75 mJy arcsec^{-2} (3σ).

If we assume that the distribution and excitation of the 8.2 & 11.3μm emitting PAHs are the same, the similarity of the 8.2μm and 11.3μm emission suggests that there can be no strong gradients in silicate absorption, or in extinction, across the object.

A two dimensional χ^2 test shows that the 10μm image is more similar to the 8.2 & 11.3μm images than to a point source, although we note that the 10μm continuum image has lower signal to noise ratio than the other images. This suggests that the continuum emitters have the *same* distribution and excitation as the PAHs and that the continuum is from small grains or PAHs, and that there is little emission from MRN type grains at 10μm. The absence of such material may be connected with the absence of detectable 1.3mm continuum emission from WL16 (André & Montmerle, 1994).

The flattened structure suggests that the mid-IR emission predominately arises either in a disk, or in bipolar lobes, although the structure is not clearly bipolar. The 2.3μm CO band-head emission seen in WL 16 has been interpreted as an edge-on Keplerian accretion disk (Chandler et al 1993), implying that the axis of such a disk lies in the plane of the sky and Dent & Geballe (1991) reported detection of high velocity CO(2 → 1) emission to the North-West of WL 16 suggesting an outflow lieing NW-SE and hence that the mid-IR emission (which lies approximately NE-SW) must lie in a disk.

What is the lifetime of the material in this disk? If we take the central object to be a 25 L$_\odot$, 2.5 M$_\odot$ Ae star then grains in the size range 0.01-3μm will be pushed out to > 10^4 AU in < 10^4 yr by radiation pressure and smaller (PAH) grains within ∼ 650 AU should spiral in to the star in < 10^4 yr by Poynting- Robertson drag. Thus the disk should have been cleared of almost all normal dust, leaving only PAH molecules and larger, coagulated particles, and the observed extended mid-IR emission may be transient, unless material is replenished from some reservoir of particles.

This 10^4 yr lifetime may be compared to a crudely estimated PAH emission phase lifetime of a pms Ae star by taking the pre-main sequence lifetime of an A star as a few 10^6 yr and noting that Brooke et al (1993) find that ~20% of Herbig AeBe stars have $3.28\mu m$ emission, leading to an expected lifetime of the PAH emission phase of ~ 2×10^5 yrs, a factor of 20 longer than the deduced lifetime of the observed disk. If WL 16 is typical of this lifetime then there must either be a reservoir of material extending out to ~ 3000 AU which will replenish the emitting material to make the PAH emission phase last for ~ 10^5 yr, or we are observing WL 16 at the end of its PAH emitting phase and seeing a transient fossil of the circumstellar disc.

3   Conclusion

Extended mid-IR PAH emission in embedded YSOs might be much more common that is generally appreciated, and PAHs and small grains lieing in a flattened distribution should be considered for incorporation into models of the spectrum and morphology of embedded YSOs.

Acknowledgements

The United Kingdom Infrared Telescope is operated by The Royal Observatories on behalf of the UK Particle Physics and Astronomy Research Council.

References

André, P., & Montmerle, T. (1994): ApJ, **420**, 837

Brooke, T.Y., Tokunaga, A.T., Strom, S.E. (1993): AJ **106**, 656

Chandler, C.J., Carlstrom, J.E., Scoville, N.Z., Dent, W.R.F. Geballe, T.R. (1993): ApJ,**412**, L71

Dent, W.R.F., Geballe, T.R. (1991): A&A, **252**, 775

Deutsch, L.K., Hora, J.L., Butner, H.M., Hoffmann, W.F., Fazio, G.G. (1995): Ap&SS **224**, 89

Emerson, J.P. (1996): this volume

Hanner, M.S., Tokunaga, A.T., Geballe, T.R., (1992): ApJ, **395**, L111

Hoffman, W.F., Fazio, G.G., Shivanandan, K., Hora, J.L., Deutch, L.K., (1994): Infrared Phys. Technol., **35**, 175

Keto, E., Ball, R., Arens, J., Jernigan, G., Meixner, M., (1992): International Journal of Infrared and Millimeter Waves, **13**, 1709

Natta, A., Krügel, E., (1995): A&A, **302**, 849

Zinnecker, H., Perrier, C., Chelli, A., (1988): *High Resolution Imaging by Interferometry*, ed. Merkle, F, ESO Conference Proceedings 29, 505

Star Formation in the Vela Molecular Clouds: Near IR Images[1]

T. Giannini[1], D. Lorenzetti[1], B. Nisini[1], L. Spinoglio[1], A. Zavagno[1], R. Liseau[2], P. Andreani[3], and A. Moneti[4]

[1] Istituto di Fisica Spazio Interplanetario CNR - Frascati (Italy)
[2] Stockholm Observatory - (Sweden)
[3] Dipartimento di Astronomia - Padova (Italy)
[4] ESA/ESTEC - (The Netherlands)

Abstract. We present the first results of a sensitive (K $\approx$ 17 mag) near IR (JHK) imaging survey of a complete IRAS selected sample of Young Stellar Objects (YSO's) belonging to a Giant Molecular Cloud (GMC) located in the Vela Molecular Ridge (VMR). The same objects have also been observed at $\lambda = 1.3$mm. We provide identifications (*i.e.* NIR counterparts) and detailed spectral energy distributions (SED's) for ten newly identified Class I objects. The envelope masses are consistent with central masses derivable from the observed luminosities. Detected multiplicity indicates that intermediate mass star formation might not represent an individual and isolated phenomenon.

1 Observations

We initiated few years ago the study of the young stellar content in the VMR constitued by four different molecular clouds (each containing $10^5 \div 10^6 \, M_\odot$) which are actively forming stars (Murphy & May 1991). The selection criteria to pick up the young objects, the ascertainment of their cloud membership, and the observational results obtained so far have been communicated in our previous papers (Liseau et al. 1992; Lorenzetti, Spinoglio & Liseau 1993)

In the following we present new results of the near IR imaging and mm-continuum observations for a sub-sample of sources which :

- belong to the VMR-D cloud
- have valid fluxes (no upper limits) in the first three IRAS bands
- have been suggested as *protostellar candidates* in our previous papers

The infrared images were obtained in February 1993 with the IRAC2 near-infrared camera on the ESO/MPI 2.2m telescope at La Silla. The adopted magnification has been 0.5 arcsec/px, thus the field of view corresponds to about 2 $\times$ 2 $arcmin^2$. The limiting magnitude of this survey is K $\approx$ 17.

The 1.3mm continuum observations have been carried out during September 1992 with the SEST telescope at La Silla. The antenna fed a ^3He-cooled bolometer of the MPIfR. The beam size is 24 arcsec (HPBW) and the chop throw is 70 arcsec.

---

[1] Based on observations collected at ESO, La Silla, Chile

2   Results and Conclusions

The results of our observations are summarized in Table 1.

Table 1. VMR-D Class I objects

Source	IRAS name	J	H	K	$F_{1.3mm}(mJy)$
IRS 62 ⋆	08328-4314	13.78 (5)	11.33 (2)	9.29 (2)	42 (12)
IRS 13	08375-4109	17.7 (20)	12.89 (2)	9.43 (2)	446 (23)
IRS 63	08393-4041	> 18.3	16.3 (10)	14.31 (4)	529 (27)
IRS 14	08404-4033	11.40 (3)	9.96 (2)	8.75 (2)	38 (12)
IRS 67	08445-4420	16.11 (5)	13.82 (2)	11.84 (2)	34 (16)
IRS 17 ⋆	08448-4343	13.98 (5)	11.94 (2)	9.21 (2)	1111 (47)
IRS 18	08470-4243	15.91 (5)	13.74 (2)	11.85 (2)	2510 (116)
IRS 19 ⋆	08470-4321	15.25 (6)	12.29 (2)	8.85 (2)	604 (19)
IRS 20 ⋆	08476-4306	14.37 (4)	12.23 (2)	10.68 (2)	388 (24)
IRS 21	08477-4359	> 18.3	15.19 (3)	12.76 (2)	121 (16)

Source	L_{bol} $(L_\odot)$	β	T_d (K)	M $(M_\odot)$	Ext./Mult.
IRS 62 ⋆	1.2 10^2	-	-	-	Y / 3
IRS 13	9.6 10^2	1.35÷1.72	34.2÷39.8	1.1÷2.5	Y / 4
IRS 63	5.1 10^2	1.13÷1.88	24.8÷36.5	1.0÷5.0	N / 1
IRS 14	2.5 10^2	1.97÷2.45	25.6÷33.0	0.3÷0.8	Y / 10
IRS 67	2.6 10^2	2.25÷2.55	25.7÷30.6	0.4÷0.8	Y / 2
IRS 17 ⋆	3.1 10^3	1.87÷2.40	22.9÷29.2	8.2÷25	Y / 1
IRS 18	5.6 10^3	1.56÷1.90	29.0÷34.5	10÷19	Y / 5
IRS 19 ⋆	1.6 10^3	1.09÷1.75	32.6÷53.0	0.7÷3.3	Y / 6
IRS 20 ⋆	1.6 10^3	1.93÷2.38	24.8÷31.2	2.9÷7.8	Y / 1
IRS 21	1.8 10^3	2.37÷2.98	24.0÷32.0	1.7÷5.7	Y / 3

- *Col.1* - Source identification according to our classification [⋆ indicates that JHK magnitudes have been obtained by integrating extended emission].
- *Col.2* - IRAS name. To select the most probable IRAS counterpart among the NIR objects in a given field, the following criteria had to be met by the NIR and/or IRAS sources: (1) The slope of the NIR SED had to be positive and a match to the IRAS SED should be obtained (Figure 1). (2) The NIR object had to be located inside the IRAS uncertainty ellipse. (3) The NIR colours should evidence the presence of an intrinsic infrared excess emission (Figure 2 displays, as an example, the two NIR colours diagram of the field associated to IRS 17).
- *Col.3-5* - NIR magnitudes with the rms-errors (1σ) in units of 0.01 mag given in parentheses.
- *Col.6* - Observed flux densities at 1.3 mm and relative 1σ errors.

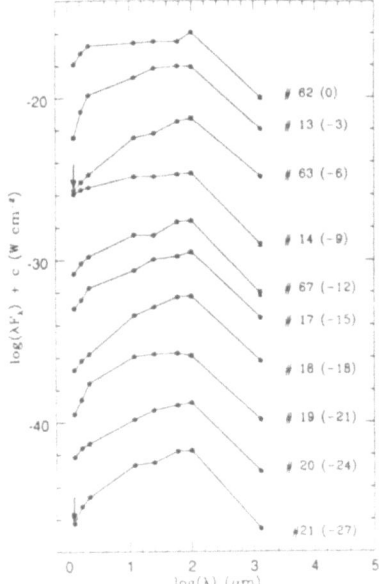

Fig. 1. SED's, scaled by the factor 10^c indicated next to each curve, of Class I sources in the VMR–D cloud. Upper limits are indicated with an arrow.

Fig. 2. [J–H] *vs* [H–K] diagram showing the colours of all the 66 individual sources detected within the 2 *arcmin* field centered on IRS17. Limits are given as open symbols indicated by an arrow. The object identified as the counterpart of the IRAS source is labelled as IRS17.

- *Col. 7* - Bolometric luminosities estimated by integrating the observed SED's (Figure 1) from 1.25 μm to 1.3mm. As expected the bolometric luminosities are dominated by the IRAS and submm emission. In addition, as a consequence of our selection, they are relatively high ($120 \leq L/L_\odot \leq 5600$) and probably reflect central masses $1 < M/M_\odot < 8$, *i.e.* protostellar candidates of intermediate mass.

- *Col.8-10* - Estimate for some relevant parameters: the dependence of the dust opacity on wavelength β ($\kappa_\lambda = \kappa_0 \, (\lambda_0/\lambda)^\beta$), the envelope total mass M_{env}, and the dust temperature T_d computed from the 60, 100 and 1300 μm flux densities. Assuming an optically thin emission at 1.3mm and an isothermal uniformly distributed population of dust grains, using the formalism of Hildebrand (1983), we have obtained the best fit to the data points by considering β, M_{env} and T_d as free parameters.

To evaluate the uncertainties related to the obtained parameters, two different fits have been obtained to the 60, 100 μm and 1.3mm emission taking into account the errors on the IRAS fluxes: the first (second) curve passes through the minimum (maximum) flux at 60 μm and the maximum (minimum) at 100 μm, both compatible with a 1σ error in the IRAS fluxes. Therefore two limiting values are given for each parameter (an upper limit at 100 μm

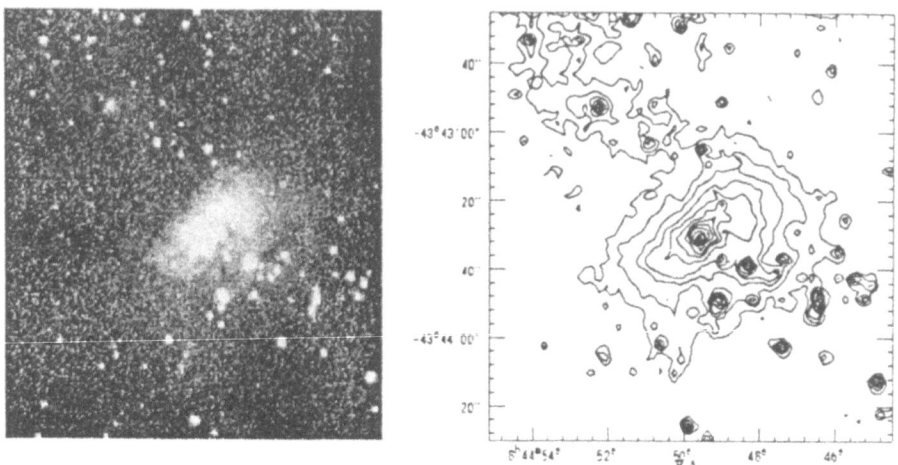

Fig. 2. K frame of the source IRS 17 along with the relative contour plot

prevents the estimate for IRS 62). Due to the large apertures, additional contributions, not related with the source, could contaminate the IRAS fluxes, leading possibly to overestimate β and M_{env}. Having this *caveat* in mind, the obtained values of β ($<\beta>\approx$ 2) are in agreement with the prescription given by Gordon (1995) for dense dark clouds; the envelope masses are consistent with the intermediate central masses derivable from the observed luminosities; the temperatures correspond to the coldest sources detectable by IRAS.

Col. 11 - This column refers to the analysis of the K frames contour plots (*e.g.* Figure 3). The flag (Y/N) indicates whether or not extended emission ($\geq$ 15 arcsec) is associated to the source(s), while the number indicates how many individual condensations, showing an IR excess, are located within the extended emission. Almost all the sources have extended emission and are composed by multiple condensations. This finding in VMR and the same result by Chen et al. (1993) for L1641, indicates that intermediate mass star formation might not represent an individual and isolated phenomenon.

References

Chen, H., Tokunaga, A.T., Strom, K.M., Hodapp, K.-W., (1992): ApJ, **407**, 639.
Gordon, M.A. (1995): A&A, in press.
Hildebrand, R.H. (1983): QJRAS, **24**, 267.
Liseau, R., Lorenzetti, D., Nisini, B., Spinoglio, L., Moneti, A. (1992): A&A, **265**, 577.
Lorenzetti, D., Spinoglio, L., Liseau, R. (1993): A&A, **275**, 489.
Murphy, D.C., May, J. (1991): A&A, **247**, 202.

Linear and Circular Imaging Polarimetry of the Chamaeleon Infrared Nebula

T.M. Gledhill[1], A. Chrysostomou,[1] and J.H. Hough[1]

Division of Physical Sciences,
University of Hertfordshire,
College Lane,
Hatfield
U.K.

Abstract. We present linear and circular imaging polarimetry observations of the Chamaeleon Infrared Nebula, a bipolar reflection nebula in the Chamaeleon I dark cloud, at near infrared (JHKn) wavelengths. The detection of both high degrees of linear polarisation and a significant degree of circular polarisation in the extended nebulosity allows us to comment on the scattering geometry and the range of particle sizes present.

1 Introduction

Measurements of linear polarisation produced by the scattering of light from dust particles in the environments of Young Stellar Objects (YSOs) have been used for many years to place constraints on scattering geometries and on the range of sizes of particles present and, to some extent, their composition. These observations, however, only allow us to sample a part of the scattering properties of the dust, with no information provided on the scattering efficiencies leading to circular polarisation. Circular polarisation is produced when polarised light is scattered. The amount of circular polarisation produced depends on the properties of the scattering medium and the degree of linear and circular polarisation present in the incident light. In regions where YSOs are found, we expect circular polarisation to be produced as a natural consequence of the higher optical depths which require that photons reaching the observer have undergone more than one scattering.

In this contribution, we present both linear and circular imaging polarimetry of the Chamaeleon Infrared Nebula, a bipolar reflection nebula in the Cha I dark cloud (Schwartz & Henize 1983, Cohen & Schwartz 1984, Scarrott *et al.* 1987). This object (hereafter ChaIRN) is located towards the centre of the Cha I cloud and is illuminated by an optically obscured source believed to be associated with IRAS 11072-7727. We discuss the results of our near-infrared polarimetry in the context of a probable geometry for the system and the local grain size distribution.

2   Observations

Observations were made using the IRISPOL imaging polarimeter at the f/15 Cassegrain focus of the 3.9m Anglo-Australian Telescope resulting in a pixel scale of 0.6 arcsec/pixel. IRISPOL is a dual beam polarimeter with both linear and circular polarimetry modes.

Linear polarimetry was obtained using standard J and H filters and a narrow K_n filter (2.0 → 2.3μm) on 22 May 1994. Circular polarimetry was obtained using the H filter during the circular mode commissioning run on 20 May 1995.

When operating in linear mode, the polarimeter incorporates a stepped half waveplate and Wollaston prism as analyzer. When in circular mode, the half waveplate is replaced with a stepped quarter waveplate to convert incident circularly polarised light to linearly polarised light which is then analyzed by the Wollaston prism. In addition, the circular mode configuration includes a constantly rotating half waveplate to elliminate incident linearly polarised light which would otherwise contaminate the observations.

3   Results

3.1   Linear Polarimetry

In Figure 1 we show greyscale images (left) and linear polarisation maps superimposed on intensity contours (right) of the ChaIRN in the three observed wavebands.

The ChaIRN is a visually striking object dominated by the bright lobe of nebulosity extending more than 1 arcmin to the east of the central source region. The central region contains a 'knot' of bright nebulosity located at the origin of our coordinate system but the source itself remains obscured. At these near-infrared wavelengths, the bipolar nature of the object is clearly evident, with a counterlobe of nebulosity extending to the west. In the longer wavelength H and K_n filters a 'spur' of nebulosity can be seen to the south east of the central region which we assume delineates the rim of the obscured western lobe/cavity.

The linear polarisation maps shown in Figure 1 reveal that ChaIRN is a bipolar reflection nebula at NIR wavelengths. The curvature of the polarisation pattern in the eastern and western lobes of the nebula, characteristic of illumination by a point source, indicates an illuminator close to, but not coincident with, the origin of our coordinate system. The polarisation pattern overlying the central regions exhibits the pattern of aligned vectors (rather than a centrosymmetric pattern) that has now become the signature of multiple scattering in an optically thick environment. Degrees of polarisation are typically 20 → 30% in the eastern lobe but rise to up to 60% along the bright northern rim and in the western counterlobe.

It is likely that a proportion of the polarisation observed towards ChaIRN is produced by dichroic extinction by aligned grains. The polarisation pattern is seen to be most centrosymmetric in K_n where extinction effects are weaker than at the shorter wavelengths.

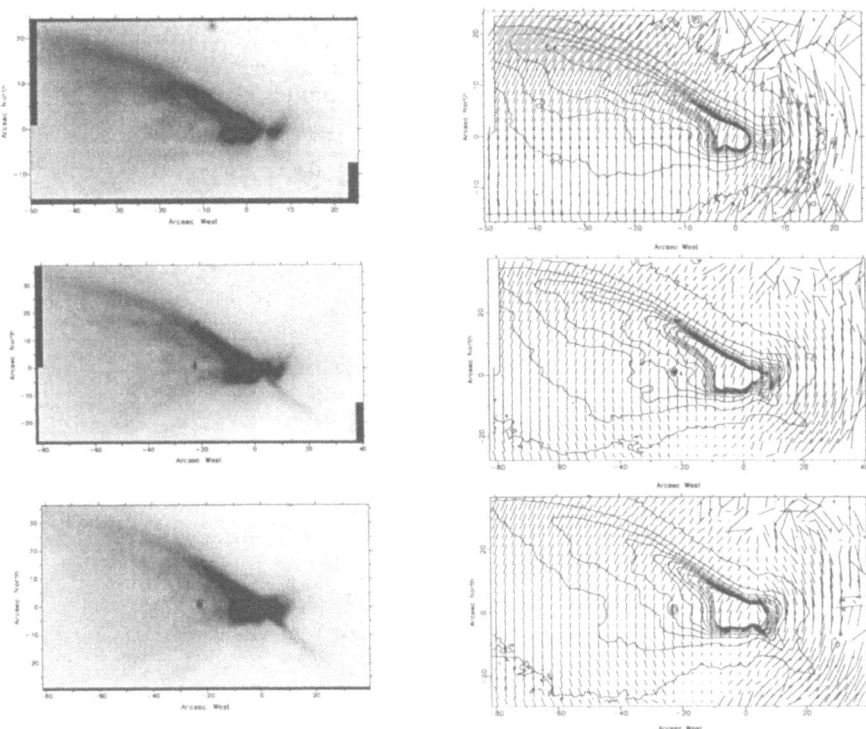

Fig. 1. On the left, greyscale surface brightness images of the Chamaeleon Infrared Nebula in the J, H and K_n wavebands (top to bottom). The coordinate system is centred on the bright knot at the apex of the nebular lobe. North is up and East is to the left. On the right, corresponding linear polarisation maps superimposed on contours of surface brightness.

3.2   Circular Polarimetry

In Figure 2 we present the first imaging circular polarimetry results for ChaIRN in the H waveband. The greyscale intensity image shows the V Stokes parameter with the same coordinate system as the linear polarimetry data. The sign reversal in V stokes between quadrants centred on the source position (changes between black and white greyscale levels) can be clearly seen, corresponding to changes in the handedness of the circular polarisation. Degrees of circular polarisation detected in the eastern nebular lobe lie in the range $-0.5 \rightarrow 0.5\%$.

The detection of circular polarisation, not only in the central source region, but in the main body of the reflection nebulosity, immediately allows us to infer two things: Firstly, throughout the nebula photons have undergone more than one scattering event (since circular polarisation can only be produced by scattering light that is already linearly polarised). These additional scatterings are most likely to occur in optically thick regions, perhaps in a circumstellar disk surrounding (and obscuring) the source. Secondly, the grains responsible for scattering the light cannot appear as Rayleigh particles at NIR wavelengths (since Mie scattering from Rayleigh particles does not induce circular polarisation).

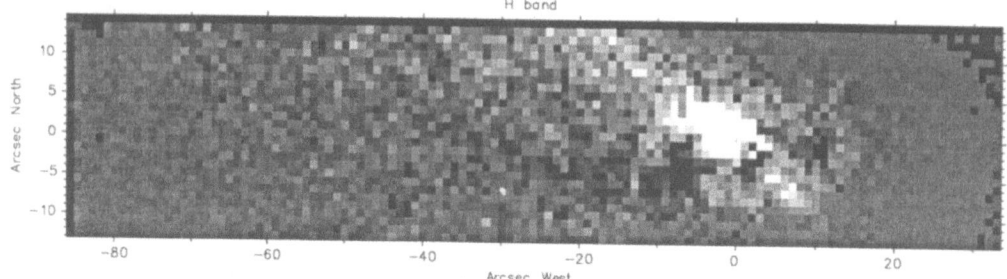

Fig. 2. A greyscale image of the V stokes parameter providing an image of the Chamaeleon Infrared Nebula in circularly polarised light.

4   Conclusion

The detection of high levels of linear polarisation in all three wavebands (up to 60%) combined with the detection of significant levels of circular polarisation in the H waveband suggests that a broad distribution of particle sizes is present in this object. Small grains ($< 0.3\mu$m) are required to produce the observed large linear polarisations at J whereas larger ($> 0.6\mu$m) particles are required to produce the observed degree of circular polarsation at H, assuming the properties of astronomical silicate grains (Draine 1985).

A more detailed analysis of our polarimetry results and their implications in terms of nebular geometry and likely grain parameters will be presented in a future publication currently in preparation.

5   References

Cohen, M. & Schwartz, R.D. 1984, AJ **89**, 277

Draine, B.T., 1985, Ap.JS **57**, 587

Scarrott, S.M., Warren-Smith, R.F., Wolstencroft, R.D. & Zinecker, H 1987, MNRAS **228**, 827

Schwartz, R.D & Henize, K.G. 1983, AJ **88**, 1665

Near-IR Speckle Imaging of Luminous Young Stellar Objects

Melvin G. Hoare[1], Andreas Glindemann[1], and Andrea Richichi[1,2]

[1] Max-Planck-Institut für Astronomie, Königstuhl 17, D-69117 Heidelberg, Germany
[2] Osservatorio Astrofisico di Arcetri Largo E. Fermi, 5, 50125 Firenze, Italy

Abstract. We present high resolution near-IR images of luminous young stellar objects using the speckle technique. Three objects are presented here which show small scale extended emission thought in most cases to be light reflected off the dusty walls of outflow cavities. The morphology can be used to constrain models of the outflow geometry and collimation mechanism.

1 Introduction

Reflection nebulosity around luminous young stellar objects (YSOs) is a common occurrence and appears to be primarily due to light reflected off the dusty walls of cavities evacuated by bipolar outflows (e.g. Tamura et al. 1991). Extinction by a dusty disk or torus perpendicular to the outflow often obscures light from the receding lobe. Monte Carlo models can be used to test if particular outflow geometries can match the observations (Whitney & Hartmann (1993); Fischer et al. 1995). Tracing the reflected light closer to the star enables a more accurate derivation of the opening angle and degree of collimation which are an important test of any model. Speckle imaging is effective in revealing subarcsec structures (e.g. Koresko et al. 1993) and it was in order to find out how common these structures are, their different morphologies, and their relation to other outflow phenomena that our study was undertaken.

2 Observations

Fifteen luminous YSOs were observed with the 3.5m telescope at Calar Alto using the MAGIC 256×256 NICMOS 3 camera. The pixel scale was set at 0.07″ per pixel in order to sample the diffraction limit at K. 1500 128×256 frames with an integration time short enough to freeze the seeing (0.06 to 0.2 seconds) were accumulated on the object interspersed with a similar number on one or more reference stars less than 1° away. The visibility amplitude was estimated from the power spectra in the usual way and the phases reconstructed from the bispectrum using the least squares method of Glindemann et al. (1992). A gaussian apodizing function was applied in order to reconstruct the images, often to a slightly lower resolution than the diffraction limit in order to bring out the faint reflection nebulosities. Artifacts from the data processing were typically at the ±1–2% level within an arcsec of the source as judged from reconstructions of reference stars against each other.

3   Results

Of the sources surveyed six show small scale reflection nebulosity and we present the results for three of these here. Fig. 1 shows the reconstructed H-band image of GL 490 (no extension was seen at K). We see faint emission extending along the NW-SE direction which begins to curve off at the ends towards the SW which is the direction of the approaching lobe of the large scale outflow from this object. The nebula is similar in size and orientation to the extended radio emission seen by Campbell et al. (1986). Haas et al. (1992) using 1D speckle interferometry and polarimetry found the near-IR halo to be much brighter and its light centre shifted to the SW. It would appear that the reflection nebula has faded and retreated closer to the star in the last few years.

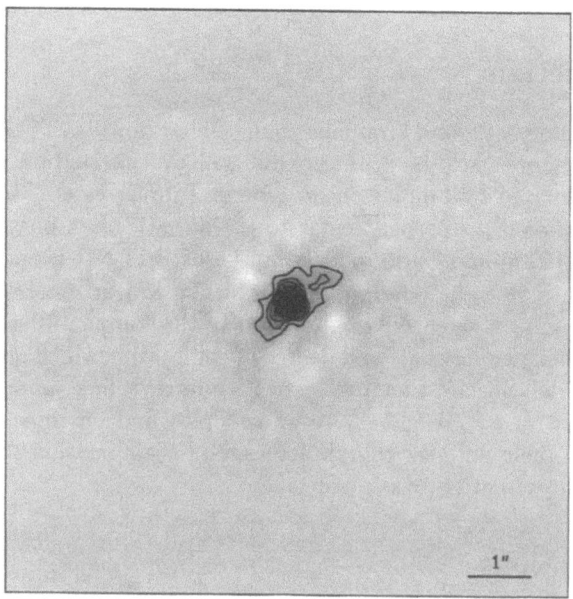

Fig. 1. Reconstructed H-band image of GL 490 with 0.23″ resolution. The greyscale ranges from -2.5% (white) to 9% (black) and the contours are at 1.5, 3, and 4.5%.

In Fig. 2 we present a K-band image of S140 IRS 1 where significant emission is seen extending to the SE. The strong linear feature gives the impression that it could be the southern wall of highly collimated SE cavity with the northern wall also present but shorter. A cavity in this direction would be consistent with the orientation of the bipolar CO flow (Minchin et al. 1993), but not with the elongated radio emission which is at a position angle of 45° and has been interpreted as a thermal jet (Hoare & Muxlow 1995). If instead the radio emission is from an equatorial wind as for S106IR (Hoare et al. 1994) then this would be somewhat more consistent with the outflow geometry.

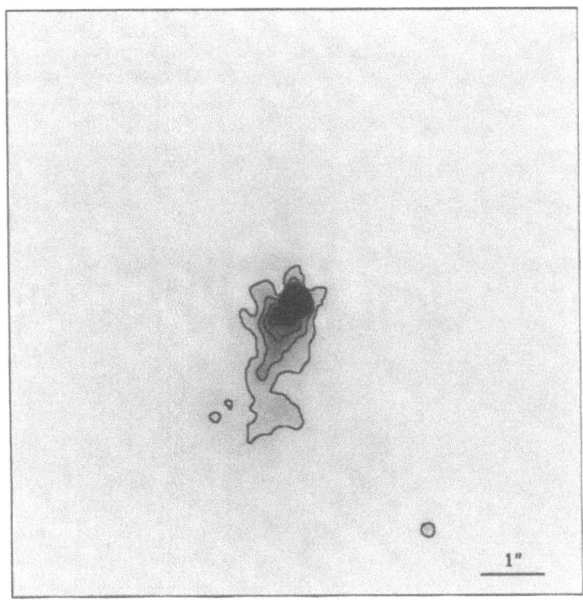

Fig. 2. Reconstructed K-band image of S140 IRS 1 with 0.19″ resolution. The greyscale ranges from -2% (white) to 8% (black) and the contours are at 1, 2.5, 4, 5.5, 7 and 8.5%.

A more complex region is S255 IRS 1 shown in Fig. 3. Tamura et al. (1991) made a low resolution K-band imaging polarimetry study of the extensive reflection nebulosity in this region. In their notation we see IRS 3 and 22 as well as IRS 1 and the two reflection nebulosities IRN 1 and 2. IRN 1 is shown to be a conical nebula with an opening angle of 35° originating from IRS 3 which polarization shows is its illuminating source. Our observations confirm the suggestion by Tamura et al. that there is a second reflection nebula, IRN 2, to the north of IRS 1. It has a spatially distinct cometary shape which does not point precisely back to IRS 1, which may mean that it is a illuminated by another, more deeply embedded source. There does appear to be reflection nebulosity to the SW of IRS 1, although the reality of the linear feature emanating from the NE is questionable.

4   Conclusions

Speckle imaging has revealed the subarcsec structure of several IR reflection nebulae around luminous YSOs which can be used to infer the outflow orientation, opening angle and collimation on scales of a few 100 AU. Improved sensitivity

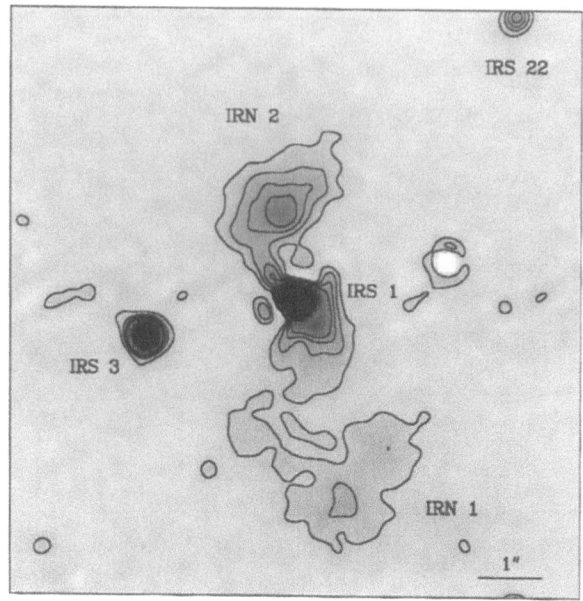

Fig. 3. Reconstructed K-band image of S255 IRS 1 with 0.29″ resolution. The greyscale ranges from -1% (white) to 4% (black) and the contours are at 0.5, 1, 1.5 and 2%. The negative feature (-3.5%) west of IRS 1 is a 'ghost' of IRS 3.

and dynamic range will come with adaptive optics and the NICMOS camera on HST. This will allow the outflow geometries in a much wider range of objects to be studied enabling the key determining processes to be identified.

References

Campbell B., Persson S. E., McGregor P. J. (1986): ApJ, 305, 336

DeWarf L. E., Dyck H. M. (1993): AJ, 105, 2211

Fischer O., Henning Th., Yorke H. W. (1995): A&A, in press

Glindemann A., Lane R. G., Dainty J. C. (1992): J. Opt. Soc. Am. A., 9, 543

Haas M., Leinert Ch., Lenzen R. (1992): A&A, 261, 130

Minchin N. R., White G. J., Padman R., 1993, A&A, 277, 595

Hoare, M. G., Drew J. E., Muxlow T. B., Davis R. J. (1994): ApJ, 421, L51

Hoare M. G., Muxlow T. B. (1995): in J. M. Paredes & A. R. Taylor eds, Radio Emission from the Stars and the Sun, ASP Conf. Ser., in press

Koresko C. D., Beckwith S., Ghez A. M., Matthews K., Herbst T. M., Smith D. A. (1993): AJ, 105, 1481

Tamura M., Gatley I., Joyce R. R., Ueno M., Suto, H., Sekiguchi M. (1991): A&A, 378, 611

Whitney B. A., Hartmann L. (1993): ApJ, 402, 605

A Large Dust Shell Observed at $10\mu m$ Around V921 Sco[*]

Pierre-Olivier Lagage[1], Sylvie Cabrit[2], Thierry Montmerle[1], and Göran Olofsson[3]

[1] CEA/DSM/DAPNIA Service d'Astrophysique, F-91191 Gif-sur Yvette Cédex, France
[2] DEMIRM, Observatoire de Paris, F-75014 Paris, France
[3] Stockholms Observatorium, S-13336 Saltsjöbaden, Sweden

Abstract. While it is generally well accepted that young pre-main-sequence stars of low mass, i.e. "classical" T Tauri stars, are surrounded by a circumstellar accretion disk, the situation is much more controversial for pre-main-sequence stars of intermediate mass, i.e. Herbig Ae/Be stars. Mid-IR imaging observations can help clarifying the situation. Here we discuss observations of V921 Sco made with the TIMMI camera on the 3.6 m telescope of the European Southern Observatory (ESO). These observations have revealed the presence of a large detached envelope, which dominates the emission at 10 μm contrary to expectations from models based on Spectral Energy Distribution (SED) fits.

1 Introduction

Accretion disk models developed to reproduce the mid-IR emission from Herbig Ae/Be stars often require two conditions which are difficult to reconcile: a large accretion rate and an inner opacity hole (Hartmann, Kenyon and Calvet 1993, Natta et al. 1993a). High-resolution imaging observations in the atmospheric window at 10 μm offer a new way of investigating potentially important sources of mid-IR emission other than accretion disks. We have devoted several observing runs to image YSO's at 10 μm, both with the Saclay CAMIRAS camera and the ESO TIMMI camera. Various sources of mid-IR emission other than an accretion disk were found such as:

1. embedded (presumably less evolved), previously unknown, companions, for example near LkHα198, LkHα234 (Cabrit et al. 1993, Lagage et al. 1993a).
2. extended emission, for example from transiently heated grains (Lagage et al. 1993b, see also Prusti et al. 1994, Deutsch et al. 1995, Moore et al. these proceedings).

In this paper, we concentrate on recent observations of V921 Sco (or CD -42°11721 or MWC 865), a relatively well studied Herbig Ae/Be star. Observations at 50 and 100 μm from the Kuiper Airborne Observatory have revealed the presence of a spatially resolved large envelope. When modelling the spectral energy distribution, the contribution of this envelope to the 10 μm flux was found negligible compared to the contribution of the disk (Natta et al. 1993a).

---

[*] Based on observations obtained at the European Southern Observatory, La Silla

2   Observations

V921 Sco was first observed in February 1994 with the TIMMI camera, built by
the Service d'Astrophysique at Saclay (Lagage et al. 1993c). This instrument is
an ESO common user instrument available on the 3.6 m ESO telescope at La Silla
(Chile). The star was further observed in June 1995. Several filters have been
used, both broad-band and narrow-band, centered on the so-called Unidentified
Interstellar Bands (UIB's) in the N atmospheric window (at 7.7, 8.6 and 11.3
μm), as well as centered on the silicate feature at 9.7 μm. The largest pixel
field of view available in TIMMI (0.6″) was used and the chopper throw was set
at 50″. The result of the observation in the 11.3 μm filter is shown in Fig. 1.
A surprisingly large envelope is revealed. The envelope is not symmetrical and
shows an inner cavity.

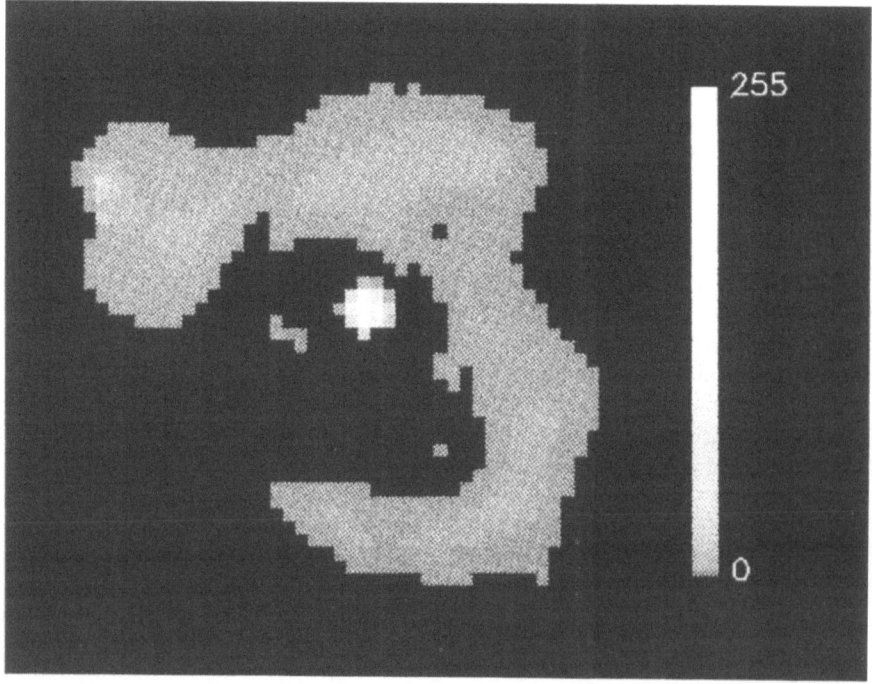

Fig. 1. V921 Sco as observed at 11.3 μm with the TIMMI instrument mounted on
the ESO 3.6 m telescope. The field of the image is 40″; the pixel resolution is 0.6″.
The image has been filtered and deconvolved according to the algorithm developed by
Pantin and Starck (submitted). On-source integration time: 828 seconds.

Photometry was done using η Sgr as reference star. The total flux (star/disk plus envelope) was found to be 110 Jy (+/- 10%). The flux measured at 12 μm with IRAS is 105 Jy, so that all the IRAS flux is contained in our frame. Surprisingly, the flux in the envelope is found to be about twice the flux from the star and its immediate vicinity.

3 Discussion

3.1 Origin of the 10 micron emission from the envelope

The spectral type and the distance of the object are not well determined. Following Natta et al. (1993a), we have adopted the parameters derived by McGregor et al. (1988), i.e. an effective temperature of 15,000 K and a distance of 2000 pc. With such parameters, normal grains in thermal equilibrium with the radiation field from the star (either silicate or graphite), would have too low a temperature to emit significantly at 10 μm.

One solution is to invoke the presence of small transiently heated grains as proposed for reflection nebulae... (Sellgren 1984). Such a possibility has also been evoked to obtain better fits to the SED of some YSO's (Natta et al. 1993b, Seibenmorgen 1993). The need for transient heating is most often required in regions where UIB's are found. We have thus looked for such features in our data. The features at 7.7, 8.6 and 11.3 μm are indeed present. The finding of UIB's around Ae/Be stars is not new; the UIB's have been found in a significant number of Herbig Ae/Be stars (20%), at least at 3.3 μm (Brooke, Tokunaga and Strom 1993).

Note that an explanation in terms of transiently heated grains is not the only possibility. A very good fit to the UIB's and underlying continuum in proto-planetary nebulae have been obtained with the coal model (Guillois et al. submitted). In this model, coal grains are in thermal equilibrium with the radiation field from the star. Given the radiation field from V921 Sco, coal grains could be hot enough to explain our observations (Papoular, private communication).

3.2 Consequences for disk model

Given that the envelope dominates the flux in the 10 μm atmospheric window, disk models for V921 Sco have to be revised. Note however that a circumstellar disk is probably present. The SED of the flux centered on the central pixels in Fig. 1 follows a power law with an index of ∼ -1.75 (λF_λ). This is not the index of a photosphere, but it is in agreement with that expected for a disk, as observed for example in T-Tauri systems. A detailed analysis is underway.

4   Conclusion

The case of V921 Sco illustrates very well the importance of mid-IR imaging observations for a better understanding of Herbig Ae/Be stars. Indeed, such observations provide specific constraints which allow to distinguish between possible disk/envelope models derived from global SED.

Acknowlegments:
We would like to thank Eric Pantin for his help in filtering and deconvolving the images.

References

Brooke, T. Y., Tokunaga, A.T., Strom, S.E. (1993): Dust emission features in 3 μm spectra of Herbig Ae/Be stars. AJ **106**, 656–671

Cabrit, S., Lagage, P.-O., Olofsson, G. (1993): High-angular resolution 10μm imaging of Herbig Ae/be stars. In *The Nature and Evolutionary Status of Herbig Ae/Be stars*, P.S. Thé, M.R. Pérez and P.G. van den Heuvel Eds, ASP conf series, 319–322

Deutsch, L.K., Hora, J.L., Butner, H.M., Hoffmann, W.F. (1995): Extended structures around YSO in mid-IR broad emission features. Ap&SS **224**, 89–92

Guillois, O., Nenner, I., Papoular, R., Reynaud, C.: Coal models for the infrared emission spectra of Proto Planetary Nebulae. ApJ, submitted.

Hartmann, L., Keynon, S.J., Calvet, N. (1993): The excess infrared emission of Herbig Ae/Be stars: disks or envelopes?. ApJ. **407**, 219–231

Lagage, P.-O., Olofsson, G., Cabrit, S., Césarsky, C.J., Nordh, L., Rodriguez Espinosa, J.M. (1993a): A deeply embedded companion to LkHα198. ApJ **417**, L79–L82

Lagage, P.-O., Césarsky, C.J., Jouan, R., Masse, P., Tarrius, A. (1993b): SubArcsec 10 μm imaging of bipolar flow sources. In *Stellar jets and bipolar outflows*, Errico L. and Vittone A.A. Eds, Kluwer Academic Publishers, 83–84

Lagage, P.-O., Jouan, R., Masse, P., Mestreau, P., Tarrius, A., Käufl, H.U. (1993c): TIMMI: a 10 μm camera for the ESO 3.6 m telescope. In *Infrared detectors and instrumentation*, SPIE Vol **1946**, 655–666

McGregor, P.J., Hyland, A.R., Hillier, D.J. (1988): Atomic and molecular line emission from early-type high luminosity stars. ApJ **324**, 1071–1098

Natta, A., Palla, F., Butner, H.M., Evans, N., Harvey, P.M. (1993a): Infrared Studies of Circumstellar matter around Herbig Ae/Be and related stars. ApJ **406**, 674–691

Natta, A., Prusti, T., Krügel, E. (1993b): Very small dust grains in the circumstellar environment of Herbig Ae/Be stars. A&A **275**, 527–533

Pantin, E., Starck, J.L.: Deconvolution of astrophysical images using the multiscale maximum entropy method. A&A submitted

Prusty, T., Natta, A., Palla, F. (1994): Extended mid- infrared emission around Herbig Ae/Be stars. A&A **292**, 593-598

Sellgren, K. (1984): The near-infrared continuum emission of visual reflection nebulae. ApJ **277**, 623–633

Siebenmorgen, R. (1993): The spectral energy distribution of star-forming regions. ApJ **408**, 218–229

Dust Emission from Bok Globules

Ralf Launhardt and Thomas Henning

Max Planck Society, Research Unit "Dust in Star-Forming Regions",
Schillergäßchen 2-3, D-07745 Jena, Germany

Abstract. We present recent results of near-infrared imaging and mm-continuum mapping of a sample of star-forming cores in Bok globules. The circumstellar masses of the embedded YSOs are in most of the cases high enough to fulfil the formal criterion of beeing class 0 protostars. Most of the objects, however, were classified as typical class I objects according to their NIR–MIR spectral energy distributions. This apparent contradiction will be discussed in this paper.

1 Introduction

Bok globules are small, opaque, and isolated clouds of gas and dust. They are the molecular clouds with the simplest structure in the Milky Way. The total number of globules in our galaxy is estimated to be some 10^5 objects (Clemens et al. 1991). Recent studies of isolated globules have shown that many of these objects are active sites of low-mass star formation, whereas others seem to be rather quiet and stable. Typical Bok globules have masses of $3\ldots50\,M_\odot$ and radii of $0.1\ldots0.5\,$pc. About $10\ldots20\%$ of all globules contain dense cores of typically $1\ldots5\,M_\odot$ and currently form solar-type stars (Launhardt 1995). Up to now, however, there is little known about the frequency and the properties of dense cores in globules in which the star formation takes place.

2 The Spectral Energy Distributions

Searching embedded protostellar cores and young stellar objects (YSOs), we performed a 1.3 mm continuum survey (ON-OFF) of ≈100 globules (Launhardt & Henning 1995). For ≈60 of these objects, we also obtained NIR images and photometry. The photometric results of these observations are best summarized in colour-colour diagrams. In Fig. 1, we show the $1.25\text{--}25\,\mu$m colour-colour diagram as well as the IRAS colour-colour diagram of the northern sample of globule cores. The 1.3 mm dust continuum fluxes are marked in both diagrams by the sizes of the dots. We use the infrared spectral index introduced by Lada (1987)

$$\alpha_\nu = \frac{d\,lg(\nu S_\nu)}{d\,lg\nu} \qquad (1)$$

From the colour-colour diagrams, it is obvious that almost all of the detected 1.3 mm sources belong to the IR class I and that mm flux, IR spectral energy distribution, and outflow activity (Yun and Clemens 1992) of the YSOs are well-correlated in a statistical way. While only two sources of the IR class II are detected at 1.3 mm, we find no class III object exhibiting mm emission. These findings are generally interpreted as an evolutionary sequence in the way that the class I sources are the youngest objects in this sequence and are still deeply embedded, while class II objects exhibit only smaller amounts of circumstellar dust mostly in the form of disks, and class III objects have only little remnants of circumstellar material. We find, however, no correlation between the NIR colour index and the mm flux within the class I group. This can be explained by the fact that most of the NIR/MIR emission is probably optically thick and, thus, gives only information on the effective temperature, while the optically thin mm dust emission is a measure of the column density of the dust envelopes. In addition, one has to take into account that in the presence of geometrically thick accretion disks as we expect them to be present around class I objects, the NIR/mm luminosity ratio strongly depends on the viewing angle.

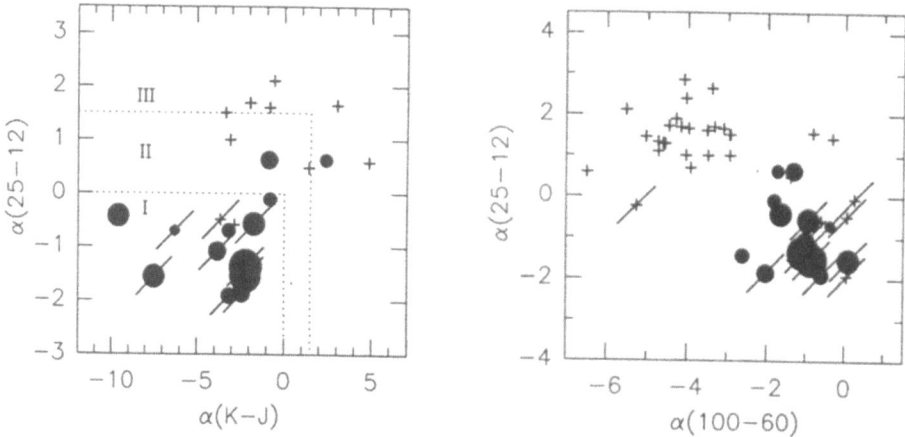

Fig. 1. Colour-colour diagrams of YSOs in Bok globules. Objects detected at 1.3 mm are marked by dots with the sizes of the dots scaled with the mm flux. Outflow sources are marked by 45° lines. Sources not detected at 1.3 mm are marked by crosses.

3 Dust Emission Maps and Circumstellar Masses

In this section, we will critically discuss the $L_{bol}/L_{1.3mm}$-ratio which was introduced as a classification criterion for protostars by André et al. (1993). This approach is based on the assumption that $L_{bol}/L_{1.3mm}$ reflects the ratio between the stellar and the envelope mass. In Fig. 2, we show the NIR images of three typical globule cores overlayed with the contour maps of the 1.3 mm dust emission. In addition, the broad band spectral energy distributions are given.

CB 244 (d = 200 pc) contains two dense condensations surrounded by an extended envelope of $\approx 4\,M_\odot$. The south-eastern core has a mass of $\approx 0.2\,M_\odot$ and is associated with a "cold" IRAS source and two faint NIR nebulae which

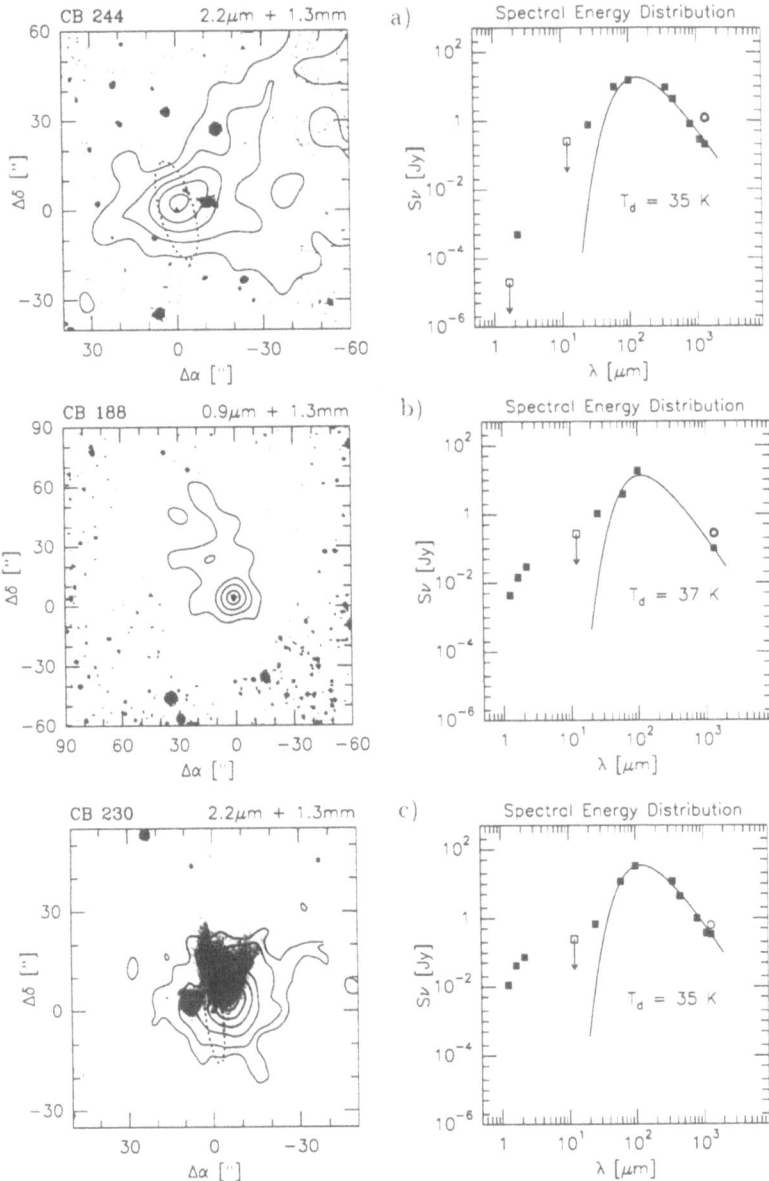

Fig. 2. a) K band image and contour map of the 1.3 mm dust continuum emission of CB 244. On the left, the boad band spectral energy distribution is shown. FIR points are from IRAS, NIR points from the diffuse objects close to the core centre. The open circle refers to the total mm flux; the filled rectangle at 1.3 mm refers to the central condensation only.

b) I band image and contour map of the 1.3 mm dust continuum emission of CB188.

c) K band image and contour map of the 1.3 mm dust continuum emission of CB 230.

might represent scattered light from a more deeply embedded object. Counting all matter within a radius of 10,000 AU around the central source as circumstellar matter, we obtain an $L_{bol}/L_{1.3mm}$-ratio of 2000 which fulfils the formal criterion of a class 0 protostar ($L_{bol}/L_{1.3mm} < 2\,10^4$).

CB 188 (d = 300 pc) contains an elongated dense core with two embedded condensations, the more massive of which is associated with a NIR and IRAS source. The total mass of the dense core ($r_{min} \approx 5000\,\text{AU}$, $r_{max} \approx 7000\,\text{AU}$) is 0.9 $M_\odot$; the mass of the compact condensation amounts to 0.15 $M_\odot$. Counting again all matter within a radius of 7,000 AU around the central source as circumstellar matter, we obtain an $L_{bol}/L_{1.3mm}$-ratio of 8000.

CB 230 (d = 450 pc) contains a dense core of 2.9 $M_\odot$ in which a young binary system is embedded. The separation of the components is 5000 AU. The western object drives a bipolar outflow and excites a bright bipolar NIR nebula. It is very probably surrounded by a massive accretion disk and a more extended shell which both cover the southern (red) lobe of the bipolar nebula. These properties would place the object in the more advanced evolutionary phase of class I. Counting only the material of the unresolved compact condensation ($r < 3000\,\text{AU}$), we obtain an $L_{bol}/L_{1.3mm}$-ratio of $1.5\,10^4$ which is still a "class 0 value". Counting all material inside $r < 9000\,\text{AU}$, this ratio decreases to 7500. Thus, CB 230 is formally also a class 0 protostar. For this object, we discuss the relevance of the $L_{bol}/L_{1.3mm}$-ratio in more detail.

A large fraction of the mm emission probably arises from an accretion disk which may be optically thick even at mm wavelengths. Moreover the $L_{bol}/L_{1.3mm}$-ratio of objects with a non-spherical geometry strongly depends on the viewing angle (Men'shchikov & Henning 1995) and L_{bol} may not immediately reflect the mass of the star. Such extremely non-spherical dust configurations are, however, typical for very young objects driving well-collimated bipolar outflows. Therefore, the physical criterion of an object beeing a class 0 protostar $M_* > M_{env}$, i.e. less than half of the final mass is accreted on the central object, can not be derived straightforward from the $L_{bol}/L_{1.3mm}$-ratio as soon as an accretion disk is formed. This may, however, happen at a very early evolutionary stage. To identify an object as a protostar beeing in its main accretion phase, one has to check additionally the presence of mass infall by analysing appropriate molecular line data.

References

André, P., Ward-Tompson, D., Barsony, M. (1993) ApJ **406**, 122–141
Clemens, D.P., Barvainis, R. (1988) ApJS **68**, 257–286
Clemens, D.P., Yun, J.L., Heyer, M. (1991) ApJS **75**, 877–904
Lada, C. (1987), In: *Star Forming Regions*, ed. Peimbert, M., Jugaku, K. J., (D. Reidel Publ. Comp. Dordrecht), 1
Launhardt, R. (1995): *Star-Formation in Bok Globules* Ph.D. Thesis, Univ. Jena
Launhardt, R., Henning, Th. (1995) in prep.
Men'shchikov, A.B., Henning, Th. (1995) A&A (in press)
Yun, J.L., Clemens, D.P. (1992) ApJ **385**, L21–L25

Density Structure of Protostellar Envelopes

F. Motte[1,2], P. André[2], and R. Neri[3]

[1] Laboratoire d'Astrophysique de Grenoble, BP53, F-38041 Grenoble Cedex 9, France
[2] Service d'Astrophysique, CEA-Saclay, F-91191 Gif-sur-Yvette Cedex, France
[3] IRAM, 300 rue de la Piscine, F-38406 St. Martin d'Hères, France

Abstract. An open question about protostellar clumps/envelopes is that of their detailed density structure. Dust continuum mapping at millimeter wavelengths is a powerful tool to tackle this problem. We present preliminary results on the radial density profiles of protostellar envelopes based on sensitive bolometer-array maps with the IRAM 30 m telescope. While the radial profiles of envelopes in isolated star-forming regions are roughly consistent with the 'standard' protostar theory ($\rho(r) \propto r^{-1.5}$ or r^{-2}), those observed in clusters appear to be, at least in some cases, steeper ($\rho(r) \propto r^{-3}$) and indicative of sharp edges.

1 Introduction

Dust continuum emission is largely optically thin at 1.3 mm making it possible to measure the mass of circumstellar material and to study its spatial distribution. With the 11" resolution of the IRAM 30 m, density structure of the protostellar envelopes surrounding Class 0 and Class I sources is accessible. Class 0 objects are at the beginning of the main accretion phase, at a stage when the envelope still contains most of the final main sequence mass; Class I sources are more evolved protostars which have already accreted a large part of their final mass but still have a substantial envelope (see André 1994 for a review).

Studies of the density structure of young Class 0 protostars can provide crucial tests of star formation models. In particular, by comparison with similar observations of pre-stellar cores (e.g., Ward-Thompson et al. 1994), they may shed light on the initial conditions for protostellar collapse.

In the standard theory of star formation (e.g., Shu et al. 1993), protostellar envelopes are expected to have outer radii $\sim 10^4$ AU and power law radial density profiles such as $\rho \propto r^{-1.5}$ in their inner regions. In this self-similar standard theory, the initial conditions correspond to singular isothermal spheres, which have $\rho \propto r^{-2}$ throughout. However, numerical models of gravitational collapse exist which do not use singular isothermal spheres as initial states. For instance, in the nonmagnetic infall calculations of Foster & Chevalier (1993), collapse is initiated in finite-size ('Bonnor-Ebert') isothermal spheres. In the magnetically-controlled models of Mouschovias and co-workers, collapse is self-initiated as a result of the formation of a supercritical core by ambipolar diffusion (e.g., Ciolek & Mouschovias 1994; see also this volume). Comparison of the predictions of these various models shows that the way the collapse proceeds (e.g., the history of the accretion rate) depends quite critically on the initial density profile.

2   Mapping Observations

In an effort to constrain protostellar models, we have started an extensive dust continuum mapping programme of a large number of Class 0 and Class I sources in nearby star-forming regions such as Taurus-Auriga and ρ Ophiuchi. In particular, 15 protostellar envelopes were mapped at 1.3 mm in March 1995, with the IRAM 30 m telescope equipped with the MPIfR 19-channel bolometer array. The dual-beam scanning mode was used with a chop throw of either 32" or 44" (in azimuth) to create high-sensitivity 'on-the-fly' maps $\sim 3' \times 4'$ in size around each source (e.g., Fig. 1). The data were reduced and restored using the new IRAM NIC software for bolometer observations (Emerson et al. 1979 ; Neri et al. in prep).

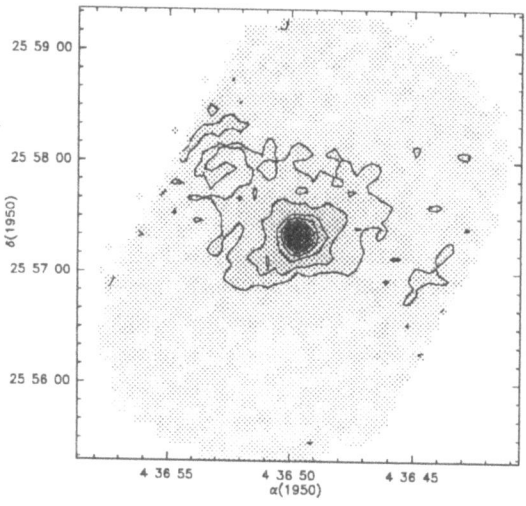

Fig. 1. Combination of two 'on-the-fly' 1.3 mm continuum maps of the 'Class 0' source L1527 in Taurus. The rms is ~ 6 mJy/11"-beam at map center; contour levels go from 30 to 360 by 30 mJy/beam.

In order to study the emission profile of each envelope, a central point source corresponding to the (possible) contribution of an unresolved disk must first be subtracted from the data. The flux of this component was estimated, whenever possible, by extrapolating 2.7 mm interferometric measurements using a dust emissivity index $\beta = 1$ for the disk (Beckwith & Sargent 1991). Assuming spherical symmetry, the millimeter emission of each protostellar envelope was then azimuthally averaged about its peak to derive a radial intensity profile (see § 4 below). For envelopes following a power law density gradient $\rho \propto r^{-p}$ and a temperature gradient $T \propto r^{-q}$ with $q \approx 0.4$ (resulting from heating by the central protostar; e.g., Butner et al. 1990), one expects specific intensity profiles of the form $I \propto \Theta^{-(p+q-1)}$ as a function of projected radius Θ (see, e.g., Adams 1991).

3   Simulations of Dual-Beam Observations

The observing technique (dual-beam mapping) and reduction method (baseline subtraction and restoration) affect the slopes of the derived flux profiles. In order to quantify these effects and to accurately compare our observational results with simple models (power laws $\rho(r) \propto r^{-p}$ or combinations of such power laws), we have developed a simulation program. For any given model envelope, this program simulates the dual-beam maps that would be observed by the various channels of the bolometer array in the on-the-fly scanning mode. (This processing includes a convolution with the beam of the telescope measured on a strong point source such as Uranus.) The simulated dual-beam maps are then reduced in the exact same way as the real data. Our simulations show that some losses do occur for 'extended' model envelopes.

Figure 2 illustrates the evolution of a single power law model through convolution with the IRAM 30 m beam and all the observation/reduction steps. The main loss of signal results from the restoration process which assumes zero emission outside the finite area of the map. For the model shown in Fig. 2 ($\rho \propto r^{-1.5}$ and $T \propto r^{-0.4}$), only the inner 15% of the profile is free of this 'edge effect'. Models with steeper slopes are less affected. Convolution with the beam (dashed line) does not affect the slope for angular radii $\Theta > HPBW$ (i.e., 11 arcsec here), in agreement with Adams (1991). The effect of baseline subtraction is negligible except at the very edge of the profile.

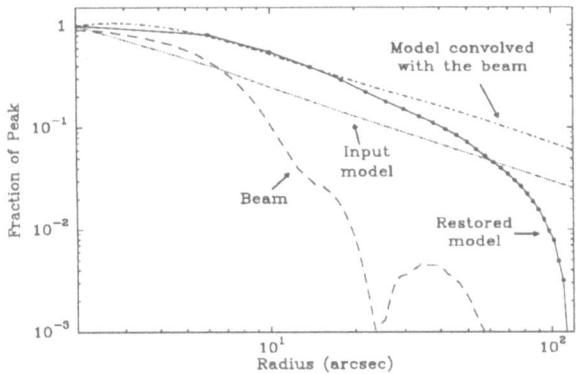

Fig. 2. Evolution of the radial profile of an input model ($\rho \propto r^{-1.5}$ and $T \propto r^{-0.4}$; dotted line) after convolution with the beam of the telescope (dashed-dotted line), simulated dual-beam observation, baseline subtraction, and restoration (solid line).

4   Preliminary Results on Protostellar Envelopes

Figure 3 shows the flux density profiles of the two Class 0 sources L1527 and VLA 1623 (solid lines) together with simulated profiles (dotted lines) and the beam profile (dashed lines). The L1527 profile (Fig. 3a) agrees well with a simulated profile corresponding to $\rho(r) \propto r^{-1.5}$ and $T(r) \propto r^{-0.4}$. On the other hand, the VLA1623 profile is much steeper and more consistent with a $\rho(r) \propto r^{-3}$ simulated profile (Fig. 3b). A sharp edge was also seen in absorption in recent deep $2\,\mu$m observations of VLA1623 (Dent et al. 1995). Note that these protostellar envelopes do not present the inner flattening observed in the prestellar cores such as L1689B (see Ward-Thompson et al., this volume).

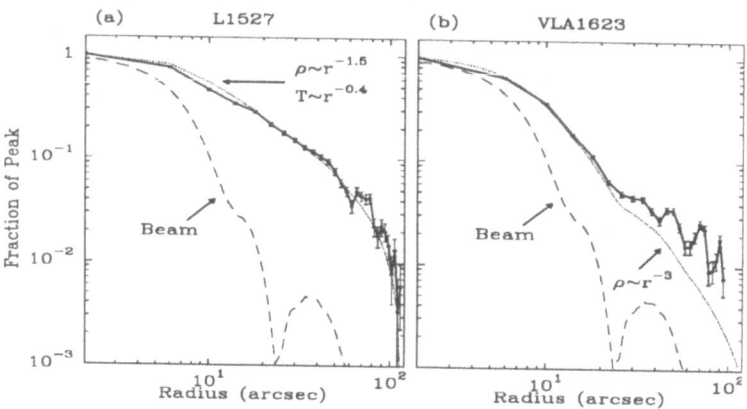

Fig. 3. (a) Azimuthally averaged flux density profile of the isolated Class 0 source L1527 in Taurus (map on Fig. 1). (b) Flux density profile of the prototype Class 0 source VLA 1623 in Ophiuchus (data averaged over a 45° sector free of confusion from other sources).

Work is in progress to determine whether the difference observed between L1527 and VLA1623 reflects a fundamental difference between protostars in isolated regions (e.g. Taurus) and protostars in clusters (e.g., ρ Oph).

References

Adams, F.C. (1991) ApJ **382**, 544
André, P. (1994) in *The Cold Universe*, Ed. T. Montmerle, C.J. Lada, I.F. Mirabel, & J. Trân Thanh Vân (Gif-sur-Yvette: Ed. Frontières), 179–192
Beckwith, S.V.W., & Sargent, A.I. (1991) ApJ, **381**, 250
Butner, H.M., Evans, N.J., Harvey, P.M. et al. (1990) ApJ, **364**,164
Ciolek, G.E., Mouschovias, T.Ch. (1994) ApJ **425**, 142
Dent, W.R.F., Matthews, H.E., Walther, D.M. (1995) MNRAS in press
Emerson, D.T., Klein U., & Haslam, C.G.T. (1979) A&A, **76**, 92
Foster, P., Chevalier, R. (1993) ApJ **416**, 303
Neri, R., Broguière, D., IRAM internal report in prep
Shu, F., Najita, J., Galli D., Ostriker, E., Lizano, S. (1993), in Levy E. H., Lumine J., eds, Protostars and Planets III. (University of Arizona Press), Tucson, 3–45
Ward-Thompson, D., Scott, P.F., Hills, R.E., André, P. (1994), MNRAS **268**, 276

What Causes the Variability of the PV Cep Nebula?

Th. Neckel and H.J. Staude

Max-Planck-Institut für Astronomie, Königstuhl 17, D-69117 Heidelberg, Germany

Abstract. We present Gunn r, I, and [SII] images of the variable PV Cep nebula, and longslit spectroscopy taken between 1989 and 1994. The variations in brightness and color are probably due to variable extinction by dust in the wind of PV Cep, implying strong variations of its mass loss rate.

1 Introduction

The Herbig AeBe star PV Cep ($L = 100$ $L_\odot$ at $d = 600$ pc) excites a strongly variable optical reflection nebula and drives a bipolar molecular outflow. On the POSS plates (1950) the nebula appears as a narrow blue streak, separated from the star by a gap of about $30''$. 1976 and later a bright and well defined fan-shaped nebulosity appeared, which is directed toward the north. Its eastern rim partially coincides with the former streak nebula. This fan-shaped structure is barely visible already on POSS. On the polar axis of the fan Herbig-Haro emission was detected at two positions and a second, weak and strongly reddened lobe was detected south of PV Cep. (Neckel et al. 1987). All these characteristics are shared with R Mon and its associated cometary nebula NGC 2261.

2 New Observations

Between 1989 and 1994 we have obtained with the 3.5 m telescope on Calar Alto broad band Gunn r and I images, deep narrow band images in the light of [SII]6716,6731 and nearby continuum, and long slit spectroscopy.

Figure 1 shows Gunn r images taken 1989 (a) and 1990 (b). Several of the filamentary structures within the fan nebula appear in both images, some only in one of them. Thus, the nebula is still variable. As can be seen from the color map I/Gunn r (Figure 1d), these structures are not due to spatially variable extinction in front of the lobe: in fact, the color changes smoothly over the nebula. Strong reddening occurs only near the star and along the rims of the fan.

The ratio between the two Gunn r images, which is shown in Figure 1c, demonstrates that between 1989 and 1990 a narrow region along the eastern border of the fan, whose northern half almost coincides with the former streak nebula, has faded, whereas a broader region at the western border of the fan became brighter. These variations can not be due to large-scale motions of the scattering material, since too high velocities would be implied. The elongated

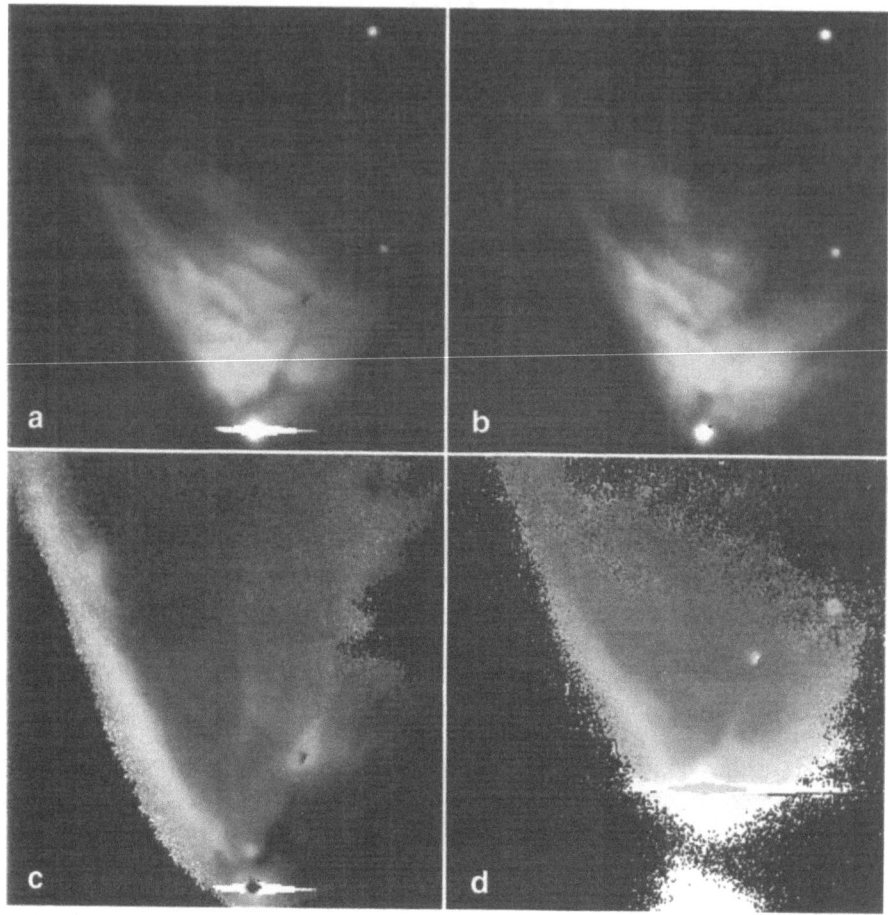

Fig. 1. (a) Gunn r image, Sep 27, 1989. (b) Gunn r image, Aug 26, 1990. (c) Ratio of the images (a) and (b). (d) Color ratio I/Gunn R (Sep 27, 1989)

structures seen in Figure 1c are directed radially away from PV Cep: this indicates that the brightness variations in the fan are due to shadow effects caused in the vicinity of the star.

Figure 2 shows a contour plot of the [SII] image after subtraction of the scattered continuum. Extended [SII] emission arises within a large region of the fan, and two isolated Herbig-Haro objects appear farther out. This extended shock-excited emission indicates that the wind emanating from PV Cep is only loosely collimated. Within the central emission region (knots 3 - 6) the radial velocities are quite uniform (-260 km/s), whereas in knots 1, 2, and 7 they range between -180 km/s and -230 km/s. The electron density is exceptionally low: it reaches a maximum of $n_e = 280$ cm^{-3} near PV Cep, and decreases to $n_e = 22$ cm^{-3} farther out.

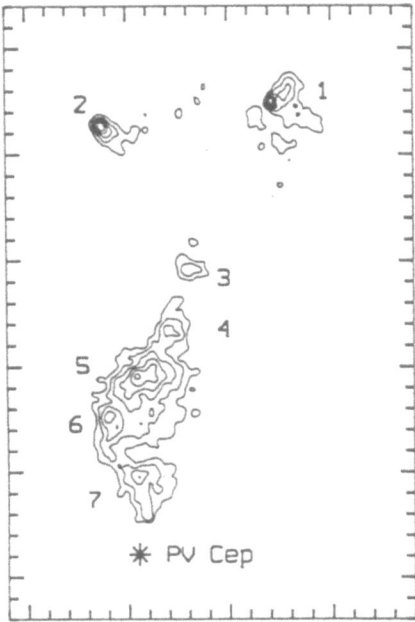

Fig. 2. Pure [SII] emission within the PV Cep nebula (continuum subtracted), Aug 5, 1990.

The spectrum of PV Cep shows a wealth of Fe emission lines (as it is characteristic for many YSOs), the forbidden Herbig-Haro lines, and a broad Hα emission line with a deep absorption at -200 km/s (Figure 3). In the light scattered by the dust in the lobe, this absorption is much broader and deeper: between -100 km/s and -500 km/s the intensity of the emission is absorbed almost completely. Thus, the wind leaving PV Cep in polar direction is much faster and denser than the wind directed toward the observer.

3   Origin of the Dust

PV Cephei is strongly reddened. It is not visible on the POSS B plate, while the streak nebula, which was illuminated by PV Cep, is obviously blue. This holds also for the central parts of the now visible fan nebula - a morphology typical of YSOs with cometary or bipolar nebulae: The YSO is surrounded by a thick dust disk, but it is free to illuminate the cavity within the surrounding molecular cloud, which has been formed by its bipolar wind. Also the low electron densities observed in the lobe confirm that the lobe is essentially an empty cavity. Since the time scale of the observed brightness variations is much shorter than the

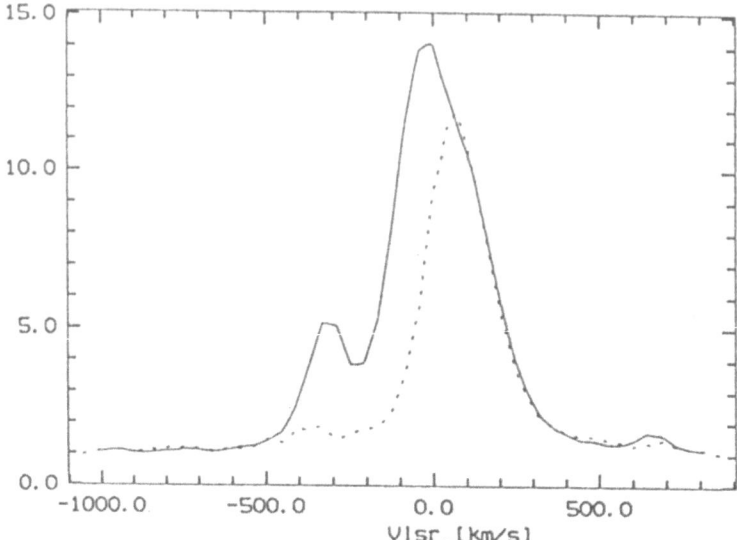

Fig. 3. The Hα emission in the direct light of PV Cep (continuous line), and in the scattered light of the reflection nebula (broken line)

time required to form the cavity, the brightness fluctuations observed in the entire fan must be caused by a transient obscuration by several magnitudes occuring in the immediate surroundings of the star. During the fifties, such kind of obscuration affected almost the entire fan, leaving free only a narrow channel through which the streak nebula was illuminated. At present, the eastern border of the fan is heavily obscured, whereas the starlight reaches its inner regions nearly unabsorbed.

The clearing process leading to the formation of the empty cavity lasts for several thousand years. It seems unlikely that huge amounts of material from the molecular cloud suddenly fall toward the star against the wind. If the obscuring dust belongs to the primary wind leaving the YSO (as observed in the Class I source IRAS 08159-3543, Neckel and Staude, 1995), the brightness variations of the PV Cep nebula reflect strong short-time fluctuations of the mass loss rate, possibly triggered by instabilities within the circumstellar accretion disk.

References

Neckel, Th., Staude, H.J. (1995): IRAS 08159-3543: Optical Detection of the Dusty, Bipolar Wind of a Luminous Young Stellar Object. Astrophys. J. **448**, 832–847

Neckel, Th., Staude, H.J., Sarcander, M., Birkle, K. (1987): Herbig-Haro Emission in two Bipolar Reflection Nebulae Astron. Astrophys. **175**, 231–237

1.3 mm Dust Continuum Observations
of Young X-ray Selected Stars in Ophiuchus

Dieter Nürnberger, Wolfgang Brandner, Harold W. Yorke and Hans Zinnecker

Astronomisches Institut der Universität Würzburg, Am Hubland, D-97074 Würzburg

Abstract. We have performed 1.3 mm dust continuum observations for a sample of 18 X-ray selected pre-main sequence (PMS) stars in the Ophiuchus star forming region with the MPIfR 7-channel bolometer array mounted at the IRAM 30m Millimeter Radio Telescope (MRT). We have detected cold dust emission from 3 of the objects and have derived 3σ upper limits (~ 21 mJy) for the remaining objects. These upper limits suggest that the cold dust masses are less than $5 \cdot 10^{-5} \, M_\odot$.
In addition, we have mapped the hierarchical triple SR 24 at $\lambda = 1.3$ mm with the 7-channel bolometer array. We detected cold dust emission only from the southern component with a peak intensity of ~ 230 mJy and a corresponding circumstellar disk mass of $\sim 0.035 \, M_\odot$ (gas + dust), while for the northern component we derived an upper limit of 10 % of the SR 24S peak flux density. The non-detection of 1.3 mm emission from SR 24N suggests a lack of cold circumstellar dust in the outer part of the disk, which might have been cleared by the close companion $0\rlap{.}''2$ away from SR 24N.

1 Introduction

The evolutionary classification scheme of Young Stellar Objects (YSOs) originally introduced by Lada (1987) and revised by André & Montmerle (1994) predicts a progressive decrease of the amount of circumstellar material from IR class I to IR class III, that is a decrease with stellar age. The classification is based on the infrared excess of the sources, as measured by the spectral index $\alpha_{IR} = d \log(\lambda F_\lambda) \, / \, d \log \lambda$ between $\lambda = 2.2$ and $10 \, \mu m$.

However, infrared observations give only poor estimates of the amount of circumstellar material around YSOs, because shortward of $100 \, \mu m$ the disk emission is generally optically thick. The IR range is also much more sensitive to the temperature and density distribution of the circumstellar material than the mm range and therefore probes only the warm regions close to the central star. Furthermore, the angular resolution presently achievable in the far IR is poor and probably insufficient to resolve the circumstellar structures.

Alternatively, observations of dust continuum emission in the millimeter range provide an excellent way to track the circumstellar evolution of YSOs. Because dusty disks around PMS stars are optically thin at wavelengths of order 1 mm, the flux is proportional to the total mass (Beckwith et al. 1986, Sargent & Beckwith 1987). Thus, it is possible to estimate the gas and dust masses around YSOs directly from the measured mm flux densities.

The sample of Rho Ophiuchi X-ray selected stars (ROX) was detected with the Einstein Observatory by Montmerle et al. (1983). Bouvier & Appenzeller (1992)

obtained low and medium resolution spectrograms, as well as visual and near IR photometry. After placing the stars in the H-R diagram they determined the stellar masses and ages based on PMS evolutionary tracks by Cohen & Kuhi (1979). Some of the ROX sources have been observed by André & Montmerle (1994) as part of a 1.3 mm continuum survey for cold circumstellar dust on a sample of over 100 YSOs in the Ophiuchus star forming region.

2   Observations and Data Reductions

The observations were performed in March 1993 at the IRAM 30 m MRT equipped with the MPIfR 7-channel bolometer array. The half power beam width of the telescope was $\sim 12''$.

The 18 ROX sources were studied by single-point photometry performing ON-OFF measurements (pointing accuracy $\sim 3''$), consisting of 10 cycles with 20 s integration time each. Strong sources were observed at least twice, fainter objects up to four times resulting in an overall (3σ) sensitivity of about 21 mJy beam^{-1}. The atmospheric transmission frequently determined by skydips was quite unstable with daily average values between 0.6 and 0.7 at zenith. Uranus was used as calibration standard adopting a brightness temperature of 95 K at 1.3 mm. For each scan a baseline, calculated from the six hexagonal channels, was subtracted.   Finally, the triple SR 24 was mapped at $\lambda = 1.3$ mm with the 7-channel bolometer array. The map data were baseline-subtracted, calibrated, restored to 'single-beam' maps, and converted into RA-DEC maps within the NOD2 software package. For deconvolution with a Gaussian maximum likelihood algorithm, maps of Uranus and Mars were used as point spread functions.

3   Results and Conclusions

We have detected cold dust emission from 3 of the objects and have derived 3σ upper limits (~ 21 mJy) for the remaining objects (**Tab. 1**). In order to compile credible statistics we extended the data set with 1.3 mm measurements obtained by André & Montmerle (1994) for 22 young X-ray selected sources in Ophiuchus (3 additional detections). For our calculations of the disk masses (gas + dust) from the 1.3 mm flux densities, we adopted a mass-averaged dust temperature of 30 K (Nürnberger 1995) and a mass opacity coefficient of 0.02 cm^2 g^{-1} (Beckwith et al. 1990). Assuming a gas to dust mass ratio of 100:1, the upper limits at 1.3 mm suggest that the cold dust masses are less than $5 \cdot 10^{-5}$ M$_\odot$. In an attempt to gain insight how cold circumstellar dust masses, traced by the 1.3 mm flux density, change with stellar mass and age (Nürnberger 1995), we have redetermined the stellar masses and ages based on the new PMS evolutionary tracks by D'Antona & Mazzitelli (1994). Comparing the new mass distribution of the stars with the old one, a shift by a factor of two to lower stellar masses is obvious. Thus, the ratio of circumstellar disk mass to stellar mass (relative disk

Table 1. Sample of X-ray selected YSOs in Ophiuchus. 1.3 mm detections are printed in bold type, upper limits (3σ) are indicated by a "$<$", values indicated by * were measured by André & Montmerle (1994). Spectral types and luminosities are taken from Bouvier & Appenzeller (1992). T_{eff} is calculated from the spectral type according to de Jager & Nieuwenhuijzen (1987). Masses and ages are based on PMS evolutionary tracks by D'Antona & Mazzitelli (1994). Bracketed values for mass and age are estimated by extrapolation. To calculate the disk masses (gas + dust) from the 1.3 mm flux densities, we adopted a mass-averaged dust temperature of 30 K (Nürnberger 1995) and a mass opacity coefficient of $0.02\,\mathrm{cm^2\,g^{-1}}$ (Beckwith et al. 1990).

Name	Sp.T.	$\log L_*/L_\odot$	$\log T_{eff}/K$	$\log t/a$	$M_*/M_\odot$	$F_{1.3mm}$ (mJy)	σ (mJy)	M_{disk} ($10^{-3}M_\odot$)	M_{disk}/M_* (%)
ROX 2	K3/M0	0.7	3.60	(4.3)	0.30	<30*	10	<4.5	<1.5
ROX 3	K3/M0	-0.2	3.60	6.0	0.48	<30*	10	<4.5	<0.9
ROX 4	K2	0.7	3.65	5.3	0.70	<28	9	<4.2	<0.6
ROX 5	K7	-0.2	3.59	5.9	0.42	<19	6	<2.9	<0.7
ROX 6	K6	0.4	3.60	(4.9)	(0.4)	31	6	4.7	1.4
ROX 7	K7	-0.2	3.59	5.9	0.42	<20*	7	<3.0	<0.7
ROX 8	K0	1.5	3.68	(4.9)	1.6	<5*	2	<0.8	<0.1
ROX 9A	M0		3.56			<22	7	<3.4	
ROX 9B	K2		3.65			<11	4	<1.7	
ROX 9C	K4		3.62			<14	5	<2.2	
ROX 9D	G?		3.72			<35	12	<5.3	
ROX 10A	K5	0.2	3.61	5.6	0.44	<25*	8	<3.8	<0.9
ROX 10B	K0	1.2	3.68	5.2	1.3	65	19	9.8	0.7
ROX 12	M0	-0.7	3.56	6.4	0.40	<19	6	<2.9	<0.7
ROX 14	B4		4.25			<15*	5	<2.3	
ROX 16	G9	0.6	3.69	5.9	1.2	<20*	7	<3.0	<0.3
ROX 20A	M5	-0.3	3.51	(4.9)	0.18	<20*	7	<3.0	<1.7
ROX 20B	M2	-0.8	3.53	6.3	0.25	<20*	7	<3.0	<1.2
ROX 21	K4/M2.5	-0.2	3.58	5.8	0.37	<35*	12	<5.3	<1.4
ROX 29	K4/K6	0.1	3.61	5.8	0.47	15*		2.3	0.5
ROX 30A	G?	-0.2	3.72	7.4	0.95	<10	3	<1.5	<0.1
ROX 30B	K4	0	3.62	5.9	0.60	<45*	15	<6.8	<1.1
ROX 30C	K4	0	3.62	5.9	0.60	40*		6.0	1.0
ROX 31	K7	-0.3	3.59	6.1	0.47	<25*	8	<3.8	<0.8
ROX 33	G0	1.3	3.75	6.1	(2.7)	<35*	12	<5.3	<0.2
ROX 34	M2.5	-0.2	3.53	5.0	0.21	60*		9.0	4.3
ROX 35A	K3	0.1	3.64	5.9	0.67	<29	10	<4.4	<0.7
ROX 35B	G4		3.72			<19	6	<2.8	
ROX 39	K5	0.3	3.61	5.3	0.42	<30*	10	<4.5	<1.1
ROX 42A	<G5	-0.5	3.72	>8	0.85	<30*	10	<4.5	<0.5
ROX 42B	M0	-0.2	3.56	5.7	0.30	<45*	15	<6.8	<2.3
ROX 42C	K6	0.4	3.60	(4.9)	(0.4)	<30*	10	<4.5	<1.3
ROX 43A	G0	0.7	3.75	6.6	1.8	<35*	12	<5.3	<0.3
ROX 43B	K5	0.2	3.61	5.6	0.44	<17	6	<2.6	<0.6
ROX 44	K3	0.3	3.64	5.7	0.61	105	11	15.8	2.6
ROX 45A	F?	0.0	3.83	>8	(1.1)	<25	8	<3.8	<0.4
ROX 45B	K?	-0.6	3.62	7.2	0.83	<19	6	<2.9	<0.4
ROX 45C	K5	-0.2	3.61	6.2	0.61	<19	6	<2.9	<0.5
ROX 45D	K0	-0.3	3.68	7.2	1.0	<19	6	<2.9	<0.3
ROX 47A	K2/K7-M0	0.2	3.60	5.4	0.37	<20*	7	<3.0	<0.8

mass) increases by a factor of two. Furthermore, plotting the relative disk mass as a function of stellar mass, there seems to be a higher fraction of circumstellar material for $M_* < 0.7\,M_\odot$ than for $M_* > 0.7\,M_\odot$. Due to the small number of detections in our Ophiuchus sample we have verified and confirmed this effect both in a sample of 32 stars in Lupus (12 detections; Nürnberger et al. 1995) and a sample of 83 stars in Taurus-Auriga (45 detections; Beckwith et al. 1990 and Osterloh & Beckwith 1995). Plotting the relative disk mass as a function of stellar age reveals that the detected 1.3 mm sources show a decrease of circumstellar matter with increasing age. It seems that there is no detection for stars older than 10^6 years. This suggests that disks around low mass PMS stars at this age become undetectable for the observer at 1.3 mm. This means that there is only little dust mass left in the form of small grains; the dust particles may have coagulated to much greater bodies (planetesimals) or may have been dispersed or accreted. In addition, we found that all 6 detected sources belong to the IR class II and most (74 %) of the non-detections are IR class III sources.

This result is consistent with the evolutionary classification scheme of YSOs, as already mentioned above.

Finally, we have mapped the hierarchical triple SR 24 at $\lambda = 1.3$ mm with the 7-channel bolometer array (Nürnberger 1995). We detected cold dust emission only from the southern component with a peak intensity of ~ 230 mJy and a corresponding circumstellar disk mass of $\sim 0.035\,M_\odot$ (gas + dust), while for the northern component we derived an upper limit of 10 % of the SR 24S peak flux density. The non-detection of 1.3 mm emission from SR 24N suggests a lack of cold circumstellar dust in the outer part of the disk, which is not surprising, as Simon et al. (1995) discovered a companion $0\rlap{.}''2$ (~ 32 AU at the distance of 160 pc) away from SR 24N. At $\lambda = 10\,\mu$m, which is an indicator of warm circumstellar dust in the inner part of the disk, both the northern and the southern component show roughly equal emission ($F_{SR24N} = 1.3$ Jy and $F_{SR24S} = 1.6$ Jy at 10 μm; Stanke & Zinnecker 1995). This means that the inner part of the disk around SR 24N is still present, but the outer part of the disk is cleared by the close companion. Such an interaction between a circumstellar disk and a close companion was recently discussed for UZ Tau by Ghez et al. (1994).

Acknowledgements We wish to thank Philippe André and Ralf Launhardt for providing us with maps of Uranus and Mars for deconvolution purposes, as well as Thomas Stanke for communicating the results of the 10 μm photometry of SR 24N/S prior to publication.

References

André P., Montmerle T.: 1994, ApJ **420**, 837

Beckwith S.V.W., Sargent A.I., Scoville N.Z., Masson C.R., Zuckerman B., Phillips T.G.: 1986, ApJ **309**, 755

Beckwith S.V.W., Sargent A.I., Chini R.S., Güsten R.: 1990, AJ **99**(3), 924

Bouvier J., Appenzeller I.: 1992, A&AS **92**, 481

Cohen M., Kuhi L.V.: 1979, ApJS **41**, 743

D'Antona F., Mazzitelli I.: 1994, ApJS **90**, 467

Ghez A.M., Emerson J.P., Graham J.R., Meixner M., Skinner C.J.: 1994, ApJ **434**, 707

de Jager C., Nieuwenhuijzen H.: 1987, A&A **177**, 217

Lada C.J.: 1987, in *Star Forming Regions*, IAU Symp. No. 115, ed. M. Peimbert, J. Jugaku, p. 1

Montmerle T., Koch-Miramond L., Falgarone E., Grindlay J.E.: 1983, ApJ **269**, 182

Nürnberger D.: 1995, Diplom thesis, Julius-Maximilians-Universität Würzburg

Nürnberger D., Chini R., Zinnecker H.: 1995, A&A, in prep.

Osterloh M., Beckwith S.V.W.: 1995, ApJ **439**, 288

Sargent A.I., Beckwith S.V.W.: 1987, ApJ **323**, 294

Simon M., Ghez A.M., Leinert Ch., Cassar L., Chen W.P., Howell R.R., Jameson R.F., Matthews K., Neugebauer G., Richichi A.: 1995, ApJ **443**, 625

Stanke Th., Zinnecker H.: 1995, A&A, in prep.

The Luminosity-mm Flux Correlation of Class I Sources Exciting Outflows

P. Saraceno[1], F. D'Antona[2], F. Palla[3], M. Griffin[4], and E. Tommasi[1]

[1] Istituto di Fisica Spazio Interplanetario, CNR, C.P. 27, I-00044 Frascati, Italy
[2] Osservatorio Astronomico di Roma, I-00040 Monte Porzio, Italy
[3] Osservatorio Astrofisico di Arcetri, Largo E.Fermi,5, 50125, Firenze, Italy
[4] Physics Dep., Queen Mary and Westfield College, Mile End Rd., London E1 4NS, England

Abstract. We discuss the possible physical explanations of the L_{bol} vs. F_{mm}^D correlation observed for Class I sources exciting outflows in nearby star forming regions. The D burning process appears to be the most likely explanation.

1 Introduction

In a paper by Saraceno et al. 1995 (hereafter SACGM) a systematic study of the millimetric emission of a sample of Class I sources with and without outflow activity is presented. The sample contains all the known Class I sources with $L_{bol} > 10 L_\odot$ in NGC 1333, L1641, ρOph L1688 and Corona Australis. In addition, 47 Class I objects of lower luminosity have been included. A luminosity of $\sim 10 L_\odot$ is expected from objects of $\sim 0.6 M_\odot$ during their accretion phase (e.g. Stahler 1988, Palla & Stahler 1993). Therefore, assuming that Class I sources represent genuine protostars, our sample should contain *all* the objects with mass $M \gtrsim 0.6 M_\odot$ currently forming within those clouds.

In SACGM it is suggested that the L_{bol} vs. F_{mm}^D diagram, where F_{mm}^D is the millimetric flux scaled at a distance D (equivalent to L_{mm}), provides a very useful diagnostics to trace different phases of protostellar evolution. The evolution of protostars is in fact controlled by two factors: the mass of the central object, which defines L_{bol}, and the mass of circumstellar material which is proportional to F_{mm}^D since the emission at millimeter wavelengths is optically thin.

Figure 1a shows this diagram for the objects in the SACGM sample plus the Class I objects exciting Herbig–Haro (HH) from Reipurth et al. (1993) scaled at the distance of 160 pc. It is clear that Class 0 (André et al. 1993), Class I exciting outflow and Class I with non detected outflow occupy different regions of the diagram. Class 0 objects are found to the right of the region indicated as the *outflow strip*, whereas non detected flows are in average located to the left of this region. This general trend is consistent with the fact that the evolution proceeds from high values of the millimeter flux, i.e. large amounts of circumstellar material, to smaller and smaller fluxes (see also André & Montmerle 1994). Class 0 are therefore the youngest objects in the diagram and Class I sources with no detected outflow are the more evolved. Class I sources exciting HH and/or molecular flows show a clear correlation between L_{bol} and F_{mm}^{160pc} that defines the *outflow strip*. This property was already pointed out by Reipurth et al. (1993) and more recently by Moriarty-Schieven et al. (1994).

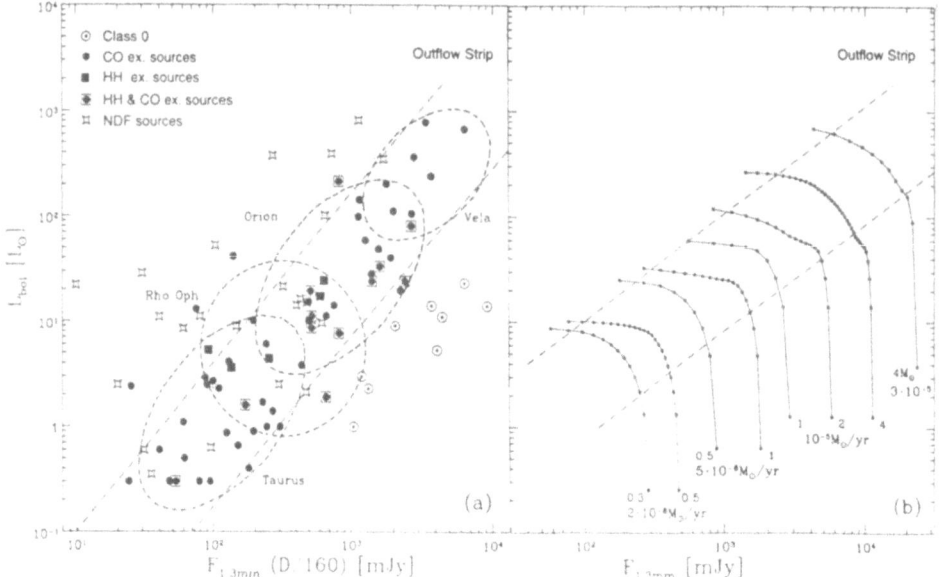

Fig. 1. (a) The L_{bol} vs F_{mm}^{160pc} diagram; (b) The computed tracks.

The important issue we want to discuss here is the origin of the observed correlation and whether this property can be used to distinguish objects in the main accretion phase (i.e. the phase in which the majority of the mass is accreted and the final mass of the star defined) from the more evolved ones. Let us discuss two possible scenarios separately.

2   Class 0 in the Main Accretion Phase

If the protostellar mass is accreted mainly during the Class 0 phase, the Class I objects on the *outflow strip* are more evolved objects with a residual circumstellar material that will be dispersed during the pre-Main Sequence (PMS) contraction. In such a case, the observed correlation between the two luminosities simply indicates that the residual circumstellar material ($\propto F_{mm}$) is proportional to the stellar mass ($\propto L_{bol}$). The range of L_{bol} of the objects in Fig. 1a is consistent with that expected for low-intermediate mass stars in the PMS phase.

   In this scenario, the low L_{bol} values of Class 0 sources are somewhat more difficult to account for by current estimates of protostellar luminosities (cf. Stahler 1988; Palla & Stahler 1993). As an example, if the formation of stars of 1-2 $M_\odot$ is occuring inside L1641 and L1630 in Orion, the predicted luminosities should be about 100 $L_\odot$ at the typical mass accretion rate of 10^{-5} $M_\odot$ yr^{-1}. Such objects are not observed (while they *should* have been, given the selection of the sample) and if they are not found in the future we will have to conclude that either Class 0 sources are not representative of the main accretion phase or that the theoretical luminosities have to be decreased by a factor ~ 10. However, in the inside–out scenario (Stahler et al. 1980) of isothermal cloud collapse this would imply a gas temperature (which determines the magnitude of the mass accretion rate) well below the observed value of 10 K.

3 Class 0 Before the Main Accretion Phase

Alternatively, we may assume that Class 0 objects are in the intial stages of the accretion process but that most of the mass is accreted while the objects are Class I on the *outflow strip*. Fig. 1b shows several evolutionary tracks originating from initial clumps of different masses and for different accretion rates. The numbers at the beginning of each track indicates the initial clump mass (ranging from 0.3 to 4 $M_\odot$) and the accretion rate (ranging from 2×10^{-6} to $3 \times 10^{-5} M_\odot yr^{-1}$). The points on each track correspond to costant time steps of 10^4 yr; at each step we computed L_{bol}, assumed to be entirely due to accretion (i.e. $L_{bol} \equiv L_{acc} = G\dot{M}M_*/R_*$). For the values of M_*/R_*, we have used those published by Palla & Stahler (1993) for an accreting object undergoing deuterium burning. The emission of the residual dust is computed as $F^{160}_{1.3mm} = \frac{M}{(160pc)^2} \kappa_{1.3mm} B_{1.3mm}$ where M is the total circumstellar mass (gas/dust = 100), $B_{1.3mm}$ is the Planck function and $\kappa_{1.3mm}$ is the dust opacity (Ossenkopf & Henning 1994).

Since each point on the tracks corresponds to the same time interval, the decrease of the distance between adjacent points indicates where protostars are spending most of the time (i.e. more objects should be observed in this area of Fig. 1a). Such concentrations of points are found to lie in the upper parts of the tracks in Fig. 1b, and delimit a computed *outflow strip*. Its location compares well with the one shown in Fig. 1a, within the uncertainties in the mass determination from the millimetric flux (which shift the tracks horizontally in the diagram) and in the distance scaling law (which changes the slope of the *outflow strip*, see SACGM). In this scenario Class I objects accrete most of their mass and burn most of their deuterium while on the *outflow strip*. Due to expected and observed accumulation of objects in this part of the diagram, SACGM suggested that the *outflow strip* represents a kind of *deuterium burning sequence* for stars of different masses.

Figure 1a shows that sources belonging to different clouds are segregated along the *outflow strip*. The most luminous sources (L_{bol} between 100 and 800 $L_\odot$) are found in the Vela molecular cloud; sources in L1641 have $L_{bol} \cong$ 5 ÷ 300$L_\odot$; intermediate luminosities ($L_{bol} \cong 2 \div 50 L_\odot$) characterize the population of ρ Oph and Perseus; finally, all the low luminosity sources belong to the Taurus molecular cloud. The comparision of Fig.1a and Fig.1b suggests that along the *outflow strip* the accretion rate (i.e. the temperature of the cloud) decrease. Since the mass of the clumps is $\propto F_{mm}$ and the mass of the star is $\propto L_{bol}$ this is an evidence that higher mass stars are formed in higher mass clumps which are warmer and with higher accretion rate.

However, the tracks of Fig. 1b cannot account for the presence of objects in the lower end of the strip. In fact, the expected luminosities of an accreting object cannot go below a few L_{bol}, as this would require $\dot{M} < 2 \times 10^{-6}$ $M_\odot$ yr^{-1}, corresponding to cloud temperatures below 10 K. It is therefore very unlikely that these sources are in the main accretion phase. The possible explanations for their location in the diagram are:

i) These objects are genuine PMS stars of $0.4 \div 1 M_\odot$. The luminosities can be evaluated from the evolutionary tracks of Palla & Stahler (1993) for M > 0.6 $M_\odot$, and from D'Antona & Mazzitelli (1994) for lower masses. The two sets of tracks yield remarkably similar isochrones, despite differences in the input physics. The estimated age is then between $10^6 - 10^7$ yr. By then, these stars should have already evolved away from the outflow strip and stopped the outflow activity (whose duration is between 10^4 and 10^5 yr, see SACGM). Therefore this interpretation seems unlikely.

ii) These objects are stars of mass in the range $0.15 \div 0.3 M_\odot$, burning deuterium at the beginning of the PMS phase. At the typical accretion rates, these stars would have an age between 10^4 and 10^5 yr and would have started the PMS phase with all the initial deuterium still unburned (Stahler 1988). Their derived age is also similar to the dynamical time of the outflow.

iii) These objects are stars of even smaller mass ($M \lesssim 0.2 M_\odot$) in early hydrostatic contraction. It is possible that such low mass stars do not go through an accretion phase, but instead contract on a Kelvin timescale from residual fragments of the parent clouds. Reasonable contraction ages ($\leq 10^5$ yr) require $M \leq 0.1 M_\odot$ (D'Antona & Mazzitelli 1994). These objects would not yet be in the D-burning phase, whose onset is at times $\gtrsim 3 \times 10^5$ yr for 0.2 $M_\odot$, but should have been able to excite a molecular outflow.

Further observations may help in discriminating between these different hypotheses. In our view, the second possibility appears the most likely one. In this case, the *outflow strip* corresponds to the locus of D-burning: in the main accretion phase for the high luminosity objects and during the first hydrostatic phases for the dimmer ones.

References

André P., Ward-Thompson D., Barsony M. (1993): ApJ 406,122
André P., Montmerle T. (1994): ApJ 420,837
D'Antona F., Mazzitelli I. (1994): ApJS 90, 467
Moriarty-Schieven G.H., Wannier P.G., Keene J., Tamura M. (1994): ApJ, 436, 800
Ossenkopf V., Henning T. (1994): A&A, 291, 943
Palla F., Stahler S.W. (1993): ApJ 418, 414
Reipurth B., Chini R., Krügel E., Kreysa E., Sievers A. (1993): A&A, 273, 221
Saraceno P., André P., Ceccarelli C., Griffin M., Molinari S. (1995): A&A submitted.
Stahler S.W., Shu F.H., Taam R.E. (1980): ApJ 241, 637
Stahler S.W. (1988): ApJ, 332, 804

PAHs as Probes of Photo-Dissociation Regions in M17 and the Orion Bar

G. C. Sloan[1], J. Bregman[1], A. S. B. Schultz[1], P. Temi[2], and D. M. Rank[2]

[1] NASA Ames Research Center, MS 245-6, Moffett Field, CA 94035-1000, USA
[2] UCO and Lick Observatory, UCSC, Santa Cruz, CA 96064, USA

1 Introduction

Polycyclic aromatic hydrocarbons (PAHs) have been proposed as the carrier of the well-known series of spectral features at 3.3, 3.4, 6.2, 7.7, 8.6, 11.3, and 12.7 μm. Here, we concentrate on the PAH features in the 3 μm regime: the main band at 3.29 μm, which arises from a C–H stretch (v=1→0) on the periphery of an aromatic ring, and the satellite band at 3.4 μm. Recent spectra by Geballe et al. (1994), Joblin et al. (1996), and Sloan et al. (1996) suggest that the satellite band is a blend of (1) a sidegroup band at 3.40 μm arising from a C–H stretch on an aliphatic (CH_3) molecule attached to a PAH molecule and (2) a hot band at 3.43 μm arising from an excited (v=2→1) counterpart of the 3.29 μm band. The sidegroup band appears to dominate the hot band in most cases.

Photodissociation regions (PDRs) lie at the interface between H II regions and molecular clouds and usually exhibit strong emission in the PAH bands. In both M17 Southwest and the Orion Bar, we can observe PDRs with a favorable edge-on geometry. Our goal is to use narrow band imaging of the PAH bands at 3.3 and 3.4 μm to probe the physics of both of these PDRs. In particular, we will compare our results to the extensive models of the Orion Bar.

We use narrow-band imaging at three wavelengths to isolate the main band (3.29 μm), the PAH pedestal (3.36 μm), and the satellite band (3.42 μm). We analyze the band strengths $F_{3.29}$ and $F_{3.42}$ and the ratio $F_{3.42}/F_{3.29}$ after subtracting the pedestal emission. We observed the Orion Bar in 1995 February and M17 SW in 1995 May at the NASA 1.5 m telescope at Mt. Lemmon/Steward Obs. with a LN_2 cooled Amber Engineering 128×128 InSb array. The pixels span 0″.78 each. A circularly variable filter wheel (CVF) isolated the bandpasses.

2 The Orion Bar

Models of the emission from CO and H_2 in the Bar by Tielens et al. (1993) show that the UV radiation drops exponentially into the Bar: $F_{UV} \propto e^{-d/d_0}$, with a 1/e folding distance $d_0 = 3''$. Observations of the PAH emission at 3.3 μm by Giard et al. (1994) result in a very different value of the folding distance: $d_0 = 9''$. They suggest that clumping within the Bar may account for this discrepancy.

Our 3.29 and 3.42 μm images show significant differences in the flux distribution in these two bands along the Bar, suggesting that the PAH material in the

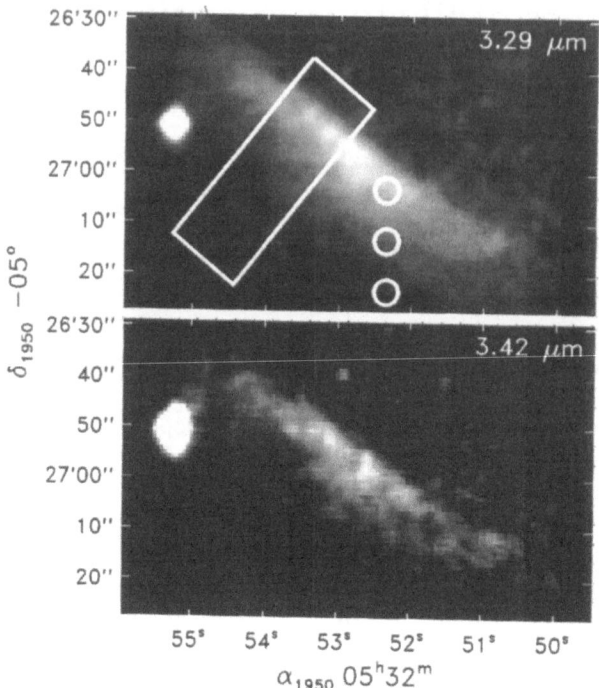

Fig. 1. Narrow-band images of the Orion Bar at 3.29 μm (top) and 3.42 μm (bottom). The box defines the region used to determine the flux profiles and ratios in Fig. 3. The three circles depict the approximate beam positions of the spectra of Geballe et al. (1989). The star to the left is θ^2 Ori A.

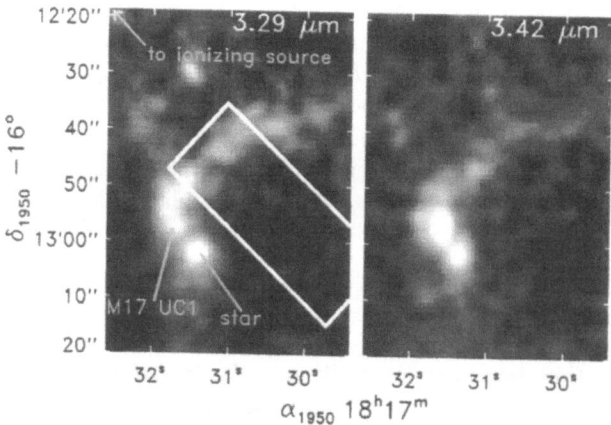

Fig. 2. Narrow-band images of M17 Southwest at 3.29 μm (left) and 3.42 μm (right). The rectangle surrounds the area used to generate the profiles in Fig. 3.

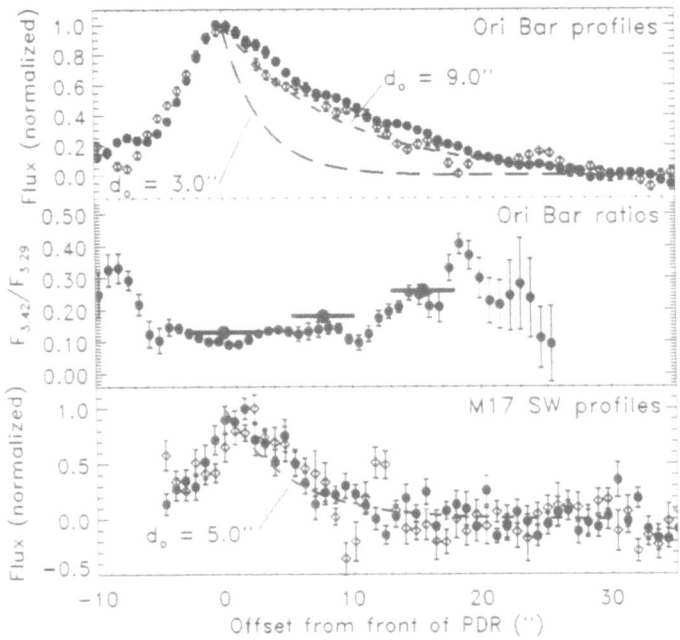

Fig. 3. Flux ratios and profiles for the Orion Bar and M17 SW, plotted as a function of distance from the front of the PDR. Top: A comparison of the 3.29 μm flux (filled circles) and 3.42 μm flux (open diamonds) from the Orion Bar with exponential drop-offs (dashed lines). The ratio $F_{3.42}/F_{3.29}$ in the Orion Bar is plotted in the middle panel, along with ratios measured spectroscopically by Geballe et al. (1989) in three discrete locations (heavy filled circles). Bottom: The 3.29 μm flux (filled circles) and 3.42 flux μm (open diamonds) in M17 SW plotted with an exponential drop-off (dashed line).

Bar has more than one component. We determine that the 1/e folding distance (d_0) is 9″ at 3.29 and 3.42 μm (confirming Giard et al. 1994). The discrepancy with d_0 derived from molecular emission suggests that the molecular emission and PAH emission are not tracing the same UV field. The flux ratio $F_{3.42}/F_{3.29}$ generally increases with distance from the front of the Bar. This was first seen in the multi-aperture spectra taken by Geballe et al. (1989), but our data extend further into the molecular region and show additional structure.

A clumpy composition in the Bar could account for the discrepancy between the UV extinction as measured from PAH features and molecular emission (Giard et al. 1994); it also produces better models of sub-mm and radio observations (Tauber et al. 1994; Hogerheijde et al. 1995). Tauber et al. (1994) suggest that the interclump region has a density of $\sim 5 \times 10^4 \, cm^{-3}$ and a filling factor of ~ 1-8%. The corresponding values as determined by Hogerheijde et al. (1995) are $\sim 3 \times 10^4 \, cm^{-3}$ and $\sim 0.5\%$. The clumps would have densities of order $\sim 10^{6-7} \, cm^{-3}$.

Our results suggest that $F_{3.42}/F_{3.29}$ would be 0.12 between the clumps, but could be as high as 0.7 within the clumps. This high ratio within the clumps makes it very unlikely that the hot band could produce the observed 3.42 μm emission; most of it must come from aliphatic sidegroups.

We suggest that the PAHs within the clumps are primitive and unprocessed, while the harsh UV field in the interclump region has stripped the PAHs there of their aliphatic sidegroups.

3   M17 Southwest

Our images show that $d_0 = 5''$, both for $F_{3.29}$ and $F_{3.42}$. Since the density is proportional to the linear folding distance and M17 SW is 4.4 times further away than the Orion Bar (2.2 kpc compared to 0.5 kpc), it follows that the Bar is 2.4 times as dense as M17 SW.

4   Conclusion

Two distinct PAH populations exist within PDRs. Primitive, unprocessed PAHs with many attached aliphatic sidegroups can survive within clumps which protect them from harsh UV radiation. In the interclump region, the radiation field has processed the PAHs, stripping most of their aliphatic sidegroups away.

The molecular emission and PAH emission do not trace the same regions within PDRs. Molecules trace the optically thickest lines of sight to the ionization front, while the PAHs trace optically thinner lines of sight. Also, the PAHs only trace conditions close to the ionization front, in regions of strong UV intensity, while more fragile molecular species like CO and H_2 trace regions with reduced UV radiation.

References

Geballe, T. R., Joblin, C., d'Hendecourt, L. B., de Muizon, M. J., Tielens, A. G. G. M., Léger, A. (1994): *ApJ*, **434**, L15

Geballe, T. R., Tielens, A. G. G. M., Allamandola, L. J., Moorhouse, A., Brand, P. W. J. L. (1989): *ApJ*, **341**, 278

Giard, M., Bernard, J. P., Lacombe, F., Normand, P., Rouan, D. (1994): *A&A*, **291**, 239

Hogerheijde, M. R., Jansen, D. J., van Dischoeck, E. F. (1995): *A&A*, **294**, 792

Joblin, C., Tielens, A. G. G. M., Allamandola, L. J., Geballe, T. R. (1996): *ApJ*, in press

Sloan, G. C., Woodward, C. E., Bregman, J. (1996): in preparation

Tauber, J. A., Tielens, A. G. G. M., Meixner, M., Goldsmith, P. F. (1994): *ApJ*, **422**, 136

Tielens, A. G. G. M, Meixner, M. M., van der Werf, P. P., Bregman, J., Tauber, J. A., Stutzki, J., Rank, D. (1993): *Science*, **262**, 86

Mid-Infrared Imaging Polarimetry of BNKL

Craig H. Smith[1], David K. Aitken[1], Toby J.T. Moore[1], Takuya Fujiyoshi[1], Patrick F. Roche[2], and Christopher M. Wright[3]

[1] School of Physics, University College, ADFA, Canberra, ACT, 2600, Australia
[2] Astrophysics, Oxford University, Keble Rd, Oxford, OX1 3RH, UK
[3] MPE, Postfach 1603, D-85740, Garching, Germany

Abstract. We present 17 μm imaging polarimetry of the BNKL region in OMC1. The polarization image shows remarkable position angle changes on small spatial scales (a few arcsec) indicating significant magnetic field complexity in this region. This field complexity can be interpreted as resulting from an accretion disk around or near IRc2, with a toroidal magnetic field in the disk, overlaid with polarization from the bi-polar outflow. Near IRc3, where the polarization seems to reach a null point, we see the effects of orthogonal overlaid fields (from the disk and the outflow) which cancel each other.

1 Introduction

The relative proximity and high luminosity of the giant molecular cloud OMC1, has allowed a more detailed study of this source than any other region of ongoing massive star formation. We are concerned primarily with the magnetic field distribution in the mid-infrared luminous core region of OMC1, generally held to be excited by the infrared source IRc2.

Large beam (20 arcsecond plus) far-infrared and submillimetre studies of OMC1 (e.g. Gonatas et al. 1990, Leach et al. 1990) show an elongated ridge of cool dust, running with a position angle $\sim 30°$. The observed polarization indicates a magnetic field essentially normal to the ridge and parallel with the local interstellar field. These observations are consistent with collapse along the field lines to form the elongated or flattened structure.

Near-infrared polarimetric observations (e.g. Burton et al. 1991, Chrysosto-mou et al. 1994) show polarization variations near IRc2, and circum-symmetric patterns attributed to scattering from the rims of the collimated outflow from the IRc2 region. From these results the inclination of the outflow is estiamted as $\sim 30°$ from the plane of the sky. However, the near-infrared polarization observations and interpretation of magnetic field distributions are made difficult by the strong scattered component and uncertainty as to the depth to which these observations penetrate the cloud.

The thermal emission from dust in the mid-infrared is more concentrated about the central sources and probes deeper into the cloud than near-infrared observations, is negligibly affected by scattering and yet maintains high spatial resolution. Studies at thermal infrared wavelengths can augment our knowledge of the magnetic field distribution as revealed at other wavelengths. With this aim we have obtained 17 μm polarimetric images of the inner 30 arcseconds of the BNKL complex in OMC1.

2   NIMPOL: a Mid-infrared Imaging Polarimeter

These observations were made with the mid-infrared imaging polarimeter (NIM-POL) constructed in the School of Physics at the University College, ADFA. This instrument is a mid-infrared (5 - 18 μm) imaging system designed from the beginning to obtain polarimetric images. It uses an Amber Engineering, 128 × 128, Si:Ga, DRO, focal plane array, which is cooled, along with all associated optics, to 20 K in a liquid He cryostat. The optics are all refractive, and the instrument provides excellent diffraction limited images on a 4 metre class telescope. A typical point spread function (PSF) shows a FWHM of $\sim$ 0.7 arcsec at 12 μm, whilst the 2nd and even third Airy maximum are seen. The pixel scale is 0.25 μm per pixel with a 32 arcsec field of view.

Digital signal processors (DSPs) are used to drive the array and coadd data, which is then downloaded to a PC for display, reduction and storage. The deep wells of the detector, coupled with fast readout (current maximum readout time is 18 msec per frame) mean that we also have the ability to obtain broad-band 10 μm images (i.e. 8 - 13 μm). The instrument currently carries broad band 10 and 20 μm filters, narrow band $\Delta\lambda = 1$ μm filters centred at 8.5, 11.5 and 12.5 μm, special 1% filters at 10.52 and 12.81 μm for observing the [S IV] and [Ne II] lines, plus a CVF with 0.25 μm resolution in the 8 - 14 μm band. The sensitivity at 12 μm is 20mJy/arcsec^2/1 σ/1 hr/$\Delta\lambda = 1$ μm.

To obtain polarimetric images a 0.2 μm spaced wire grid analyser (on a KRS5 substrate) is mounted inside the cryostat, while CdSe or CdS half waveplates are mounted in the optical path to rotate the plane of polarization. Images are obtained at various rotations of the waveplate, and differences in these images provide the polarimetric information. As differences are used there is a need for very high signal to noise images, i.e. S/N of many hundreds per pixel is required to obtain acceptable polarimetric images. We have also produced an excellent on-line polarimetric data reduction package.

The polarimetric images were obtained in January 1995, at the United Kingdom Infrared Telescope (UKIRT), a 3.9m telescope at Mauna Kea Observatory, Hawaii. Fig. 1 shows an image at 17 μm ($\Delta\lambda \simeq 2$ μm) of the BNKL region in OMC1, smoothed to the diffraction limit of about 1 arcsec. Fig. 2 shows the same region, but with the polarization vectors overlaid on the image.

The polarimetric image is most notable for the strong variations in position angle (θ) on quite small spatial scales (only a few arcseconds). The degree of polarization also varies dramatically from around 12 % near BN to effectively 0% near IRc3. At a first glance this polarization image appears rather chaotic.

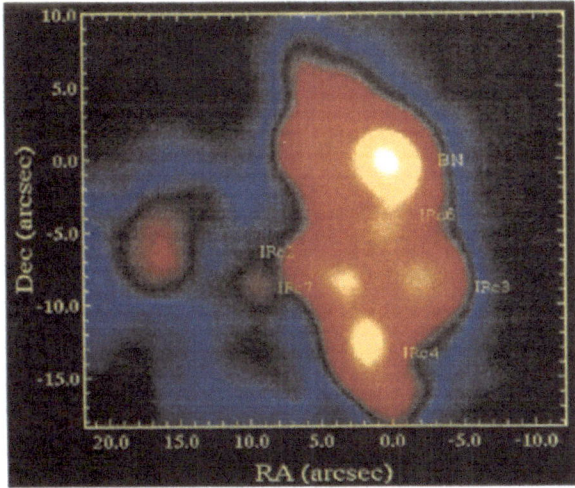

Fig. 1. A 17 μm image of the BNKL region in OMC1. The compact sources are labelled.

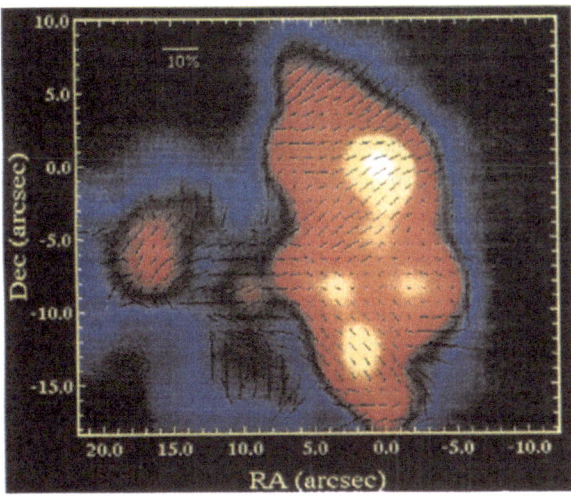

Fig. 2. The same image as Fig. 1 with the polarization vectors overlaid. In this source the polarization is in absorption so the polarization vector direction also indicates the magnetic field direction.

3   Discussion

Polarization at 17 μm is due to emission or absorption from aligned dust grains, and no matter what the mechanism that actually aligns the grains, the average orientation of a non-isotropic distribution of grain spins will be determined by the ambient magnetic field. So observing the polarization distribution relates almost directly to the ambient magnetic field. The only problem being a 90° ambiguity between emissive and absorptive polarization. In fact, in the case of the BNKL region, the observed position angle complexity might be explained by a combination of sources displaying emissive and absorptive polarization giving relative 90° position angle changes in the same magnetic field environment. However, 8 - 22 μm spectro-polarimetry of most of the compact sources in the BNKL region (Aitken et al. in preparation) is able to separate the absorptive and emissive components (which have different spectro-polarimetric signatures). These observations show that, while an emissive polarization component is present in this region, it is dominated by absorptive polarization, even at 17 μm, and has little effect on the observed position angle, which therefore represents the local magnetic field direction projected on to the plane of the sky.

To explain the rather complex polarization structure in the BNKL region we have developed a simple, yet effective model for the magnetic field distribution. We interpret our results as the overlaying of two magnetic field components. The first is a toroidal field in the accretion disk that surrounds IRc2 (or whatever the central source). Toroidal fields in disks are in fact the normal state as determined by mid-infrared spectro-polarimetry (Aitken et al. 1991). Near IRc7 and IRc4 we see the front edge of the disk (which is inclined by about 30° from the plane of the sky) and near IRc6 and IRc3 we are seeing the far side of the disk. However, the far side of the disk is seen through the optically-thin, bi-polar outflow material, which is aligned with the large scale field (~118°). Fig. 3 shows the proposed geometry of the disk and outflow. Near IRc3 the polarization in the disk is very nearly orthogonal to that in the outflow, cancelling and producing an essentially null point in the polarization. The strong polarization near BN represents the large scale field free from disk polarization and disordering effects of the disk, whereas the strong polarization near IRc4 represents the disk field unobscured by outflow material.

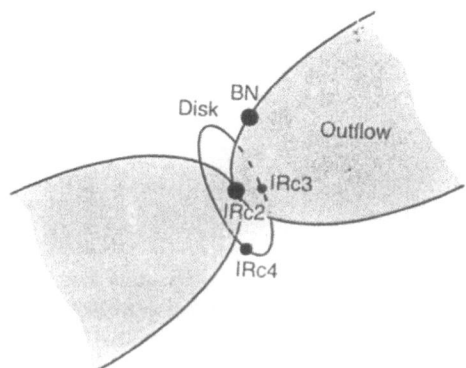

Fig. 3. A model of the complex polarization structure with a toroidal magnetic field in an accretion disk, overlaid with an orthogonal field in the bi-polar outflow.

References

Aitken D.K., Wright C.M., Smith C.H., Roche P.F., (1991) MNRAS, **262**, 456

Burton M.G. et al., (1991) MNRAS **375**, 611

Chrysostomou A., Hough J.H., Burton M.G., Tamura M., (1994) MNRAS, **268** 325

Gonatas et al., (1990) ApJ, **357**, 132

Leach R.W., Clemens D.P., Kane B.D., Barvainis R., (1990) ApJ, **370**, 257

Star Formation in the Northern Region of NGC 6334

Mauricio Tapia[1,2] and Paolo Persi[2]

[1] Royal Greenwich Observatory, Madingley Rd., Cambridge, CB3 0EZ, England
[2] Instituto de Astronomía, UNAM, Apdo. Postal 877, Ensenada, Mexico
[3] Istituto Astrofisica Spaziale, CNR, C.P.67, 00044 Frascati, Italy

Abstract. Sub-arcsec images in the J, H, K and H_2 and nearby continuum centred in the far infrared source NGC 6334 I are presented. The presence of a loop-shaped $2\,\mu m$ reflection nebula originating from the mid-IR source IRS1 and extending some $25''$ to the west is established. IRS1 is composed of at least four very red near-IR components cohabiting within $3''$. Three knots of H_2 emission coinciding with recently discovered NH_3 masers were found. Two of these are aligned along the CO bipolar outflow. From the photometry carried out of the 213 sources found, the presence of an deeply embedded ($A_V > 70$) very young (age $< 10^6$ years) cluster of approximate dimension $70''$ was detected and measured photometrically.

1 Introduction

The far–IR source NGC 6334I lies in the northeastern edge of the extended molecular cloud complex NGC 6334. This site of recent star formation contains three mid–IR sources, OH, H_2O and CH_3OH masers, a compact H II region, $\sim 3''$ size and a 1mm continuum peak emission. A near–IR survey of the region conducted by Straw et al. (1989) to a limit of K=13.5 mag, revealed the presence of a compact young stellar cluster of ~ 1 pc . A high–velocity bipolar outflow was discovered by Bachiller & Cernicharo (1990) centred near the IRS1-IRS2 complex of which excites the compact H II region. Recently, Kraemer & Jackson (1995) found three NH_3 (3,3) masers in NGC 6334I, two of which occur near the ends of the high–velocity CO outflow. They also detected an extended thermal $NH_3(3,3)$ source in the centre.

In this paper, we present the results of deep imaging with sub–arcsec resolution in the J, H, K broad–bands and in the H_2 $v = 1 \rightarrow 0$ S(1) line and nearby continuum of the region in order to study in detail the interaction between the bipolar outflow and the ambient gas in NGC 6334 I, as well as the properties of the embedded young stellar cluster

2 Observations

Our observations were made on 16 June 1994 with the 2.5m Dupont telescope of Las Campanas Observatory, Chile and a NICMOS3 array camera (Persson et al. 1992). A mosaic made from six overlaping frames covering approximately

100×170 square arcsec was constructed in each of the J, H and K, H_2 and [2.2 μm] filters. The scale was 0.35 $''$/pix, and the measured point–spread function had a FWHM of around 0.9$''$, constant across the field. The total exposure times were 480s for J, 300s for H and 100s for K. The limiting magnitudes in the photometric passbands were $J = 18.3$, $H = 17.8$ and $K = 16.2$. The effective area covered in the three colours is 17000 square arcsec. The narrow–band filters are centred at $\lambda = 2.125\,\mu m$ ($\Delta\lambda = 0.024\,\mu m$), which includes the H_2 $v = 1 \rightarrow 0$ S(1) line, and at $\lambda = 2.20\,\mu m$ ($\Delta\lambda = 0.11\,\mu m$) for measuring the neighbouring continuum. In each filter, a mosaic of 92×99 square arcsec in size was constructed using four overlapping frames. The exposure times in the line and continuum filters were 360s and 240s, respectively. The image of the area under study is shown in Fig. 1a.

3   Results

3.1   Nebular features

The H and especially the K images are characterized by the presence of a loop-shaped nebula extending over $\sim$ 25$''$ (0.2 pc) in the west-east direction with IRS1 at its western apex and a blueish, presumably field, star located by chance projection at its eastern edge (Fig. 1a). The nebula is roughly perpendicular to the CO outflow observed by Bachiller & Cernicharo (1990). The most likely explanation for this reflection nebula is that it is caused by radiation from the ionizing star that escape freely towards the lower density region. On the opposite side, the UC H II region has a steep surface brightness gradient. The dense molecular material traced by the thermal $NH_3(3, 3)$, CO, and HC_3N molecular lines may prevent the escape of the radiation from IRS1 in this direction. Similar IR nebula, though aligned in the direction of high velocity outflows, have been found associated with the massive YSOs GL 2591 and are interpreted in terms of radiation scattered by dust entrained in the bipolar outflow. In addition a second, more extended and diffuse infrared nebula is observed some 30–40$''$ to the northwest of IRS1 in our K–image (see Fig. 1a).

The source IRS1 appears complex in the near-infrared. It consists of at least four components of different JHK colours, as shown by the contour plots in Fig. 2. IRS1E is extended at $\lambda > 2.0\,\mu m$ and is the reddest source with $J - H > 5$ and $H - K = 3.9$. It dominates all emission longward of 2 μm, and is coincident with the 30 μm peak. This source is the recently formed O-B0 type star responsible for the ionization of the compact HII region and of pumping the nearby H_2O and OH masers. IRS1W is about 1$''$ to the southwest of IRS1E and is the brightest component in J and H ($J - H = 2.3$) while the component IRS1SW is some 1.5 magnitudes fainter in J and H. A fourth very red ($H - K \geq 3.6$) peak (labelled IRS1SE) is aligned with the infrared tail, as shown in Fig.2. No near-IR source brighter than about $K = 17.5$ was detected at the positions of IRS2 and the H_2O maser.

We found molecular hydrogen emission knots which coincide with the $NH_3(3,3)$ masers. Two of the H_2 knots resemble symmetric bow shocks and are aligned

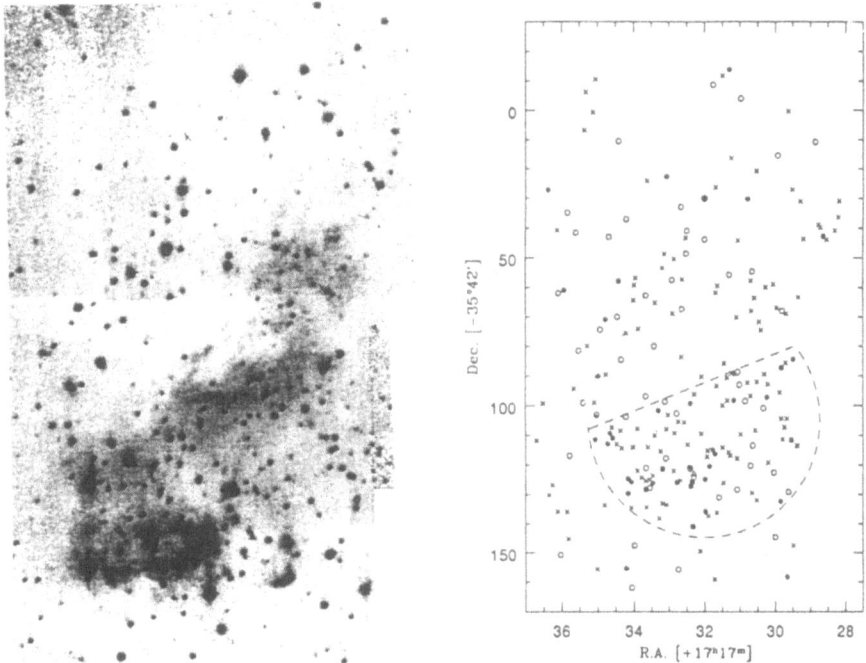

Fig. 1. (a) K image of NGC 6334I. The scale 0.35 ″/pix and has dimensions 115×195 square arcsecs. (b) Spatial distribution of all the sources in our survey. The open circles are field stars with JHK colours of reddened late spectral type. Asterisks are early type stars according to their JHK colours and filled squares are sources with near-IR excesses. Crosses are stars detected in H and K only. The position of the peaks of far-IR emission are indicated by filled circles. Open squares mark the position of the radio HII regions and the triangle, that of the H_2O maser. The large segment of circle (broken line) marks the aparent boundary of the embedded cluster. North is at top, east to the left.

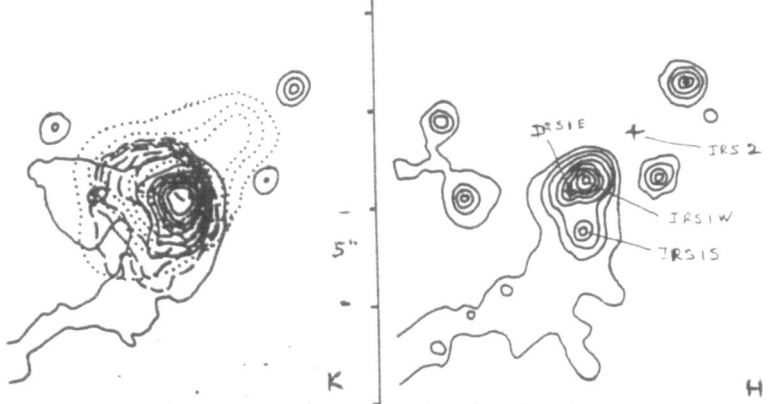

Fig. 2. H, K (solid), 30 μm (dashed) and 6cm (dotted) contour maps of central NGC 6334I. The location of the near-infrared components are indicated. The 30 μm map is from Harvey & Gatley (1983) and the 6cm map is from De Pree et al. (1995)

with the *thermal* NH$_3$ central structure, and with the HC$_3$N peak roughly at the centre of the CO bipolar outflow. Finally, a third peak is located at the eastern edge of the tail-shaped nebula. This is the first case where an accurate positional coincidence between NH$_3$(3, 3) masers and H$_2$ v= 1 → 0 S(1) line emission has been found. The precise spatial coincidence between these two phenomena (any possible offset is less than 1″) suggests that the two types of emission require similar physical and chemical conditions.

3.2 Embedded cluster

A total of 213 sources were detected and measured in at least H and K. Due to the very high extinction in the area, only 87 objects were detected in the three passbands J, H and K, as most of the objects are fainter than our limit at 1.25 μm ($J = 18.3$). Unlike other sites of suspected embedded infrared clusters, there seems to be only a small fraction of sources with significant near-infrared excess in the direction of NGC6334 I and these include the well known object Irs 1 which is a recently formed B0 type star (Harvey & Gatley 1983).

Recent formation of massive stars in NGC 6334I is evident, nevertheless, from the presence of the luminous embedded near- and mid-IR source IRS1. Fig. 1b shows the spatial distribution of all sources detected at H and K in our images. A larger concentration of sources is seen in the area delimited by the segment of a circle of radius 80″. This could be due to two factors or a combination of both: variable projected star density and patchy extinction. The latter is evident from the molecular and millimeter observations, but the observed concentration occurs near one of the maxima of the molecular density, at the position of IRS2. This clearly suggests the presence of an embedded cluster of which IRS1 and IRS2 are members. A confirmation of this can be drawn by comparing the distribution of late and early type stars, as deduced from their JHK colours. The ratio of the number density of late type stars within the of the cluster area to outside its boundaries N_{in}/N_{out} is about six times smaller than that of early type stars, and about twenty times smaller than that of the stars with near-IR excess. The age of the embedded cluster seems to be less than one million years.

References

Bachiller R., Cernicharo J. (1990): A&A **239**, 276

Davis C.J., Eislöffel J., (1995) A&A **245**,

De Pree, C.G., Rodríguez, J.F., Dickel, H.R., Goss, H.N., (1995) ApJ **447**, 220

Harvey P.M. and Gatley I. (1983): ApJ **269**, 613

Kraemer K.E., Jackson J.M., (1995): ApJ **439**, L9

Persson, S.E., West, S.C., Carre, D.M., Sivaramakrishnan, A., Morphey, D.C. (1992): PASP **104**, 204

Straw S.M., Hyland A.R., McGregor P.J., (1989): ApJ Supp **69**, 99

Morphology of the Star Forming Region Associated with HH25-26[1]

Elisabetta Tommasi, Dario Lorenzetti, and Brunella Nisini

Istituto di Fisica Spazio Interplanetario, C.N.R., C.P. 27, I-00044 Frascati (Italy)

Abstract. We present near-infrared images of the star forming region associated with the Herbig-Haro (HH) objects HH25-26 in the H,K bands and through a narrow-band filter centered on the 2.122 μm ν = 1-0 S(1) transition of molecular hydrogen. Shocked molecular hydrogen emission is observed in the direction of the two HH objects, moreover knots of H_2 emission are also detected along the axis of the associated molecular outflow, both in the blue and in the red lobe. The reached sensitivity allows us to detect infrared young objects not revealed in previous surveys of the region, making possible the identification of the molecular outflow driving source.

1 Introduction

HH25-26 are two low excitation Herbig-Haro objects located few arcmin south of NGC2068, within the L1630 dark cloud. Close to them, other two HH objects, HH24 and HH27, are also present, at about 2 arcmin north and east respectively. A low sensitive 2.2 μm survey (K < 10.7 mag, Strom et al. 1976) detected two sources in the vicinity of HH25-26, one of which, SSV59, has been suggested as the exciting source of the two HH objects, being positioned between the two (Cohen & Schwartz 1983).

CO observations (Edwards & Snell 1984, Gibb & Heaton 1993) detected two molecular outflows in the region: one associated with HH25-26 and the other situated north of this region, probably related with HH24. HH25 and 26 are both located in the blue lobe of the outflow but not aligned with its axis. SSV59 is 50 arcsec apart from the geometrical centre of the outflow. Based only on geometrical considerations, the system of the two HH and the IR source are unlikely associable with the molecular outflow.

In this contribution, we use high sensitivity broad and narrow-band NIR images to link together the different observations in a coherent picture for the morphology of the region.

2 Observations

The broad-band K and the narrow-band H_2 (λ = 2.12 μm, $\Delta\lambda$ = 0.039) images were taken on October 22, 1994, using the IRAC2 camera on the ESO 2.2 m telescope. The camera is equipped with a 256x256 pixels NICMOS detector.

[1] Based on observations collected at ESO, La Silla, Chile and at TIRGO, Gornergrat, Switzerland (TIRGO is operated by CAISMI-CNR Arcetri, Firenze, Italy)

The image scale is 0.49″/pixel which gives a total FOV of about 2x2 arcmin². A sky frame was constructed from frames taken in different positions and then subtracted from the image. The image was then flat-fielded using a differential dome flat. The limit of sensitivity in the K image is about 17 mag. A continuum-free H_2 line image was obtained by subtracting from the narrow-band image an appropriately scaled K image.

Two other broad-band images (in the H and K filters) were taken on November 8, 1994, at the telescope TIRGO (Gornergrat, Switzerland), using the infrared camera ARNICA. The scale of the images is 0.95″/pixel and the total FOV is 4x4 arcmin². The reduction technique was similar as for the ESO images, except that we used a median image of sky frames for flat-fielding. The sensitivity limits are about 17 and 16 mag for the H and K images respectively.

3   Results and Discussion

Figures 1 and 2 show the ESO K image and the continuum subtracted H_2 image respectively. In both figures, the CO molecular outflow contours taken from Gibb & Heaton (1993) are superimposed along with the positions of the known objects.

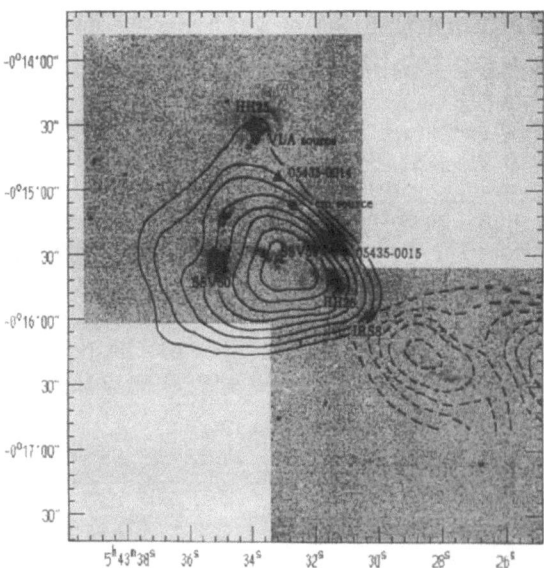

Fig. 1. ESO K image with superiposed the CO molecular outflow from Gibb & Heaton (1993) and the positions of known objects. The coordinates of the VLA and of the 1.3 cm sources are from Bontemps et al. (1995) and from Verdes-Montenegro et al. (1995) respectively.

We associated the brightest IR continuum sources in the field with the IR sources SSV59 and SSV60 detected by Strom et al. (1976), although the position they give for SSV60 is off by 15″ in declination. Table 1 reports the coordinates and the TIRGO H and K photometry of the detected sources. One of them, IRS8, is positioned exactly at the center of the CO outflow.

As regard to the line emission, H_2 is observed toward the two HH objects but also a chain of line emission spots aligned with the molecular outflow is revealed (knots A, B and C).

From the obtained results, we suggest the following picture for the morphology of the region:

1. IRS8 is the reddest star of the field according to its H-K color ([H-K]=2.4 mag) suggesting it as a young object. Giving also its extremely favourable position, it appears as the best candidate to be the exciting source of the CO outflow.
2. The proper motion of HH26, measured by Jones et al. (1987), is directed in the opposite direction with respect to IRS8. It thus seems that HH26 neither is excited by IRS8 nor is associated with the CO outflow. More likely, HH26

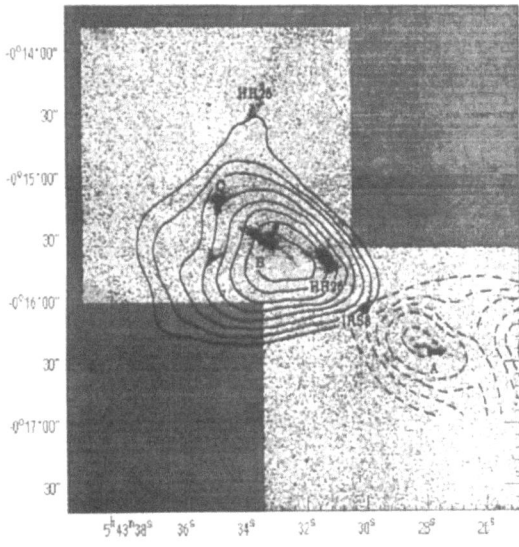

Fig. 2. The continuum subtracted H_2 image with superiposed the CO molecular outflow from Gibb & Heaton (1993) and the position of IRS8.

and HH25 could be both excited by the cm/IRAS source laying in the middle position between the two.

3. The new detected knots of infrared line emission are displaced exactly along the axis of the outflow and IRS8 is aligned with them. Moreover, the knots A and B, which lay almost at the same distance from IRS8, are coincident with the peaks of the blue and red lobe of CO emission respectively. Therefore the molecular H_2 and CO emission are likely evidences of the same phenomenon. This occurrence seems to be a quite common feature related to matter flow from embedded sources (*e.g.*, Davis & Eislöffel 1995, Hodapp & Lada 1995) being the H_2 emission originated from shock excitation due to the interaction of the flow with the ambient gas.

Table 1. Photometry of sources

IRS	$\alpha(1950)$ h m s	$\delta(1950)$ o ′ ″	H	K (mag)	[H−K]
1	5 43 30.4	−0 13 53	16.6(40)	15.0(10)	1.6(40)
2	5 43 34.0	−0 14 34	13.3(20)	12.28(3)	1.0(20)
3	5 43 34.1	−0 14 44	> 17	15.2(20)	> 1.8
4	5 43 38.8	−0 14 52	16.3(30)	15.3(20)	1.0(40)
5	5 43 39.5	−0 15 22	15.4(20)	14.8(10)	0.6(20)
6(SSV59)	5 43 31.4	−0 15 24	10.7(20)	8.77(1)	1.9(20)
7(SSV60)	5 43 35.0	−0 15 34	9.7(20)	8.61(1)	1.1(20)
8	5 43 30.3	−0 15 56	14.3(20)	11.94(2)	2.4(20)
9	5 43 35.7	−0 16 06	15.9(30)	15.4(20)	0.5(40)
10	5 43 38.1	−0 16 35	12.8(20)	11.92(2)	0.9(20)
11	5 43 26.7	−0 17 00	15.2(20)	14.9(10)	0.3(20)

Note to the table:

the errors on the magnitudes are given in units of 0.01 mag.

References

Bontemps, S., André, P., Ward-Thompson, D. (1995): A&A **297**, 98

Cohen, M., Schwartz, R.D. (1983): ApJ **265**, 877

Davis, C.J., Eislöffel, J. (1995): A&A **300**, 851

Edwards, S., Snell, R.L. (1984): ApJ **281**, 237

Gibb, A.G., Heaton, B.D. (1993): A&A **276**, 511

Hodapp, K.-W., Ladd, E.F. (1995): ApJ in press

Jones, B.F., Cohen, M., Wehinger, P.A., Gehren, T. (1987): AJ **94**, 1260

Strom, K.M., Strom, S.E. , Vrba, F.J. (1976): AJ **81**, 308

Verdes-Montenegro, L. et al. (1995): in preparation.

New Millimetre Observations of Pre-stellar Cores

D. Ward-Thompson[1], P. André[2], and F. Motte[2,3]

[1] Royal Observatory, Blackford Hill, Edinburgh, UK
[2] Service d'Astrophysique, CEA, Saclay, F-91991 Gif-sur-Yvette Cedex, France
[3] Laboratoire d'Astrophysique de Grenoble, BP53X, F-38041 Grenoble Cedex, France

Abstract. Results are outlined of a submillimetre study of pre-stellar cores. The pre-stellar phase is one in which a dense core in a molecular cloud is gravitationally bound, but contains no embedded luminosity source. This takes place prior to the protostellar Class 0 phase, which is the main infall phase of star formation. The observations of pre-stellar cores are compared to different models of star formation. The radial density profiles are inconsistent with a singular isothermal sphere but consistent with predictions of a magnetic support model, indicating that the cores are undergoing ambipolar diffusion prior to the main protostellar collapse phase.

1 Introduction

The earliest protostellar stage observed so far appears to be the main infall and outflow phase of star formation, in which a central hydrostatic protostar has formed, but has not yet accreted the majority of its final main-sequence mass. At this stage, most of the mass is still in the form of a dense circumstellar envelope. The best observational candidates for this phase are the recently identified Class 0 objects (André, Ward-Thompson & Barsony 1993; André 1995).

Prior to this protostellar stage, the pre-protostellar phase of star formation (hereafter referred to as the pre-stellar phase for brevity) may be defined as a phase in which a gravitationally bound fragment has formed in a molecular cloud, and evolves towards progressively higher degrees of central condensation, eventually leading to protostellar collapse. Theoretically, this evolution of magnetically subcritical cloud cores is thought to occur through the progressive loss of magnetic support by ambipolar diffusion (Ciolek & Mouschovias 1994 and references therein – hereafter CM; Mouschovias, this volume). During this quasi-static phase, which precedes dynamical collapse, no self-luminous YSO is present in the centre, but the core is gravitationally bound.

2 Data

The first submillimetre continuum maps of pre-stellar cores were made by Ward-Thompson et al. (1994). Figure 1(a) shows an 800-μm map of the pre-stellar core L1689B taken from Ward-Thompson et al. They found that pre-stellar cores have larger FWHM sizes than, but comparable masses to, the protostellar envelopes

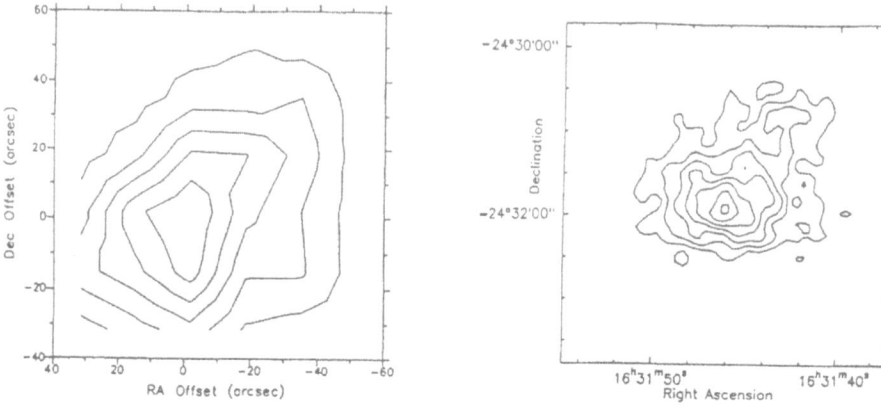

Fig. 1. (a) 800-μm JCMT contour map of pre-stellar core L1689B. (b) 1.3-mm IRAM contour map of the same core at higher resolution.

surrounding Class 0 protostars. This is consistent with pre-stellar cores being the precursors of Class 0 protostellar clumps, the latter already having begun to collapse, as witnessed by their higher luminosities.

This study showed that the radial density profiles of pre-stellar cores are relatively steep towards their edges (roughly $\rho \propto r^{-2}$) and flatten out near their centres, becoming less steep. This is apparently inconsistent with the scale-free r^{-2} power-law density distributions characterizing singular isothermal spheres, which are generally taken as initial states in the standard protostellar models (e.g. Shu 1977; Shu, Adams & Lizano 1987).

Obtaining further observational constraints on the initial conditions for collapse is crucial for protostellar evolutionary models. In particular, if most low-mass stars form from finite-sized clumps instead of singular isothermal spheres, then the mass-infall rate is expected to be time-dependent (e.g., Foster & Chevalier 1993) rather than constant, as generally assumed.

New millimetre dust continuum data were taken (André, Ward-Thompson & Motte, 1995) for one of the pre-stellar cores from the Ward-Thompson et al. (1994) sample, L1689B, using the IRAM 30-m telescope equipped with the MPIfR 7- and 19-channel bolometer arrays. These new 1.3 mm data, which have 11-arcsec resolution and were obtained using the on-the-fly mapping technique, supplement the 800-μm data, which were taken at JCMT with 20-arcsec resolution using a point-by-point ('on-off') mapping technique and a single-channel bolometer. As a result, we are now able to rule out any telescope, or technique-dependent, effects which might have affected our previous conclusions on pre-stellar cores. Furthermore, the better sensitivity and spatial sampling of the new data allow us to make a detailed comparison with the predictions of ambipolar diffusion models.

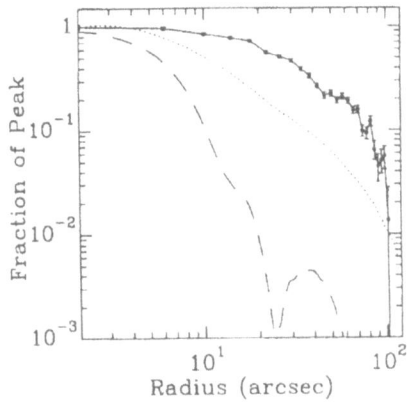
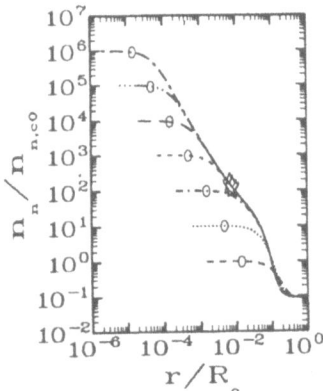

Fig. 2. (a) Azimuthally averaged radial flux density profile of L1689B (solid line) compared to the IRAM beam (dashed line) and a simulated isothermal $\rho \propto r^{-2}$ profile (dotted line). (b) Predicted theoretical radial density profiles at various times for a magnetically supported pre-stellar core undergoing ambipolar diffusion.

Figure 1(b) shows the co-added 1.3-mm isophotal contour map of L1689B smoothed to 15-arcsec resolution. The rms is $\sim$ 3 mJy per 11-arcsec beam. In this map, the L1689B core appears as an east-west elongated source (aspect ratio $\sim$ 0.7). In addition, a secondary component is clearly visible $\sim$ 40 arcsec to the north-west of the main core. Its presence probably explains the apparent SE-NW elliptical shape seen in the 800-μm map taken at lower angular resolution. The two datasets are consistent when smoothed to the same resolution.

3   Discussion

The radial density profile of the core may be derived from the observed intensity profile, assuming there is nó central heating source. If $\rho(r) \propto r^{-p}$ and $T(r) \propto r^{-q}$ are the density and temperature distributions as a function of radius r, and $I(\theta) \propto \theta^{-m}$ is the observed intensity profile as a function of projected radius θ, then p, q and m are related by the simple equation m=p+q−1. With no central heating source there is no temperature gradient, and hence q=0. Therefore the equation simplifies to m=p−1.

   Figure 2(a) shows the azimuthally averaged flux density profile of L1689B, measured from Fig. 1(b) as a solid line. Note that we have measured the difference between the profiles in the major and minor axes of the core and found them not to be significantly different in form. For comparison a simulation was made of a dual-beam observation of a scale-free $\rho(r) \propto r^{-2}$ isothermal sphere profile. The dotted line in Fig. 2(a) shows its radial flux density profile and the IRAM beam is shown as a dashed line. The observed radial flux density profile shows a flat inner region up to 20 arcsec from the center with $\rho \propto r^{-1}$ to $r^{-1.5}$, and a steeper region beyond 20 arcsec with roughly $\rho \propto r^{-2}$. This is very similar to the profile observed for this core in the 800-μm data (Ward-Thompson et al. 1994). Hence we have here shown that the observed radial density profile does indeed flatten towards the centre, and this is not an effect of telescope, instrumentation or observing technique. The profiles we observe are inconsistent with a single, scale-free power-law.

Figure 2(b) shows theoretical radial density profiles (Basu & Mouschovias 1995) of a pre-stellar core which is undergoing ambipolar diffusion. The profiles show a time sequence of evolution of the core from bottom to top, with each successive step having an order of magnitude greater central density than the previous one. The core spends the majority of its time in the lowest state. In this state the central part of the profile is flat, and it steepens to r^{-2} in the outer part. The supercritical core generally forms at a relatively early stage, when a substantial region in the center still has a flat density distribution (e.g. CM).

Hence the data appear to agree reasonably well with theoretical predictions for radial core density profiles during the ambipolar diffusion phase (CM and references therein). In all theoretical models of this phase, dense cloud cores are initially magnetically supported (i.e. subcritical). Then the drift of the neutral gas through the magnetic field lines allows the central mass-to-magnetic flux ratio, M/Φ, to increase, which eventually leads to the formation and collapse of a 'supercritical core'.

Thus we see that the data appear to show that pre-stellar cores are consistent with the ambipolar diffusion phase prior to the main protostellar collapse. However, there may be other models with which the data are also consistent. For instance, a recent hydrodynamic model (Bonnell et al. 1995), which takes no account of the magnetic field, can produce core morphologies which are qualitatively similar to those we see – cf: Fig. 1(b) with Fig. 1 of Bonnell et al. However, this hydrodynamic code appears to predict a greater degree of central condensation than we observe, so we feel it is unlikely that there will also prove to be quantitative agreement with models which ignore magnetic fields. Further work is required to see how well ambipolar diffusion models can exactly predict the evolutionary states of individual observed pre-stellar cores.

References

André, P. (1995) Ap&SS **224**, 29
André, P., Ward-Thompson, D., Barsony, M. (1993) ApJ **406**, 122
André, P., Ward-Thompson, D., Motte, F. (1995) A&A in prep
Basu, S., Mouschovias, T.Ch. (1995) ApJ **452**, 386
Ciolek, G.E., Mouschovias, T.Ch. (1994) ApJ **425**, 142 – CM
Bonnell, I.A., Bate, M.R., Price, N.M. (1995) MNRAS in press
Foster, P.N., Chevalier, R.A. (1993) ApJ **416**, 303
Shu, F.H. (1977) ApJ **214**, 488
Shu, F.H., Adams, F.C., Lizano, S. (1987) ARA&A **25**, 23
Ward-Thompson, D., Scott, P., Hills, R.E., André, P. (1994) MNRAS **268**, 276

Young Stellar Objects in L1641:
a Submillimeter Continuum Study

A. Zavagno[1,2], S. Molinari[1], E. Tommasi[1], P. Saraceno[1], and M. Griffin[3]

[1] Istituto di Fisica dello Spazio Interplanetario, CNR, CP 27, 00044 Frascati, Italy
[2] Observatoire de Marseille, 2 Place Le Verrier, 13248 Marseille Cedex 4, France
[3] Queen Mary and Westfield College, Phys. Dept., Mile End Rd., London E1 4NS, England

Abstract. We present JCMT observations of the 350 - 1300μm continuum emission of a sample of 10 Class I young stellar objects in the L1641 molecular cloud. These observations, together with 60 and 100μm IRAS data, are used to derive and discuss the properties of the circumstellar matter, to which the submillimeter emission is usually attributed. Finally, we attempt to discuss these results in an evolutionary scheme.

1 Introduction

During the past decade, many attempts have been made to classify, in evolutionary terms, young stellar objects (hereafter YSOs) using their characteristic observed properties (Hillenbrand et al. 1992; Hodapp, 1994 and references therein). For highly embedded YSOs, millimeter facilities allow observational studies of the first steps of star formation by giving access to the continuum emission due to the circumstellar dust. As the young star evolves, the amount of circumstellar material decreases (accretion/dispersion) and, consequently, the submillimeter emission diminishes. For this reason, the far IR and submillimeter emission are expected to trace an evolutionary sequence.

The purpose of this work is to study the submillimeter properties of the observed continuum and to identify a possible new evolution indicator, using a sample of 10 Class I sources in the Lynds 1641 molecular cloud situated in the southern part of Ori A at a distance of 480pc (Strom et al., 1989).

2 Observations

We carried out continuum observations from 350μm to 1.3mm with the UKT14 common user bolometer receiver on the 15-m James Clerk and Maxwell Telescope (JCMT) in October 93 and March 94, of the 10 most luminous Class I (Lada & Wilking, 1984) sources in the L1641 molecular cloud from the study of Strom et al. (1989). The FWHM beamwidth varied between 18 and 20$''$, depending on the filter. The observed sources are: S06, S11, S22, S31, S32, S55, S59, S72, S85 and VLA1 (not studied by Strom et al., 1989, see Harvey et al., 1986 for the infrared measurements). IRAS data at 60 and 100μm have also been used (Strom et al., 1989) to obtain a complete spectral energy distribution between 60μm and 1.3mm for each object.

3   Data Analysis

In order to derive submillimeter properties (dust temperature and emissivity, circumstellar mass) for the 10 observed YSOs, we adopted two different methods. The first one is a least-squares fit using the Hildebrand formalism (Hildebrand, 1983). For this we assume optically thin greybody emission, between $60\mu m$ and $1.3mm$, from an isothermal sphere (see also Reipurth et al., 1993). The dust emissivity, κ, is assumed to have a power law variation with wavelength: $\kappa = \kappa_0(\lambda_0/\lambda)^\beta$, with $\kappa_0 = 0.1$ cm^2/g and $\lambda_0 = 250\mu$m. The second approach has been to develop a spherical model, assuming the dust envelope to be a series of concentric dust shells, following temperature and density power laws. Details about these methods can be found in Zavagno et al. (1995, in preparation). Figure 1 presents a typical result of the least-squares fitting procedure, for S06.

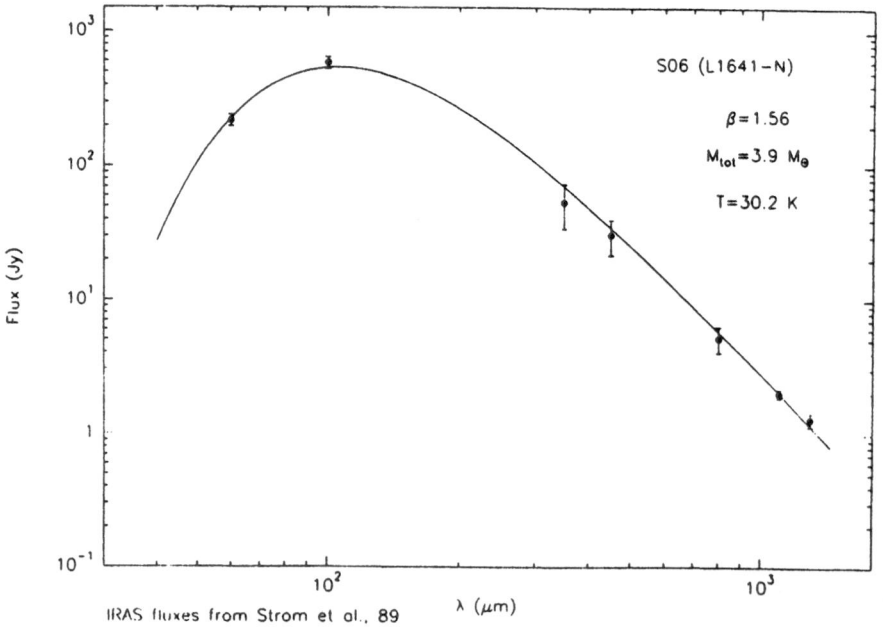

Fig. 1. S06 least-squares fit result

4   Results - Discussion

We derived mass, temperature and β values for 9 of the 10 sources of our sample. For S31, no fit was possible, as the IRAS fluxes appear inconsistent (too high) with submillimeter measurements.

The derived **temperatures** and **masses** are, respectively, in the range 25 - 40K and 0.5 - 4 $M_\odot$, indicating that we are dealing with cold low mass objects.

The β **values** are found to be in the range 0.9 - 1.9, for the whole sample. However, there is a trend towards higher β values for sources having higher $(L_{bol}/L_{1.3mm})$ values, as can be seen in the $(L_{bol}/L_{1.3mm})$ versus β diagram (see Fig. 2). Search for a linear correlation between the two quantities gives a correlation probability of 95.5%. The function $Log(L_{bol}/L_{1.3mm}) = 1.32\beta + 2.635$ is the line shown in Fig. 2.

The $(L_{bol}/L_{1.3mm})$ ratio has been proposed as an evolutionary indicator by André (1994). If this is true, the correlation seen in Fig. 2 means that β values change as the source evolves and could imply dust evolution (intrinsic properties and/or geometrical distribution). On the other hand, it is also possible that variations of the β value (due to different dust properties and/or distribution) may cause variations of the $(L_{bol}/L_{1.3mm})$ ratio, without invoking evolutionary explanations. Both effects are probably present and our data do not allow us to separate the two possible contributions. This point will have to be checked in the next future. However, we emphasize that Dent et al. (1995) have also noted a change of the β value among sources in different evolutionary stages and evidence for a flatter submillimeter continuum spectra has been found for Class 0 sources (see Ward-Thompson et al., 1995).

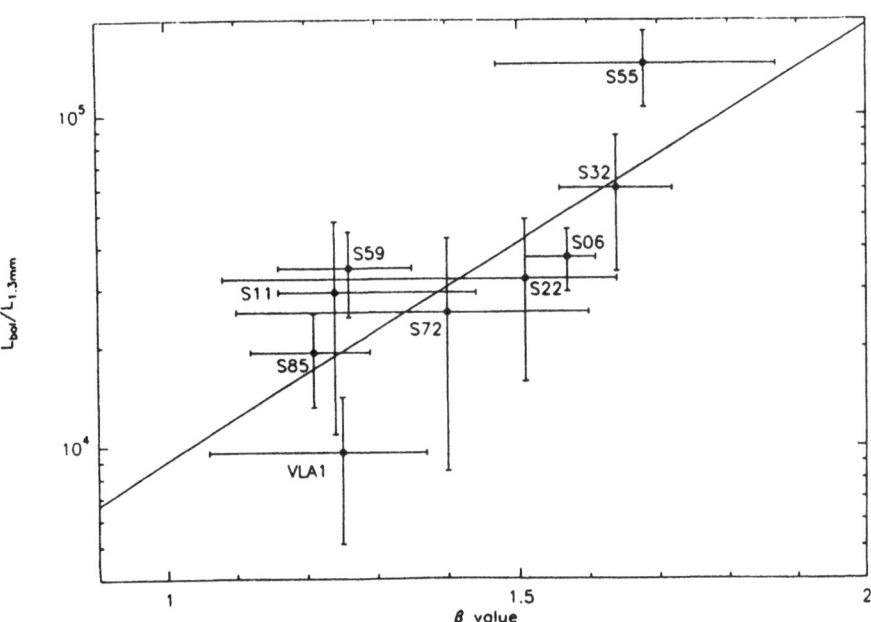

Fig. 2. β value versus $(L_{bol}/L_{1.3mm})$. The line represents the linear correlation between $Log(L_{bol}/L_{1.3mm})$ and β (see text)

In order to obtain some indication on the κ_0 value, we used the relation between the near IR spectral index (α) and the A_V value (see Strom et al., 1989). Using the A_V value and assuming the circumstellar material to be distributed in a sphere, it is possible to derive a total mass associated with each object (see Walker et al., 1990). Doing so, we obtain an independant mass estimate that can be compared with the previous one (least-squares fit).

A good agreement between the two mass determinations is found for sources associated with a high (i.e. >1.3) β value, supporting the choice of κ_0. This result is also reinforced by the fact that, taking the Hildebrand κ_0 value and extrapolating it to $1.3mm$ with a mean β value of 1.5, we obtain a $\kappa(1.3mm)$ value comparable to the one given by Ossenkopf & Henning (1994) for dense cores, assuming a gas-to-dust ratio of 100. A discrepancy is found for sources associated with lower β values, indicating a possible problem when taking the same κ_0 value for all sources.

5   Conclusions - Perspectives

This observational study of 10 Class I sources in the L1641 cloud has revealed, using a least-squares fit procedure, that more evolved sources tend to be associated with higher β values. If confirmed on a larger statistical sample, this relation might provide a new evolutive tracer (the β value) to the study of YSOs. The possible connection between the evolution and changes in dust distribution and/or properties will be tested in the next future using new high resolution facilities in the IR and submillimeter range.

6   Aknowledgement

A. Zavagno is grateful for the support of an ESA fellowship.

References

André, P., Ward-Thompson, D., Barsony, M. (1993): ApJ 406, 122
Dent, W.R.F., Matthews, H.E., Ward-Thompson, D. (1995): Ast. Sp. Sc. 224, 85
Harvey, P.M., Marshall, J., Lester, D.F., Wilking, B.A. (1986): ApJ 301, 346
Hildebrand, R. H. (1983): Q. Jl R. astr. Soc 24, 267
Hillenbrand, L.A., Strom, S.E., Vrba, F.J., Keene, J. (1992): ApJ 397, 613
Hodapp, K.W. (1994): ApJS 94, 615
Lada, C.J., Wilking, B.A. (1984): ApJ 287, 610
Ossenkopf, V., Henning, T., (1994): A&A 291, 943
Reipurth, B., Chini, R., Krügel, E., Kreysa, E., Sievers, A. (1993): A&A 273, 221
Strom, K.M., Newton, G., Strom, S.E., Seaman, R.L., Carrasco, L., Cruz-Gonzalez,
    I., Serrano, A., Grasdalen, G.L. (1989): ApJS 71, 183
Walker, C.K., Adams, F.C., Lada, C.J., (1990): ApJ 349, 515
Ward-Thompson, D., Chini, R., Krügel, E., André, P., Bontemps, S. (1995): MNRAS
    274, 1219

Part II

Observations of Dust Factories

3-D Structure of the Bipolar Dust Shell of η Carinae

Douglas G. Currie[1], Daniel M. Dowling[1], Edward J. Shaya[1], J.Jeff Hester[2], The HST WF/PC Instrument Definition Team, and The HST WFPC2 Instrument Definition Team

[1] Physics Department, University of Maryland, College Park, MD 20972
[2] Department of Physics and Astronomy, Arizona State University, Tempe, AZ 85287-1504

Abstract. We present a symmetric bipolar 'double-flask' model for the homunculus of η Carinae derived from narrow band HST WFPC2 doppler imaging of H-α line and ground based spectra. The combination of high spacial (0.1″) resolution imaging and high spectral resolution (R=10,000) ground based spectra allow us evaluate three different models of the 3-D structure of η Carinae. We compare our 'double-flask' model to the 'double-sphere' and 'spherical-caps' models proposed in recent papers. Only our model is consistent with the observed morphology and the emission spectra.

Our recent work on the astrometric expansion of the homunculus show small, but significant deviations from pure linear expansion. In this paper we consider only the general linear expansion (0.64 %/year). Deviations from our presented symmetric model will be the subject of future papers.

1 Introduction

The homunculus of η Carinae originated in the "Great Eruption" of η Carinae in the late 1830s though 1840s (Currie et al, 1995a). At that time a large amount of material (Davidson, 1987) was ejected at high velocity. This material cooled and formed the expanding gas and dust cloud we see today. From the ground, the homunculus of η Carinae appears as an ellipsoid roughly 18″ x 10″ with major along position angle 132°. HST WFPC2 images resolve the ellipsoid into two roughly equal lobes (SE and NW) and an equatorial 'skirt.' Within the skirt, along the major axis is an oar-like shaped region called the 'paddle.' Our recent astrometric work (Currie et al, 1995a) shows that the plane-of-the-sky motions in the homunculus are generally radial, linearly increase in magnitude with distance from the central star. Within the homunculus (and NN/NS knots) the measured expansion is 0.64%/year giving an average ejection date of 1839, in excellent agreement with the historical great eruption which peaked in 1843. The velocities of the outer condensations have be shown by Walborn(1978,1988) to be consistent with earlier eruptions, dating back as far as the 1300s.

The three dimensional shape of the homunculus has been the subject of many papers over the years. While many models have been proposed, the most successful have been bipolar models with the SE lobe of the homunculus expanding toward the observer and NW lobe receding. Two of the most recent models have been the double-sphere (Duschl et al 1994, Atiken et al 1995, Frank, et al 1995) and the bipolar-caps model (Allen and Hillier, 1993; Currie et al 1994).

For the present analysis we use the distance to η Carinae of 2.2 kpc (Allen and Hillier, 1993). We determine this distance compared to the canonical value of 2.5 kpc(Davidson, 1987) for symmetry within the lobes: At 2.2 kpc the lobes are basically symmetric (i.e. between radial and astrometric velocities) about a axis passing through the central star. For distances greater then 2.2 kpc the symmetry is reduced to point-symmetry about the star. While point-symmetry is known to exist, we prefer a symmetric bipolar shell model as it has more secure theoretical basis in the models of (e.g. Frank et al, 1995).

In this paper we address only the overall structure of the homunculus and leave the deviations from the symmetric model for future papers. Deviations include possible acceleration in the SE lobe (Currie et al, 1995a) and holes in the shell of the homunculus (Allen and Hillier, 1993).

2   Observations

High spacial resolution(0.1″/pixel), narrow band imaging of the H-α line was obtained with the HST WFPC2 wide field camera in filters F658N (January 1994) and F656N(August 1994). Two filters are needed to study the reflection of the H-α line due to high red-shifts (greater the 1000 km/s or 20$\overset{\circ}{A}$) in the NW lobe.

Spectroscopic data cubes (x by y by λ) were obtained from multiple long slit spectra observed at the AAT in March 1986. Detailed description of the observations and processing used in producing these cubes are described by Hillier and Allen (1992). For the present analysis we used Hillier and Allen's cube ETA5 (7026-7435$\overset{\circ}{A}$). This data cube has x-y pixel dimensions of 0.7″/pixel with spectral dispersion of 0.7$\overset{\circ}{A}$/pixel. The seeing for the observations was about 2″. The cube has been orientated such that the major axis of the homunculus (PA=132°) is on the x-axis of the cube, with the central star in the center of the x-y plane. Thus x-λ 'cuts' through the cube are equilivent to long slit spectra on the major axis of the homunculus of η Carinae.

3   Astrometric Analysis

Images of η Carinae, obtained with the HST Wide Field/Planetary Camera in October 1990 (Hester et al, 1991), April 1991, and December 1993 (WFPC2), have been used to perform a detailed study of the proper motion of the homunculus of η Carinae (Currie et al, 1995a). This analysis yields the plane-of-the-sky astrometric velocities which range from tens of kilometers per second to over 1000 km/sec with estimated uncertainties on the order of 40 km/sec.

Our primary conclusion from these astrometric measurements is that the motion of the homunculus of η Carinae is largely radial, increasing linearly with distance from the central star. We measure an average radial expansion rate of

0.64% per year. Our direct measurements imply a single eruptive event centered in 1838.6±0.8 years (standard deviation of mean) or ±4 years when one includes some corrections in the error estimate for the correlated motions and possible relative plate scale errors. This agrees well with the historical "Great Eruption" which peaked in 1843. The motion of the individual fragments indicates "times of ejection" for the fragments occurred over an interval of less than 20 years.

4   Three Dimensional Structure

To deduce the 3-D shape of η Carinae we use spectroscopic evidence in combination with high resolution imaging. In cube ETA5 there are forbidden emission lines of [NiII], [CaII] and [FeII] as well as reflected emission lines of HeI. For the present analysis we will use the observed emission line shape to construct a model for the 3-D shape of the homunculus.

Following the analysis of Allen and Hillier (1993) have analyzed the [NiII] 7378Å,[NiII] 7412Å, [CaII] 7291Å,and [CaII] 7324Å lines in spectral cube ETA5. The strongest forbidden emission line in the data cube is [NiII] 7378Å. We will base our modeling below primarily on this line. The weaker lines have been used to verify our analysis.

To reveal the fine details of the shape of the [NiII] line we have applied several processing techniques. First we select a plane in the data cube representing an x-λ slice along the homunculus major axis with λ-spectral range of 7378±30Å and subtract the continuum. We then apply a 'continuum' normalization to this image. To 'continuum' normalize we divide each column (spectra) by its average value. This procedure produces an image in which all the major [NiII] peaks have the same magnitude. While this procedure tends to wash out any dim features, it does allow major structures to be displayed at the same level. In Fig. 1 we present the continuum subtracted image at two different stretches (saturated linear scale and square root scale) and the 'continuum' normalized image.

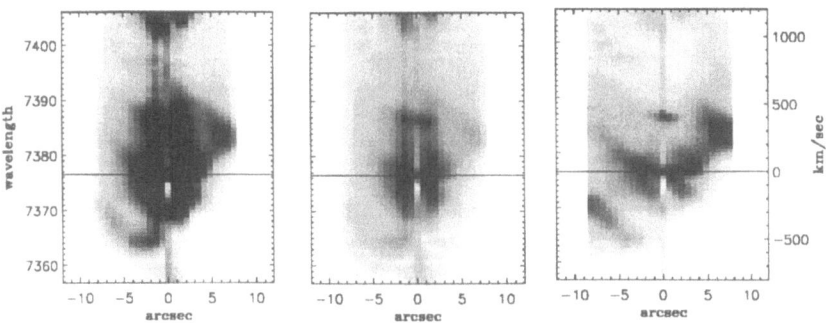

Fig. 1. [NiII] 7378Å emission line across the major axis of the homunculus. Here we the 'continuum subtracted' image at two different stretches (left and center) and the 'peak normalized' spectra

In Fig. 1 we see all the basic features of the homunculus of η Carinae. The SE lobe is blue shifted, NW lobe red shifted, and a small but significant blue shift at the location of the 'paddle' (<3″NW of the central star). Through our analysis below, it has become apparent that this [NiII] line displays *both* emission and reflection components in the SE lobe. The [CaII] lines do not show reflected components, however they are contaminated with weak HeI and [NiII] lines. Notable in the [CaII] lines is the emission from the 'paddle' in the skirt region (Allen and Hillier, 1993).

From our astrometric analysis we can convert x-y locations on the sky to v_x and v_y velocities. Doppler shifts from emission lines give v_z. Finally, the assumption that all the material in the homunculus originated from the central star (rather then the arbitrary line between the central star and earth), constrains the z spacial position. Thus we have the full 6-dimensional phase space of the homunculus. The velocities v_x and v_y do however depend on the distance to η Carinae. We have chosen to use a distance to η Carinae of 2.2 kpc, as this distance produces axially symmetric models and is consistent with observed doppler shifts (Allen and Hillier, 1993). The canonical distance of 2.5 kpc to η Carinae introduces only mild asymmetries to our proposed model.

Emission line shapes in v_x-v_z plane can now be used to create a model to match the line shape in velocity space. Our approach in modeling the shape was to use a fourier representation (i.e. sum of low order cosine functions).

5   Comparison of Models

We can now compare our model to two recently proposed models: the double-sphere and bipolar-caps. For this comparison we consider the combination of the emission line shapes and the model onto the plane of the sky.

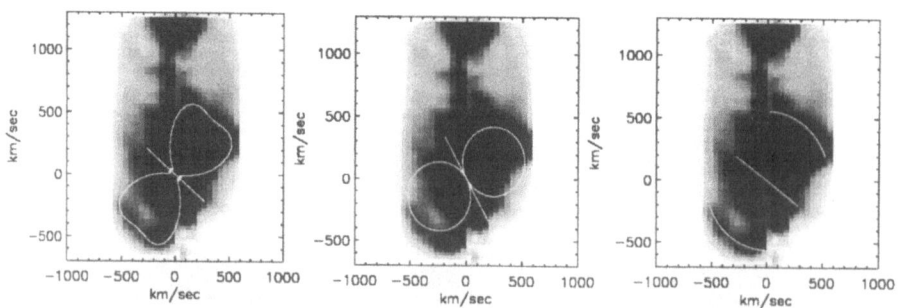

Fig. 2. Models for the 3-D shape of the homunculus plotted on top of the observed spectra taken along the major axis of the homunculus. From left to right are the double-flask, double-bubble, and bipolar-caps

We compare the each of the three models to the observed emission shape. In Fig. 2 we plot the model emission line shape on top of the [NiII] 7378Å line.

Only the our double-flask model is consistent with the observed spectrum. The double-sphere model has a major problem: Notice in Fig. 2 that the double-sphere model predicts significant blue shifted emission in the NW lobe. This blue shift can not be removed by rotation in the spacial-spectral plane because the symmetry axis is constrained by the plane-of-the-sky projection above (Fig. 3). Nor is this a distance effect: If the distance to η Carinae is greater the 2.2 kpc, the doppler shift would remain unchanged. Similar problems exist for the bipolar-caps model. The bipolar-caps provides a good representation the SE lobe emission. The problem occurs in fitting the reflection line in the NW (Currie, et al 1995b). A thin wall is need to produce the structure NW of the star along the slit. Only reflection off this wall can produce the reflection line shape in the NW lobe. The existence of the wall was proposed by Allen and Hiller (1993) with some reservations.

axis with respect to the plane of the sky for the double-bubble and bipolar-caps models.

Only our double-flask can fit the observed line profiles and projection onto the plane of the sky. We have also modeled the reflection line shapes predicted by these models. The results for the reflection lines support our general conclusion on the homunculus (Currie, et al 1995b).

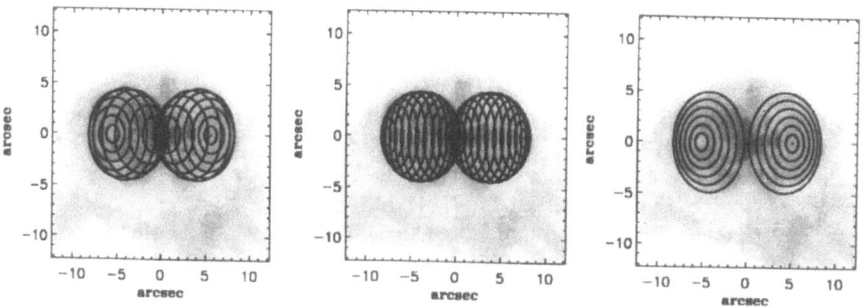

Fig. 3. Projection of models onto the plane of the sky. Notice all models can be adjusted to fit the observed image (upper left). Models are the double-flask(upper right), double-bubble(lower left), and bipolar-caps(lower right)

For projection onto the plane of the sky we necessarily assume that the homunculus is largely symmetric between and SE and NW lobes with a symmetry axis passing through the central star. In Fig. 3 we plot the x-y (plane of sky) projections for each model. For clarity we have not plotted the projection of the skirt (equatorial disk) for each model. Notice in Fig. 3 all three models can accurately produce the general shape of the homunculus and skirt as seen from HST. The importance of this test is that it constrains the angle of the symmetry

6 Acknowlegements

We would like to thank John Hillier for the use of his spectral data cubes and advice on how to use them. This research has made use of the Simbad database operated at CDS, Strasbourg, France. This research was conducted by the WFPC Instrument Definition Team supported in part by NASA Grant No. NAS5-1661.

Copies of this paper and figures are available from Douglas G. Currie (currie@khaos.umd.edu).

References

Atiken D.K., Smith C.H., Moore T.J.T., and Roche P.F. (1995): Mid-infrared studies of eta Carinae. II. Polarimetric imaging at 12.5 mu.m and the magnetic field structure. MNRAS **273**,359–366

Allen D.A. and Hillier D.J. (1993): The shape of the Homunculus Nebula around Eta Carinae. PASA, **10** 338–341

Currie D.G., Dowling, D.M., Shaya, E.J., Hester, J.J., Scowen, P.S. Groth, E.J., WF/PC IDT, WFPC2 IDT (1994): Hubble astrometry and the 3D structure of eta Carinae. BAAS **26**, 914

Currie D.G., Dowling, D.M., Shaya, E.J., Hester, J.J., Scowen, P.S. Groth, E.J., Lynds, R., O'Neil, E.J., WF/PC IDT, WFPC2 IDT (1995a): Astrometric Analysis of the Homunculus of η Carinae with the Hubble Space Telescope. To be submitted to AJ

Currie D.G., Dowling, D.M., Shaya, E.J., Hester, J.J., Scowen, P.S. Groth, E.J., Lynds, R., O'Neil, E.J., WF/PC IDT, WFPC2 IDT (1995b): The 3-Dimensional Shape of η Carinae I: The Symmetry Plane. In preparation

Davidson, Kris (1987): The relation between apparent temperature and mass-loss rate in hypergiant eruptions. ApJ,**317** 760–764

Duschl, W.J., Hofmann, K.h., Rigaut, F., Weigelt, G. (1994): Morphology and Kinematics of η Carinae. Preprint

Frank A. Balick B. and Davidson K. (1995): The Homunculus of eta Carinae: an Interacting Stellar Winds Paradigm. ApJ,**441** L77–L80

Hester, J.Jeff, Light, R.M., Westphal, J.A., Currie, D.G., Groth, E.J., Holtzman, J.A., Lauer, T.R., O'Neil, E.J.(1991): Hubble Space Telescope Imaging of η Carinae. AJ, **102**, 654-657

Hillier D.J. and Allen D.A. (1993): A spectroscopic investigation of eta Carinae and the Homunculus nebula I. Overwiew of the spectra. A&A **262**, 153–170

Walborn, N.R., Blanco B.M. and Thackery A.D.(1978): Proper motions in the outer shell of eta Car. ApJ, **219**, 498–503

Walborn, N.R., Blanco B.M. and (1988): Third-epoch Proper Motions in the Outer Shell of eta Carinae. PASP, **100**, 797–800

Dust Around Post-AGB Sources with 21 μm Feature

Ryszard Szczerba[1,2], Kevin Volk[1], and Sun Kwok[1]

[1] University of Calgary, Department of Physics and Astronomy, 2500 University Drive N.W., Calgary, AB T2N 1N4, Canada
[2] Polish Academy of Sciences, N. Copernicus Astronomical Center, 87-100 Toruń, Rabiańska 8, Poland

Abstract. We present fits to the spectral energy distributions of two post–AGB sources with the 21 μm feature: IRAS 22272+5435, where the 30 μm band was also detected, and IRAS 05113+1347, where observations around 30 μm are not yet available. Results of modeling suggest that all 21 μm emitters also have the 30 μm band. We show that magnesium sulfide could be responsible for the observed 30 μm feature. We also find that the stellar effective temperatures necessary for modeling of post–AGB objects are rather higher than those inferred from the spectral classification.

1 Introduction

Proto–planetary nebulae (PPNe) are objects in transition between asymptotic giant branch (AGB) stars and planetary nebulae, which emit a significant fraction of their total energy in the infrared. They are very intriguing sources because of rapid changes in the stellar temperature and, consequently, in the physical conditions inside the ejected shell and the remnant stellar atmosphere. Nine PPNe are known to display the 21 μm emission band (Kwok et al. 1995) which was discovered first in four objects by Kwok et al. (1989). The 21 μm band was believed to be characteristic only for post–AGB phase of evolution but Henning et al. (1995) reported a few young stellar objects (YSO) with the 21 μm feature. Therefore, it seems that the appropriate conditions responsible for excitation of this band exist in two different stages of stellar evolution.

Four of the post–AGB sources known to have 21 μm feature in their spectra were observed with the Kuiper Airborne Observatory (KAO) and a spectacular, very broad emission band around 30 μm was detected for all of them (Omont 1993; Cox 1993). Analysis of the spectral energy distribution (SED) for the sample of post–AGB sources from Kwok et al. (1995) suggests that 30 μm band always appears together with the 21 μm one (Szczerba et al. 1995b – hereafter Paper II). Thus it is possible that the 30 μm feature will also be detected in YSO, and investigation of of post–AGB objects with these features could help with the identification of the dust grain producing these features and bring us closer to understanding the nature of the physical and chemical processes involved in its excitation. Here we present analysis of the spectral energy distributions for two post–AGB objects with the 21 μm band: one observed by the KAO (IRAS

22272+5435, hereafter IRAS 22272) with detection of the 30 μm band and one
not observed by the KAO (IRAS 05113+1347, hereafter IRAS 05113).

Detailed description of the model used and dust properties involved can be
found in Szczerba et al. (1995a) – hereafter Paper I. Briefly: the frequency-
dependent radiative transfer equation is solved for a dusty envelope under assumption of a spherically–symmetric geometry taking into account dust size
distribution and quantum heating effects for very small dust particles. Because
the sources are C–rich we assumed that dust is a mixture made of policyclic
aromatic hydrocarbons (PAH) and amorphous carbon (AC) grains. In addition,
to provide a good fit to the spectral energy distribution around 21 and 30 μm
features an empirical opacity function (hereafter EOF) was constructed (see Paper I for details).

2   IRAS 22272+5435

The best fit to the spectral energy distribution of IRAS 22272 is shown by heavy
solid line in Fig. 1. Replacing EOF by opacity function of MgS (see Begemann
et al. 1994) and assuming that pure MgS grains have the same temperature distribution as PAH grains with radius 10 Å (which have the highest temperature
distribution as found from the best fit model) we estimated emission from the
shell if it is optically thin in far–infrared: heavy dotted line shows the case of
spherically symmetric grains and heavy dashed one represents results for continuous distribution of ellipsoids. We can see that observed emission at 30 μm can
be explained by pure MgS grains if they are not only spherical. The required
amount of sulphur is $n(S)/n(H) = 4\,10^{-6}$, well below the abundance estimated
for planetary nebulae. Note that our best fit needed the star effective temperature (T_{eff}) about 5300 K, while spectral classification (G5 Ia pec – Hrivnak
1995) suggests a temperature below 5000 K. It is worth noting that Začs et
al. (1995) estimate $T_{eff} = 5600 \pm 250$ K for this source in a good agreement with
our estimate.

3   IRAS 05113+1347

In Fig. 2 we present two fits to the spectral energy distribution of IRAS 05113:
heavy solid line – the best fit with $T_{eff} = 5300$ K and heavy dotted line – the best
fit with $T_{eff} = 4900$ K. In the far–infrared the quality of the fits is very similar
while in the visual and UV the fit with lower temperature is able to explain only
observational data which were not corrected for interstellar extinction. Note that
according to the spectral type (G8 Ia pec – Hrivnak 1995) the temperature should
be around 4600 K. The heavy dashed line in the far–infrared represents the fit
which was obtained when that part of EOF responsible for the hypothetical 30
μm band was removed. As can be seen this fit cannot explain the behavior of the
energy distribution for $\lambda > 22\,\mu$m. Therefore we conclude that the 30 μm band
is also present in this source.

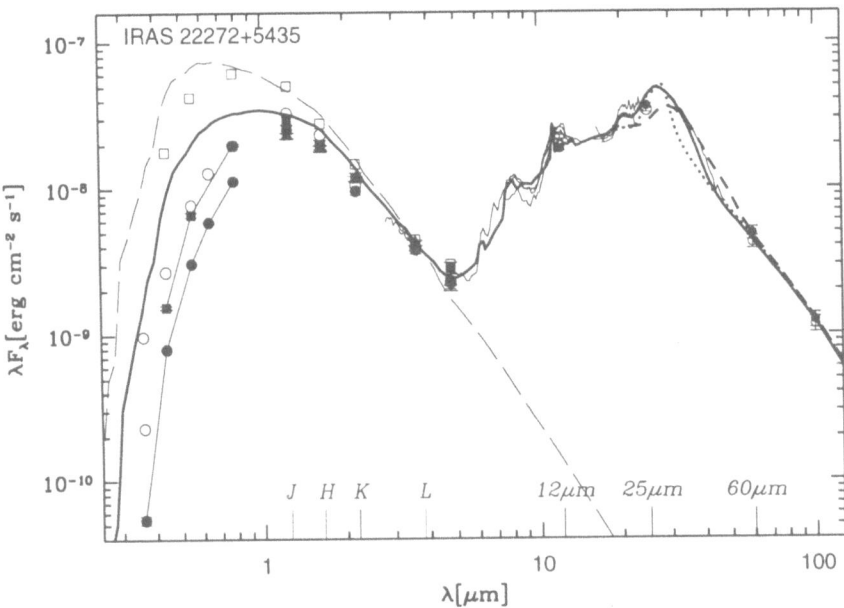

Fig. 1. Fit to the spectral energy distribution of IRAS 22272 – heavy solid line. Heavy dotted and dashed lines represent estimated emission by MgS grains (see text for details). Thin long–dashed line shows input energy distribution of the central star taken to radiate according to model atmosphere calculations for $\log(g) = 0.5$ and $T_{eff} = 5300\,\mathrm{K}$. Observational data are described elsewhere – see Paper I. Note that open symbols in the visual range of the spectrum represent extinction corrected data.

4   Conclusions

Results of our modeling of SED for post–AGB objects seem to suggest that 30 μm band is connected to the 21 μm one. This implies that carriers of these features are chemically connected. However MgS could be responsible for the 30 μm band but probably not for the 21 μm one. We need more precise opacity data for MgS in the visual and UV range to check if, in reality, it could be the carrier of the 30 μm feature. We have also found some discrepancy between the stellar temperature necessary for modeling of SED and the temperature inferred from the spectral classification. It is possible that the effective temperature value for post–AGB supergiants differs from that of "real" supergiants because post–AGB stars presumably have extremely low mass stellar envelopes.

Acknowledgments. One of us (R.Sz.) express his gratitude to Canadian Institute of Theoretical Astrophysics for a postdoctoral fellowship.

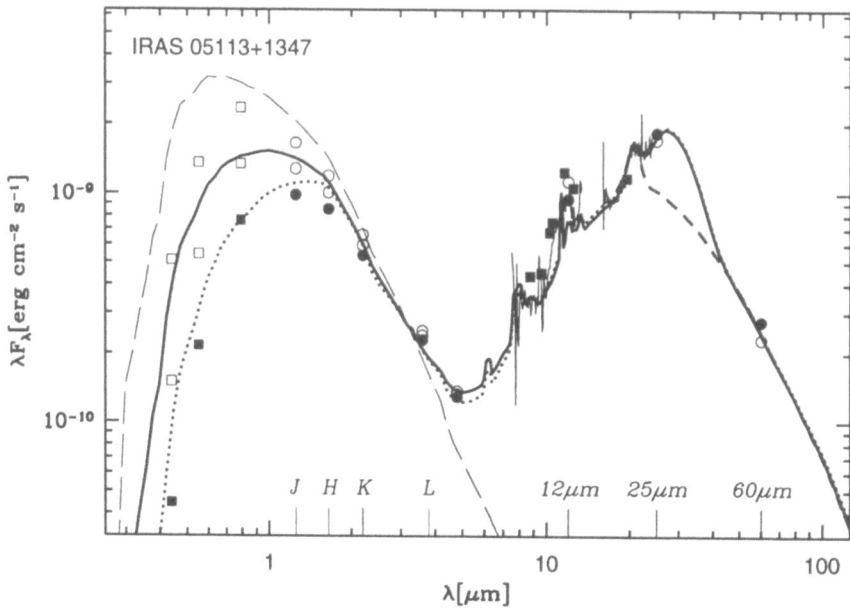

Fig. 2. Fits to the spectral energy distribution of IRAS 05113: heavy solid line – the best fit with $T_{eff} = 5300$ K; heavy dotted line – the best fit with $T_{eff} = 4900$ K. The heavy dashed line in the far–infrared represents the fit obtained for an EOF which does not contain the 30 μm band. Thin long–dashed line shows the input energy distribution of the central star taken to radiate according to a model atmosphere calculation for $\log(g) = 0.5$ and $T_{eff} = 5300$ K. Observational data are described elsewhere – see Paper II. Open symbols in the visual range of the spectrum represent extinction corrected data.

References

Begemann B., Dorschner B., Henning T., Mutschke H., 1994, ApJ, 423, L71

Cox P., 1993, Far-Infrared Spectroscopy of Solid State Features. In: Kwok S. (ed) Astronomical Infrared Spectroscopy: Future Observational Directions. ASP Conf. Ser. 41, p. 163

Henning Th., Chan S.J., Assendorp R., 1995, A&A submitted

Hrivnak B.J., 1995, ApJ, 438, 341

Kwok S., Hrivnak B.J., Geballe Th. R., 1995, ApJ, in press

Kwok S., Volk K., Hrivnak B.J., 1989, ApJ, 360, L23

Omont A., 1993, Circumstellar Matter: Cool IRAS Sources. In: Kwok S. (ed) Astronomical Infrared Spectroscopy: Future Observational Directions. ASP Conf. Ser. 41, p. 87

Szczerba R., Cox P,, Volk K., Omont A., Kwok S., 1995, A&A, submitted

Szczerba R., Volk K., Kwok S., Cox P., 1995, in preparation

Začs L., Klochkova V.G., Panchuk V.E., 1995, MNRAS, 275, 764

Part III

Observational Results Based on IRAS, COBE or Balloon Borne Platforms

HIRES IRAS Images of the Serpens Core

Mary Barsony and Robert L. Hurt

Physics Department, University of California, Riverside, Riverside, CA 92521 USA

Abstract. The nearby Serpens cloud core (d=310 pc) is the site of on-going star-formation. Recent mm/submm mapping of the thermal radiation in this region has led to the discovery of heretofore undetected emission peaks lacking near-infrared (NIR) counterparts (Casali, Eiroa, & Duncan 1993). A follow-up detailed, multi-transition study of these dust emission peaks in the mm/submm transitions of H_2CO determined the presence of associated, dense ($10^6 \leq n_{H_2} \leq 10^{6.5}$ cm^{-3}) gas, with several sources exhibiting self-absorbed line profiles perhaps best interpreted in the context of infall models (Hurt, Barsony, & Wootten 1996).

To further elucidate the nature of these sources, we present IRAS images, processed with the HIRES maximum correlation method. These images have point source spatial resolutions of $\sim 30''$, $30''$, $40''$, and $60''$ at the four IRAS bands of 12 μm, 25 μm, 60 μm, and 100 μm, respectively. Due to the improved flux upper limits we can obtain from the HIRES-processed data, we find the resultant spectral energy distributions (SED's) and luminosities of several Serpens sources to be consistent with that expected for Class 0 protostars (cf. Barsony 1994). We also report the discovery of a new IRAS source in this region.

1 HIRES Processing

We have used the HIgh-RESolution (HIRES) image processing algorithm as implemented in the software program YORIC available at IPAC[1] to produce new IRAS 12, 25, 60, & 100 μm images of the central $5' \times 6'$ Serpens core region. This program uses the Maximum Correlation Method (MCM; see Aumann, Fowler, & Melnyk 1990) to *construct* images in the four IRAS wavelength bands. The MCM algorithm as applied to IRAS data differs fundamentally from other image restoration techniques in that it combines image creation with resolution restoration, rather than restoring resolution from an already existing, initially blurry image.

The IRAS survey data consist of the responses of 62 rectangular detectors to celestial sources, with flux measurements read out typically every $0.5'$ along the scan direction. Detector orientations relative to a given source varied between subsequent tracks, resulting in highly non-uniform point-source response functions. Data pre-processing, known as "laundering," corrects for known systematic effects, such as baseline subtraction, de-striping, and cosmic ray hits. After data "laundering," the MCM algorithm constructs the image recursively,

---

[1] The Infrared Processing and Analysis Center (IPAC) is funded by NASA as part of the InfraRed Astronomical Satellite (IRAS) extended mission under contract to the Jet Propulsion Laboratory (JPL).

starting with an initial image consisting of a uniform, flat, positive-definite flux distribution. The IRAS detector response to the current image is then simulated and the result is statistically compared with the actual IRAS detector responses; multiplicative correction factors are then calculated for each image pixel and applied to the current image to arrive at the next iteration.

Typical "default" HIRES processing consists of 20 iterations in each IRAS band and is often sufficient for many sources. Given the high signal to noise and close separation of the Serpens core sources, we found substantial improvements in image resolution after 50 iterations for the 12 & 25 μm maps and after 150 iterations for the 60 & 100 μm maps. There is no clear demarcation to indicate when the "correct" number of iterations has been reached, such that one has produced the highest resolution image consistent with the IRAS detector data, but has not "over-processed" the image. A qualitative indication of "over-processing" is the introduction of unwanted artifacts, such as striping or "ringing" around strong point sources. For reference, we found substantially no change in our images, with no artifacts, for up to 100 iterations in the 12 & 25 μm bands, and for up to 200 iterations in the 60 & 100 μm bands.

2   Results for the Serpens Core

Of particular interest in the Serpens core are a number of submillimeter sources which have been identified as protostellar candidates: FIRS1, SMM2, SMM3, SMM4, & S68N (Hurt, Barsony, & Wootten 1996). The locations of these sources are indicated in Figure 1, which shows the contour plot of our HIRES processed 25 μm IRAS image overlaid on the greyscale map of the 1.1mm emission in the region (from Casali, Eiroa, & Duncan 1993). Also indicated in Figure 1 are the locations of three bright NIR sources, IRS53, SVS2, & SVS20, and the recently discovered "FU Ori"-type object (Hodapp 1995).

Except for FIRS1, the 25 μm emission peaks do not coincide with the 1.1mm continuum peaks. In the region near S68N, for example, the location of the 25 μm peak must be a result of source confusion between emission from S68N, IRS53/SMM5, and the recently reported FU Ori source. In order to quantitatively explore the implications of source confusion on our derived IRAS flux upper limits for the Serpens mm-continuum sources, we produced model IRAS detector responses to an assumed point source input model, HIRES processed the model IRAS detector responses, and compared the resultant "point-source model" HIRES images with the HIRES-processed images obtained with no *a priori* assumptions about the source brightness distribution. The point-soure model was iterated (in source positions and/or fluxes) until we produced the best match between the resultant HIRES-processed images. Via such comparisons, we found we could reproduce the observed 25 μm peak location assuming either that the three known sources in the area (S68N, IRS53, and the FU Ori source) each contribute equally to the observed 25 μm flux, or assuming only a single point source to be present at the 25 μm peak location. Although the point-source model is clearly non-unique, it does set quantitative upper limits on the flux contributions of any source producing 25 μm emission in the region.

We combined our best-determined IRAS flux upper limits from the point-source models at each wavelength with previously published mm/submm contin-uum fluxes (CED; McMullin *et al.* 1994) to derive spectral energy distributions [SED's] for the five Serpens protostellar candidates. Fitting these SED's with modified, single-temperature blackbodies, we find all five sources with $23 \text{ K} \leq \text{T} \leq 27 \text{ K}$ and $0.02 \leq \tau_{250 \ \mu m} \leq 0.10$, assuming an opacity law varying as $\lambda^{-1.5}$. We estimate circumstellar masses by assuming the 1.3mm emission to arise from sources with $20''$ diameters and a 1.3mm dust emissivity of $\kappa_{1.3mm} = 0.01 \text{ cm}^2 \text{ gm}^{-1}$ (André 1994). In this way, we find the range of circumstellar masses sur-rounding the Serpens protostars to lie in the range $0.6 \text{ M}_\odot \leq \text{M} \leq 3 \text{ M}_\odot$, in good agreement with values of these masses previously inferred from H_2CO ob-servations (Hurt, Barsony, & Wootten 1996). Integrating under the SED results in the determination of a source's bolometric luminosity. We find $L_{bol} = 46 \text{ L}_\odot$ for FIRS1, the most luminous source in the region, with luminosities of $9 \text{ L}_\odot$ for SMM4, $8 \text{ L}_\odot$ for SMM3, and $6 \text{ L}_\odot$ each for SMM2 and S68N. We find an $8 \text{ L}_\odot$ upper limit for the pre-outburst bolometric luminosity of the recently reported FU Ori source.

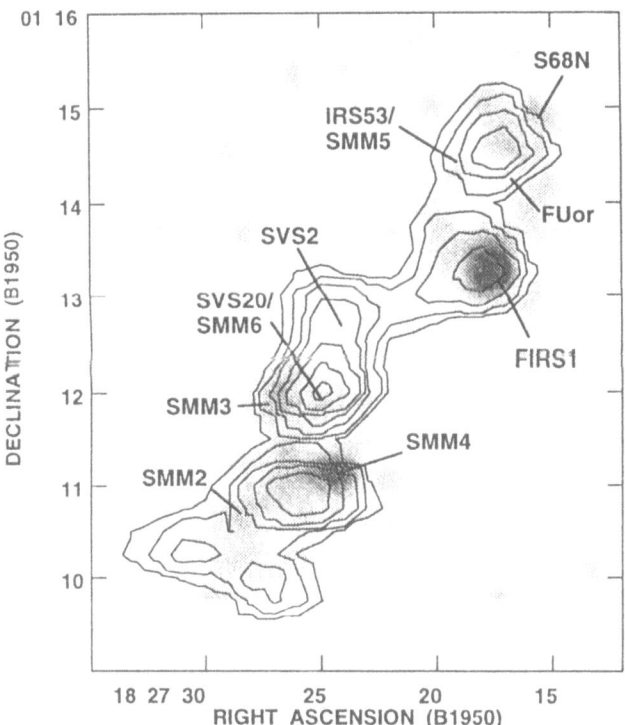

Fig. 1. Comparison of the 25 μm HIRES-processed IRAS map (contours) with the 1.1-mm emission (greyscale) from Serpens. The 1.1mm data are from Casali, Eiroa, & Duncan (1993).

Based on a comparison of the amount of circumstellar mass found around each Serpens candidate protostar, with limits on the central heating source's mass inferred from our IRAS HIRES-derived bolometric luminosities, we find SMM4 to be definitively a Class 0 protostar, with FIRS1, S68N, SMM3, and SMM2 highly likely to be members of this class as well. Definitive determinations of the evolutionary status of these objects awaits pending ISO observations.

References

André, P. (1994): in *"The Cold Universe,"* Proc. XIIIth Rencontres de Moriond, eds. T. Montmerle, C.J. Lada, F. Mirabel, & J. Tran Thanh Van, in press.

Aumann, H.H., Fowler, J.W., & Melnyk, M. (1990): AJ, **99**, 1674

Barsony, M. (1994): in *"Clouds, Cores, and Low Mass Stars,"* Proc. 4th Haystack Conf., eds. D.P. Clemens & R. Barvainis, PASPC, **65**, 197

Casali, M.M., Eiroa, C., & Duncan, W.D. [CED] (1993): A& A, **275**, 195

Hodapp, K.W. (1995): IAU Circ. No. 6186

Hurt, R.L., Barsony, M., & Wootten, A.H. (1996): ApJ, **456**, in press

McMullin, J.P., Mundy, L.G., Wilking, B.A., Hetzel, T., & Blake, G. (1994): ApJ, **424** 222

A Catalogue of Massive Young Stellar Objects: A Description

S.J. Chan and T. Henning

Max Planck Society, Research Unit "Dust in Star-forming Regions", Schillergäßchen 2-3, D-07745 Jena, Germany

Abstract. We compiled a catalogue of massive young stellar objects, which contains about 250 objects. Massive young stellar objects are compact and luminous infrared sources. In this catalogue, we have summarized important and characteristic observational data and provide the most comprehensive data base presently available.

Keywords: Catalogues and dictionaries – Stars: formation of – ISM: dust – Infrared radiation – Radio lines: molecular – Radio continuum

1 Introduction

During the last decade, a huge amount of new data concerning the formation and early evolution of massive stars has been obtained at infrared, submillimetre/millimetre, and radio wavelengths. Thus, it should be the right time to update the catalogue of massive young stellar objects (YSOs) which was compiled about ten years ago (Henning et al. 1984). Massive young stellar objects are compact and luminous infrared sources. The stellar core is still surrounded by optically thick dust cocoons (cf. Henning 1990). The spectral energy distributions of these objects are characterized by strong infrared continuum radiation with its maximum at about 100 μm. They have roughly a blackbody energy distribution from 2 to 20 μm with a colour temperature of about 300 - 700 K. The objects are frequently associated with energetic outflows revealed by the presence of broad molecular emission lines and with maser sources. Our main goal is to collect the observational data of these sources as complete as possible. This updated version of our earlier catalogue will provide comprehensive information on infrared and radio flux densities, molecular line data, the association with maser sources, and outflow phenomena.

2 Selection Criteria

The first sky survey from space at infrared wavelengths was performed by the Infrared Astronomical Satellite (IRAS). This sky survey has turned out to be a powerful tool for the systematic study of the star formation phenomena because this process takes place in molecular cloud cores which have high optical depths. The study of the IRAS data showed that these objects have a well-defined range of colour indices. A general feature of the spectral energy distribution of all the massive YSOs absent in most of the other classes of infrared sources is that the fluxes in the IRAS bands increase with wavelengths. Therefore, the IRAS colours and fluxes in the IRAS bands will be used as the main selection criteria for this new catalogue.

The objects in the catalogue were selected from the PSC based on the following criteria:

1. IRAS flux density qualities ≥ 2 in the four IRAS bands;
2. $F_\nu(12\ \mu m) \leq F_\nu(25\ \mu m) \leq F_\nu(60\ \mu m) \leq F_\nu(100\ \mu m)$;
3. $F_\nu(100\ \mu m) \geq 1000$ Jy;
4. IRAS colours (including uncertainty 0.15) within the following colour boxes:
   $-0.15 \leq R(12/25) \leq 1.15,\ -0.15 \leq R(25/60) \leq 0.75,\ -0.35 \leq R(60/100) \leq 0.35$;

   where $R(\lambda_i/\lambda_j) = \log\ (\lambda_i F_\nu(\lambda_j)/\lambda_j F_\nu(\lambda_i))$ (Henning et al. 1990);
5. IRAS 'idtype' (type of objects) $\neq 1$; objects are not associated with galaxies or late-type stars;
6. $b \leq 10^o$.

The items (5) and (6) are to avoid the contamination by galaxies and/or evolved objects.

254 objects are found to fit the above selection criteria. Figs. 1a and 1b show the colour-colour diagrams of the YSOs in our catalogue. Objects without enough information/data to confirm their YSO nature are marked by filled circles. Furthermore, the selection criterion by Wood & Churchwell (1989) for ultra-compact H II regions is shown by a dashed line. Almost all objects of our sample fall within the colour interval given for the ultra-compact H II regions.

3   Content of the Catalogue

This catalogue is divided into three subcatalogues which are the subcatalogue (1) of basic data, the subcatalogue (2) of flux densities and dust feature information as well as the subcatalogue (3) of maser and molecular line data. Each subcatalogue contains one or more tables. The subcatalogue (1) contains the basic information about the selected objects such as IRAS name, name of the object, the position of the object, the spectral type of the object, other name of the object and the (kinematic) distance of the object. In the subcatalouge (2), the flux densities from the near-infrared to the radio range are assembled (J, H, K bands, IRAS bands, 350 μm, 800 μm and 1.3 mm bands, 2 cm and 6 cm bands). The information on dust features (such as ice, silicate, PAH) come from the IRAS Low Resolution Spectrometer Atlas as well as other literature. In the subcatalogue (3), the maser sources (H_2O, OH, CH_3OH) and NH_3, HCO^+, and CS molecular line data towards these objects are reported. The CO outflow velocity will be given if the object is found to be associated with an outflow.

The information (detected, non-detected or observed) on the above molecules are also summarized in two tables. There are 14 confused objects

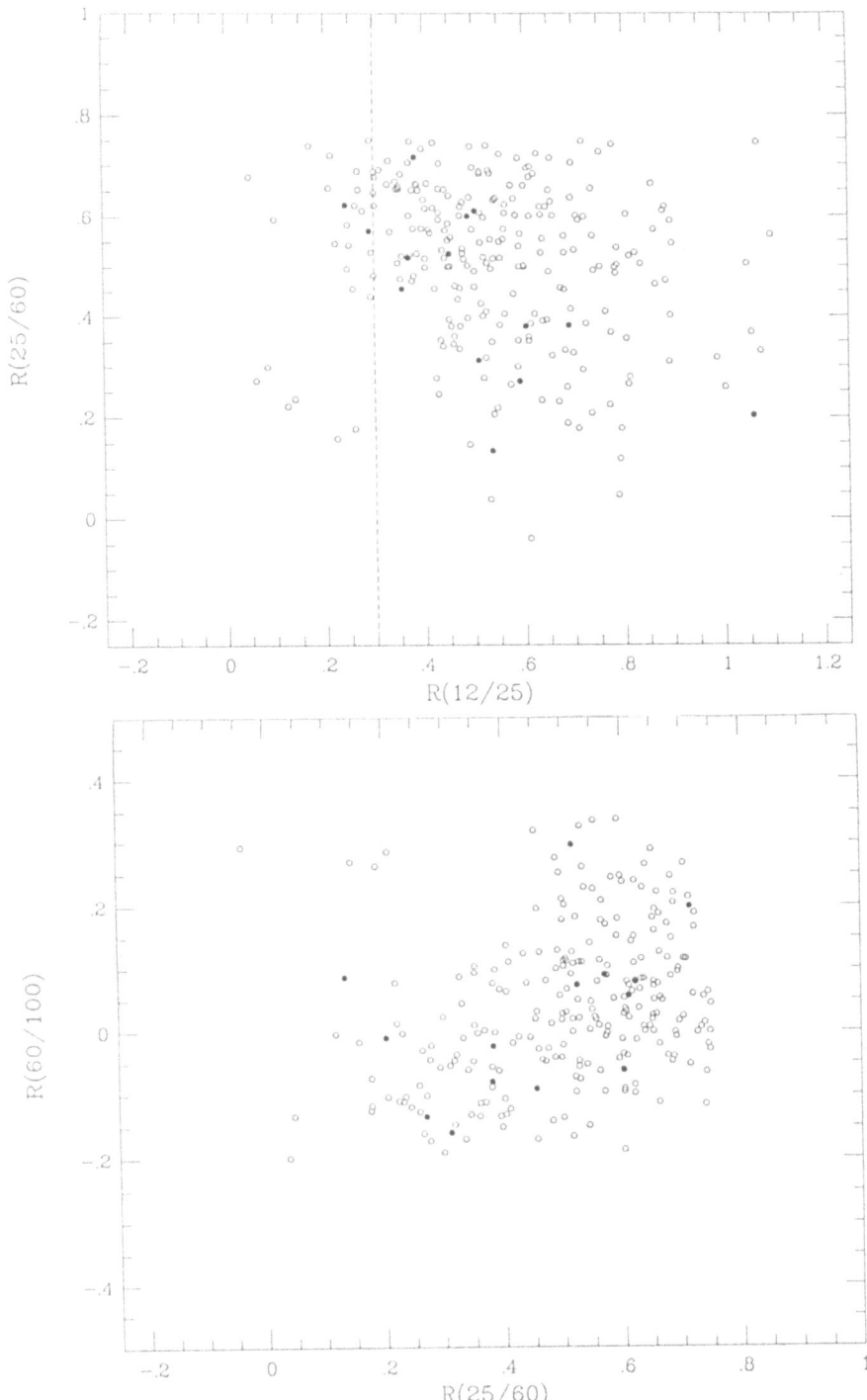

Fig. 1. The 12/25/60 μm colour-colour diagram (a) and the 25/60/100 μm colour-colour diagram (b) of selected massive young stellar objects with good and moderate photometric fluxes in the IRAS 12, 25, 60, 100 μm bands.

which may be YSOs or young planetary nebulae because there is not enough information available in order to confirm the actual character of these objects. However, from their locations in the IRAS colour-colour diagrams, they should be good candidates for YSOs. More observations for these objects are necessary. Full information on the catalogue can be found in Chan et al. (1995). Tables are available in electronic form from the CDS data centre in Strasbourg via anonymous ftp (130.79.128.5).

4   Conclusion

We compiled a catalogue of about 250 massive young stellar objects. In this catalogue, we summarized important and characteristic observational data and provide the most comprehensive data base presently available for these objects.

References

Chan S.J., Henning Th., Schreyer K., 1995, A&AS (in press)
Henning Th., Friedemann C., Gürtler J., Dorschner J., 1984, Astron. Nachr., 305, 67
Henning Th., 1990, Fundamentals of Cosmic Physics, 14, 321
Wood D.O.S., Churchwell E., 1989, ApJ, 340, 265

The S135 Star Formation Region

Carsten Kömpe[1] and Gilles Joncas[2]

[1] Universitäts-Sternwarte, Schillergäßchen 2, D-07745 Jena, Germany
[2] Université Laval, Dépt. de Physique, Ste-Foy, Québec, G1K 7P4, Canada

Abstract. Data covering an extended wavelength range have been combined to investigate the relation between the different physical components that make up the S135 gas/dust complex. The IRAS data reveal the general spherical shape of the dust complex that breaks up into one main and two smaller components. Young embedded stars appear to be present. The main component was fully mapped in the $J=1-0$ ^{13}CO and ^{12}CO lines allowing to estimate the gas mass (1100 $M_\odot$). The well developed ionization front of the H II region is associated with the main component. A high resolution (20″) map of the dense gas near IRAS 22200+5831 has been obtained in the CS $J=2-1$ line. DRAO 21 cm data revealed both the continuum emission of the ionized gas and H I line emission of atomic hydrogen. A tentative evolutionary scenario for the S135 complex is proposed.

1 Motivation

There are numerous physical processes that play a role in the formation of stars in molecular clouds, and their mutual interactions are clearly complex. These processes include the formation and fragmentation of molecular clouds, the ionization of the gas by earlier generations of stars and shock waves of different origins. The result of the interaction of these processes is a complicated pattern of gas and dust in very different physical conditions: hot ionized gas, warm H I gas, and cold molecular gas and dust particles. These interactions are ultimately responsible for stars being formed rather in one location of a molecular cloud than in another. However, there is currently no theory available that can make any predictions as to where and when what kind of stars will form within a given molecular cloud.

Star formation regions are thus very complicated objects. In order to make progress in their understanding, it is important to study the interactions of their different gaseous and dusty components. As a first step, it is necessary to determine their spatial distributions and physical parameters. To do this, one would have to use line and continuum data with wavelengths ranging from nano- to centimeters.

Some time ago, we have begun this kind of investigation with the Sharpless H II regions S142 (Joncas et al. 1985, 1988), S247 (Kömpe et al. 1989), and S187 (Joncas et al. 1992). We are continuing our series of multi-wavelength investigations of star formation regions here with the presentation of our analysis of the gas and dust associated with the H II region S135.

2 Brief Review of the S135 H II Region

S135 is an optically visible H II region (Fig. 2a) located at RA(1950) = 22^h 20^m and DEC(1950) = 58° 31'. It lies in the second quadrant (l = 104.6°) of the Galaxy and slightly above the galactic plane (b = +1.2°). The nebula is ionized by the star BD+57 2513 of spectral type O9.5 V (Georgelin et al. 1973). A realistic value for its distance is 1.9 kpc that is consistent with distances found by Georgelin et al. (1973) and Lahulla (1985). S135 is thus located at the inner edge of the Perseus arm of the Galaxy. The 21 cm continuum emission of the nebula was first observed by Israel (1977). His results (n_e=1200 cm^{-3}, $\emptyset$=0.3 pc) classify S135 as a compact H II region of class II (Habing & Israel 1979). The velocity field of the ionized gas was studied by Pişmiş et al. (1986) who find that it is consistent with a blister model. The authors also note some striking similarities between S135 and the better known S140 H II region.

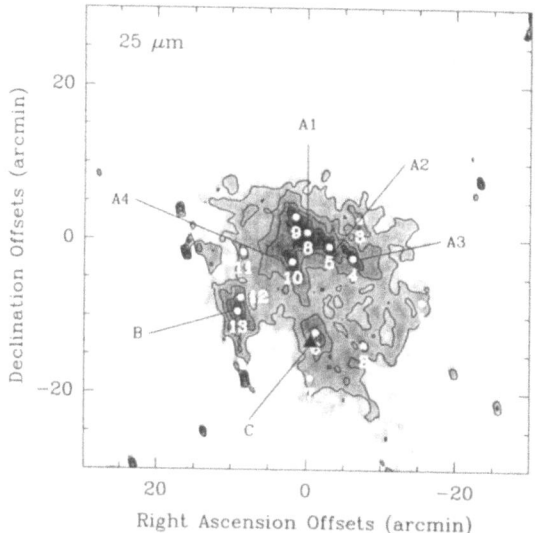

Fig. 1. IRAS contour plot and grey-scale map at 25 µm of the area around S135. The reference position of the map is the central position of the H II region at RA(1950) = 22^h 20^m, DEC(1950) = 58° 31'. The lowest contour level is 2 MJy/sr; the values of the higher levels are respectively twice that of the preceding lower level. The positions of IRAS point sources are indicated by white dots and coded by numbers (1: 22180+5820, 2: 22190+5817, 3: 22190+5833, 4: 22192+5828, 5: 22196+5830, 6: 22198+5818, 7: 22190+5812, 8: 22200+5831, 9: 22201+5833, 10: 22202+5828, 11: 22210+5929, 12: 22211+5823, 13: 22211+5821). The triangle indicates the position of the compact continuum source 25P45 (Joncas & Higgs 1990).

3 Data and Results

Dust distribution. IRAS maps are a good starting point in order to get an understanding of any dust/gas complex. The Infrared Processing and Analysis Center

(IPAC) at Pasadena provided high resolution IRAS data of the S135 region at 12, 25, 60, and 100 μm. As an example, we show the map at 25 μm (Fig. 1). The spatial resolution ranges from 2' at 100 μm to 1' at 12 μm. The IRAS maps show that the dust complex has a spherical shape and breaks up into one major component (A) and two smaller components (B, C). Component A breaks up again into four fragments (A1 to A4). Note that there is an IRAS point source associated with each of the fragments A1 to A4 and each of the small components B and C pointing to the presence of embedded young stars.

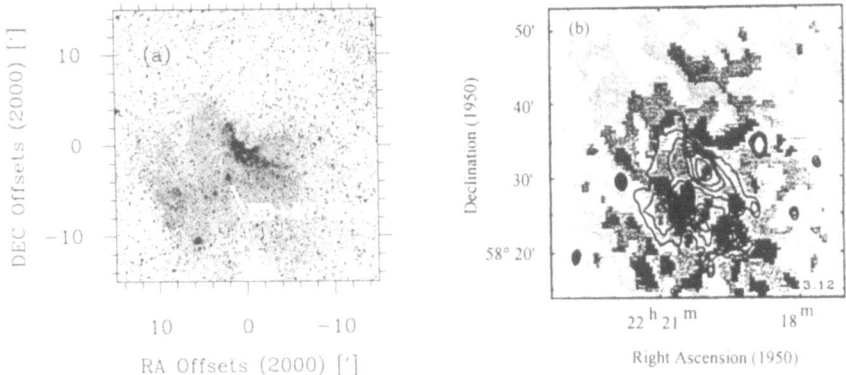

Fig. 2. (a) Reproduction of the POSS red print of the S135 H II region. The exciting star of the nebula, BD+57 2513, is indicated. (b) Superposition of contour plots of the 21 cm continuum radiation and grey-scale maps of the 21 cm line emission for a velocity of -23.12 km/s. The intensities of the maps are given in brightness temperature. The contour levels of the continuum data are at 2.3, 3.0, 4.0, 5.0, and 7.0 K. The grey-scale levels of the line data are at 6, 12, 18, 24, 30, 36, and 42 K.

Molecular gas. We have used the 2.5 m telescope of the Observatoire de Bordeaux, France, to map a region of about 40' × 20', covering component A, in the ^{13}CO and ^{12}CO $J=1-0$ lines with a resolution of 4'. There is general agreement between the molecular-line and the IR maps, but there are some differences in detail. From these data, we estimate the gas mass of component A as 1100 $M_\odot$. The average kinetic temperature is about 7 K. Of particular interest is fragment A1 that is located at the center of fragment A near the ionization front and is associated with IRAS 22200+5831. An area of 180'' × 160'' fully covering fragment A1 was mapped at high spatial resolution (20'') in the CS $J=2-1$ line using the IRAM 30 m telescope on Pico Veleta, Spain. Six additional CS $J=3-2$ spectra were measured towards the central part of the mapped area. The CS map shows that the IR source lies close to an elongated clump of dense matter.

Ionized gas. The distribution of the ionized gas can be inferred from the recombination-line emission of various ions in the optical wavelength range and from the free-free continuum emission. The optical emission can be seen on the red POSS print (Fig. 2a). It reveals a remarkably well developed ionization front that indicates the location where the ionized gas interacts with the adjacent molecular cloud. Continuum data at 21 cm were taken from the DRAO galactic-plane survey which was carried out with the DRAO synthesis telescope at Penticton, Canada, providing a spatial resolution of about 1.1′. (Joncas & Higgs 1990). Fig. 2b shows a contour map of the continuum data. The ionized gas is clearly distributed throughout the whole dust complex revealing two intensity maxima: the main peak coinciding with the optically visible ionization front, and a second one located to the south-east of the first.

Atomic hydrogen. Atomic hydrogen can be studied from the 21 cm H I line. The data were also obtained from the DRAO galactic-plane survey and are shown in Fig. 2b. Patches of H I emission are located in the general area of S135. Most of the atomic hydrogen is probably located at the outer edges of the molecular/dust clouds. Some patches closely follow the outline of the continuum emission and probably represent dissociated molecular hydrogen.

4   Discussion

Taking into account all the data available to us, we tentatively suggest that the evolution of the S135 complex took place in three main episodes: (1) Stars began to form in the center of the molecular cloud which has a roughly spherical shape. This is evidenced by the location of the ionizing star of the nebula. (2) These young stars ionized the gas in their neighborhood and ultimately clear the central part of the complex of gas and dust. This is indicated by the presence of diffuse emission seen on the POSS print and the paucity of dust in this region. (3) A second star formation episode took place in the periphery of the complex and may have been triggered by the first generation stars. The presence of IR point sources in the dust components A, B, and C are evidence of the presence of young embedded stars that are younger than the ionizing star. Four of these point sources are located close to the ionization front of S135 and may be an indication of sequential star formation.

Acknowledgement
We thank Dr. J. Gürtler from the Universitäts-Sternwarte Jena for helpful discussions and suggestions. CK acknowledges financial support from the BMBF (Förderkennzeichen 05 2JN13A).

References

Georgelin, Y.M., Gorgelin, Y.P., Roux, S. (1973): A&A **25**, 337
Habing, H. J., Israel, F. P. (1979): ARA&A **17**, 345
Israel, F.P. (1977): A&A **61**, 377
Joncas, G., Dewdney, P.E., Higgs, L.A., Roy, J.R. (1985): ApJ **298**, 596
Joncas, G., Durand, D., Roger, R. S. (1992): ApJ **387**, 591
Joncas, G., Higgs, L. A. (1990): A&AS **82**, 113
Joncas, G., Kömpe, C., de La Noë, J. (1988): ApJ **387**, 591
Kömpe C., Joncas G., Baudry A., Wouterloot J. G. A. (1989): A&A **221**, 295
Lahulla, J.F. (1985): A&AS **61**, 537
Pişmiş, P., Hasse, I., Moreno, M.A. (1986): Rev. Mex. Astron. Astrofis. **13**, 131

Star Formation in the Cepheus Flare Molecular Clouds

M. Kun

Konkoly Observatory of the Hungarian Academy of Sciences,
P.O.Box 67, H-1525 Budapest, Hungary

1 The Clouds of the Cepheus Flare

The Cepheus region at $100° \lesssim l \lesssim 120°$ and $b \gtrsim 10°$ contains a large number of dark clouds (Lynds 1962; Taylor, Dickman & Scoville 1987; Clemens & Barvainis 1988). Although most of them received attention in molecular line studies, and these studies usually assume nearly the same distance (300–400 pc) for the clouds, the region as a whole is seldom mentioned as a nearby giant molecular cloud complex, having a significance in large-scale studies of star formation comparable to Taurus or Orion.

The region as a whole was first studied by Lebrun (1986). Mapping of an area of 190 square degrees in ^{12}CO revealed the Cepheus molecular cloud complex at an average latitude of $+15°$ to be a most prominent feature of the local interstellar medium. At a probable average distance of 400 pc the mass of the cloud complex is about 2×10^5 M$_\odot$.

2 Star Formation in the Cepheus Flare

The dark clouds of the Cepheus Flare are potential sites of low-mass star formation. A number of dark clouds of the complex have already been included in star formation studies and observational results clearly show that low and intermediate mass star formation is widespread in this region.

There is relatively little information available on the optically visible pre-main sequence stellar population of the region. The *Third Catalog of Emission-Line Stars of the Orion Population* (Herbig & Bell 1984) contains only ten objects in the Cepheus Flare. A few clouds of the region have been searched for Hα emission objects as classical T Tauri star candidates. Ogura and Sato (1990) report the detection of 69 Hα emission line stars and list 49 suspected Hα objects in the region of Lynds 1228. Kun and Prusti (Kun and Prusti (1993)) found 12 emission stars within the area of Lynds 1251.

3 The Present Survey

I performed a more or less unbiased survey for young stellar object candidates in the Cepheus Flare molecular cloud complex. This survey included:

1. *Objective prism search for Hα emission stars with the 60/90/180 cm Schmidt telescope of Konkoly Observatory.* Thus we expect to find classical T Tauri stars. As spectroscopic identification of these stars is beyond the scope of this survey, we may regard these stars as T Tauri candidates. T Tauri stars have a characteristic far infrared flux distribution, therefore IRAS detection of the Hα emission stars gives a support to the reality of their T Tauri nature. In addition to looking for positional associations in the IRAS Point Source Catalog and Faint Source Catalog, IRAS detector scans are being examined at the position of each Hα emission star.

2. *A search for point sources having young stellarlike colours in IRAS catalogues.* The selection criteria applied during the present work are essentially the same as were used by Beichman et al. (1986). After selecting the sources on the basis of flux criteria the digitized POSS plates are searched for optical counterparts of the object selected. Galaxies can be eliminated during this step. Next, the spectral type and magnitude of the probable stellar counterparts have to be found if the star is included in any available catalog. For stars brighter than about V ≈ 13^m spectral types are derived from blue-sensitive objective prism plates taken with the Schmidt telescope.

Both types of search embrace the same part of the sky between $100°$ $\lesssim$ $l \lesssim 120°$ and $+10° \lesssim b \lesssim +22°$. Weak-line T Tauri stars and cold protostellar objects not detected by IRAS shortward of $100 \mu m$ cannot be detected by this survey. The catalogue of the YSO candidates obtained during the survey will be published elsewhere (Kun 1996).

4   Results

4.1   Hα Emission Stars

155 Hα emission stars were found until now by visually inspecting the plates several times. There are 36 previously known emission objects among them, as our survey included the fields covered by earlier studies. 11 of these stars coincide in position with IPSC objects and 17 with IFSC sources. Calibrated IRAS data sets (IRDS) available from the IRAS data base server of the SRON (Assendorp et al. 1995) are used to look for far infrared fluxes of the remaining stars. This work is still in progress. 51 positions were examined till now using the Groningen Image Processing System (GIPSY) software and 35 stars stars were found to have weak infrared fluxes.

Distribution of the Hα emission stars overlaid on the $100 \mu m$ optical depth image of the region is displayed in Fig. 4.1. Several groups of these objects can be recognized. If we consider them to be young solar-type star candidates, their groups mark the probable regions of low-mass star formation within the cloud complex.

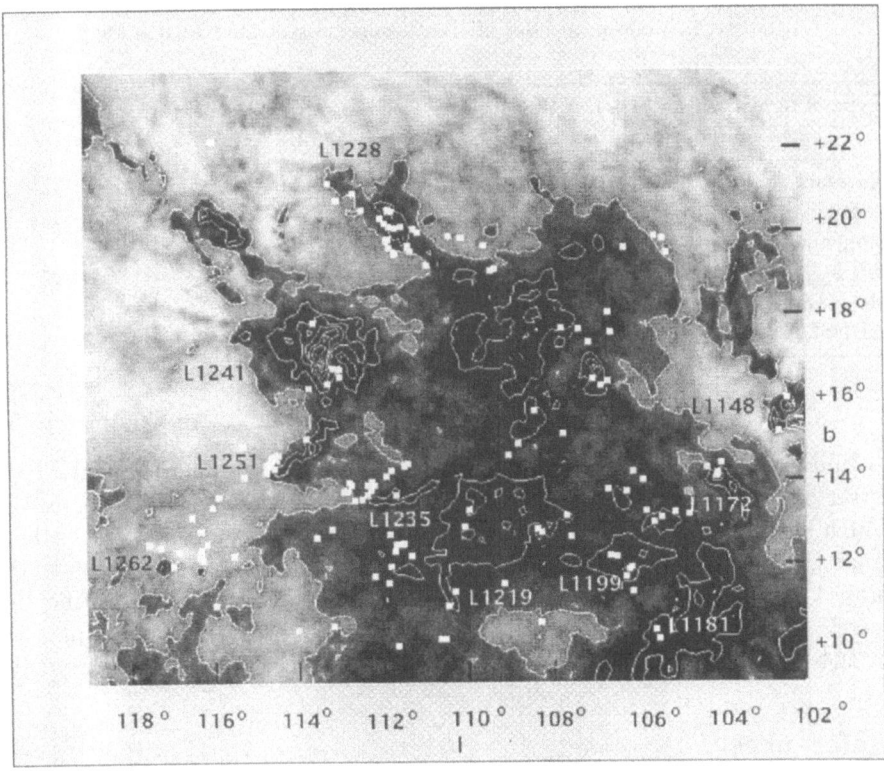

Fig. 1. 100μm optical depth image of the Cepheus Flare indicating the dust column density distribution in the region. The most prominent Lynds dark nebulae are indicated. White dots mark the Hα emission stars found with the Schmidt telescope of Konkoly Observatory.

4.2 IRAS Sources

191 IRAS sources from both Catalogs were found to satisfy our flux criteria. Their optical identification is still in progress. Till now, 40 of them have been identified with faint uncatalogued galaxies. 28 sources coincide in position with Hα emission stars found during this survey, and 15 with known young stellar objects. A number of IRAS sources coincide with apparently "ordinary" stars. Their IRAS colours are similar to the emission stars. They may be pre-main sequence objects as well.

Table 1 lists the mean IRAS properties of three subsamples of stars detected by IRAS: Hα emission stars and nonemission stars having infrared counterparts either in PSC or FSC, as well as Hα emission stars with ADDSCAN fluxes. IRAS luminosities were calculated adopting a distance of 350 pc.

Table 1. Mean properties of IRAS sources associated with stars

Type of object	F(12) Jy	F(25) Jy	T_d(12-25) K	L_{IRAS} $L_\odot$
Hα stars in PSC and FSC	1.27	2.53	183	3.6
Nonemission stars in PSC and FSC	0.49	1.16	181	1.3
Hα stars with ADDSCAN fluxes	0.57	0.89	193	1.06

Acknowledgements The IRAS data were obtained using the IRAS data base server of the Space Research Organisation of the Netherlands (SRON) and the Dutch Expertise Centre for Astronomical Data Processing funded by the Netherlands Organisation for Scientific Research (NWO). The IRAS data base server project was also partly funded through the Air Force Office of Scientific Research, grants AFOSR 86-0140 and AFOSR 89-0320. This work was supported by the Hungarian grants OTKA No. T4341 and No. T7438.

References

Assendorp R., Bontekoe T.R., de Jonge A.R.W., Kester D.J.M., Roelfsema P.R. and Wesselius P.R. 1995, A&AS 110, 395-403
Beichman C.A., Myers P.C., Emerson J.P., Harris S., Mathieu R., Benson P.J., Jennings R.E., 1986, ApJ 307, 337
Clemens D.P. & Barvainis R., 1988, ApJS 68, 257
Herbig G.H. & Bell K.R., 1984, Lick Obs. Bull. 1111
Kun M., 1996, in preparation
Kun M. & Prusti T., 1993, A&A, 272, 235
Lebrun F., 1986, ApJ 306, 16
Lynds B.T., 1962, ApJS 7, 1
Ogura K. & Sato F., 1990, PASJ 42, 583
Taylor D.K., Dickman R.L. & Scoville N.Z., 1987, ApJ 315, 104

An Overview of the COBE Infrared Datasets

David Leisawitz

Code 631, NASA Goddard Space Flight Center, Greenbelt, MD 20771, USA

Abstract. Characteristics of the *DIRBE* and *FIRAS* instruments on NASA's Cosmic Background Explorer (*COBE*) are summarized. Data products likely to be useful in studies of star formation and the interstellar medium are described. The *DIRBE* sky survey is compared with the *IRAS* survey, which has been used in many such studies over the past decade.

1 Instrument and Sky Survey Characteristics

Two *COBE* instruments – the Diffuse Infrared Background Experiment (*DIRBE*) and the Far Infrared Absolute Spectrophotometer (*FIRAS*) – yielded important new information about discrete regions of star formation and diffuse infrared radiation from interstellar dust and gas.

Following Boggess et al. (1992), Table 1 summarizes characteristics of the *FIRAS*, *DIRBE*, and Infrared Astronomical Satellite (*IRAS*) instruments and their respective sky surveys. The *FIRAS*, which is best known for its precise measurement of the cosmic microwave background spectrum, also measured the far–infrared ($\lambda > 105\ \mu$m) thermal dust continuum spectrum and emission in several important interstellar cooling lines. *FIRAS* mapped 95% of the sky and had a 7° beam. Orion and perhaps a few other star formation regions notwithstanding, the *FIRAS* data are suitable mainly for studies of diffuse emission (*e.g.*, Bennett et al. 1994; Reach et al. 1995). A *FIRAS* spectrum centered on the Orion complex ($\ell, b = 209°, -19°$) contains considerably stronger dust continuum and [C II] 158 μm line emission than a comparison spectrum taken 10° away.

Although the *DIRBE* and *IRAS* experiments had different scientific objectives, both instruments were used to measure diffuse infrared emission over practically the whole sky. The instruments were multiband photometers and had overlapping spectral coverage at 12, 25, 60, and 100 μm. The *DIRBE* also included near–IR bands at 1.25, 2.2, 3.5, and 4.9 μm, far–IR bands at 140 and 240 μm, and measured polarized intensities at 1.25, 2.2, and 3.5 μm. While the *DIRBE* was designed to detect diffuse radiation, *IRAS* was dedicated to the measurement of point source fluxes. Thus, the nearly wavelength–independent spatial resolution of the *DIRBE* (0°.7 × 0°.7 instantaneous field of view) is an order of magnitude lower than that of *IRAS*. Except for the somewhat shorter long–wavelength cutoffs at 25 and 60 μm, the *DIRBE* spectral response was similar to that of *IRAS* in the four commonly–held photometric bands. By making redundant celestial observations over a broad range of solar elongation angles and covering a wide wavelength range, the *DIRBE* has contributed to a better

Table 1. Instrument Characteristics

	FIRAS	DIRBE		IRAS
Wavelength bands	0.5–10 mm 0.1–0.5 mm	1.25 μm [a] 2.2 μm [a] 3.5 μm [a] 4.9 μm 8–17 μm	17–26 μm 43–70 μm 85–116 μm 120–200 μm 200–300 μm	8–15 μm 18–30 μm 45–78 μm 85–115 μm
Spectral resolution	$\Delta\nu > 0.2\,\mathrm{cm^{-1}}$ $(\nu < 20\,\mathrm{cm^{-1}})$ $\Delta\nu > 1.\,\mathrm{cm^{-1}}$ $(\nu > 20\,\mathrm{cm^{-1}})$	$\lambda/\Delta\lambda = 1 - 10$		$\lambda/\Delta\lambda = 2 - 3$
Field of view	7° circular diameter	0°.7 square		$(0'.76 - 3') \times 5'$
Instrument type	Polarising Michelson interferometer	Multiband filter photometer/polarimeter		Multiband filter photometer
Flux collector	Flared horn	Off–axis Gregorian telescope, 19 cm primary		57 cm Ritchey–Chretien
Look direction	On spin axis[b]	30° off spin axis[b]		Scans at constant ε
Solar elongation range	94°	$64° \leq \varepsilon \leq 124°$		$75° \leq \varepsilon \leq 105°$
Sky coverage	95%	100%		96%
Temperature at bolometers	1.55 K	1.55 K		2.6 K
Detector	Composite bolometers	Photovoltaics bands 1–4 Photoconductors bands 5–8 Composite bolometers bands 9, 10		Photoconductors
Calibration	Absolute using on–board reference sources	Absolute using on–board reference sources		TFPR model
Sensitivity	rms noise per FOV in 10 months for 3–20 cm⁻¹ $\Delta T = 0.24\,\mathrm{mK}$ $\Delta\nu I_\nu = 1\,\mathrm{nW\,m^{-2}\,sr^{-1}}$	rms noise per FOV in 10 months[c] Band $\quad$ νI_ν $\qquad$ (nW m⁻² sr⁻¹) 1.25 μm $\quad$ 1.0 2.2 μm $\quad$ 0.9 3.5 μm $\quad$ 0.6 4.9 μm $\quad$ 0.5 12 μm $\quad$ 0.3 25 μm $\quad$ 0.4 60 μm $\quad$ 0.4 100 μm $\quad$ 0.1 140 μm $\quad$ 11.0 240 μm $\quad$ 4.0		dimmest discernible 0°.5 source Band $\quad$ νI_ν $\qquad$ (nW m⁻² sr⁻¹) 12 μm $\quad$ 5.0 25 μm $\quad$ 2.4 60 μm $\quad$ 1.0 100 μm $\quad$ 2.1

[a] Linear polarisation also measured.
[b] COBE spin axis is approximately normal to the Sun, directed away from the Earth.
[c] Based on instrument dark noise in orbit; actual performance is reduced by sky confusion noise.

understanding of the zodiacal emission which interferes with Galactic mid–IR intensity determinations.

The DIRBE included on–board thermal and cryogenic reference sources and chopped the sky signal against a stable, zero–flux reference, enabling absolute calibration of the diffuse sky brightness, whereas the IRAS measurements were not absolutely calibrated. The zero point is not a matter of concern to analysts of extended interstellar dust emission who must in any case subtract a zodiacal emission baseline from the data. However, there is also a "gain" discrepancy between the DIRBE and IRAS photometric scales such that the IRAS brightness appears to be about 10–30% higher at 60 and 100 μm than the DIRBE brightness (Wheelock et al. 1994); the agreement is better at 12 and 25 μm. Ongoing efforts to characterize the discrepancy thus far have not yielded a simple prescription that one would hope could be followed to mitigate the problem.

Although confusion precludes using the DIRBE data to study individual star formation regions in the inner Galaxy, many such regions were detected as point sources and can be studied if they are sufficiently isolated. A few relatively nearby star formation regions were resolved by the DIRBE.

Table 2. *DIRBE* Data Products[a]

Product	Description	Number of files	MB/file	Format	Distribution medium[b]
Weekly Sky Maps	Weekly–averaged inten- sities for 10 bands (1.25, 2.2, 3.5, 4.9, 12, 25, 60, 100, 140, and 240 μm) and Stokes Q and U parameters for 1.25, 2.2, and 3.5 μm bands; 1 file per week	41	28	FITS BINTABLE	c, t
Annual Average Sky Maps	Average of intensities for all 10 bands over the entire cryogenic mission; 1 file per band	10	10	FITS BINTABLE	n, c, t
$\epsilon = 90°$ Sky Maps	Intensities for all 10 bands for the special case of Solar elongation $\epsilon = 90°$; 1 file per band	10	13	FITS BINTABLE	n, c, t
Galactic Plane Maps	Galactic plane subset of the $\epsilon = 90°$ Sky Maps; limited to a 6 month interval	1	6	FITS BINTABLE	n, c, t
Beam Profiles	Effective two-dimensional shape of *DIRBE* beam for each detector	16	0.5	FITS IMAGE	n, c, t
Color Corrections	Color correction factors needed in case source spectrum is not $\nu I_\nu = $ constant	1	0.04	ASCII	n, c, t
System Responses	Normalised system spectral response functions for each wavelength band	1	0.07	ASCII	n, c, t

[a] *DIRBE* Time–ordered data are also available.

[b] Data available from the NSSDC on tape (t), CD–ROM (c), or over the network (n).

Table 3. *FIRAS* Data Products Germane to ISM Studies[a]

Product	Description	Number of files	MB/file
Line Emission Maps	"Maps" of the intensity and the associated uncertainty in each of the follow- ing spectral lines: CO (115.27, 230.54, 345.80, 461.04, 576.27, and 691.47 GHz), O_2 (424.75 GHz), [C I] (492.23 and 809.44 GHz), H_2O (556.89 and 1113.3 GHz), [N II] (1461.1 and 2459.4 GHz), [C II] (1900.5 GHz), [O I] (2060.1 GHz), and [Si I] (2311.7 GHz). One file for ν<630 GHz; another for higher frequencies.	2	0.7
Interstellar Dust Maps	"Maps" of T_{dust} and $\tau_{1800\ GHz}$ are included in the high–frequency Line Emission Maps file.		
Continuum Spectrum Maps	Spectral sky maps from which the line emission has been subtracted; 1 file covers the wavelength range 476μm – 5mm, the other covers 104μm – 5mm.	2	8
Spectral Sky Maps	Calibrated, destriped, combined spectra covering 95.4% of the sky; 1 file per channel/scan mode.[b] Ancillary files contain calibration model parameters, errors associated with the calibration model and destriping process, and covariance matrices.	14[c]	12

[a] All of the products are available from the NSSDC over the network or on tape; CD–ROMs will be made, too. All of the data are in FITS binary tables.

[b] The *FIRAS* instrument had two output ports ("left" and "right") and two frequency bands (60 – 630 GHz and 630 – 2910 GHz), and a Mirror Transport Mechanism (MTM) that could operate at two speeds ("slow" or "fast") and over two scan lengths ("short" or "long"). Some data products represent output port and MTM mode combinations.

[c] Only the LRES, HIGH, RHSS, RHFA, LHSS, and LHFA products include high–frequency (far–infrared) data.

2   COBE Data Products

Well documented data products were prepared for use by the research community. Information about and access to the data products, instrument Explanatory Supplements, and analysis software can be obtained via the *COBE* Home Page at the URL address

http://www.gsfc.nasa.gov/astro/cobe/cobe_home.html

Tables 2 and 3, respectively, describe the *DIRBE* and *FIRAS* data products that are applicable to studies of star formation regions. The *FIRAS* products include dust continuum spectra and spectral line emission maps. The *DIRBE* products include absolute brightness maps of the full sky at 1.25, 2.2, 3.5, 4.9, 12, 25, 60, 100, 140, and 240 μm. Zodiacal light–subtracted maps are scheduled to be released in 1996. With a browsing tool accessible from the *COBE* Home Page one can readily assess the visibility of an unresolved or small extended source in the *DIRBE* data and see its spectral energy distribution.

The NASA Goddard Space Flight Center (GSFC) is responsible for the design, development and operation of the Cosmic Background Explorer (*COBE*) and preparation of the *COBE* data products. Scientific guidance is provided by the *COBE* Science Working Group. The data products are archived at the National Space Science Data Center at NASA/GSFC.

References

Bennett, C.L., Fixsen, D.J., Hinshaw, G., Mather, J C., Moseley, S.H., Wright, E.L., Eplee, R.E., Jr., Gales, J., Hewagama, T., & Isaacman, R.B. 1994, *ApJ*, **434**, 587–598

Boggess, N.W., Mather, J.C., Weiss, R., Bennett, C.L., Cheng, E.S., Dwek, E., Gulkis, S., Hauser, M.G., Janssen, M.A., Kelsall, T., Meyer,S.S., Moseley, S.H., Murdock, T.L., Shafer, R.A., Silverberg, R.F., Smoot, G.F., Wilkinson, D.T., & Wright, E.L. 1992, *ApJ*, **397**, 420–429

Reach, W.T., Dwek, E., Fixsen, D.J., Hewagama, T., Mather, J.C., Shafer, R.A., Banday, A.J., Bennett, C.L., Cheng, E.S., Eplee, R.E., Jr., Leisawitz, D., Lubin, P.M., Read, S.M., Rosen, L.P., Shuman, F.G.D., Sodroski, T.J., & Wright, E.L. 1995, *ApJ*, **451**, 188–199.

Wheelock, S.L., Gautier, T.N., Chillemi, J., Kester, D., McCallon, H., Oken, C., White, J., Gregorich, D., Boulanger, F., Good, J., & Chester, T. 1994, *IRAS Sky Survey Atlas Explanatory Supplement*, JPL Publication 94–11 (Pasadena: JPL)

Submillimeter Continuum Emission in the Orion A Cloud, Observed with PRONAOS

I. Ristorcelli[1], A. De Luca[2], M. Giard[1], F. Pajot[2],
J.P. Torre[2], G. Serra[1], and J.M. Lamarre[2]

[1] C.E.S.R., CNRS-UPS, BP 4346, F31029, Toulouse cedex, France
[2] IAS, bât 120, Campus d'Orsay, 91405 Orsay CEDEX, France

Abstract. We present submillimeter observations performed with the french balloon-borne experiment PRONAOS, during its first flight on September 1994. A map containing the M42 nebula has been drawn up over 10'x50', with angular beams varying from 2' to 3.5', and in four photometric bands (from 180 to 1100μm). Around the emission area of the BN/KL objects, three other brightness peaks have been observed. One is about 5' × 5' in size, and can be interpreted as the dust emission of a very cold molecular cloud (about $12 \pm 2K$), with a total mass of 15 $M_\odot$ over 3.5'. The emission spectra of the other sources show a variation of the emissivity coefficient, which could reveal a dependence of the grain emission characteristics with the external radiation, inside a same molecular cloud.

1 Instrument Characteristics

PRONAOS is a french balloon-borne experiment devoted to submillimeter astronomy. It consists of four main sub-systems : a two-meter Cassegrain telescope, a stabilised gondola and two focal instruments, a Multiband Photometer (SPM) and a heterodyne Spectrometer (SMH), planned to fly alternatively on the experiment. The SPM instrument has been designed to realise simultaneous measurements in four bands from 180 to 1100 μm and its characteristics have been described in Lamarre et al. (1994). It uses He^3 cooled bolometers (300mK) and dichroic filters inside a cryostat; the warm optics ensures, in particular, the beam switching and the in-flight calibration by means of two blackbodies. The absolute accuracy of this calibration is better than 10%. The sensitivities reached during the flight as well as the field of view in each photometric band are given in table 1. The electrical signal provided by each bolometer is amplified, filtered, demodulated and synchronously transmitted by telemetry to the ground station. The telescope (manufactured by Matra Marconi Space-Toulouse) is made of a 2m diameter primary mirror, consisting of six identical panels in carbon fiber and covered by a gold coating. A servo-controlled loop system between capacitive sensors and actuators on the panels allows, all along the flight, to correct any misalignment introduced by thermal or gravity effects, and maintains the nominal alignment of the panels. The payload pointing is provided by the gondola using a two-stage system. The first one ensures the azimuthal stabilisation within a relative precision of 30 arcseconds, thanks to a magnetometer and a gimbal servoed with gyrometers and inertial wheels. The fine pointing is obtained with an inertial platform which drifts are corrected thanks to the pointing of a star sensor that can be oriented off-axis from the telescope. The final relative pointing accuracy is better than 5 arcseconds. The PRONAOS-SPM experiment was launched on September 17, 1994, from the NASA launching site of Fort-Sumner (New Mexico, USA). Unfortunately, the failure of the star sensor during daytime, together with a much too low ceiling altitude at night, did not allow to perform the observations planned. Only very partial observations were possible. We report here the results obtained in the region of the Orion M42 nebula. The observations have been done with a cross-elevation scanning at a constant speed of $20 arcsec/s$ and with a beam switching amplitude of 5.35 arcmin. The covered map corresponds to $50 \times 10 arcmin^2$.

Table 1. Instrument characteristics

Band	1	2	3	4
$\Delta\lambda\ (\mu m)$	180-240	240-340	340-540	540-1100
FOV(arcmin)	2	2	2.5	3.5
NET ($mKs^{1/2}$)	1.2	0.9	0.5	0.35
NEB ($MJysr^{-1}s^{1/2}$)	85	92	21	11

1.1   Results

The contours map of the signal measured in the second photometric band, for instance, is given in figure 1. Four brightness peaks are visible and have been identified to the following sources:

1) The brightest peak corresponds to the BN/KL objects area.

2) An extended emission region over around $5' \times 20'$, 15' SE from BN/KL coincides with an IRAS $100\mu m$ band flux enhancement and a CO emission peak; it has been identified by Castets et al. (1990) with a compressed gas area due to a shock driven into the neutral gas by the expansion of the HII region.

3) The third brightness peak in the PRONAOS map is located at $10'W \times 4'N$ from BN/KL. It is associated with the edge of an ionized region (which emission peak at $100\mu m$ is about 20'W from BN/KL).

4) An extended peak emission (over $5' \times 5'$) is clearly observed at 30'NW from BN/KL, particularly in the second photometric band (240 - $340\mu m$). This area coincides neither with a brightness enhancement of the 13CO emission nor of the $100\mu m$ IRAS band emission.

The emission spectrum associated with each of these sources is displayed on figure 2. In order to avoid dilution effects due the different beam values in the wavelength bands, the data have been averaged over 3.5'. The extended emission on component (2) has been averaged over $5' \times 8'$. The spectra are compatible

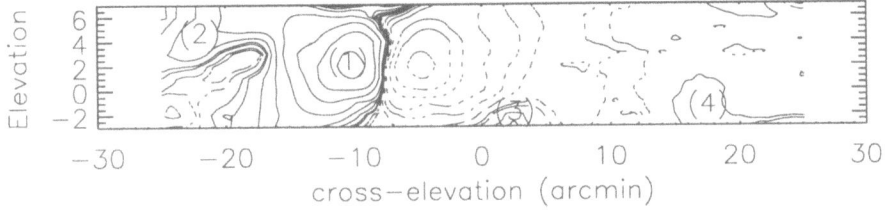

Fig. 1. Contours of the measured signal in the second photometric band. Large negative signal (dashed contour) is due to the beam switching.

with a dust emission and can be schematically modeled by assuming an average dust temperature Td and an emissivity power law $Q_{em} \alpha \nu^\beta$. The best fits have been determined by a least square method and the derived values of Td, β and the optical depth τ (in the longer wavelength band), are given in table 2. For the sources (2) and (3), the fit has been deduced from the PRONAOS-SPM measurements together with the IRAS 100μm emission. On the BN/KL spectrum, we have overplotted the data of Harper et al. (1974), Chini et al.(1984) and Mezger et al.(1990) at respectively 90, 1000, and 1300 μm, corresponding to large beams of 3.5', 3.9' and 3'. The frequency index β for BN/KL is similar to that found by Gordon et al. (1987) (β=1.2). It remains quite different from the values obtained on others warm interstellar clouds associated with HII regions ($\sim$2) (Gordon et al. (1988)). The 4th source can be associated to dust emission with a mean very cool temperature of $12 \pm 2K$. Its total luminosity over 3.5' is $0.3 L_\odot$. We have estimated the corresponding mass using the ratio τ_λ / NH adopted in the dust model of Desert et al. (1990). We derive a total gas mass of about $15 M_\odot$.

Table 2.

Source	RA_{50}	DEC_{50}	T(K)	β	λ ;τ
BN/KL (1)	5:32:48	-5:25:30	83.	1.0	628μm; 4.1e-3
Source (2)	5:33:24	-5:32:00	31.	2.2	628μm; 2.3e-4
Source (3)	5:32:01	-5:18:32	33.	1.9	628μm; 1.2e-4
Source (4)	5:31:27	-5:05:49	12.	1.9	645μm; 4.e-4

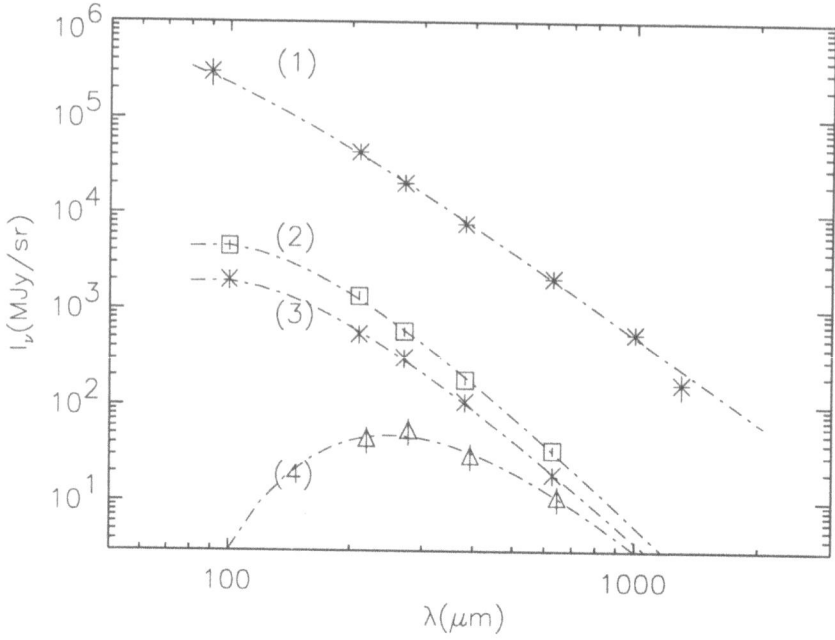

Fig. 2. Emission spectrum of the identified sources. The fluxes measured with PRONAOS-SPM are given at effective wavelengths determined in each photometric band in accordance with the spectrum. We have overplotted on the BN/KL spectrum, the data of Harper et al. (1974), Chini et al.(1984) and Mezger et al.(1990) at respectively 90, 1000, and 1300 μm. The fluxes deduced from IRAS data at 100 μm have also been added to the spectrum of sources (2) and (3). The dashed curves correspond to the fit described in the text.

References

Castets, A., Duvert, G., Bally, J., Langer, W.D., Wilson, R.W. (1990): A&A, 234, 469

Chini, R.,Kreysa, E.,Mezger, P.G.,Gemund, H.P.(1984): A&A, 137,117

Desert, F.X.,Boulanger, F.,Puget, J.L. (1990) A&A, 237,215

Gordon, M.A. (1987): ApJ,316,258

Gordon, M.A. (1988): ApJ,331,509

Harper, D.A.,(1974): ApJ,192,557

Lamarre, J.M., Pajot, F., Torre, J.P., Guyot, G., Bernard, J.P., De Luca, A., Giard, M., Mangin, J., Peytureaux, R., Puget, J.L., Serra, G., Boulanger, F., Desert, F., Ristorcelli, I., Jegoudez, G., Lagardere, H., Leblanc, L., Pons, R., Recouvreur, G., Barthelemy, M., Bourguignon, C., Crussaire, J.P., Dambier, G., Delille, L., Korczak, E., Lepeltier, J.P., Leriche, B., Lizambert, C., Narbonne, J., Plaisant, C., Reigner, J., Renault, J.C., Rioux, C. (1994): Infrared Phys. Technol., Vol. 35, No. 2/3, 277

Mezger, P.G., Wink, J.E., Zylka, R.(1990): A&A, 228, 95

Catalogue of IRAS Loops in the IInd Galactic Quadrant

Tóth L. Viktor[1], Kiss Csaba[1] and Moór Attila[1]

Eötvös L. University, Dept. of Astronomy, Budapest,
Ludovika tér 2., H-1083, Hungary

Abstract. A catalogue of bright loops on ISSA (Infrared Sky Survey Atlas) maps of the IInd galactic quadrant has been compiled. More than 150 loops has been identified. A significant portion of them are located at high galactic latitudes.

1 Introduction

Loops on radio continuum maps have been long since known. However shell or arc like local density enhancements of the interstellar gas may be formed by various processes, most of the radio loops are explained as 2D projected views of bubble shaped supernova remnants (SNR's). The association of the radio continuum spurs with old SNR's was suggested by Hanburg et al. (1960). The galactic supernovae may occur at a rate sufficent to produce a swiss-cheese morphology of the cold interstellar gas, with hot coronal gas in the connecting bubbles (see eg Cox and Smith (1974). The amount of coronal gas in the galactic disk may be estimated calculating the total volume of bubble interiors. Detection of X-ray emission of the hot gas is possible for the nearby bubbles (see eg. Bunner et al. (1972)) only. It is however possible to detect the far infrared (FIR) emitting shells of the swept up gas from larger distances. The cold interstellar matter (ISM) in these shells may be warmed up by new stars formed by the trigger of the large pressure of the bubble interior (see eg. Kun et al. (1987)). Due to the similarly low optical depth HI (21cm) mapping is the other effecive tool exploring the bubble structure of the galactic disk (see eg. Heiles (1979). The best survey so far which could be used for this purpose was provided by Burton and Hartmann (1994). We report here a large scale FIR survey (using IRAS data) in the IInd galactic quadrant.

2 Input Data

The visual search for loop like features has been carried out on the computer readable ISSA (see Wheelock et al. (1994)) images. All the 100micron sky intensity plates and a number of the 12, 25 and 60micron plates of the IInd galactic quadrant has been analysed using the Skyview (R3.1.5 beta) software package of the IPAC (Infrared Processing and Analysis Center, CALTECH). Loops by our definition must show an excess FIR intensity as compared to it's surrounding. The curve must be at least 60% of a complete ellipse. It may consist of a set of bright extended spots, or may be a difuse ring or part of a ring. The parameters of the loops has been obtained fitting ellipses to them. The loop interior has been defined by a concentric smaller ellipse.

3   Discussion

The Catalogue of IRAS loops in the IInd galactic quadrant containes 156 items. The distribution of the loops is shown in Fig. 1.

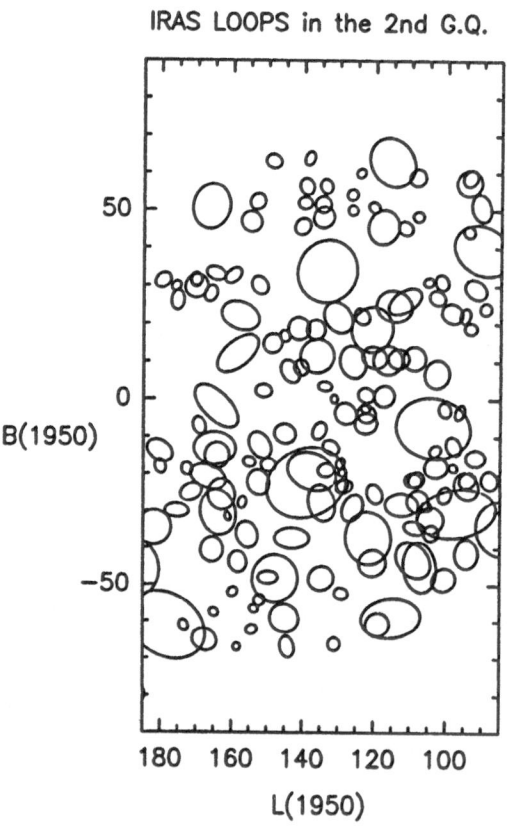

Fig. 1. The distribution of the FIR (100micron) loops. Each of the 156 loops are represented by an ellipse.

As it is seen on Fig. 1 there are large loops at $b_{II} > 50°$ As it is shown in Fig. 2 the loops have a characteristic size, R_{eff} of 3 degrees in effective radius

$(R_{eff} = \sqrt{a \times b})$ where a and b are the main axes of the fitted ellipse. The smallest loops have R_{eff} of about 1 degrees, and the largest R_{eff} is 11 degrees. It is not easy to confirm larger loops due to confusion by smaller scale features. The total volume of hot gas in the dense disk has been estimated assuming (i) all the loops are bubbles contain coronal gas, (ii) the bubbles have a homogenous distribution. Then the fraction of hot gas is $\approx 10\%$. The value depends strongly on the galactic scale height and the viewing distance one defines in the FIR disk.

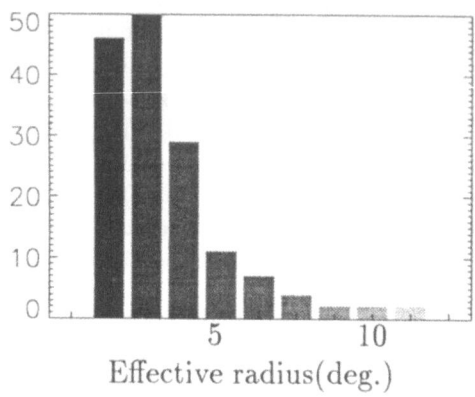

Fig. 2. The size distribution of the IRAS loops.

We have compared the distribution of the FIR loops to the distribution of HI gas (Burton and Hartmann (1994)), and the distribution of the 1420 MHz continuum radiation (Reich (1982), Reich and Reich (1984)). There are many radio continuum - FIR and HI - FIR associates. We note that the there are several high latitude loops on the HI channel maps at velocities which characterise nearby gas.

There are usually one or more OB associations seen towards the loops near the galactic equator, and the physical connection may be assumed in most of these cases.

It is an interesting task to explain the origin of the loops found at very high galactic latitudes, but it is not impossible. Loop like features may be formed by the following processes: (i) expanding stellar and/or SN wind blown bubble (see eg. deGeus (1988) discussing Loop I a nearby example); (ii) like (i), but in a flattened cloud, or cloud layer forming and expanding ring; (iii) two expanding SN bubbles forming an expanding ring in their intersection plane ; (iv) nonradiative/radiative collision of a cloud with a thick gas layer resulting a hemispheric/cylindrical shell (see Tenorio-Tagle et al. (1986)); (v) impact of a high velocity cloud (HVC) (see eg. the NCP Loop Meyerdirks). Real 3D bubble is expected only in the first case. The shear and Corriollis force then distort the growing shell into an ellipsoid, which closes up after an epiciclic period (Tenorio-Tagle and Palous (1987)).

There are obviously no OB associations at high galactic latitudes. The large number of loops at such coordinates may be a result of a large number of single B type stars at high z, or an intensive bombardement of the dense disk by HVC's.

4   Results

There are at least 156 IRAS loops in the IInd galactic quadrant. The dense disk as traced by its cold gas/dust has a Swiss cheese structure. The filling factor of the bubble interiors' hot gas in the dense disk is about 10%. The number of high latitude loops may suggest a high HVC impact rate.

The catalogue is available at the WWW page:

http://enlil.elte.hu/staff/IRASLoops.html

The parameters listed in the catalog's columns are: (1) Cat.no.; (2) (3) galactic coordinates $l(1950)$ and $b(1950)$; (4) (5) main axes of the ellipse fitted to the loop (5) size R_{eff} in degrees; (6) excess intensity ΔI of the shell relative to the loop's interior; (7) remarks: listing some of the associated objects.

Acknowledgement : This research was partly fund by the grant: "Csillagok és intersztelláris anyag kölcsönhatása..." of the OTKA Hungarian Research Fund.

References

Bunner, A. N. et al. (1972): ApJ **172** L67.

Burton, W.B. and Hartmann, D. (1994) Astrophysics and Space Sci. **217**. 189.

Cox, D.P., Smith, B.W.(1974): ApJ **189**, L105

Elmegreen, B.G. (1985): ApJ **299** 196

deGeus (1988) *Ph.D Thesis*, Leiden.

Hanburg, B.R., Davies, R.D., Hazard, C. (1960): Observatory, **80**. 191.

Heiles, C. (1979): ApJ, **229**, 533.

Meyerdirks, H., Heithausen, A. Reif, K. (1991): A&A, **245**, 247.

Kun, M., Balázs, L.G., Tóth, I. (1987): AsSpSc, **134**, 211

Reich, W. 1982, A&AS, **48**, 219

Reich, P., and Reich, W. (1986): A&AS, **63**, 205

Tenorio-Tagle, G. et al. (1986): A&A, **170**, 107.

Tenorio-Tagle, G. and Palous, J. (1987): A&A, **186**, 287.

Wheelock S. L. et al. (1994): *IRAS Sky survey Atlas Explanatory Supplement*, JPL Publication 94-11 (Pasadena:JPL)

Triggered Core Formation in Nearby Clouds

Tóth L. Viktor[1], Horváth András Jr.[1,2]

[1] Eötvös L. University, Dept. of Astronomy, Budapest,
   Ludovika tér 2., H-1083, Hungary
[2] Szechenyi István College, Dept. of Physics, Győr,
   Hungary

Abstract. We present the results of observational and numerical studies of medium sized interstellar clouds. High pressure events played an important role in the formation of the present structure of the studied clouds in the Cep-Cas void. The triggred core formation may be a general process in low and intermediate mass clouds which then leads to low mass star formation.

1 Introduction

Interstellar clouds usually may not free-fall collapse, but instead they need 5 to 15 times the free-fall time to loose the portion of their turbulent, rotational, magnetic energy, required for dense core formation. (See eg. by Elmegreen 1992, or by Mouscovias in this volume.) Since a frequent passage of shock fronts is expected in our galactic disk, clouds will form cores due to this trigger. Compression by stellar or SN winds may form cometary clouds and trigger a cloud collapse directly or may act indirectly speeding up the diffusion of a strong magnetic field.

We analyzed 7 dark clouds towards the Cep-Cas void, a SNR as described by Grenier et al. (1989). The results were compared to our hidrodynamical simulations of shock cloud encounters (see Horváth and Tóth(1995)).

2 Input Data and Data Processing

The distribution of the far infrared emission was studied at the 12, 25, 60, and 100 micron IRAS bands using ISSA (Infrared Sky Survey Atlas) plates A statistical investigation (factor analysis) indicated a multiple bubble system of the void between the Cepheus Flare and Cassiopeia giant molecular clouds. Several shells were already identified on the 100micron images alone. At least 3 of these IRAS loops are well seen in the distribution of the nearby HI clouds in the channel maps of the Leiden/Dwingeloo survey Burton and Hartmann (1994). The swept out gas forms clumps appearng as bright spots on both the IRAS and the HI intensity maps. Some of these clumps are associated with mapped molecular clouds. The distance of the clouds in the region is between 300 and 800pc (Grenier et al. (1989) and Kun and Prusti (1993)) which allows a good spatial resolution of their morphology.

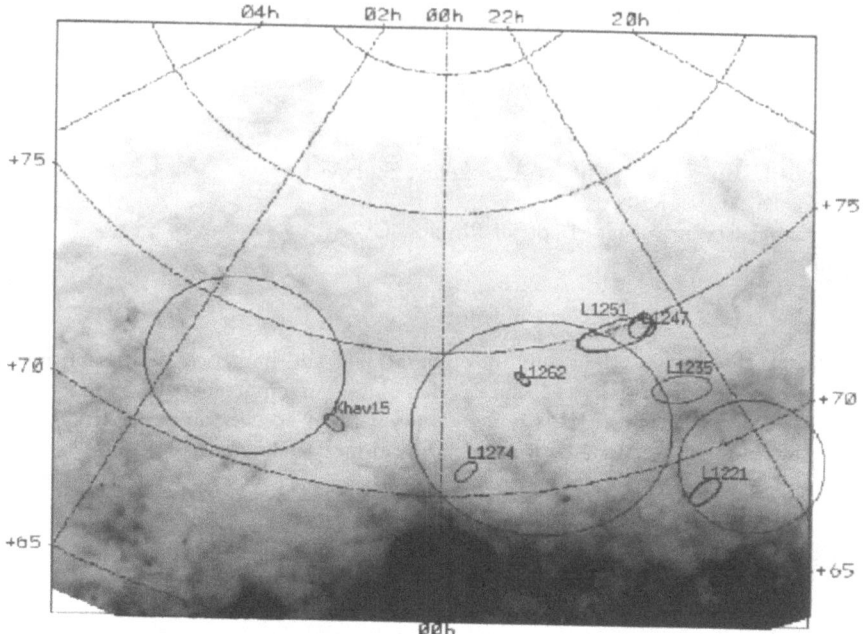

Fig. 1. IRAS 100 micron mosaic image of 6 ISSA plates. The small ellipses are 2× enlarged representations of the ^{13}CO clouds. The elongated clouds point into the centres of 3 IRAS loops (shown as large ellipses). The cometary cloud morphologies were resulted probably by the passage of the shocks which created the loops.

The distribution of molecular gas in the dark clouds were analyzed using ^{13}CO(1-0) maps of L1221(by Umemoto et al. (1991)), of L1235 (by Snell (1981)) and of KH15, L1274 and L1262 (by Fukui), and of L1251 (by Sato et al. (1994)) and H$_2$CO map of L1247 (by Tóth and Walmsley (1996)). Ellipses were fitted to the integrated intensity contour levels (assuming optically thin radiation) to measure the displacement of the center of gravity for the different column density regions. The directions of the clouds are shown on Fig. 1, where the ellipses representing the ^{13}CO clouds has been enlarged by a factor of 2.

High resolution radio spectral line observations were carried out towards L1251 and L1247. The distribution of the H$_2$CO absorption and NH$_3$ (1,1) and (2,2) emission was mapped using the Effelsberg-100 telescope in May 1993, April, 1995. (by Tóth and Walmsley (1995), Tóth and Walmsley (1996)). The extent of the molecular clouds was measured by the formaldehyde 6cm line. The dense core formation efficiency was tested with the ammonia line measurements. The spatial resolution of the survey was about 1 arcminute. Cores were mapped with a 3σ limit of ammonia column density: $N(\mathrm{NH_3}) = 1.25 \times 10^{14} \mathrm{cm}^{-2}$.

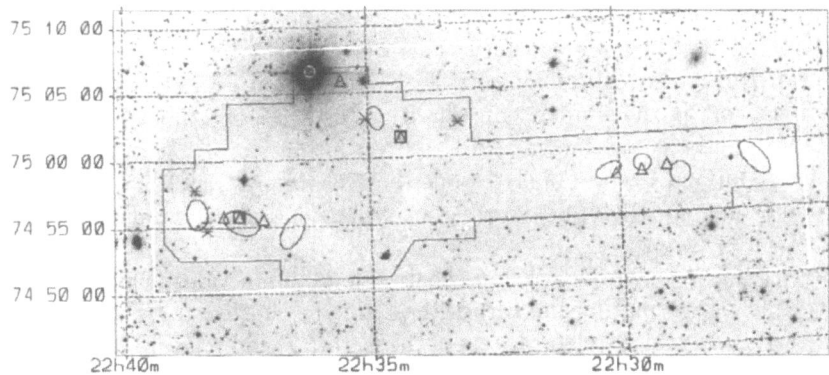

Fig. 2. L1251: POSS red print indicating the positions observed, and the cores found, with IRAS point sources overlaid. We marked the region where ammonia was observed (black contour). The cores are represented with ellipses. The IRAS sources are marked according to their types: star with circle: SAO red giant foreground star; star: T Tauri like object with Hα emission; triangle: cold objects (class 1 or 60μm only); box: CO outflow driving source.

The cloud morphology was compared to hydrodynamical simulations of collision of clouds of similar mass with low velocity shock fronts. We applied a hydrocode assuming He, H$_2$, and optically thin CO cooling of real gas mixture of HI, HII, H$_2$, and He. Detailes on the model are given in Horváth and Tóth(1995).

3   Results

All the listed clouds except L1247 were found to have cometary structures i.e. flattened, and the densest regions displaced towards one end of the clouds. The directions given by the offset of the column density peak from the geometrical centre, were used to test the multibuble view drawn by the IRAS and HI data. The cloud directions indicate 3 centres of disturbance (see Fig. 1). All 3 are centres of IRAS loops (see Tóth et al. in this volume). From the degree of elongation and the maximum estimated densities we concluded that these clouds were possibly shocked by slow (5 to 30 kms^{-1}) shock fronts.

The clouds L1251 and the neighbouring L1247 on the flank of the Cepheus Flare GMC showed a large contrast. There is $\approx$ 50% of the molecular mass concentrated in 8 dense ammonia cores in L1251 (see on Fig. 2), while there is no ammonia core in the similarly opaque L1247 which is $\approx$4 times less massive than L1251. There is an efficient (SFE$\approx$ 10%) formation of low mass stars in L1251 (see Kun and Prusti (1993)), while there is no any ClassI or Class0 type IRAS pointsource in L1247. Low mass star formation is apparent also in L1221, in L1235 and in L1262.

These clouds are examples of slow shock – cloud interactions, which resulted in a triggered core formation and turned these small clouds to star forming nebulae. The low mass young stellar objects of the Cep-Cas void tends to concentrate around the shocked clouds. No spontaneous low mass star formation has been found i.e. at least in this region the low mass star formation is triggered.

Acknowledgement : This research was partly fund by the grant: "Csillagok és intersztelláris anyag kölcsönhatása..." of the OTKA Hungarian Research Fund. We also received support from the German-Hungarian Technological and Scientific Cooperation Project no. 121.

References

Burton, W.B. and Hartmann, D. (1994) Astrophysics and Space Sci. **217**, 189.

Elmegreen, B.G. (1985): ApJ **299**, 196

Fukui, Y. at al.(1995): *private comm.*

Grenier, I., et al. (1989): ApJ **347**, 231

Horváth, A.Jr. and Tóth, L.V. (1995): in *Shocks in Astrophysics*, eds. Millar, T.J. and Raga, A.C., publ. Kluwer in press.

Kun, M. and Prusti, T., 1993, A&A, **272**, 235.

Sato et al. (1994): ApJ **435**, 279.

Snell, R.L. (1981): ApJS **45**, 121.

Tóth, L. V. and Walmsley, C.M. (1995) A&A in press.

Tóth, L. V. and Walmsley, C.M. (1996) in prep.

Umemoto, T. et al. (1991): ApJ **377**, 511

Searching for New Young Stars in the IRAS Point Source Catalog

H. J. Walker[1], T. L. Lim[2], B. M. Swinyard[1], P. J. Richards[1], and R. J. Emery[1]

[1] CLRC Rutherford Appleton Laboratory, Chilton, Didcot, Oxon, OX11 0QX, UK
[2] Dept of Physics, Queen Mary & Westfield College, Mile End Road, London, E1 4NS

Abstract. Using the "colours" of known T Tauri stars, the IRAS Point Source Catalogue was searched for sources with similar "colours". Almost 3500 IRAS sources matched our criteria. We have examined the global properties of the sample; there are several clusters of sources and just over 200 sources more than 40 degrees away from the galactic plane. Due to the fact that very few sources in the sample have optical counterparts we suspect that our sample contains sources with dense dust shells (i.e. sources younger than the classic T Tauri stars). We have selected several groups of sources for further investigation; a control sample in Taurus, a sample in Cygnus, one in Aquila, and the high latitude sources.

1 Introduction

A sample of known T Tauri stars, with fluxes in the IRAS Point Source Catalog (PSC, 1985) at 12, 25 and 60 μm were used to define a set of "colours" for T Tauri stars. The flux ratios were:

$$0.17 < S_{12}/S_{25} < 1.51$$
$$0.00 < S_{25}/S_{60} < 1.62$$

These ratios were very similar to those used by other groups, e.g. Harris et al. (1988). The IRAS PSC was then searched for sources with flux ratios in this range. No other constraint was imposed except that the sources must have good quality flux measurements at all three wavelengths. We found 3489 sources matched our criteria.

2 Results

Since we were uncertain about the evolutionary stage of our candidates, we decided to refer to them as young stars, rather than the more specific term of T Tauri stars. We suspect they must have denser dust shells than the T Tauri stars used to define the flux ratios, since very few of our sample have optical counterparts and those used to define the flux ratios were all optically identified stars. We have effectively selected a particular set of warm dust shells around a central star. Surprisingly few sources were associated with entries in the almost 40 catalogues selected by the IRAS Science Team for cross-referencing (see the IRAS PSC Explanatory Supplement for more information). Most of the associations were to compact HII regions or dark nebulae (which was reassuring).

The young star candidates, when plotted in galactic coordinates, are strongly concentrated towards the galactic plane. Away from the plane, two conspicuous concentrations of sources are visible, in Orion and in the Large Magellanic Cloud, and Taurus is also noticeable. When the data in the plane are put into 5° bins, more sources are seen above the plane from 60° to 120°, and more below the plane from 220° to 300°, reflecting the presence of Gould's belt. There were many sources at high galactic latitudes (more than 200), which was surprising, since star formation is largely concentrated in the plane. We selected several regions for further study; Taurus, a sample in Cygnus, one in Aquila, and the high latitude sources.

Taurus: This is our control sample; the area of sky covered is 169 square degrees and contains 43 sources. Compared to the list of Harris et al. (1988) we have 29/38 of their known sources in our sample, and 10/12 of their new sources (plus 4 new sources not in their list).

Cygnus: This area covers 188 square degrees and contains 137 sources of which 20 have associations with the IRAS catalog set. The sources are well distributed throughout the region, but a few, small clusters are noticeable.

Aquila: This is a smaller area of 63 square degrees, containing 62 sources (8 associations). These sources are concentrated in the galactic plane, but they do not not show any clustering.

High Latitude Sources: These sources have | b | > 40°. Orion, Taurus and the Large Magellanic Cloud have been explicitly excluded from the sample. There are 203 sources in this sample. They show no particular concentration, but are uniformly scattered around the high latitude sky.

We are currently trying to identify our IRAS sources on sky survey plates. We would like to find sources with optical counterparts from our sample, to follow up optically. Hα emission, lithium and mass loss are signs of a young star, and these (mainly) need optical spectra. The nature of most of the candidates will have to be confirmed at wavelengths other than the optical, and it will be the properties of their circumstellar gas and dust which will be used. Walker et al. (1989) showed that some galaxies had IRAS flux ratios (colours) similar to T Tauri stars, the high latitude sample (in particular) will be carefully checked for contamination by this type of object.

References

Harris, S., Clegg, P. and Hughes, J. (1988): MNRAS **235**, 441

Joint IRAS Science Working Group (1985): *The IRAS Point Source Catalog* (U.S. GPO, Washington D.C.)

Walker, H. J., Cohen, M., Volk, K., Wainscoat, R. J. and Schwartz, D. E. (1989): AJ **98**, 2163

Part IV

Vega-Type, T-Tauri, λ Bootis and Herbig Ae/Be Stars

Vega-Type Systems

Pawel Artymowicz[1,2,3]

[1] Stockholm Observatory, Stockholm University, S-133 36 Saltsjöbaden, Sweden
[2] Space Telescope Science Institute, 3700 San Martin Dr, Baltimore MD 21218, USA
[3] pawel@astro.su.se

Abstract. Vega and β Pic-type stars are common main-sequence stars exhibiting a large infrared excess – the thermal radiation of solid grains orbiting the stars. In this review we discuss them as individuals and as a group. We show that all the objects are divided into two categories: one with gas-poor, dust-dominated disks equally or less 'dusty' than that of β Pic, and a second category of gas-dominated disks with larger 'dustiness' (ratio of bolometric IR excess to stellar luminosity). The first category contains maturing planetesimal or planetary systems similar to β Pic. Such systems descend from objects in the second category, which is similar to late-stage protostellar T Tau and post-T Tau disks and/or Herbig Ae/Be systems. Some of the Vega-type systems belong to the class of λ Boo stars. We interpret this within the framework of planetary system formation. We emphasize the manifold role the stellar radiation pressure on dust in the common A-type systems plays in: (i) setting an upper limit to the dustiness of the gas-free systems; (ii) removing the small grains from such disks (submicronic collisional debris, transiently heated silicates, and polycyclic hydrocarbon grains) which, among others, causes color-neutral scattering in the visible; (iii) placing typical disk particles on elliptic orbits, and (iv) repelling the interstellar medium, which therefore has no direct bearing on the evolution of the best studied Vega-type systems.

1 Decade of Discovery

1.1 Nature and Frequency of the Vega Phenomenon

Little more than a decade ago very few astronomers would think a perfectly normal, main-sequence, seemingly non-variable, isolated (remote from star forming regions), single – in short – boring[1] star can be of much interest except perhaps as a photometric standard. Vega, β Pic, and other now-famous stars were, in fact, used as standards. It all changed with the surprising discovery of a large infrared excess of the emitted radiation over the photospheric stellar flux, during the calibration of IRAS satellite in 1983. The telltale excess was demonstrated to be the thermal radiation of solid grains with sizes much larger than 1 μm , i.e., very much larger than typically found in the interstellar medium. The grains could not be produced in an outflow, they had to be in orbit around the star. The requirement of no stellar mass loss (which, if present, could imply dust-forming winds) was incorporated into the definition of Vega-type stars. (A decade of discovery that followed was aptly reviewed by Backman & Paresce

---

[1] However, β Pic has been known for some time as a spectroscopic 'shell star', exhibiting narrow circumstellar lines.

1993.) As a result of studies of the Vega-excess phenomenon (or simply Vega phenomenon), we now know that many normal main-sequence stars in the Galaxy are surrounded by rings or disks of dust. Many observed systems contain A-type dwarfs. Later types are somewhat underrepresented due to their smaller luminosity. In fact, more than 1 out of 5 main-sequence stars with type B, A, or F, are now estimated to possess circumstellar dust (Patten & Willson 1991). At least 50% (!) of sun-like G-type stars are expected to have circumstellar dust (Aumann & Good 1991). Individual systems' ages are varied, and we believe we are not merely witnessing 'excesses of the youth', phenomena found only in young stellar objects, e.g., the pre-main sequence stars. There is a growing consensus that Vega phenomenon signifies common occurrence of planetary systems around stars (sect. 5).

1.2   Prototypes of the Vega-Excess Phenomenon

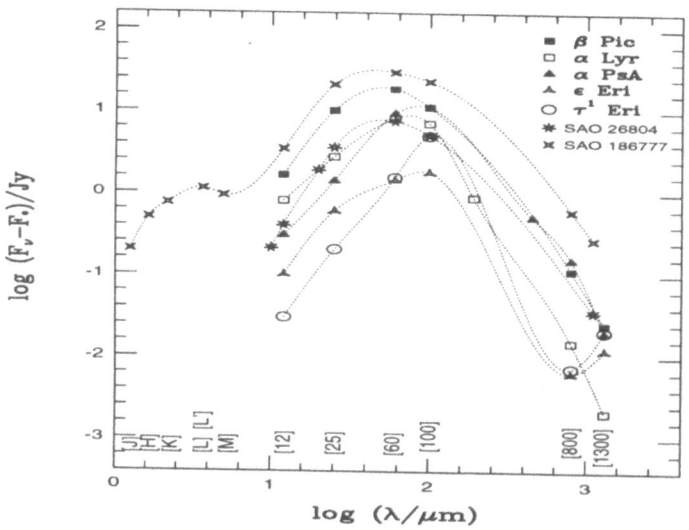

Fig. 1. Infrared excess of radiation from several Vega-type systems. Near-IR and IRAS band positions are indicated in square brackets by their name or wavelength in μm at the bottom of the figure. (Data from: Gillett 1985, Chini et al. 1991, Zuckerman & Becklin 1993a, Sylvester et al. 1994, and Skinner et al. 1995.)

Historically, four objects gained the status of prototypes for Vega phenomenon (Gillett 1986, Backman & Paresce 1993): α Lyr (Vega itself, A0V star 8.1 pc away), α PsA (Fomalhaut, an A3V star 15 pc away), β Pic (the best studied representative; an A5V star 17 pc away), and ϵ Eri (K5V star 3.3 pc from us). Their infrared spectra (with photospheric components removed) are reproduced in Fig. 1. The dotted lines smoothly connecting observational points are just a visual guide to the shape of the spectrum – they do not represent any particular model (models fitting the data have been constructed among others by Gillett 1986, Artymowicz et al. 1989; Backman et al. 1992). Notice that among the prototypes β Pic has not only the largest absolute flux but also the largest absolute and relative excess over the extrapolated photospheric contribution. It is thus not surprising that this star attracted most of the attention, especially after Smith & Terrile (1984) have for the first time imaged a disk-like structure in the scattered (visible/near-IR) light. A wealth of data has since been obtained on β Pic, as summarized by Norman & Paresce (1989), Backman & Paresce (1993), Lagrange (1995), and Artymowicz (1994). In comparison, the progress in understanding the remaining 3 objects has been slow, no doubt due to inability of current coronographic and other imaging techniques to separate the faint disk glow from the extended, bright stellar image. Literally hundreds of IRAS-excess sources have been scrutinized for evidence of light scattered by their disks, so far without success. It is known that edge-on orientation of β Pic disk contributes significantly to its observability, as does the sheer area of grains it possesses (see about parameter τ_{dust} below). The exceptional status of β Pic might be promoted by unusually large visible albedo of its dust grains. There is little doubt that the uniqueness "problem" is technical and that other systems have dust circling around the stars in β Pic-type disks and not, e.g., spherical shells. Spatially resolved mid-IR observations of SAO 26804 by Skinner et al. (1995) illustrate that point clearly. They have shown the disk inclination to be intermediate between edge-on (as in β Pic) and pole-on (as in Vega). Earlier, the substantially different extents of α PsA in two perpendicular directions of IRAS slow scans at 60 μm (Gillett 1986) provided arguments for a similar orientation of its disk. However, a direct spectroscopic proof that the disks is in the expected Keplerian rotation is still lacking, even for the best-studied system of β Pic.

The spectral energy distributions of most systems in Fig. 1 are close to a superposition of Planck curves with temperature T contained in a range from 30 K to 250 K (e.g., Gillett 1986). Peak temperature, or single-temperature model's T, varies in a narrow range from $T \approx 70$ K for α PsA and $T \approx 90$ K for α Lyr to $T \approx 110$ K for β Pic. The corresponding black-body grain separations from the respective stars are subject to much more variation due to different stellar luminosities: in ϵ Eri they range from about 1 to 50 AU, in α PsA from 30 to 140 AU, in α Lyr from 40 to 200 AU, and in β Pic from 15 to more than 400 AU. All the temperatures are below the silicate thermal evaporation limit of 1500 K. Moreover, typical disk temperatures ($T < 110K$) allow water ice to remain unaffected against thermal evaporation (sublimation). It is a remarkable fact that the Vega-type stars have a very small fraction of solid grains at temperatures in excess of about 200 K, which gives rise to the central clearing or gap in the disk. For instance, in β Pic the dust density inside the gap radius (10 to 40 AU, as inferred from IR excess modeling) is at least one order of magnitude smaller than in the outlying disk. The origin of the gaps may be (non-exclusively) understood in terms of either the clearing dynamical action of planets, or the ice evaporation boundary.

2   The Disk of β Pic

2.1   The Dust Component

β Pic is the best-studied example of a Vega-type system due to the availability of imaging in B,V,R,I,[12 μm], high resolution spectroscopy in the mid-IR, and broad-band spectroscopy in far-IR up to the mm region. Such a comprehensive data set is required to address the basic issues of mean dust albedo, composition, and size distribution (for reviews see Backman & Paresce 1993; Artymowicz 1994). To that end, *simultaneous* modeling of thermal and scattered radiation must be performed, using theoretical $T(r)$, temperature–radius relations, for specific assumed types of grains. Results show that β Pic particles have a wide range of sizes, extending from submicron dust to at least mm-sized sand. Most of disk surface area is contained in 1–20 μm particles, sometimes called "mid-sized" grains. Warm silicates responsible for the emission feature (described below) are strongly dominated by particles with radius of order 2 μm. One group of good candidate materials are microscopically translucent silicates *very much different* from the so-called "astronomical silicates" found in ISM or any other materials containing a large percentage of dark carbon or iron. Common olivine and pyroxene-based rocks fit the bill. It is tempting to think that not coincidentally, magnesium/iron silicates are the most common *solid* materials in the universe (for example, consider the planet you are on). Alternative material is a slightly darkened ice (H_2O or other). The mean albedo is rather high, $\gtrsim 0.6$, and polarization of order 15% (oriented perpendicular to disk plane; Wolstencroft et al. 1995). Both the polarization and size range are similar to those of the zodiacal light disk – an extremely tenuous Vega-type disk filling the planetary region around the sun[2]. Telesco & Knacke (1991), Aitken et al. (1993), and Knacke et al. (1993) have recently found the 10 μm silicate emission feature from amorphous and perhaps partly crystalline silicates, strongly resembling the analogous feature in comet Halley's dust. The emission comes from a small amount ($\sim 10^{-11}M_\odot$) of dust heated to $T \sim 300$ K in the disk gap region.

The β Pic disk is geometrically thin, and over a range of radii (80–400 AU) has a roughly constant opening angle $\sim 10°$. The ground-breaking 12 μm sub-arcsecond resolution imaging by Lagage & Pantin (1994) has directly pinned down the size of the gap – its midpoint radius (since it may be a rather gradual transition zone) is of order 15 to 20 AU, consistent with the evaporation radius of 'dirty' water ice (but see Artymowicz 1994 about the rapid *photoevaporation* of ice). It also revealed pronounced and puzzling asymmetries between NE vs. SW disk extensions. Much smaller but also currently unexplained asymmetries

---

[2] Kuiper belt of comets outside the Neptun's orbit may provide a better analogy with Vega-type systems, but in any case our system is dust-poor in comparison.

plague the distant disk parts as well (Kalas & Jewitt 1995). It is sometimes thought but far from being proven that asymmetries are created by planets (cf. articles by Pantin & Lagage, Roques, and Artymowicz in Ferlet & Vidal-Madjar 1994). The outer limit of the system is unknown, the disk fades below the detectable level beyond $\sim 10^3$ AU from the star.

2.2   Gas Around β Pic

The rarefied gaseous component of the disk is not the primordial massive nebula but probably the result of thermal and/or collisional evaporation of solid bodies within ~ 1 AU from the star (for review see Lagrange 1995, Beust et al. 1994). Peculiar temporal variation of narrow circumstellar absorption lines of singly and multiply ionized metals shows that ~ 1 km-sized bodies, called Falling Evaporating Bodies (FEB), pass close the the star and disintegrate very much like the near-grazing comets around the sun. It is an open question whether the time-stationary gaseous subsystem is produced by FEBs alone or mainly by the dust disk. Ongoing work concentrates on the way planet or planets perturb planetesimals (i.e., undifferentiated, small, planet-building bodies; e.g., comets) onto plunging, elliptic orbits. The solar abundance ratios between elements such as Ca, Na, Fe, Mn, Zn (Lagrange et al. 1995) support the association of β Pic with our planetary system, in they characterize meteoritic and cometary matter. Many recent results on gas and dust in β Pic and other systems can be found in the materials from a 1994 topical meeting (Ferlet & Vidal-Madjar 1994).

3   Other Vega-Type Stars

3.1   Lists, Statistics and the Two Classes of Vega-Type Systems

The number of Vega-type stars and candidate systems grew rapidly in the 1980s. For instance, Aumann (1985) presented the list of 36 stars within 25 pc from the sun, from which 12 turned out to belong to the Vega class. The list of Sadakane & Nishida (1986) contains 12 stars. Backman et al. (1987) listed data for 26 most prominent systems from their sample of 134 stars with 25 μm (and mostly also 60 μm and 100 μm) excess detected by IRAS. Most (15) of the systems have color temperature from 60 to 85 K, and only one has $T > 110$ K. Walker and Wolstencroft (1988) presented their list of IRAS stars with flux ratios similar to those of the 4 prototypes. More than 20 Vega-type stars have been found, at distances > 25 pc. The statistics of the dust shell temperatures computed from the spectra clearly showed the peak at temperatures $T = 100 \pm 20$ K. Only in two cases $T \approx 600$ K; these two atypical results exclude icy composition and favor silicate composition of dust. In other cases, the broad-band spectroscopy is consistent with the possibility that the dust consists primarily of water ice, stable at $T < 110$ K. In the context of the planetary hypothesis, ice sublimation boundary could play an indirect role by defining the region of efficient planet formation. Whether the gap clearing is mechanical or thermal, it is important that the gap radii seem less similar than the typical or the maximum temperatures in the main disk parts. Thus, a purely dynamical *and* temperature-independent gap clearing seems inadequate.

Recently, Sylvester et al. (1994) have significantly improved the mid-IR coverage of 14 Vega-type stars (see also Walker & Butner 1995). In Fig. 1 we have included spectra of two objects studied by Sylvester et al. (1994) and Skinner et al. (1995): SAO 26804, probably a K2V star at a distance of $\sim$ 25 pc, and SAO 186777, a B9V star at a distance of 230 pc. The first system, like the objects from Aumann's list, lies in solar neighborhood and emits IR spectrum similar to Vega (no significant 12 μm excess). The second is a luminous B9V star located far away, yet emitting more thermal excess than β Pic not only in all IRAS bands including the 12 μm band, but also showing a definite near-IR excess between 1 and 5 μm.

Superficially, it seems that in order to subdivide the now large and varied group of Vega-type stars into more manageable subsets we should define a subclass of strictly 'Vega-type' stars without noticeable 12μm IRAS excess and another subclass of 'β Pic-type' stars with such an excess. The two SAO stars considered above would then fall into different categories. However, the strength of 12 μm excess is not necessarily an important fact. Taking into account luminosity difference of the host stars and detection limits, the disks in the categories thus defined need not be much different physically. For instance, SAO 26804 might have a small, currently undetectable, $\lesssim$ 0.1 Jy near-IR excess. To establish the essential physical similarity or dissimilarity of disks, we compare the total absorbing area of (or solid angle covered by) the dust.

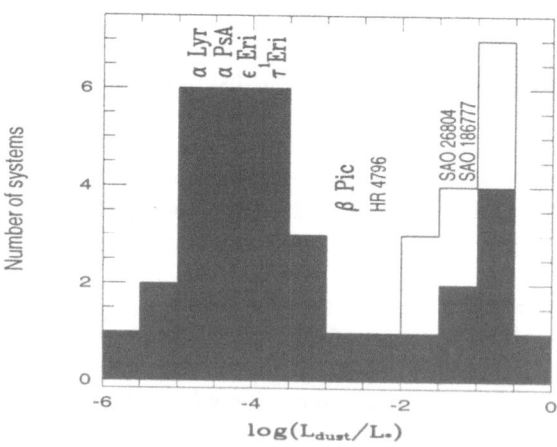

Fig. 2. The distribution of dustiness, or dust coverage, parameter $\tau_{dust} = L_{dust}/L_*$ among 34 Vega-type systems (black-shaded). Solid-line histogram includes, in addition, 7 suspected mass-losing systems not belonging to Vega class.

The 'dustiness' parameter $\tau_{dust} = L_{dust}/L_*$, is the bolometric IR excess divided by the bolometric stellar luminosity (the division cancels the pronounced effect of L_* on the normalization of the observed spectrum; τ_{dust} was denoted as τ by Backman & Gillett 1987). Conveniently, τ_{dust} also informs about the dust coverage factor of the sky seen from the star. Comparison of data published by Walker & Wolstencroft (1988), Backman & Paresce (1993), Sylvester (1994), and Skinner et al. (1995), lead to interesting conclusions regarding the dustiness of the known disks and the nature of the systems (Artymowicz et al. 1996, in preparation). In Fig. 2 we show the black-shaded histogram of the distribution of τ_{dust} among 34 Vega-excess systems, with two bins per each dex of τ_{dust}. Stellar names above the histogram indicate the bins into the systems fall. For comparison, the the upper envelope (solid line) presents the distribution of 41 stars including those found in section B of Table I of Walstencroft & Walker (1988), for which there is evidence of non-Vega-type dust formation in mass outflows, and/or non-main-sequence classification of stars (giant/supergiant).

Strangely, the values $0.001 < \tau_{dust} < 0.01$ are avoided! The only two stars in that range are β Pic and HR 4796A. The latter A0V star was studied by Jura (1991), Jura et al. (1993), and Stauffer et al. (1995), who concluded that this star exhibiting β Pic-type excess forms a wide physical pair, and probably has the same age, with an M dwarf HR 4796B.

Estimates of the age of the M dwarf based on its large lithium abundance and detailed PMS isochrone model fitting, vary between the ~ 3 Myr and 8 ± 2 Myr. If indeed HR 4796A is a main-sequence star, it must be very young (much younger than β Pic, according to Paresce 1991), and its gaseous disk may still be disappearing.

Since there is no obvious observational selection effect concerning the intrinsic dustiness of the systems of order 10^{-3} (only the cutoff at very low values) we believe that Vega-type systems are divided into distinct and non-overlapping classes: little or moderately dusty (left peak, with all the prototype systems) and extremely dusty (right peak, at $\tau_{dust} > 0.0032$). The two groups contain both nearby ($d < 20$ pc) and faraway ($d \gg 20$ pc) stars of different spectral types, with predominance of hot distant stars in the two rightmost bins. We give a brief account of the theory (outlined at the present meeting but preceding the construction of the histogram!) providing the physical reason for the dichotomy in sect. 4.4. In a nutshell, the main group, containing the prototypes, is the one in which the dusty disks are optically thin along any line of sight, and largely gas-free. The mass of the other disks must be dominated by gas, optical thickness may be large, and the systems may be very young (similar to, or identical with, T Tauri or optically thin post-T Tauri disks). We shall call the two classes gas-poor and gas-rich Vega-type systems, although so far there are no tight observational limits on gas contents, especially in disks which may be viewed far from edge-on.

3.2   Related Systems

A close connection between gas-rich Vega-type systems and post-T Tauri disks around binary stars was illustrated recently by one prominent system   SAO

179815 (HD 98800), which was known earlier as the Vega/β Pic system with one of the largest dust covering factors (Zuckerman & Becklin 1993b), but now should be regarded as a weak-line T Tau disk system, in which there are two (possibly not physically connected) pairs of stars. One of them, a single-lined spectroscopic binary with a K5V spectrum and period 262 days (Torres et al. 1995), fits within the 1–2 AU radius gap in a dusty disk predicted earlier on the basis of IR modeling. The high frequency of binary stars in the Galaxy tells us that the story may be relevant to other gas-rich systems.

Another class of related objects are the Herbig Ae/Be stars, more massive cousins of T Tau stars, predecessors of B and A-type Vega systems. Spectroscopic similarities with β Pic are discussed in Ferlet & Vidal-Madjar (1994). It is intriguing that these stars show signs of time-dependent accretion of gas similar to that of β Pic, but making a direct analogy between them and β Pic-type systems is risky. E.g., some of Ae/Be stars may be dominated by spherical envelopes rather disks and they are much younger on average.

3.3 Silicates and Very Small Grains in Vega-type Systems

Infrared spectroscopy brought evidence of the presence in disks of amorphous silicate grains: the broad and appropriately centered peak near 9.7 μm is present in β Pic, 51 Oph, SAO 179815, SAO 184124, and some other stars that, however, do not include Vega (Skinner et al. 1992, Fajardo-Acosta et al. 1993, Sylvester et al. 1995, Walker & Butner 1995).   There is evidence for very small grains: amorphous carbon, polycyclic aromatic hydrocarbons (PAHs, which are really big molecules), transiently heated grains less than 10 Å in size (surviving and radiating for a brief moment at temperatures > 1500 K following absorption of a single UV photon), and the 'usual' small submicronic grains (Sylvester et al. 1994, 1995; Walker & Butner 1995). Importantly, the spectra in which PAHs and other small grains dominate over the broad 10 μm silicate emission attributed to larger grains, belong to the gas-rich Vega-type disks from sect. 3.1 (SAO 186777 provides a good example).

4   Physics of Dust

4.1 Radiation Pressure

Dust grains with a radius b in the interval from zero (or very small) to the so-called blow-out limit b_{rad} are removed from the circumstellar disk by the net repelling force of combined gravity and radiation pressure around the stars more luminous than the sun. This limit is respected by, and often coincides with, the empirically determined typical grain sizes in the relevant Vega-type systems (Backman & Paresce 1993). For instance, $b_{rad} \approx 2$ μm for β Pic . The absence of large numbers of submicron grains offers the explanation for the gray color of the disk between B and I bands. The ratio of radiation pressure to gravity equals $\beta \approx b_{rad}/b$ for larger particles. If the disk is gas-rich then the grains with $b < b_{rad}$

are kept in the disk by the gas drag (and additional possible shadowing effect in dense disks). In this way, the presence of submicronic debris of interparticle collisions, and the very small grains discussed in the previous section is expected in spite of the sometimes large radiation pressure around Vega-type stars.

4.2   Interactions with Interstellar Dust

Do internal or external factors control the evolution of Vega-type systems? In the former case, the Vega/β Pic phenomenon is a matter of nature, rather than nurture, and the current state of the system gives insight into its past. In the latter, the dustiness of the system would depend on the galactic environment. Two current theories argue for the first case (for review see Backman & Paresce 1993). Lissauer & Griffith (1989) proposed that "sandblasting" by ISM dust (typical size 0.1 μm) during the rapid passages through atomic clouds depletes circumstellar disks; β Pic happens to have a small velocity with respect to the clouds and is thus spared the sandblasting. In another theory, Whitmire et al. (1992) proposed that nearby Vega-type stars all have recently passed through nearby molecular clouds and were sandblasted, which created the observable population of dust grains in disks from larger meteoroids. In fact, the ISM grains are not a serious threat to disks around A-type stars. They neither create nor destroy them: rather, ISM grains respectfully give way to approaching A-type stars because they are repelled by radiation pressure from several to a few dozen times stronger than they are attracted by gravity (Artymowicz 1994).

4.3   Internal Sandblasting

Close to 1 Earth mass ($10^{-6} M_\odot$) of solid particles have been directly detected in IR/mm-range near β Pic. One can show that this considerable mass will and disappear from the disk in just a few Myr. Internal collisional erosion in disks can be orders of magnitude stronger than ISM counterpart, if any. The main reason for large fluxes of dust within Vega-type disks is connected with the dust size distribution that must peak near the radiative cutoff $b = b_{rad}$, because of general steep power-law dependence above that characteristic size. Dominant grains, stable against the blow-out but not by a wide margin, find themselves on elongated orbits the moment they are born (in meteoroid collisions, or ejected from comets) because part of the gravity is canceled by the radiative force, and their initial velocity is "too large" for them to remain on nearly circular orbits. That enhances the estimated mass flux $\dot{M}$ throughout the collisional cascade of: planetesimals $\longrightarrow$ meteoroids $\longrightarrow$ typical disk dust $\longrightarrow$ sub-b_{rad} dust leaving the system. Estimates show that $\dot{M} \sim 10^{-12} M_\odot/$yr, i.e., that a whole planetary system can be crushed to dust and removed from β Pic in only 10^8 yr. This, in turn, means that β Pic : (i) cannot be too old (and have a disk all the time), (ii) will unavoidably "clean" itself substantially within the next few times 10^8 yr. Less dusty systems remove solid material less rapidly, but they may be older.

4.4 Dust Avalanches

A *gedankenexperiment* about the interaction of radiation pressure with dust around the A-type stars shows the possibility of dust 'avalanche'. It happens when a submicronic chip from the inner edge of the β Pic-type (gas-poor) disk is accelerated to the velocity of ~ 10 km/s by radiation pressure and while heading out strikes an unfortunate disk particle ($b > 1$ μm), either catastrophically shattering or at least cratering it. This creates a fresh supply of, say, N_β fresh submicron chips to be accelerated by radiation. The result is an exponentially growing avalanche, or chain reaction, that can destroy the disk if the logarithm of the final number N of outflowing grains per one inital grain, $\log N \sim N_\beta \tau_{dust}(z/r)^{-1} \gtrsim 10^2$, where τ_{dust} stands for the total solid angle occupied by dust ($\tau_{dust} = 2.4 \times 10^{-3}$ for β Pic), and z/r its opening angle in radians ($\sim 10^{-1}$ for β Pic). From the physics of grain collision it follows that $N_\beta \sim 10^3$ (a plausible but very rough estimate, since N_β depends on velocity of collision). If that is true then we see that β Pic, although not subject to immediate avalanche instability, may not be far from the theoretical *upper limit on the dustiness*: $\tau_{dust} \lesssim 10^2 N_\beta^{-1}(z/r) \sim 10^{-2}$. This estimate is consistent with the observed cutoff on τ_{dust} (Fig. 2). Are some of the asymmetries in β Pic caused by small avalanches touched off by, say, collisions between planetesimals? Notice that without the exponential proliferation of grains in an avalanche, even a giant collision of two 10^3 km radius planetary bodies would not result in observable amount of collisional debris; it would also be extremely unlikely.

5   Vega-Type Systems and λ Boo Systems: On the Way to Planetary Systems?

Vega-type systems are probably an intermediate or an end-stage in an evolutionary sequence leading from the forming star to the main-sequence star with a planetary system. They descend from the solar nebulae and their more massive analogs. The condensation and settling of dust grains toward the equatorial plane of the disk opens the way to further buildup of meteoroids and planetesimals, and leaves the gaseous part of the disk depleted in refractory elements such as Mg, Fe, Si, Ca, Na. The volatile CNO elements stay mostly in the disk and are eventually partly accreted onto the star, and partly photoevaporated, mechanically and/or magnetically flushed from the disk into the ISM. A thin, accreted, veneer of metal-poor gas can remain on the surface of an A star for an unkown, but probably considerable period of time of order 10^8 yr, because an A star has no surface convection zone promoting the mixing. It has been proposed that the class of chemically peculiar type A main-sequence stars known as λ Boötis stars be identified with a stage of planetary system formation around stars (Artymowicz 1994). They show appropriate selective metal depletion pattern. Vega is known to exhibit a weak λ Boo symptoms, and β Pic is a suspected λ Boo star, although there is also some evidence to the contrary (cf. contribution to this meeting by Gerbaldi et al., and Holweger & Rentzsch-Holm). Both possibilities regarding β Pic are understandable. Since each Vega-type star must undergo the period of λ Boo style chemical surface anomalies at some stage, perhaps there exists a large overlap of the two classes, mainly between the gas-rich Vega stars and those λ Boo stars, which exhibit the cold IR excess (due to the same elements that are missing from the atmosphere, now locked in a Vega-type disk of circumstellar grains).

Is there a direct evidence of planets in any of the Vega-type systems? Not yet, we know very little about most of them. The best-known system, β Pic, most certainly has a disk of solid bodies from dust to planetesimals, which *are* seen (the latter as FEB) as directly as possible. Indirectly, we can argue that a collection of bodies with substantial surface escape speed, of order 1000 km diameter, are needed to gravitationally stir the disk to its observed thickness. Even larger, truly planetary bodies, are needed to explain the episodic infall of planetesimals, and perhaps also the large radial span of the disk by invoking the process of Kuiper belt and (much more extended) Oort cloud formation from solid material condensing in a relatively small initial region. Finally, the inner disk gap may be created and maintained by the planet(s). But it should be kept in mind that the presence of Earth-size and larger planets may not be a necessary or even common feature of Vega-type planetesimal systems, because under some conditions (e.g., low density of the primordial disk) planetary accumulation may be halted at the level of bodies not larger than the Moon.

6   Conclusions

A decade after discovery, we still have much to learn about Vega-type systems. Hopefully the ISO satellite will open a new era in this study. Among others, ISO will improve the statistics discussed in sect. 3.1, either supporting or contradicting our claim that the two proposed categories of Vega-type systems are almost non-overlapping on the histogram of dustiness τ_{dust}. Further studies may also establish how fast they are able to clean themselves of gas and dense dust, and how to tie quantitatively their dustiness and other features to their age and the possible existence of planets. It will be interesting to see how closely systems like β Pic resemble our solar system, past or present.

Acknowledgments. I thank the organizers for inviting me to present this review, and the informative meeting they have prepared. I thank Mark Clampin for his hospitality during my visit to STScI related to β Pic, and the support by the Visitor Program at STScI. Partial support for this work has been provided by the NFR (Swedish Natural Science Research Council) research grant.

References

Aitken, D.K., Moore, T.J.T, Roche, P.F. et al. (1993): MNRAS, **265**, L41
Artymowicz, P. (1994): in *Circumstellar dust disks and planet formation*, (Editions Frontieres: Gif sur Yvette), pp. 47, 335

Artymowicz, P., Burrows, C., & Paresce F. (1989): ApJ, **337**, 494

Aumann, H.H. (1985): PASP, **97**, 885

Aumann, H.H., Gillett, F.C., Beichmann, C.A., et al. (1984): ApJ, **278**, L23

Aumann, H.H., & Good, J.C. (1990): ApJ, **350**, 408

Backman, D.E., Gillett, F.C., & Witteborn, F.C. (1992): ApJ, **385**, 670

Backman, D. E., & Paresce, F. (1993): in *Protostars and Planets III*, Eds. E. H. Levy, J. I. Lunine, & M. S. Matthews (Tuscon: Univ. Arizona Press), p. 1253

Beust, H., Vidal-Madjar, A., Ferlet, R., Lagrange-Henri, A.M. (1994): Ap&SS, **212**, 147

Chini, R., Krügel, E., Shustov, B., Tutukov, A., & Kreysa, E., (1991): A&A, **252**, 220

Fajardo-Acosta, S.B., Telesco, C.M., & Knacke, R.F. (1993): ApJ, **417**, L33

Ferlet, R., & Vidal-Madjar, A. (Eds.) (1994): *Circumstellar dust disks and planet formation*, (Editions Frontieres: Gif sur Yvette)

Jura, M. (1991): ApJ, **383**, L79

Jura, M., Zuckerman, B., Becklin, E.E., & Smith, R.C. (1993): ApJ, **418**, L37

Kalas, P., & Jewitt, D. (1995): AJ, **110**, 794

Knacke, R. F., Fajardo-Acosta, S. B., Telesco, C. M. et al. (1993): ApJ, **418**, 440

Lagage, P. O., & Pantin, E. (1994): Nature, **369**, 628

Lissauer, J.J., Griffith, C.A. (1989): ApJ, **340**, 468

Lagrange, A.-M., (1995): Ap&SS, **223**, 19

Lagrange, A.-M., Vidal-Madjar, A., Deleuil, M., Emerich, C., Beust, H., & Ferlet, R. (1995): A&A, **296**, 499

Norman, C.A., & Paresce, F. (1989): in *The Formation and Evolution of Planetary Systems*, Eds. H. Weaver & L. Danly (Cambridge: University Press), p. 151

Patten, B.M., & Willson, L.A. (1991): AJ, **102**, 323

Paresce, F. (1991): A&A, **247**, L25

Sadakane, K., Nishida, M. (1986): PASP, **98**, 685

Skinner, C.J., Barlow, M.J., & Justtanont, K. (1992): MNRAS, **255**, 31P

Smith, B.A., & Terrile, R.J. (1984): Science, **226**, 1421

Stauffer, J.R., Hartmann, L.W., & Barrado y Navascues (1995): ApJ, **454**, 910

Sylvester, R.J., Barlow, M.J., and Skinner, C.J. (1994): Ap&SS, **212**, 261

Sylvester, R.J., Barlow, M.J., and Skinner, C.J. (1995): Ap&SS, **224**, 405

Telesco, C.M., & Knacke, R.F. (1991): ApJ, **372**, L29

Torres, G., Stefanik, R.P., Latham, D.W., & Mazeh, T., (1995): ApJ, **452**, 870

Venn, K.A., & Lambert, D.L. (1990): ApJ, **363**, 234

Walker, H.J., & Wolstencroft, R.D. (1988): PASP, **100**, 1509

Walker, H.J., & Butner, H. (1995): Ap&SS, **244**, 389

Whitmire, D.P., Matese, J.J., Whitman, P.G. (1992): ApJ, **388**, 190

Wolstencroft, R.D., Scarrott, S.M., & Gledhill, T.M., (1995): Ap&SS, **224**, 395

Zuckerman, B., & Becklin, E.E. (1993a): ApJ, **414**, 793

Zuckerman, B., & Becklin, E.E. (1993b): ApJ, **406**, L25

Far-Infrared Spatial Observations of Herbig Ae/Be Stars and Low Mass Stars

H. M. Butner[1]

Dept of Terrestrial Magnetism, Carnegie Institution of Washington, Washington, DC 20015, USA

Abstract. Thermal emission from dust provides us with a detailed look at the overall temperature and density structure of a star-forming region, from the cold outer envelope to the hot inner disk. Much of our knowledge of the disk and envelope properties for young stellar objects (YSOs) comes from a comparison of the observed spectral energy distribution to the spectral energy distribution predicted from radiative transfer calculations for various YSO source models. However, as many recent studies have pointed out, spectral energy distributions alone are not sufficient to distinguish among many possible YSO source models. High resolution far-infrared spatial observations provide a critical test for these models. We will illustrate how far-infrared observations can be used to separate different YSO source models; each which can produce the same spectral energy distribution but which produce different far-infrared spatial distributions. We will focus on recent studies of Herbig Ae/Be stars and compare to what is seen for low mass stars.

1 Introduction

In the mid-1980s, the advent of IRAS as well as improved ground based photometry at both near-infrared and millimeter wavelengths allowed the first complete spectral energy distributions to be constructed for large number of young stars. The stars were found to fall into three classes:

Class I — these sources had a rising infrared excess towards far-infrared wavelengths, and they were often associated with embedded young stars.

Class II — these sources had a substantial infrared excess, but the flux was flat or falling at long wavelengths. The list included many T Tauri stars.

Class III — these sources had a small or modest infrared excess, and included many naked or weak-lined T Tauri stars.

Spectral Energy distributions (SEDs) were quickly recognized as powerful tools for testing star formation models. This was first done by Adams, Lada, and Shu (1987) for embedded young stars. They combined the Terebey, Shu, and Cassen (1984) "inside-out" collapse model with a radiative transfer code. They could predict the expected the spectral energy distributions for various source models within the Terebey, Shu, and Cassen framework. Adams, Lada, and Shu compared these model SEDs to real SEDs for some nearly low mass embedded sources. They found that for reasonable physical parameters, like the mass infall rate and stellar luminosity, of the Terebey, Shu, and Cassen picture, they were able to match the overall SED. In particular, for L1551 IRS

5, they were able to get excellent agreement between a model SED and the observed SED. Since L1551 IRS 5 is often used as a prototype of Class I low mass young stellar objects (YSOs), it was an important test of the Terebey, Shu, and Cassen picture. Following this success, Adams, Lada, and Shu (1988) matched the observed SEDs of T Tauri stars with the SEDs predicted from disk source models. Again, an excellent match for many of the SEDs was obtained within a reasonable range of disk parameters.

This work led to the realization that, as had been suggested previously by others, the spectral energy distribution provided a clue to the source's overall geometry (disk, envelope, or some combination therein). Furthermore, if embedded Class I sources have envelope dominated SEDs, matched by collapse models and Class II sources have disk dominated SEDs, matched by disk models, then it is logical to assume further that the SED also reflects what evolutionary stage the star is at in its pre-main-sequence life.

2 Spatial Observations

SEDS, however, are not produced by a single source geometry alone. Instead, the situation is more complicated with many different source models capable of reproducing the same overall SED. For example, for many of the "disk-like" SEDs seen around some T Tauri stars, envelope only models have been proposed (Kenyon et al. 1993). In other cases, the portion of an SED that in one model might originate from a disk has been explained by envelope emission instead. L1551-IRS 5's mid-infrared emission was explained by Adams, Lada, and Shu (1987) as due to disk emission. In contrast, Butner et al. (1991) were able to explain the emission as instead arising from the inner envelope by invoking a shallower dust opacity law. Kenyon et al. (1993) achieved much the same effect with a non-spherical geometry for the envelope rather than an additional disk component. The reason for the inability to associate a SED with a unique source model is that the SED is simply a sum of different emission contributions. Each physical component in the source has a characteristic temperature, leading to a contribution to the overall emission at each wavelength. By adjusting the relative strength of different components, one can produce the same total emission sum (i.e. the SED).

Spatial information provides an additional constraint on source models. For example, if one model requires a disk and the other a very extended envelope to match the SED, then spatial observations can resolve the situation. This is because the disk model predicts an unresolved source whereas the extended envelope model will produce an extended emission region. By carefully selecting the appropiate wavelength, one can separate different source models from each other. Each wavelength regime will have its strengths in terms of the size scales and temperatures that will contribute most. In particular, far-infrared observations (50 to 200 microns) probe the thermal emission of the warm inner envelope and the outer portions of the disk, a critical regime for assessing the "inside-out" collapse model.

When further combined with a SED, spatial observations set much better limits on many of the important physical source properties than either the SED or a single spatial observation will alone. The properties that can be studied include:

(1) the overall geometry,

(2) the dust opacity law,

(3) the envelope properties (inner radius, total column density, density gradient, temperature profile),

(4) disk properties (optical depth, temperature profile, and if certain assumptions are made, the disk mass).

(5) the physical nature of the central source: star, disk, or some combination of the two.

However, to achieve this flexibility, it is necessary to have high spatial resolution. IRAS had too small an aperature to allow studies of this type to be done in much detail. The Kuiper Airborne Observatory, which has a moderate aperature (0.9m) can achieve sufficient spatial resolution to be interesting, particularly in studies of the envelope density gradient on size scales of a few thousand AU.

The basic technique is to obtain high spatial scans or maps of the source, and to also measure the telescope's point source (or beam) profile. The extended emission, even for extended sources, will often only be 50% broader than the diffraction limit of the telescope. This means that one has to know both the source and beam profiles accurately if one is to pull the necessary spatial information from the profiles. Second, as inspection of any of the papers in this area will reveal, the sources are often rather round or flattened structures, not much different in appearance from the beam profile, except larger. One way to analyze such data would be to use Maximum Entropy Methods(MEM). However, since we are interested more in what we can say about underlying source models, it is often better to convolve the results of a radiative transfer model with the beam profile. The model profiles can then be compared directly with the observed profiles. Since this can be done at all the wavelengths for which one has data, the technique is more robust for assessing physical properties than MEM alone would be.

Among the earliest work to use SED and spatial data was the studies of Ladd et al. (1991). They studied a number of nearby embedded low mass stars, and compared their observations with the predictions of the Terebey, Cassen, and Shu models. Using the Yerkes Far-Infrared Camera, they mapped out the cores (and the equivalent point source profiles) at several far-infrared wavelengths (60, 100, and 160 microns). Their results suggested good agreement with the star formation theories, with density gradients $\alpha = 1.5$ matching the observations. However, at that time, they did not achieve the diffraction limit of the Kuiper Airborne Observatory. In addition, they primarily did a consistency check of the Terebey, Shu, and Cassen theory with the data. They did not explore how tightly their observations actually constrained the density gradient to be $\alpha = 1.5$ as required by the "inside-out" collapse model as opposed to values of the density gradient of $\alpha = 1$ or 2.

To take full advantage of spatial observations required diffraction limited observations where possible combined with models that explored a wide range of parameter space. Some of the first work of this type was done using the University of Texas Scanning Photometry and the KAO on two sources L1551 IRS 5 (Butner et al. 1991) and NGC 2071 (Butner et al. 1990). L1551 IRS 5, as one of the original test sources for the Terebey, Shu, and Cassen picture, was found to be well matched by a density gradient $\alpha = 1.5$, consistent with the theoretical predictions. In addition, the observations placed limits on the density gradient as lying with a range of 0.3 of the expected value, an excellent match to the model.

Several concerns with the standard model of Adams, Lada, and Shu (1987) arose, however, in a detailed examination of the data. First, to achieve a good match at far-infrared wavelengths, the total envelope column density had to be increased. This caused the mid-infrared flux to drop dramatically due to the much higher optical depths of the envelope. Unless the dust opacity or the geometry were altered, no emission from the central region of the source could be seen. However, if one altered the dust opacity at far-infrared wavelengths a modest amount, then the observed mid-infrared flux could be recovered. More detailed studies by Butner et al. (1994) found that no disk was required to explain the mid-infrared emission. In fact, for the 1-D case considered by Butner et al., the disk was invisible unless the dust optical properties had a much larger far-infrared to mid-infrared extinction ratio than assumed by Draine and Lee type dust models. It turned out that the millimeter wavelength regime was the best place to look for evidence of disk emission. If one had a large ratio of flux within an interferometric observation relative to a single dish, a disk was likely (Butner et al. 1991, 1994). NGC 2071, a more massive embedded YSO, provided a nice confirmation of the $\alpha = 1.5$ Terebey, Shu, and Cassen prediction. Here again there were surprises. To match the spatial extent at 50 microns, a rather large cavity was necessary inside the envelope—even if the multiple sources in the beam were taken into account.

3   Herbig Ae/Be Stars

Unfortunately, most low mass stars are not bright enough that they would be expected to produce extended far-infrared emission at 100 microns. In Taurus, only L1551 IRS 5 was expected to be resolvable, and it was only just doable. Thus, a more luminous class of objects was needed. The Herbig Ae/Be stars (henceforth HAeBe stars), which are moderate massive YSOs between 2 and 8 solar masses, were ideal candidates for spatial studies. There were a number of nearby objects identified. Hillenbrand et al. (1992) searched the literature for available photometry, supplemented by additional observations. They provided a complete SED (including from U to Q band, IRAS data, and 1.3 millimeter observations) for more than 50 HAeBe stars. Hillenbrand et al. had identified 3 groups analogous to the traditional Class I, Class II, and Class III classification scheme for low mass YSOs (called Group II, I, and III respectively). They modeled the Class II sources as disk-only systems, similar to the T Tauri stars. Since then, this interpretation has been challenged. Hartmann et al. (1993) explored a large number of different source models and found envelopes might work as well under the correct conditions. These sources are, thus, a natural choice for far-infrared studies.

Consider the case of LkHα 198, a nearby A star placed in the "Class I-like" HAeBe stars by Hillenbrand et al. Natta et al. (1992) found that the SED could be modeled very nicely by a disk, from 1 micron out to 1 millimeter. Far-infrared observations, however, revealed a very extended emission region. This emission was much too extended to come from any sort of traditional disk model. Envelope-only models could, however, easily explain the extended far-infrared emission as arising from a very shallow density gradient envelope. Unfortunately, an envelope-only model failed to match the observed SED, with a mis-match in the mid-infrared. Unlike L1551 IRS 5, the optical depths of the envelopes were not high enough and the emission too extended to permit modifications of dust opacity to solve the problem. Instead, a second very warm component was needed. A disk could fit this role quite naturally. Thus, the basic parameters of the Natta et al. LkHα 198 models included a very flat density gradient for the envelope ($\alpha = 0.5$), a relatively large inner cavity (3 arcseconds or 0.01 pc), and a very modest total envelope column density ($A_V = 4.5$). The central source had to include a disk. The disk had a rather flat temperature profile ($T(r) \propto r^{-0.5}$), similar to that seen around many T Tauri stars (Beckwith et al. 1990). The disk mass was between 0.01 and 0.27 solar masses. Thus, Natta et al. found that only a combined disk+envelope model matched both the available SED and spatial data.

Natta et al. (1993) extended the spatial studies to include a larger sample of Class I-like HAeBE stars. In all cases, the emission was at 100 microns was extended. Detailed models of the sources had a similar result as was found for LkHα 198. Two components, a disk (for the mid-infrared) and an envelope (for the far-infrared) were required to explain both the spatial extent of the far-infrared emission and the overall SED. However, among the various HAeBe stars, a variety of density gradients were found. Some, like LkHα 198, R Mon, and CD-42 11721, had very shallow density gradients. Others, like R Cr A, and V645 Cyg, had the $\alpha = 1.5$ to 2.0 density gradients expected from theory. However, the envelope's optical depths were so low, that it is likely that the stars were in the final stages of dispersing their envelopes. This seemed confirmed by the additional result that the source models required large inner cavities (between 0.01 and 0.1 pc in size) to match the 50 and 100 microns observations.

In a related study, Di Francesco et al. (1994) observed a sample of Class II-like HAeBe stars (the Group I or disk-like stars from the Hillenbrand et al. paper). These stars had been identified as possible examples of pure disk systems. If that was correct, and the far-infrared emission arose from the disk also, it was expected that the stars would be unresolved. In 5 out of 6 cases, the 100 microns emission was found to be very extended. Only in the case of AB Aur was the emission consistent with arising from a disk (or small envelope). From this initial study of Di Francesco, it appears that envelopes, not disks, are responsible for the far-infrared emission. Given the very extended emission seen around most of them, the density gradients will likely to turn out to be quite shallow. Again this is consistent with an interpretation that the HAeBe stars have finished the bulk of their accretion and are in the process of dispersing their parent envelopes.

4   Visible Versus Embedded Sources

One of the interesting issues raised by the far-infrared studies of the past few years is the relative role of visible and embedded sources. In the case of the HAeBE stars, it is possible that the far-infrared emission is dominated not by the visible star but instead by invisible Class 0/Class I companions. If such companions exist, they could be quite luminous in the far-infrared. Consider the case of LkHα 198. Lagage et al. (1993) discovered a mid-infrared source 6 arcseconds away from the optical star LkHα 198. It was very faint at 2 microns, and had evidence for a deep silicate absorption at 10 microns. They suggested that it was possible that the IR source rather than the known optical HAeBe star LkHα 198 was responsible for the very extended emission reported by Natta et al. (1992, 1993). In Natta et al., the 100 microns emission was very extended and thus, the modeling process led them to the conclusion that LkHα 198 was surrounded by a very diffuse, flat density gradient envelope. The Lagage et al. suggestion, if true, could dramatically alter the interpretation of the far-infrared emission around LkHα 198 as well as other HAeBe stars.

To investigate that problem further, Butner and Natta (1995) decided to see what sort of source models would be consistent with the available data. They investigated three additional source models.

(1) a deeply embedded star—this case was found to produce very steep far-infrared profiles. This was caused by the simultaneous requirement that the mid-infrared region emit strongly (hence emission arose from close to the star) and that the total optical depth be large (due to the large silicate absorption feature seen). This type of model was ruled incapable of producing the type of extended far-infrared emission observed.

(2) a deeply embedded star+disk—this case was found to not be ruled out by the available mid-infrared and far-infrared data. The disk can produce the strong mid-infrared emission, and then the envelope can have its inner edge anywhere. These type of models have a large inner cavity (to produce the required 100 microns size) and should be easily visible at millimeter wavelengths as a ring-like object centered on the IR source. Butner and Natta argued that the lack of 800 microns emission centered on the IR source tended to argue against this interpretation.

(3) Multiple sources in the beam—here Butner and Natta considered the role of both the optical star and the IR source, each with equal luminosities. If the IR source is sitting in the center of a dense cloud, and the optical star is at the edge, so that it effectively does not dominate the far-infrared emission, then the situation effectively reverts to (2) above. If the IR star is at the center of a dense

cloud, but the optical star is in the center of a diffuse region, then the optical star can contribute substantially to the total IR emission. The profile KAO would see on average would lie between the density gradients of the dense and diffuse regions. I.e. if the IR source contributed to the total 100 microns emission, then the emission due to the optical star must have even a flatter density profile than the single star models alone had in order to match the observed profiles.

These results led to the following general statements that

(1) If embedded sources are to produce large far-infrared emission regions, the envelopes need to have large inner holes.

(2) An optically visible source surrounded by a flat density gradient envelope can dominate the observed size of the overall far-infrared emission region, provided it is responsible for a sizeable fraction of the total IR emission.

5   Comparison with Low Mass Stars

While prospects to do 100 microns high resolution work on nearby low stars is extremely limited, by moving out to 200 microns, observers can achieve much greater flexibility. Using the Kuiper Airborne Observatory and the Yerkes 60-element Far-Infrared Array Camera, a number of low mass stars in Taurus have been examined. Initial results (Butner et al. 1994, 1995) suggest that a range of density gradients may exist. L1551 IRS 5 and L1489 IR have the expected $\alpha = 1.5$. Other sources, L1527 and L1551 NE (both close to Class 0), appear to have much flatter density gradients ($\alpha = 0.5$). If verified by additional modeling, then these profiles pose an interesting challenge to the traditional Terebey, Shu, and Cassen picture. Class 0 sources are thought to be extremely young. Hence one expects to see either $\alpha = 1.5$ or $\alpha = 2.0$. A flatter density gradient is difficult to reconcile with theory.

6   Summary

Far-infrared spatial studies have yielded considerable information on the density gradients around young stars. Both HAeBe stars and low mass stars tend to show a wide range of density gradients, not just the simple $\alpha = 1.5$ expected from theory. In addition, HAeBe stars are surrounded by envelopes with large inner cavities, and the total envelope optical depths are a few A_V. In the cases studied to date, the emission appears to be dominated by the optical stars rather than deeply embedded companions. These techiques can be extended to other wavelengths provided one has adequate sensitivity and spatial resolution.

In addition to the studies mentioned, additional HAeBe and low mass stars have been observed. These samples should enlarge the total data base to 20 HAeBe stars and 10 or so low mass stars. We also anticipate that when SOFIA, which is a joint US-German collaboration to replace the KAO with a 2.5 meter IR telescope on a 747, comes on line in 5 years, it will greatly increase our spatial resolution and our sample size at far-infrared wavelengths.

References

Adams, F. C., Lada, C. J. and Shu, F. H. (1987): ApJ **312**, 788

Adams, F. C., Lada, C. J. and Shu, F. H. (1988): ApJ **326**, 865

Beckwith, S. V. Sargent, A. I., Chini, R. S., and Güsten, R. (1990): AJ, **99**, 924

Butner, H. M., Evans, N. J. II, Harvey, P. M., Mundy, L. G., Natta, A., and Randich, M. S. (1990): ApJ **364**, 172

Butner, H. M., Evans, N. J. II, Lester, D. F., Levreault, R. G., and Strom, S. E. (1990): ApJ **376**, 164

Butner, H. M., Moriarty-Schieven, G. H., Ressler, M. E., and Werner, M. W., (1994): **Airborne Astronomy Symposium on the Galactic Ecosystem: From Gas to Stars to Dust** (ASP, San Francisco), 235-238.

Butner, H. M., Moriarty-Schieven, G. H., Ressler, M. E., and Werner, M. W., (1995): **Circumstellar Matter 1994** (Kluwer: Dordrecht), 77-81.

Butner, H. M., and Natta, A., (1995): ApJ **440**, 874

Butner, H. M., Natta, A., and Evans, N. J. II, (1994): ApJ **420**, 326

Di Francesco, J., Evans, N. J. II, Harvey, P. M., Mundy, L. G., and Butner, H. M. (1994): ApJ **432**, 710

Hartmann, L., Kenyon, S. J., and Calvet, N. (1993): ApJ **407**, 219

Hillenbrand, L. A., Strom, S. E, Vrba, F. J. and Keene, J. (1992): ApJ **397**, 613

Kenyon, S. J., Calvet, N., and Hartmann, L., (1993): ApJ **414**, 773

Ladd, E. F., Adams, F. C., Casey, S., Davidson, J. A. Fuller, G. A., Harper, D. A., Myers, P. C., and Padman, R. (1991): ApJ **366** 203

Lagage, P. O., Olofsson, G., Cabrit, S., Cesarsky, C. J., Nordh, L., and Rodriguez Espinosa, J. M. (1993): ApJ **417**, L79

Natta, A., Palla, F., Butner, H. M., Evans, N. J. II, and Harvey, P. M. (1992) ApJ **391**, 805

Natta, A., Palla, F., Butner, H. M., Evans, N. J. II, and Harvey, P. M. (1993) ApJ **406**, 674

Terebey, S., Shu, F. H., and Cassen, P. (1984): ApJ **286**, 529

λ Boo Stars in the Orion OB1 Association

Michèle Gerbaldi[1] and Rosanna Faraggiana[2]

[1] Institut d'Astrophysique (CNRS), 98bis, Bd Arago, F-75014 Paris, France and Université de Paris Sud XI, France
[2] Dipartimento di Astronomia, via G.B. Tiepolo 11, I-34131 Trieste, Italy

Abstract. The origin of the photospheric underabundances detected in the Pop I A-type stars, the λ Boo stars, is interpreted as the signature of the last evolution of a star before to reach the main sequence. In order to test this hypothesis, UV observations of λ Boo candidates in the young Orion OB1 Association have been made.

1 Introduction

Among the early type stars the small group of λ Boo stars are particularly interesting because of the pattern of their abundances have been recently interpreted as representative of the last phases of the evolution towards the main-sequence of stars having mass around 2 solar masses.

The proofs for the youth of λ Boo stars are the following:

a) They are Pop I stars.

b) The atmospheric abundance pattern mimics that of the ISM (Venn and Lambert (1990)). It has been interpreted as the result of the last phases of accretion of material still surrounding the star; the accreted gas is depleted in heavy elements which remain locked into dust grains. The radiation pressure keeps the grains away from the star (Waters et al. (1992)). The CNO abundances are not affected, due to the low condensation temperatures of these elements which therefore remain in the gas phase.

c) Spectral indications for the presence of gas and/or dust in an envelope have been found for some λ Boo stars. The dust component has been detected by IRAS around 3 of these stars. The non-detection of IR excess for the other λ Boo stars could be simply due to the low IRAS sensitivity limit. Circumstellar absorption in the CaII line has been found in five of the eleven λ Boo stars studied by Holweger and Rentzsch-Holm (1995).

d) A search of such stars in two young Associations (Orion OB1 and Lacerta OB1) have been carried out by Gray and Corbally (1993) and Levato and Malaroda (1995). They detected 4 new λ Boo candidates. This result together with the fact that no λ Boo stars have ever been found in an old cluster supports once more the arguments mentioned above.

The specific abundance pattern which characterize these stars are the result of a selective accretion, according to Waters et al. (1992) interpretation: the mechanism of gas, but not dust accretion proposed by these authors should take place during the late phases of the pre-main sequence evolution and should stop when the star, approaching the ZAMS, is surrounded by a disc containing too low an amount of material.

The accretion on the surface of a λ Boo star occurs just before the star arrives on the ZAMS, and the very low number of known λ Boo stars may indicate that mixing due to meridional circulation works rapidly and efficiently and that the λ Boo episode of stellar life is quite short, these peculiarities being quickly destroyed. Theoretical investigations of an accretion/diffusion model by Charbonneau (1993) and by Turcotte and Charbonneau (1993) give a strong support to this scenario.

How to test this scenario: by analyzing the properties of the λ Boo stars according to their position on the HR diagram, compared to tracks of evolutionary models. With the above hypotheses, the highest anomalies must be found in the objects nearest the ZAMS. The λ Boo stars so far analyzed in detail were field stars; those for which the bolometric luminosity L could be measured, have been placed on the HR diagram together with tracks for pre-main sequence evolution (Gerbaldi et al. (1993)). Comparing the position of 6 λ Boo stars relatively to pre-main sequence tracks and their abundance deficiencies, it follows that the highest peculiarities are observed in stars nearest the ZAMS, i.e. at the end of their pre-main sequence phase. Such conclusion is in favor of the above scenario.

But a first critic to this sample, is that it is very small; the second one is that we have no information on the age of the stars. So, dedicated observations of λ Boo stars, members of young clusters, are needed, in order to settle down this conclusion on a sample of stars homogeneous in age, except the spread due to the star formation mechanism in the cluster. We present here the observations we did with the satellite IUE in the ultraviolet domain of λ Boo candidates in the Association Orion OB1.

2   The Observations

Why do we need ultraviolet observations ?

The character "λ Boo" has been attributed on the basis of low resolution spectroscopic observations in the visible range only, and the list of stars classified "λ Boo" can vary from one author to another. We have shown, Faraggiana et al. (1989, 1990), that the λ Boo stars present UV criteria at low resolution, which proved to be very efficient in selecting stars which share with λ Boo itself more than a weak Mg 4481 line which may produce a mis-classification. The UV observations allow also a measure of the "intensity" of the "λ Boo phenomena" by the value of the ratio CI1657/AlII1670, the C having a "normal" abundance and the Al being depleted compared to the solar abundances.

We have observed the very few number of early A-type stars members of Orion Associations which have been classified λ Boo by Gray and Corbally (1993) or Levato and Malaroda (1995) or for which an indication of a slight metal deficiency had been found.

The stars we have observed are listed in the Table below. From the value of their photometric indices in the Strömgren system, we computed E(b-y), T_{eff} and log g using the Moon and Dworetsky codes (1985).

Table 1. Parameters of the observed λ Boo stars

HD	E(b-y)	T_{eff}	log g
36203	0.014	8640	4.35
37078	-0.004	8820	4.21
290799	0.004	8000	4.26
294253	0.045	10540	4.51

From the Wu et al. (1991) Atlas (as well as the list of standard stars of Gray and Garrison (1989) we selected normal stars also observed by IUE and having a similar T_{eff} and log g.

Table 2. Parameters of the comparison stars

HD	E(b-y)	T_{eff}	log g
23643	0.013	8250	4.12
97603	0.005	8300	3.89
102647	[-0.026]	8660	4.34
216956	[-0.026]	8790	4.34
222661	0.003	10510	4.29

3 Results

The star HD 290799 has been classified as a λ Boo star by Gray and Corbally (1993) and Levato and Malaroda (1995). The comparison stars are HD 23643 and HD 97603. In HD 290799 the ratio CI1657/AlII1670 is clearly greater than one. Moreover on the long side of λ1600 a strong absorption feature is clearly detectable, the spectrum at shorter wavelength being under-exposed. So this star is a "classical λ Boo star".

The star HD 36203 has been classified as a "metal weak star" by Gray and Corbally but is not quoted in the Levato and Malaroda paper. The comparison star is HD 102647. It is clear than in HD 36203, the ratio CI1657/AlII1670 is grater than in the comparison star, which indicates a slight metal deficiency, but nothing more.

The star HD 37078 has been classified as a "metal weak star" by Gray and Corbally but is not quoted in the Levato and Malaroda paper. The comparison star is HD 216956. In the spectrum of HD 37078 we notice that the ratio

CI1657/AlII1670 is similar to that of the comparison star, but in general all the features are much weaker than in the comparison spectrum. A definite deficiency in Si can be noticed because the λ1400 bound-free absorption feature is drastically reduce in HD 37078. So, in this target at least one element is clearly deficient : the Si, but the metal deficiency in HD 37078 is different from that observed in HD 36203.

The star HD 294253 has been classified as a λ Boo by Levato and Malaroda (1995), but is not in the list of Gray and Corbally (1993). The comparison star is HD 222661. In the spectrum of HD 294253 all the features are weaker than in the comparison star which indicates a metal deficiency; the AlII1670 blend is absent but the CI1657 one has the same intensity as in the comparison star. The flux distribution indicates a weak metallicity; the λ1600 depression is not observed, but it is not expected at that T_{eff} (Allard and Koester (1992); Holweger et al. (1994). This star is the hottest metal weak star so far detected among this group of objects.

4   Conclusion

Classical λ Boo stars have been detected in the Association Orion OB1 as well as metal weak stars. No counterpart for these last ones has been detected among the field stars. We are pursuing our observations with IUE and we intend to make a fine analysis of the most interesting objects based on high resolution optical spectra.

References

Allard, N.F., Koester, D. (1992) Astron. Astrophys. **258** 464

Charbonneau, P.:1993, Ap.J. **405** 720

Faraggiana, R., Gerbaldi, M. (1990) ESA SP-310, 287

Faraggiana, R., Gerbaldi, M., Boehm, C. (1989) Astron. Astropys. **209** 233

Gerbaldi, M., Zorec, J., Castelli, F., Faraggiana, R. (1993) IAU Coll. 138, ASP Conf. Series No. 44, p. 413

Gray R.O., Corbally C.J. (1993) A.J. **106**, 632

Gray, R.O., Garrison, R.F. (1989)AjJS **70**, 623

Holweger, H., Koester, D., Allard, N.F. (1994) Astron. Astropys. **290** L21

Holweger, H., Rentzsch-Holm, I. (1995) Astron. Astropys., in press

Levato, H., Malaroda S. (1995) The MK Process at Fifty Years, eds. C.J. Corbally, R.O. Gray, R.F. Garrison ASP Conf. Series No. 60, p. 93

Moon, T.T., Dworetsky, M.M. (1985) Mon. Not. R.A.S. **217**, 305

Turcotte, S., Charbonneau P. (1993) Ap.J. **413**, 376

Venn, K., Lambert, D. (1990): Ap. J. **363**, 234

Waters, L.B.F.M., Trams, N.R., Waelkaens, C. (1992) Astron. Astrophys. **262**, L37

Wu, C., Crenshaw, D.M., Blackwell,Jr., J.H., Wilson-Diaz, D., Shiffer,III, F.H., Burstein, D., Fanelli, M.N., O'Connell,R.W. (1991) IUE NASA Newsletter No. 41

Comets as a Source of the Dust in the β Pictoris Disk

J. Mayo Greenberg and Aigen Li

Laboratory Astrophysics, University of Leiden, Postbus 9504, 2300 RA Leiden, The Netherlands
e-mail: mayo@rulhl1.LeidenUniv.nl; agli@strw.LeidenUniv.nl

Abstract. The infrared continuum and the 10 μm spectral energy distribution of the dust in the disk of β Pictoris may be derived by assuming that the dust is continually replenished by comets orbiting close to the star. The basic, initial dust shed by the comets is taken to be the fluffy aggregates of interstellar silicate core-organic refractory mantle dust grains whose size distribution is like that observed for Comet Halley. The primary heating provided by the organic refractory mantle absorption of the stellar radiation, is sufficient for some of the particles to crystallize the initially amorphous silicates. The dust grains are then distributed in the disk by radiation pressure. In steady state the disk then consists of a mixture of crystalline silicate aggregates and aggregates of amorphous silicate core-organic refractory mantle particles with variable ratios of organic refractory to silicate mass. The temperature distribution of a radial distribution of such particles provides an excellent match to the silicate 10 μm (plus 11.2 μm) spectral emission as well as the excess continuum flux from the disk.

1 Introduction

Since IRAS detected the so-called "Vega Phenomenon"; i.e., that many main sequence stars have large infrared (IR) excesses over the black body emission of their photosphere (Aumann et al, 1984; Aumann 1985; Gillet 1986; Walker & Wolstencroft 1988), much attention has been paid to the characteristics of circumstellar dust orbiting "Vega-type stars" (Becklin & Zuckerman 1990; Backman & Paresce 1993; etc.). It is commonly believed that this dust is heated to about 100 K and that its thermal emssion contributes to the observed IR excess. Among Vega-type stars, β Pictoris (an A5 star, HD 39060, HR 2020) is the most extensively studied, not only because it has the largest IR excess, but also because its resolved nearly edge-on protoplanetary disk structure (Smith & Terrille 1984; Lagage & Pantin 1994) shows evidence of planet formation and provides clues for us to understand the formation process of our solar system.

Based on imaging (e.g., Paresce & Burrows 1987; Golimowski et al 1993; Lecavelier des Etangs et al 1993; etc) and infrared photometry observations (e.g., Telesco et al 1988; Chini et al 1991; Backman et al 1992; etc), many theoretical studies have been carried out (Artymowicz et al 1989; Backman et al 1992; Chini et al 1991; Nakano 1991; etc). All these models contain many uncertainties and differ in assumptions regarding the nature of the dust, such as composition, size, number density distribution, temperature, etc. For example, thermal modelling by Backman et al (1992) required particles as small as 1 μm, however, much larger particles are required for optical light scattering data (Paresce & Burrows 1987), thermal modelling (Chini et al 1991) and orbital stability against radiation pressure and Poynting-Robertson drag (Aitken et al 1993).

Spectroscopic data are undoubtedly powerful means to study the disk structure and dust properties. Recently, the sucessessful discovery of the silicate feature by Telesco & Knacke (1991) and the subsequent confirmation by Knacke et al (1993) provided valuable information needed to help understand better the dust composition, size, and distribution. It provides information about the connections among interstellar dust, protoplanetary/planetary disk dust and comet dust. In particular, the fact that the observed crystalline silicate feature at 11.2 μm is similar to that of comet Halley and comet Levy supports the assumption that comets may be sources of dust particles in the disk. However, in their detailed modelling, Knacke et al (1993) failed to reproduce the silicate feature basing on the Draine-Lee (Draine & Lee, 1984) silicate particles. This led them to suggest that "the silicate mineral composition is different from that incorporated in the Draine & Lee constants". Their attempts to improve the fits by other silicate mineral compositions were also unsuccessful.

In all the above models, the dust has been taken to consist of compact particles. In view of the similarity of the emission spectrum to that of comets and the fact that comet spectral emissivity is well matched by fluffy aggregates (Greenberg & Hage 1990) it is natural to expect that the dust in the β Pictoris disk is also fluffy. The fluffiness could occur because of aggregation of primary interstellar dust particles in the first stage of star formation. This is as suggested by Krügel & Siebenmorgen (1994) (also see: Greenberg 1985; Siebenmorgen 1993; Henning & Stognienko 1995), but the problem with β Pictoris is that one requires a substantial fraction of the silicates to emit at 11.2 μm which corresponds to well heated — to $\sim 1000K$ — dust and, since this occurs only rather close to the star it would be an unstable situation because the particles would then be modified so as to be ejected by radiation pressure. An alternative to this is that comets have already been abundantly formed in the disk of β Pictoris and some of these comets have orbits which bring them close to the star where they shed dust replenishing that which has already been blown out. These ideas lead us to believe that the dust grains in β Pictoris disk are certainly different from Draine & Lee silicates and are probably fluffy aggregates very similar to comet dust (Greenberg 1982a; Greenberg 1989).

Assuming comet dust replenishment, we present a dust disk model which reproduces the 10 μm silicate emission along with the continuum very well.

2   Dust Replenishment Source

Backman et al (1992), considering removal and destructive processes of dust grains in the β Pictoris disk, found that all particles would be lost in $\leq 10^8$ years which is shorter than the lifetime of β Pictoris which is about $\leq 2 \times 10^8$ years (Paresce 1991). They conclude that there must be continuous replenishment of the disk particles. As a source comet-like bodies are proposed by Telesco et al (1988), Telesco & Knacke (1991) and Knacke et al (1993). Similar suggestions have been conjectured for other systems (Weissman 1984; Harper et al, 1984; Matese et al, 1987; Beust et al, 1990). Obviously this would imply a strong connection between interstellar dust, comet dust and protoplanetary disk dust. On the basis of Greenberg's cyclic core-mantle interstellar dust theory (Greenberg 1978; Hong & Greenberg, 1980; Chlewicki & Greenberg, 1990), Greenberg developed a comet model in which the basic idea is that the proto-solar nebula interstellar dust particles were completely preserved during aggregation into comets (Greenberg 1982a; 1989). In contrast to the separate silicate and graphite dust model (Mathis et al, 1977; Draine & Lee, 1984), Greenberg's core-mantle dust in diffuse clouds consists of silicate cores mantled by organic refractory material resulting from ultraviolet photoprocessing of ice accreted on the small silicate particles. The silicate particles produced in the atmosphere and blown out of cool evolved stars (Schutte et al 1989) provide the core for the mantle ice accretion and processing. In protostar molecular clouds additional outer ice mantles dominated by water are accreted on the silicate core-organic refractory mantle dust (Greenberg 1982b). When the protostar dust grains aggregate into comets, the dust properties are almost completely preserved and one obtains a very fluffy porous structure (Greenberg & Gustafson 1981; Greenberg 1982a; Greenberg 1989). This high prorosity (as high as 0.975) for comet dust, whose volatiles have evaporated, can explain the observations of comet Halley very well (Greenberg & Hage, 1990; Greenberg & Li, 1995a).

When the comets which have formed in the same parent cloud as β Pictoris come close to the star in their orbits around it, some comet dust particles would be sputtered out and stay in the disk. Exposed to the strong β Pictoris radiation, the outer ice mantles would evaporate, leaving the disk dust grains as fluffy silicate core-organic refractory mantle aggregates. In the inner region of the disk, the dust temperatures are so high that some of the organic refractory mantle also evaporates and some silicates are even crystallized. The strong radiation pressure blows the small particles out, so that the crystalline silicate particles become distributed all over the entire disk. On the other hand, because of the strong rigidity of such aggregates (Greenberg & Gustafson, 1981), the fluffy structure can survive.

3   Dust Properties

As to the theoretical modelling of β Pictoris's spectral energy distribution, the major dust properties are dust composition, optical constants, dust size distribution and the radial number density distribution.

3.1   Dust composition and optical constants

Knacke et al (1993) have pointed out that the dust grains in the disk must consist of a mixture of minerals rather than a single one. As discussed in section 2, there are two main kinds of dust grains in the β Pictoris disk: (1) porous aggregates of amorphous silicate core- organic refractory mantle interstellar dust; (2) porous aggregates of crystalline silicates grains.

Following Greenberg & Li (1995b), we use the optical constants of amorphous olivine $MgFeSiO_4$ obtained in the laboratory by the Jena group (Dorschner et al, 1995) which was considered to be most representative of the true astronomical silicates. The optical constants of organic refractories were derived from a combination of astronomical (Murchison meteorite) and laboratory spectra of residues of ultraviolet photoprocessed ices (Greenberg & Li 1995b; Greenberg et al 1995; Schutte 1988). The optical constants of crystalline silicate are obtained from Mukai and Koike (1990).

3.2 Dust size distribution

There is controversy about the dust size in the earlier works on β Pictoris as mentioned in Section 1. Telesco & Knacke (1991) pointed out that the particles must be be less than 10 μm to give the silicate emission. In fact, because of the emission/absorption properties at 10 μm the particles really have to be significantly smaller than 10 μm to show an emission feature (Greenberg & Hage 1990). However, in order to fit the mm-observation, Chini et al (1991) argued that the typical size of dust grain radii was between 10 μm and at least 1 mm. Aitken et al (1993) reported that 2 μm grains could reproduce the silicate emssion as observed by themselves, using Draine & Lee silicates. The optical imaging by Paresce & Burrows (1987) implied that the starlight scattering particles were much larger than 1 μm. Modelling both the scattered starlight and the far IR emssion (Artymowicz et al 1989) indicated that the particles were approximately between $1\mu m$ and 20 μm. Furthermore, in another paper, Artymowicz (1988) showed that particles between 10^{-2} and 1 μm would be blown out by radiation pressure leaving only the large particles as a stable population. Inconsistent with this was Backman et al's (1992) thermal modelling which showed that the particles reponsible for the IR emission were near 1 μm. One should keep in mind that all the size distributions mentioned above were model dependent and dust mineral dependent. So it is not surprising to obtain such widely ranging different values.

We note that each of the above discussions was focussed on different aspects of the emission by the β Pictoris dust particles; either continuum or spectral but not both simultaneously. The problem is to reconcile the two pictures. Can one, at the same time, have small (sub micron or micron size) particles and large (up to mm size) particles in the same framework? A possible answer to this lies in the properties of very porous aggregates of sub micron size interstellar dust where the smaller aggregates have many of the absorption/emission properties of their component particles while the large aggregates act, indeed, like large particles.

As discussed in section 2, if comets are the dust sources of the β Pictoris disk, we will assume as a first approximation that the disk dust has a dust size distribution similar to that of comet Halley which was measured by the spacecrafts Vega 1, Vega 2 and Giotto approaching close to comet Halley (see Figure 3(a) in Greenberg & Hage, 1990). Although we can not say that this is exactly the case in the β Pictoris disk, it is a reasonable starting point. Note that very fluffy interstellar dust aggregates can resemble either their basic units (sub micron size interstellar dust) or their aggregate properties depending on the wavelength observation.

3.3   Dust radial number density distribution

Just as for the size distribution, previous models yield widely different dust volume number density distributions, partly due to the poor quality of observational data. With the availability of relatively high resolution direct imaging data of optical light scattering (Paresce & Burrows 1987 and Artymowicz et al 1989), and in particular with the availability of unprecedented infared imaging of the inner dust disk (Lagage & Pantin 1994), we are able to investigate the volume number density distribution more accurately and place stronger constraints on it.

In this study we use a power law density distribution, $n(r) \propto r^{\gamma-1}$, where r is the distance from the central star. In the outer region of the disk plane $(a > 100 AU)$, we adopt $\gamma \approx -1.7$, namely, $n(r) \propto r^{-2.7}$ derived from optical light scattering data (Artymowicz et al 1989; Backman & Paresce 1993). In the inner region, we keep γ as a free parameter to be controlled by the observed disk surface brightness distribution.

4   Spectral Energy Distribution

In order to model the spectral energy distribution, the knowledge of the dust temperature of each size at each location $T_d(r, a)$ is required, where r is the distance from the dust to the cental star and a is the dust size. We calculate the dust temperatures on the basis of the dust energy balance,

$$\omega \int_0^\infty \pi a^2 Q_{abs}(a, \lambda) B(T_*, \lambda) d\lambda = \int_0^\infty \pi a^2 Q_{abs}(a, \lambda) B(T_d, \lambda) d\lambda \qquad (1)$$

where ω is the dilution factor, $\approx (\frac{R_*}{2r})^2$; T_* ($\approx 8200 K$) and R_* ($\approx 6.23 \times 10^{-3} AU$) are the temperature and radius of the central star, respectively; $B(T, \lambda)$ is the Planck function; $Q_{abs}(a, \lambda)$ is the absorption efficiency. Note that the stellar flux is approximated by a blackbody at temperature T_*.

The absorption efficiency $Q_{abs}(a, \lambda)$ is obtained from Mie theory, assuming both the fluffy aggregates and the individual particles in the aggregates are spherical. The Maxwell-Garnett effective medium theory (Maxwell-Garnett 1904; Bohren & Huffman 1983) is applied to calculate the effective dielectric function first of the individual core-mantle particles and then of the aggregates (Greenberg & Hage 1990; Hage & Greenberg 1990). The core of the individual particle is regarded as the inclusion and the mantle as the matrix. Thus the effective dielectric function of an individual core-mantle particle ϵ_{cm} is

$$\epsilon_{cm} = \epsilon_m[1 + \frac{3f(\frac{\epsilon_c - \epsilon_m}{\epsilon_c + 2\epsilon_m})}{1 - f(\frac{\epsilon_c - \epsilon_m}{\epsilon_c + 2\epsilon_m})}] \tag{2}$$

where ϵ_c and ϵ_m is the dielectric function of the core and mantle, respectively; f is the volume fraction of the core. In Greenberg's comet model, $f \approx 2$, which is consistent with interstellar observations (Greenberg 1982a; Greenberg & Hage 1990). In the case of the β Pictoris disk, we adopt $f \approx 1.6$, taking into account that the inner region dust is so hot that some, even all of the organic refractory mantle has evaporated. One will find that such an amount is reasonable when looking at the dust temperature distribution.

With the individual core-mantle particles considered as inclusions and vacuum as matrix, the average effective dielectric function of the aggregates is,

$$\epsilon_{av} = 1 + \frac{3(1 - P)(\frac{\epsilon_{cm} - 1}{\epsilon_{cm} + 2})}{1 - (1 - P)(\frac{\epsilon_{cm} - 1}{\epsilon_{cm} + 2})} \tag{3}$$

where the porosity P is taken as 0.975 (Greenberg & Hage 1990; Greenberg & Li 1995a).

Note that the above approach for ϵ_{av} is only valid for $x = \frac{2\pi a^i}{\lambda} \ll 1$ where a^i ($\sim 0.1\mu m$) is the size of individual particle (Greenberg & Hage 1990; Hage & Greenberg 1990). For shorter wavelengths, e.g., $\lambda \leq 1\mu m$, the aggregates are considered to be similar to a cloud of independent particles, therefore,

$$Q_{abs}(a, \lambda) \approx 1.0 - e^{\tau_{abs}} \tag{4}$$

$$\tau_{abs} \approx \frac{3}{4} N R C_{abs}^i \tag{5}$$

where N is the number of particles per unit volume; R is the radius of the aggregates; C_{abs}^i is the absorption cross section of individual particle (Greenberg & Hage 1990).

In Fig.1 we plot the dust temperatures as a function of the distance from the central star. It is clear that the dust in the inner several AU region is hot enough to be crystallized — it has been shown in laboratory studies that amorphous silicate can be annealed when heated to $\sim 700 - 1200 K$; for example, amorphous magnesium silicate was converted to crystalline olivine by heating to 1270 K for one hour (Day & Donn 1978). Glassy silicate particles can be crystallized by heating to 875 K for 105 hours (Koike & Tsuchiyama 1992). On the other hand, most of the dust grains giving rise to the silicate emission is at about 150 K which is consistent with observations (Walker & Wolstencroft 1988).

Following Backman et al (1992) and Knacke et al (1993) we treat the β Pictoris disk as optically thin, circularly symmetric and wedge shaped, namely, the disk thickness is proportional to r. Furthermore the disk is considered as

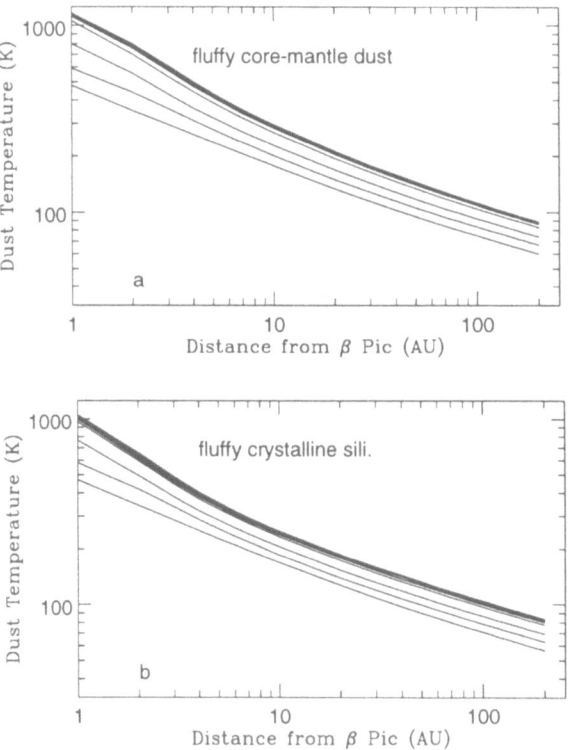

Fig. 1. The radial distribution of dust temperatures in the β Pictoris disk plane. The lines from top to bottom correspond to the dust with mass $10^{-13}, 10^{-12}, 10^{-11}, \ldots, 10^{-7}, 10^{-6} g$, respectively. **(a)** — The dust temperature radial distribution of porous aggregates of silicate core-organic refractory mantle particles. **(b)** — The dust temperature radial distribution of porous aggregates of crystalline silicate particles.

strictly edge-on because the tilt effect is negligible (Artymowicz et al 1989). Therefore the flux density emitted by the dust received at the Earth is

$$F(\lambda) = \int_0^{r_{disk}} \int_{a_1}^{a_2} \pi a^2 Q_{abs}(a, \lambda) n(a) rn(r) B(T_d, \lambda) da \frac{2\pi r dr}{D^2} \qquad (6)$$

where n(a) is the dust size distribution and n(r) is the dust volume density distribution; D ($\sim 16.6pc$) is the distance from the Earth to β Pictoris ; r_{disk} is the disk extent in this study. Although the disk actually extends to 2000 AU

(Smith & Terrile 1987), most of the silicate band emssion originates in the inner region, so $r_{disk}= 200$ AU is large enough.

Fig.2 shows the calculated spectral energy distribution in which we assume 20% silicates crystallized and the dust number density distribution is $n(r) \sim r^{\gamma-1}$: $\gamma = 2, 1AU \leq r \leq 20AU; \gamma = 0.5, 20 \leq r \leq 45; \gamma = -0.5, 45 \leq r \leq 100; \gamma = -1.7, 100 \leq r \leq 200$ which is consistent with observation (Lagage & Pantin 1994) and it shows a dust depleted inner region as observed by Lagage & Pantin (1994) and Lecavelier des Etangs et al (1993). This is also consistent with the expected effect of planets pertubation in the dust disk (Roques et al 1994; Lazzaro et al 1994). As shown in Fig.1, most of the infrared emission ($\lambda \leq 20\mu m$) is emitted by the dust within the inner 100 AU disk, which does not place strict constraints on γ for the outer 100 AU disk. We adopt $\gamma = -1.7$ following Artymowicz et al (1989).

Our model result gives the best match to date to the observations: it not only fits the total silicate emssion shape, but also reproduces the crystalline feature exactly. Although our model result is a little lower in the short wavelength wing than the observation, it is not a real problem if one realizes that an A5 star is not a good blackbody. The silicate emission data shown in Fig.2 is a 8200 K blackbody function substracted one whereas the true substracted distribution could be in error at the short wavelength side (Cohen, private communication; Cohen et al 1992). Furthermore, the terrestrial atmospheric ozone feature at 9.4 – 9.8 μm may contaminate the 9.7 μm amorphous silicate peak and cause uncertainties in the short wavelength wing.

The total dust mass required to produce the observed silicate emission is $\sim 7.5 \times 10^{22}g$ while Knacke et al (1993) derived it to be $\sim 3.0 \times 10^{22}g$ within 30 AU of the star assuming a two-component disk model. Telesco et al's (1988) simple estimation led to $\sim 10^{22}g$. Aitken et al's (1993) estimation is $\sim 3.0 \times 10^{21}g$, assuming that single size $2\mu m$ Draine & Lee silicate dust grains produced their observed silicate emission, but they thought that the real dust mass may be much more and comparable to $\sim 7.0 \times 10^{23}g$ of SAO 179815, a K5 star (Skinner et al 1992) which has similar dust excess emission and at a similar distance. One should bear in mind that the dust mass estimation is also dependent on the dust model: dust composition, size, and density distribution. It does not make much sense to compare the derived dust mass among different model calculations. However, we believe that the most reliable way of approaching the amount of dust mass should be that which can best reproduce the observations.

We have also calculated the spectral energy distribution of 2 μm single size Draine & Lee (1984) silicate grains which was claimed to be the best match to the observation of Aitken et al (1993). This result is also plotted in Figure 2. Clearly, it does match the entire shape in the 10 μm region, but, of course, it fails to reproduce the 11.2 μm crystalline silicate feature, and it also has no way to fit the emission at longer wavelengths. Furthermore, we have tried the case of Draine & Lee silicate core-organic refractory mantle particles + a fraction of crystalline silicate particles with comet Halley size distribution, but no good match could be obtained.

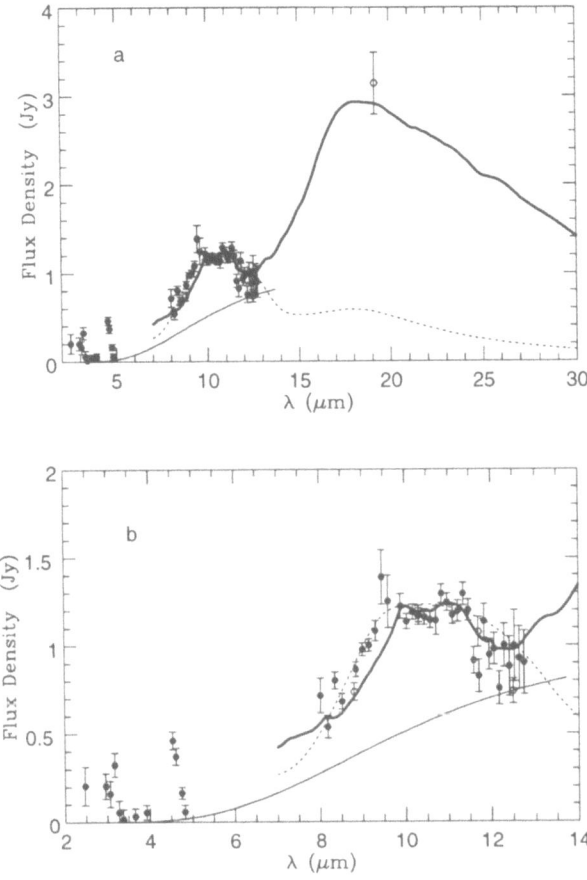

Fig. 2. (a) — The near and mid infrared emission spectra of the dust in the β Pictoris disk. The full points are the observational data from Knacke et al (1993) and the open points are from Telesco & Knacke (1991). The thin solid line is the dust thermal continuum (Knacke et al 1993). The thick solid line is the theoretical prediction of our model (see text). The dotted line is calculated from a single size $2\mu m$ Draine & Lee (1984) silicate grains. **(b)** — The same as **(a)** but detailed in the $10\mu m$ region.

5   Discussion

The 10 μm and 20 μm emssion or absorption silicate features due to the Si-O stretching and bending vibrations are ubiquitous in interstellar dust and circumstellar dust both of late type giants (Schutte et al 1989) and of young stars, e.g. T Tauri stars (Cohen et al 1985). Most of the interstellar dust sources exhibit a broad, structureless 10 μm feature which is characteristic of amorphous silicates. To date, the only interstellar dust source which exhibits an 11.2 μm structure is AFGL 2591, a heavily obscured molecular cloud source which is not representative of ordinary interstellar dust (Aitken et al 1988; Aitken et al 1989). This

11.2 μm structure is attributed to crystalline olivine although the exact anneal-ing mechanism is not yet clear. But circumstellar dust in some AGB stars does show an 11.2 μm structure. It is still uncertain whether this structure is due to crystalline silicate (Little-Marenin & Little 1990) or PAH (Polycyclic Aromatic Hydrocarbon) (Sylvester et al 1994), except for the case of some carbon stars in which it is attributed to SiC.

Recent observations have shown that Vega-type stars also display broad sil-icate 10 μm emission (Coulson & Walther 1995; Skinner et al 1992; Skinner et al 1995; Fajardo-Acosta et al 1993; etc.) and, in particular, β Pictoris and SAO 206462 (Coulson & Walther 1995) show a prominent 11.2 μm emission feature. The 11.2 μm stucture in SAO 206462 was identified as PAH emission because a series of PAH emssion features at 3.3 μm, 6.2 μm, 7.7 μm, 8.6 μm as well as 11.2 μm which are characteristics of PAH C=C stretching and C-H vibration modes, were observed (Coulson & Walther 1995). Thus dust in the β Pictoris disk could be similar to ordinary circumstellar dust and PAH molecules can not automatically be ruled out as a source of its 11.2 μm structure (Dorschner & Henning 1995). This holds also for SiC. But the reasons against this are : (1) the entire silicate emission feature is much broader than that of circumstellar dust (and interstellar dust); (2) the 11.2 μm PAH feature is too narrow compared with the observation; and (3) there is no evidence for the presence of any PAH emissions at other wavelengths than 11.3 μm. On the other hand, SiC was also ruled out (Knacke et al 1993). Obviously, besides the criterion of the dust re-plenishment source as discussed in Section 2, these arguments also support our idea that comets replenish the dust in the β Pictoris disk.

It is well recognised that silicates are an important component of comet dust. Many comets show an 11.2 μm crystalline silicate emission as well as the broad 10 μm emission (see Hanner et al 1994, for a review), e.g., comet Halley (Bregman et al 1987; Campins & Ryan 1989); Bradfield (Lynch et al 1989); Levy 1990 XX (Lynch et al 1992); and Mueller 1993a (Hanner et al 1994), etc. Figure 3 shows a comparison of the silicate excess emission (the dust thermal emission was subtracted) of both the observation and the theoretical calculation of the β Pictoris disk dust, comets Halley, Levy 1990 XX and Mueller 1993a. The similarity of these spectra is evident as has been pointed out by Knacke et al (1993). It reinforces our idea that comets replenish the dust in the β Pictoris disk. One may ask if the silicate component of comet dust in the β Pictoris disk has been crystallized in the protostar nebula. If t yes, it would not change our model fit and even simplify the model condition — no longer requiring strong radiation and crystallization. However, for solar comets, it seems rather unlikely that the crstalline silicate originates in the presolar nebula. For more detail see Greenberg & Li (1995c, in preparation).

Observations of the gaseous lines in β Pictoris also imply that comets may be the dust source. Spectral lines of CaII, MgII, FeII, AlIII, AlII, CIV, CI and CO show strong time variation and are almost always redshifted relative to the stellar spectra. These can be explained to result from some comets orbiting the star and falling inwards as a consequence of planet perturbation. These redshifted

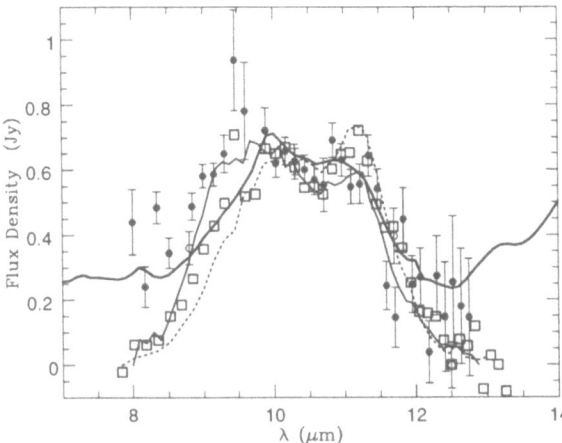

Fig. 3. The silicate excess emission (i.e., with the dust thermal continuum subtracted) of β Pictoris and some comets (all the comets data are normalized to the 10.3 μm data point), points — the observational data of the dust of the β Pictoris disk; thick solid line — our model result; open squares — comet Mueller 1993a (Hanner et al 1994); thin solid line — comet Halley (Campins & Ryan 1989); dotted line — comet Levy 1990 XX (Lynch et al 1992).

temporal variations lines are produced by the evaporation of the dust grains shed from comets (Beust et al 1990; Vidal-Madjar et al 1994; Lagrange et al 1987; etc.). Moreover, the fact that the deduced evaporation rate of the comet-like body in β Pictoris is comparable to that of such solar system comets as comet Halley, comet Bradfield, further supports our suggestion (Beust et al 1989).

The fluffiness of comet dust structure is necessary. Mukai & Koike (1990) have shown that, with a size distribution of comet Halley, which was deduced from the observed dust mass distribution on the basis of the assumption of compact dust, even pure crystalline silicate particles failed to produce an 11.2 μm structure. Furthermore, large compact particles act like black bodies and their equilibrium temperatures may be too low to cause evaporation, let alone crystallization. As already mentioned in Section 3.2, the fluffy aggregates of dust can avoid such problems because their optical properties are more like that of individual small particles (at least for the smaller aggregates) and they are much more absorbing than the eqivalent compact ones (Greenberg & Hage 1990).

Finally, we note that ice mantles may be present in the outer disk region, say, several hundred AU's away from the star where it is so cold that the subli-mated ice mantle may reaccrete on the core-mantle particles or have never been sublimated. If the water vapor condenses on the outer surface of the fluffy aggre-gates closing up the pores, the presence of ice mantles in the outer region may explain the high albedo (~ 0.6, $100AU \leq r \leq 500AU$) derived from optical light scattering data (Artymowicz et al 1989) while the albedo of the initially porous comet dust is quite low (~ 0.2).

6   Conclusions

We have developed a model for the dust in the β Pictoris disk. The basic idea is that the dust in the disk plane is continually replenished by comets orbiting the star where the dust may be quickly swept out by radiation pressure or spiral onto the star as a result of Poynting-Robertson drag. We assume that the initial dust consists of fluffy aggregates of interstellar amorphous silicate core-organic refractory mantle dust grains. The dust size distribution is assumed similar to that of comet Halley as observed by spacecrafts. When comets come close to the star, some dust aggregates are heated sufficiently to sublimate or evaporate the organic refractory mantle and even the amorphous silicate core is crystallized so that there will be a new kind of dust — porous aggregates of crystalline silicates in the disk. Exposed to the strong radiation pressure, both the initial aggregates of core-mantle dust and the new ones — aggregates of crystalline silicates — would be distributed over the entire disk plane. Based on such a dust model, we calculated the infrared spectral energy distribution and reproduced the observations very well.

Acknowledgements: We are grateful for support by NASA grant NGR 33-018-148 and by a grant from the Netherlands Organization for Space Research (SRON) for research on the organic refractories in space. We thank Dr. S.B. Fajardo-Acosta and Dr. R.F. Knacke for providing us with the β Pictoris observational data, Dr. T. Kozasa for sending us the optical constants of crystalline olivine and Dr. M. Hanner for sending us the silicate emission data of comet Mueller 1993a. One of us (AL) wishes to thank the World Laboratory for a fellowship.

References

Aitken, D.K., Roche, P.F., Smith, C.H., James, S.D. & Hough, J.H., 1988, MNRAS 230, 629
Aitken, D.K., Smith, C.H. & Roche, P.F., 1989, MNRAS 236, 919
Aitken, D.K., Moore T.J.T., Roche, P.F., Smith, C.H. & Wright, C.M., 1993, MNRAS 265, L41
Artymowicz, P. 1988, ApJ (Letters) 335, L79
Artymowicz, P., Burrows, C., & Paresce, F., 1989, ApJ 337, 494
Aumann H.H. et al., 1984, ApJ (Letters) 278, L23
Aumann H.H., 1985, PASP 97, 885
Backman, D.E., Gillett, F.C. & Witteborn F.C., 1992, ApJ 385, 680
Backman, D.E. & Paresce, F., 1993, in: Protostars and Planets III, eds. Levy E.H., Lunine J.I. & Mathews M.S., Tucson: Univ. Arizona Press, P.208

Becklin, E.E. & Zuckerman, B., 1990, in: Submillimeter Astronomy, eds. Watt, G.D. & Webster, A.S., Kluwer Academics Publishers, P.147

Beust, H., Lagrange-henri, A.M., Vidal-Madjar, A., Ferlet, R., 1989, A&A 223, 304

Beust, H., Lagrange-henri, A.M., Vidal-Madjar, A., Ferlet, R., 1990, A&A 236, 202

Bohren, C.F., Huffman, D.R., 1983, Absorbtion and Scattering of Light by Small Particles, Wiley, New York

Bregman, J., Campins, H., Witteborn, C., Wooden, D.H., Rank, D.M., Allamandola, L.J., Cohen, M. & Tielens, A.G.G.M., 1987, A&A 187, 616

Campins, H. & Ryan, E., 1989, ApJ 341, 1059

Chini, R., Krügel, E., Shustov, B., Tutukov, A. & Kreysa, E., 1991, A&A, 252, 220

Chlewicki, G. & Greenberg, J.M., 1990, ApJ 365, 230

Cohen, M. & Witteborn, F.C., 1985, ApJ 294, 345

Cohen, M., Witteborn, F.C., Carbon, D.F., Augason, G., Wooden, D., Bregman, J. & Goorvitch, D., 1992, AJ 104, 2045

Coulson, I.M. & Walther, D.M., 1995, MNRAS 274, 977

Day, K.L. & Donn, B., 1978, ApJ (Letters) 222, L45

Dorschner, J., Begemann, B., Henning, Th., Jäger, C. & Mutschke, H., 1995, A&A, 300, 503

Dorschner, J. & Henning, Th., 1995, Astronomy &Astrophysics Review, in press

Draine, B.T. & Lee, H.M., 1984, ApJ 285, 89

Fajardo-Acosta, S.B., Telesco, C.M. & Knacke, R.F., 1993, ApJ (Letters) 417, L33

Gillett, F.C., 1986, in: Light on Dark Matter, ed. Israel, (Dordrecht, Reidel), P.61

Golimowski, D.A., Durrance, S.T. & Clampin, M., 1993, ApJ (Letters) 411, L41

Greenberg, J.M., 1978. in: Cosmic Dust, McDonnell, J.A.M.(ed.), Wiley, P.187

Greenberg, J.M. & Gustafson, B., 1981, A&A 93, 35

Greenberg, J.M., 1982a, in: Comets, Wilkening, L.L.(ed.), University of Arizona Press, P. 131

Greenberg, J.M., 1982b, in: Submillimetre Wave Astronomy, Beckman, J.E. and Phillips,J.P.(eds.), Cambridge University Press, P. 261

Greenberg, J.M, 1985, Physica Scripta 11, 14

Greenberg, J.M., 1989, in: Interstellar Dust (IAU Symp. 135), eds. Allamandola, L.J. & Tielens, A.G.G.M., Kluwer, Dordrecht, P.345

Greenberg, J.M. & Hage, J.I., 1990, ApJ 361, 260

Greenberg, J.M. & Li, A., 1995a, submitted to Planetary & Space Science

Greenberg, J.M. & Li, A., 1995b, A&A, in press

Greenberg, J.M., Li, A., Mendoza-Gómez, C.X., Schutte, W.A., Gerakines, P.A., de Groot, M., 1995b, ApJ (Letters) 455, L177

Greenberg, J.M. & Li, A., 1995c, in preparation

Hage, J.I. & Greenberg, J.M., 1990, ApJ 361, 251

Hanner, M.S., Hackwell, J.A., Russell, R.W. & Lynch, D.K., 1994, ICARUS 112, 490

Hanner, M.S., Lynch, D.K., & Russell, R.W., 1994, ApJ 425, 274

Harper, D.A., Loewenstein, R.F. & Davidson, J.A., 1984, ApJ 285, 808

Henning, Th. & Stognienko, R., 1995, A&A, in press

Hong, S.S., Greenberg, J.M., 1980, A&A 88, 194

Knacke, R.F., Fajardo-Acosta, S.B., Telesco, C.M., Hackwell, J.A., Lynch, D.K. & Russell, R.W., 1993, ApJ 418, 440

Koike, C. & Tsuchiyama, A., 1992, MNRAS 255, 248

Krügel, E. & Siebenmorgen, R., 1994, A&A 288, 929

Lagage, P.O. & Pantin, E., 1994, Nature 369, 628

Lagrange, A.M., Ferlet, R., Vidal-Madjar, A., 1987, A&A 173, 289

Lazzaro, D., Sicardy, B., Roques, F. & Greenberg, R., 1994, ICARUS 108, 59

Lecavelier des Etangs, A., Perrin, G., Ferlet, R., Vidal-Madjar, A., Colas, F., Buil, C., Sevre, F., Arlot, J-E., Beust, H., Lagrange-henri, A-M., Lecacheux, J., Deleuil, M. & Gry, C., 1993, A&A 274, 877

Little-Marenin, I.R. & Little, S.J., 1990, AJ 99, 1173

Lynch, D.K., Russell, R.W. & Campins, H., 1989, in: Interstellar Dust (IAU Symp. 135), eds. Allamandola, L.J. & Tielens, A.G.G.M., Kluwer, Dordrecht, P.417

Lynch, D.K., Hanner, M.S. & Russell, R.W., 1992, IRACUS 97, 269

Matese, J.L., Whitmire, D.P., Lafleur, L.D., reynolds, R.T. & Cassen, P.M., 1987, BAAS 19, 829

Mathis, J.S., Rumpl, W., Nordsieck, K.H., 1977, ApJ 217, 425

Maxwell-Garnett, J.C., 1904, Phil.Trans.R.Soc., London, 203A, 385

Mukai, T. & Koike, C., 1990, IRACUS 87, 180

Nakano, T., 1991, in: Origin and Evolution of Interplanetary Dust, eds. Levasseur-Regourd, A.C. & Hasegawa, H., Kluwer Academic Publishers, P.421

Paresce, F. & Burrows, C., 1987, ApJ (Letters), 319, L23

Paresce, F., 1991, A&A 247, L25

Roques, F., Scholl, H., Sicardy, B. & Smith, B.A., 1994, ICARUS 108, 37

Schutte, W.A., 1987, The evolution of Interstellar Organic Grain Mantles, PhD Thesis, Leiden

Schutte, W.A. & Tielens, A.G.G.M., 1989, ApJ 343, 369

Siebenmorgen, R., 1993, ApJ 408, 218

Skinner, C.J., Barlow, M.J. & Justtanont, K., 1992, MNRAS, 255, 31P

Skinner, C.J., Sylvester, R.J., Graham, J.R., Barlow, M.J., Meixner, M., Keto, E., Arens, J.F. & Jernigan, J.G., 1995, ApJ 444, 861

Smith, B.A. & Terrile, 1984, Science 226, 1421

Smith, B.A., 1987, BAAS, 19, 829

Sylvester, R.J., Barlow, M.J. & Skinner, C.J., 1994, MNRAS 266, 640

Telesco, C.M. & Knacke, R.F., 1991, ApJ (Letters) 372, L29

Telesco, C.M., Becklin, E.E., Wolstencroft, R.D. & Decher, R., 1988, Nature 335, 51

Vidal-Madjar, A., Lagrange-henri, A-M., Feldman, P.D., Beust, H., Lissauer, J.J., Deleuil, M., Ferlet, R., Gry, C., Hobbs, L.M., McGrath, M.A., McPhate, J.B. & Moos, H.W., 1994, A&A 290, 245

Walker, H.J. & Wolstencroft, R.D., 1988, PASP 100, 1509

Weissman, P.R., 1984, in: Protostars and Planets II, eds. Black, D.C. & Matthews, M.S., Tucson, Univ. Arizona Press, P.895

Magnetic Fields of T Tauri Stars

Eike W. Guenther and James P. Emerson

Physics Dept., Queen Mary & Westfield College, Mile End Rd., London E1 4NS, UK

Abstract. In this paper, new evidence for the presence of strong magnetic fields in T Tauri stars is presented. Four optical flares in Weak Line T Tauri stars (WTTSs) but none in the classical T Tauri stars (cTTSs) were found by monitoring the stars with a multi-object spectrograph, and 3 magnetic field strength in WTTS and one upper limit in a cTTSs were derived by applying the Zeeman-broadening method to high resolution optical spectra.

1 Introduction

Classical T Tauri stars (cTTSs) are young, low mass pre-main sequence stars, with a circumstellar accretion disk. Large observational and theoretical efforts have been invested in trying to understand how cTTSs can keep their low rotation rates while accreting matter. Currently, the most popular model is that strong magnetic fields ($\approx 1kG$) couple star and disk, so that angular momentum is transported outward, while matter is flowing inward. Although magnetic fields are thus highly important for the structure and evolution of T Tauri stars (TTSs), very few attempts to measure the field have actually been made, and evidence for the fields is thus rather indirect at present. In this paper, new evidence for the presence of strong magnetic fields in TTSs is presented.

2 Flares

One of the most important arguments for the presence of strong magnetic fields in TTSs is the observations of X-Ray flares. From the observation of a giant X-ray flare on the cTTS LkHα92, Preibisch et al. (1993) was able to derive a minimum value of the magnetic field strength of 210 G. However, the authors had to assume that flares on TTSs are scaled up versions of solar flares. If this is the case, it must be possible to detect these flares in the optical regime, and they must closely resemble flares on the sun or flare-stars. Some surveys for flares in the optical regime have already been made (see Gahm 1989). However, Gahm (1989) concluded from a statistical analysis of broad band photometry that the events seen on WTTSs were flares, whereas the events on cTTSs were not (slow rise-time, and different colours) Since the spectra of flares can be expected to be dominated by emission lines, it would be much better to have complete spectroscopic time series of these events to distinguish flares from other events. Progress in this field has hitherto been severely hindered because of the rarity of these events.

However, multi-object spectrographs with large fields of view, like FLAIR on the UK Schmidt, allow one to monitor many stars simultaneously, thus increasing the probability of observing flares. Since some fibers can be placed onto photometric standard stars, the data from such instruments are even spectrophotometric. In January 1995 17 cTTSs and 18 WTTSs and one Herbig Ae/Be star of the Chamaeleon star forming region were monitored for 16 hours. The spectral region around $H\alpha$ (5350-6870 Å) was chosen. The spectral resolution was 2.5 Å, and the time resolution about 4 minutes. The data reduction of the resulting 5040 spectra brought up at least 4 clear flares in WTTSs but none in cTTSs or the Herbig Ae/Be star. The properties of the flares where exactly as expected: (1) The increase in flux is much larger in $H\alpha$ than in the continuum. For example, in J1149.8-7850 the continuum flux increased by 25%, whereas the the flux in $H\alpha$ increase by 320%. (2) The rise-time is much smaller than the decay-time. Typical rise-times are about 10 minutes, whereas decay-times are about one hour (a figure showing spectra of the rise-time and the light-curve of one flare can be found in Guenther (1995)). Although a number of events were seen in the cTTSs as well, none of these shows the strong increase within minutes and decrease in an hour characteristic of a flare. These events in the cTTSs are definitely not like the flares seen in WTTSs, and are more likely to be due to sudden variations of the accretion rate. Although the cTTSs have a strong $H\alpha$ emission line, it can be shown that if the events are like those seen in the WTTSs, and if they are superimposed on the emission line, they could have easily been detected. The results of this run thus somewhat deepens the mystery in that there are no, or at least very few, optical flares in cTTSs. There are three possible explanations for the non-detection of flares in the cTTSs: (1) Just by chance no flare occurred in a cTTSs, although their frequency is the same as in WTTSs. The probability for this is 0.125. (2) It could be that the flares originate very close to the star. If much of the $H\alpha$ emission comes from further out, flares could not be observed if $H\alpha$ is optically thick. (3) There are really no, or very few optical flares in cTTSs, because either the majority of cTTSs lack fields (the few flares seen in the X-rays might came from unusual sources), or the flares on TTSs are really not solar-like, and do not emit in the optical regime.

3   Direct Measurements

Obviously, the best way to prove, whether TTSs have magnetic fields or not, are direct measurements but measuring the fields has some difficulties. Since the magnetic field of a TTS will have its maximum magnetic flux density near the surface of the star and will decrease rapidly outwards, the strongest fields can be expected close to the star. Thus, magnetic field measurements should be carried out using photospheric lines, rather than emission lines. Since the photospheric spectrum of cTTSs is veiled, it will be easier to detect the fields in WTTSs, especially if they are slowly rotating. If the magnetic field structure is very complicated, observation of Stokes V will show no signal, because the signal from each magnetic north pole is canceled out by a magnetic south pole.

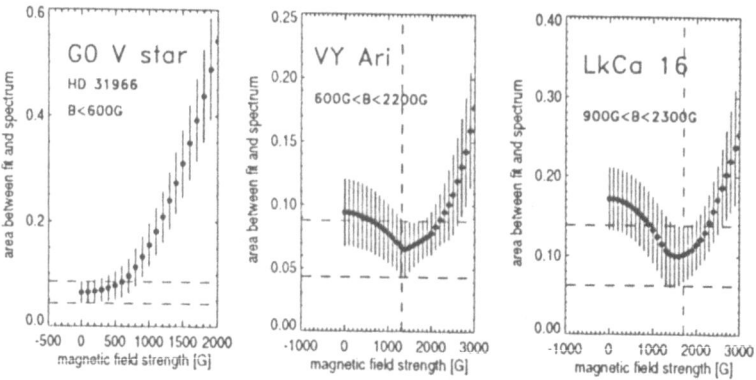

Fig. 1. Left: The magnetic field strength for the main sequence star HD 31966, which was found as expected to be zero.

Middle: The magnetic field strength for the magnetic template stars VY Ari was found to be 1320 G, with excellent agreement with the literature.

Right: Derived magnetic field strength for LkCa16 of $fB = 1700^{+600}_{-600}$ G.

However, the presence of many regions of opposite polarity will still lead to a broadening of the Stokes I-profile (Mathys, G., 1992), and the magnetic field strength can be measured if the non-magnetic broadening (rotation, turbulence) is exactly known. The best way to do this, is to measure the line-width of lines with very low sensitivity to the magnetic field. Magnetic fields of TTS can thus best be detected by the enhanced broadening of Zeeman sensitive lines over Zeeman insensitive lines in the photospheric lines of WTTS. However, the non-magnetic broadening of the lines could in principle be different for different lines. For example, if the turbulence in the stellar atmosphere has a strong gradient. Although, it can be expected that this effect is very small, because the amplitude of the the vertical component of the micro-turbulence decreases from just 2.5km/s to 1km/s when going from $\tau = 1$ to $\tau = 10^{-3}$ (Hollweger 1967), and the macro-turbulence is of the order of 2km/s only (Stix 1989), it is probably safer to use lines originating in the same layers of the atmosphere (Robinson 1980, Guenther & Mattig, 1991), for example by using lines of the same ion, of about the same equivalent width, in about the same spectral region. Using the Zeeman-broadening method on WTTSs, Basri et al. (1992) detected in this way a magnetic field of 1000 ± 500 gauss in the WTTS TAP35, and found an upper limit of 700 G in Tap 10.

Encouraging as this work is, the detection of a magnetic field in a single WTTSs does not necessary imply that kilogauss magnetic fields are a common feature of all TTSs, and it is thus necessary to observe a much larger sample which should include cTTSs as well. In this paper, the first results from a survey for fields on cTTSs and WTTSs are presented. In November 1994, spectra

with a resolution of $\lambda/\Delta\lambda = 55000$ in the range between 5260 and 9240 Å of 18 WTTSs and cTTSs together with 15 spectroscopic template stars, and 4 magnetic template stars were taken. The signal to noise ratio of the spectra is one-hundred or higher. For the reasons described above, the results presented here are only derived by using FeI lines of about the same equivalent width. In order to test the method, the magnetic field strength of the magnetic template stars and non-magnetic template stars were derived. As expected, we detected no fields in the non-magnetic main-sequence stars but did in the magnetic template stars. Two examples are shown in Fig. 1 (left, middle). The Figure shows the average area between the fit and observed line profiles. The best fit is achieved when the area between the observed spectrum and the fit is minimised. The best fit for HD 31966 is at $fB = 0^{+600}$ G and for VY Ari at $fB = 1400^{+700}_{-700}$ G), which is in excellent agreement with the value of $B = 2000G$, $f = 0.66$ from the literature (Bopp et al. 1989). The Figure (right) shows a similar measurement for the WTTS LkCa16. A magnetic field is definitely detected. The Table gives an overview of the measurements for all TTS measured so far. The errors are somewhat overestimated. Magnetic fields with strength of more than 1kG were detected in all three WTTSs but no field could be detected in the cTTSs.

Table 1.   Magnetic field strength

star	$W_\lambda\ H\alpha$ [Å]	$W_\lambda\ HeI$ [Å]	veiling	$Lx/10^{23}$ [W] [*]	fB [G]
LkCa 15	23.8	0.24	0.43 ± 0.13	< 0.408	< 2000
LkCa 16	1.8	0.08	0.04 ± 0.07	$1.369^{+0.775}_{-0.327}$	1700^{+600}_{-600}
LkCa 7	1.7	0.02	0.08 ± 0.08	$0.731^{+0.211}_{-0.169}$	2500^{+500}_{-2500}
HBC 427	0.5	0.00	0.18 ± 0.09	$0.717^{+0.177}_{-0.170}$	2500^{+300}_{-1000}

* from Neuhäuser et al. 1995

References

Basri, G., Marcy, G. W., Valenti, J.A. (1992): ApJ **390**, 622

Bopp, B.W., Saar, S.H., Ambruster, C., Feldman, P., Dempsey, R., Allen, M., Barden, S.P. (1989): ApJ **339**, 1059

Gahm, G.F. (1989): In *Flare Stars in the Star Clusters, Associations and in the Solar Vicinity*, L.V. Mirzoyan ed., IAU 137

Guenther, E. Mattig, W. (1991): A&A **243**, 244

Guenther, E. (1995): AAO newsletter, No. 73

Hollweger, H. (1967): Z. Astrophys. **22**, 265

Mathys, G. (1992): IAU Colloquium 138

Neuhäuser, R., Sterzik, M.F., Torres, G., Martín, E. (1995): preprint A&A.

Preibisch, Th., Zinnecker, H., Schmitt, J.H.M.M. (1993): A&A **279**, L33

Robinson, R.D., Worden, S.P., Harvey, J.W. (1980): ApJ **236**, L155

Stix, M. (1989): *The Sun*, Springer Heidelberg, p. 142

Lambda Bootis Stars and 'Dusty' A Stars

Hartmut Holweger and Inga Rentzsch-Holm

Institut für Astronomie und Astrophysik, D-24098 Kiel, Germany

Abstract. We derive calcium abundances for two classes of A stars that show direct or indirect evidence for the presence of circumstellar (CS) gas and/or dust: λ Boo stars and A stars with infrared excesses. We argue that λ Boo stars are pre-main-sequence objects, and that Vega also belongs to this class. By contrast, the surface composition of other A stars with dust disks, including β Pic, is surprisingly normal. In λ Boo stars a correlation exists between the deficiency of dust-phase elements, stellar rotation, and surface gravity, suggesting that the surface anomaly develops during the final contraction towards the ZAMS, and that the rotation rate decreases rather than increases during this phase.

1 Introduction

In this paper we discuss two groups of A stars that may be related to the final phase of star formation: (i) a class of metal-poor A stars known as λ Bootis stars, whose deficiency probably is due to accretion of depleted gas, (ii) normal A stars with IR excesses arising from circumstellar dust.

The results reported here are based on an analysis of 11 of the 16 known λ Boo stars, and of 7 B9.5-A6 IV-V stars that are confirmed *IRAS* sources (Cheng et al. 1992). Details are given elsewhere (Holweger and Rentzsch-Holm 1995, paper I); here we present a summary and supplementary considerations.

2 λ Boo Stars – Accretion, Rotation and Stellar Evolution

The deficiency of 'metals' is a characteristic property of λ Boo stars. In paper I we find that metallicity (as measured by [Ca/H]) and rotation are correlated, as illustrated in Fig. 1. The deficiency of Ca in λ Boo stars diminishes towards high $v \sin i$ values. It is argued that CS lines are preferably seen in CS disks viewed nearly equator-on ($\sin i \approx 1$). In Fig. 1 just these stars show the correlation most clearly. Presumably the other λ Boo stars, if also viewed equator-on, would fit into this correlation. The dependence of surface abundances on rotation can be explained by models of accretion and rotationally induced circulation (Turcotte and Charbonneau 1993): in the rapid rotators meridional mixing efficiently removes the accreted matter from the stellar surface.

As pointed out in paper I, there is circumstantial evidence that λ Boo stars are in the pre-main-sequence phase of evolution, supporting previous speculation (Gerbaldi et al. 1993). Also shown in Fig. 1 is Vega. Recent high-S/N

spectroscopy (Gulliver et al. 1994) has confirmed earlier suspicion that Vega is a rapid rotator seen pole-on ($i \approx 5°$). If viewed at a more typical inclination, Vega will nicely fit into the correlation of [Ca/H] and rotation of λ Boo stars. Figure 2 shows a new aspect of the λ Boo phenomenon. The gravities vary from star to star by as much as a factor of 6, but the spread in mass is much smaller – about 1.5–2.2 $M_\odot$ (see Gerbaldi et al. 1993 and paper I). This implies significantly different radii. Are we observing newly formed stars during their final contraction

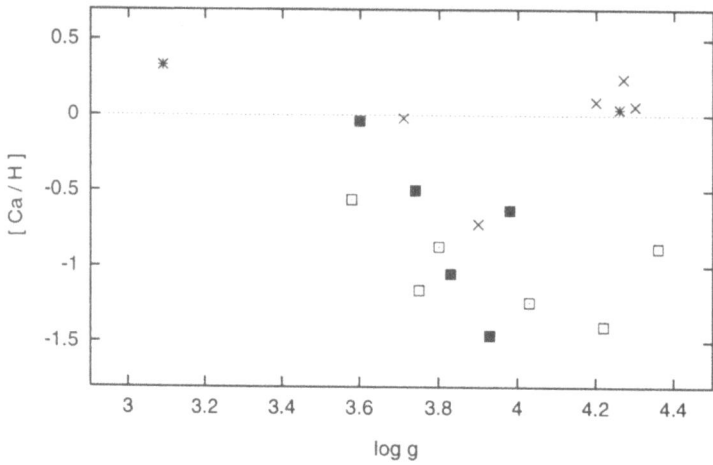

Fig. 1. Photospheric abundance of calcium vs. projected rotational velocity for λ Boo stars (squares) and dusty normal A stars (crosses and asterisks). Stars with CS gas are denoted by filled squares and asterisks, respectively. β Pic is located at (132, +0.02). The three crosses connected by a dashed line represent the rapid rotator Vega as viewed almost pole-on from Earth ($v \sin i = 22\,kms^{-1}$), at a more typical inclination of $i = 45°$ ($v \sin i = 173\,kms^{-1}$), and equator-on ($v = 245\,kms^{-1}$).

towards the ZAMS? However, if the evolution of λ Boo stars indeed proceeds from low to high gravity, Fig. 1 then implies that their rotation slows down, with substantial loss of angular momentum. A similar, more severe problem is encountered with disk accretion models of T Tau stars; one possible solution is transfer of angular momentum from the star to the CS disk by magnetic stresses (Cameron and Campbell 1993). Is this model applicable also to the probably much less massive disks of λ Boo stars?

3 Silicon in λ Boo Stars: Differential Depletion

Separation of gas and dust during star formation as an explanation for non-solar metallicity is an old idea (see Turcotte and Charbonneau 1993). Accretion of gas depleted in dust-phase elements, as invoked for λ Boo stars (Venn and Lambert 1990), will lead to a general deficiency of refractory elements, while volatiles like

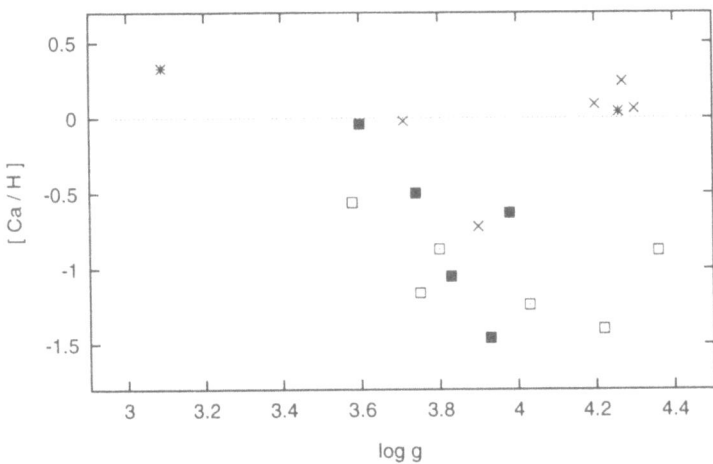

Fig. 2. Calcium abundance vs. surface gravity for λ Boo stars and dusty normal A stars. The symbols refer to the same stellar samples as in Fig. 1.

C stay with H. Although naive in view of the complexity of grain physics and chemistry, this model explains the principal abundance features (Fig. 3).

In four of the λ Boo stars silicon behaves like carbon (Fig. 4). Thermal processing just above the condensation temperature of Si may have removed this element from the grains. Incomplete condensation of Si is an alternative. Indeed, recent work on interstellar clouds suggests that only 35 % of the available Si is incorporated into dust-grain cores (Sofia et al. 1994).

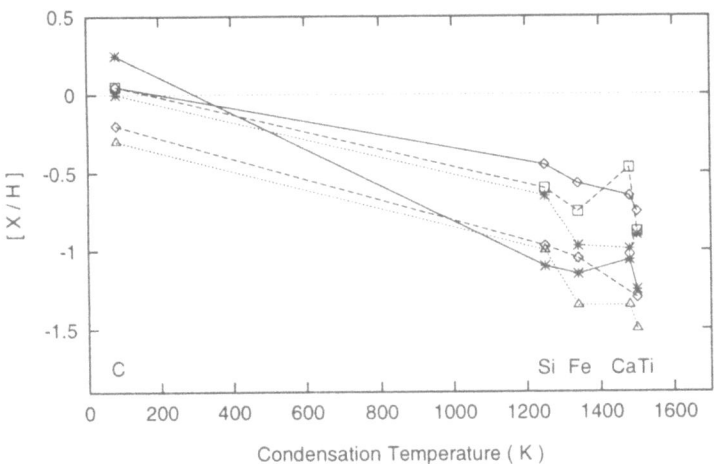

Fig. 3. The surface composition of λ Boo stars indicates accretion of CS matter depleted in dust-phase elements. The sample includes HR 12, 1570, 4881, 7400, 7736 and 8437 (C, Si, Fe, Ti: see Holweger and Stürenburg 1993, same ref. as Gerbaldi et al. 1993)

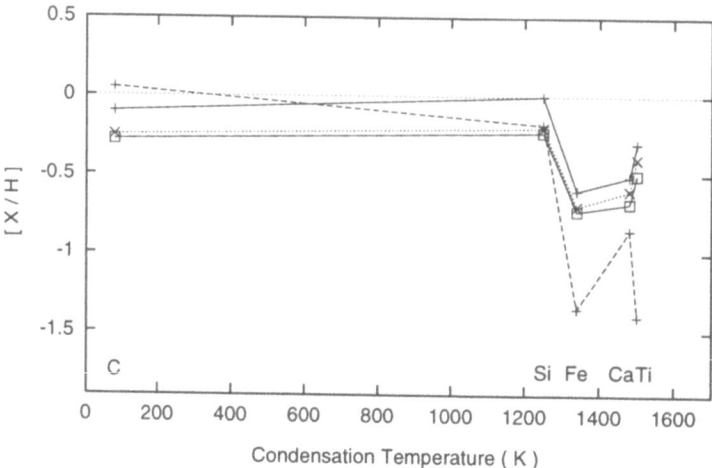

Fig. 4. In some of the λ Boo stars silicon does not follow the depletion pattern of the more refractory elements (HR 541, 7764C, 7959, 7960)

4   Vega – a λ Boo Star Unlike Other 'Vega-type' A Stars

According to Sect. 2 the combination of metal-deficiency and rotation in Vega is typical for λ Boo stars. Further support that Vega belongs to this class comes from Fig. 2: Vega ($\log g = 3.9$) also fits well into the contraction sequence.

By contrast, in other dusty A stars including β Pic ($\log g = 4.2$) calcium is normal or even slightly enhanced (51 Oph, $\log g = 3.1$). The abundance signature of accretion is not seen in these stars.

References

Cameron, A.C., Campbell, C.G. (1993): Rotational evolution of magnetic T Tauri stars with accretion discs. A&A **274**, 309–318

Cheng, K.-P., Bruhweiler, F.C., Kondo, Y., Grady, C.A. (1992): Newly identified main-sequence A stars with circumstellar dust. ApJ **396**, L83–L86

Gerbaldi, M., Zorec, J., Castelli, F., Faraggiana, R. (1993) Parameters T_{eff} and $L/L_\odot$ for a group of λ Bootis candidate stars and their evolutionary status. In: Dworetsky, M.M., Castelli, F., Faraggiana, R. (eds.), IAU Coll. 138, ASP Conf. Ser. **44**, 356

Gulliver, A.F., Hill, G., Adelman, S.J. (1994): Vega: a rapidly rotating pole-on star. ApJ **429**, L81–L84

Holweger, H., Rentzsch-Holm, I. (1995): High-resolution spectroscopy of λ Bootis stars and 'dusty' normal A stars: circumstellar gas, rotation, and accretion. A&A, in press

Sofia, U.J., Cardell, J.A., Savage, B.D. (1994): The abundant elements in interstellar dust. ApJ **430**, 650–666

Turcotte, S., Charbonneau, P. (1993): Particle transport and the λ Bootis phenomenon. II. An accretion/diffusion model. ApJ **413**, 376–389

Venn, K.A., Lambert, D.L. (1990): The chemical composition of three Lambda Bootis stars. ApJ **363**, 234–244

Modelling of the IR Intensity Maps for HAEBE Stars with Algol-Like Minima

Vladimir Il'in[1], Natalia Krivova[2] and Alexander Men'shchikov[1]

[1] Max Planck Society, Research Unit "Dust in Star-Forming Regions", Schillergäßchen 3, D-07745 Jena, Germany
[2] Astronomical Institute, St. Petersburg University, 198904 St. Petersburg, Russia

Abstract. A numerical radiative transfer code is used to compute intensity maps of the circumstellar dust shells around the HAEBE stars with Algol-like minima in the spectral region from near to far infrared. Parameters of the shell models were determined from fitting of infrared spectra, colour–magnitude diagrams and circumstellar extinction curves. Main features of the shell images are discussed and their dependence on the model parameters is considered.

1 Introduction

While it is well known that Herbig Ae/Be (HAEBE) stars are surrounded by gas and dust, our understanding of these dusty shells still remains incomplete. The shell models are based on observed spectral energy distributions in infrared (IR), "circumstellar" (CS) extinction curves in ultraviolet (UV) and, for HAEBE stars with Algol-like minima (UX Ori, WW Vul, etc.), on colour–magnitude diagrams showing a "blueing" effect.

We refrain from commonly used method of fitting of one kind of the observational data only (usually IR spectra), since such an approach does not allow one to constrain even the parameters of the simplest models (see, e.g., Thamm et al. 1994; Men'shchikov & Henning 1995a). We believe that only a fitting of all the data may give a confidence in the derived values of the parameters.

During several last years, images of the shells around brightest HAEBE stars have been obtained at different wavelenghts from visual to far IR (see, e.g., Nakajima & Golimowski 1995; Di Francesco et al. 1994). In this paper, we briefly discuss whether these data could be used to constrain the shell parameters.

2 Model

To reduce a large number of free parameters we adopt the following simple model: the star is surrounded by a spherical shell with an optical thickness $\tau_V < 1$, power law density distribution $\rho(r) \sim r^{-\alpha}$, and minimum and maximum radii R_{in} and R_{out}, respectively. The dust grain model used is a usual MRN-like one: a mixture of silicate and graphite grains having a power-law size distribution $n(a) \sim a^{-q}$ with the minimum and maximum sizes a_{min} and a_{max}. A simple power-law $\rho(r)$

and the same sizes of silicate and graphite grains for the whole shell are justified by the absence of any detailed observational constraints.

Since the stellar parameters are usually well determined, our model has eight parameters: $\tau_V, \alpha, R_{in}, R_{out}$, and $n_{si}/n_{gr}, q, a_{min}, a_{max}$. Fitting of IR spectrum can constrain α, τ_V, R_{in}; colour–magnitude diagrams help to estimate τ_λ and Λ_λ, albedo of grains (see Krivova & Il'in 1995). Extinction curves in UV can give information on $a_{min}, q, n_{si}/n_{gr}$, but at the moment only for one HAEBE star with Algol-like minima these data are available.

For our modelling we used the radiative transfer method and the code described by Men'shchikov & Henning (1995b). The method is usable in both spherical and disk geometry with an arbitrary number and distribution of dust grain components (chemical composition, sizes).

3   Results

We have computed intensity maps of the shells at different wavelengths from visual to far IR, using the values of the model parameters derived by fitting IR spectra and colour–magnitude diagrams (see Krivova & Il'in 1995).

Figure 1 gives the distribution of specific intensity I_ν vs. impact parameter b. It is easy to follow the change of the distributions with wavelength. At $\lambda \approx 0.6$–$1.25\ \mu m$, we have $I_\nu(b) \sim b^{-(1+\alpha)}$ in agreement with a simple theoretical model of the shell with a density gradient and single scattering (see, e.g., Weintraub et al. 1992); at $\lambda \sim 3\ \mu m$ only very hot grains in the inner layers are seen; at $\lambda \sim 10$–$20\ \mu m$ the intensity distribution becomes wider and for $\lambda > 60\ \mu m$ the radiation comes from the whole shell.

Figure 2 shows the size of image area which produces half of the total flux at a given wavelength. The transition from scattered light to the thermal emission of dust at 2–3 μm is clearly seen. To observe the shells at shorter wavelengths one needs to use coronographic techniques (or lunar occultations).

Let us now consider how the maps change with variations of the model parameters. We selected the following characteristics of the maps: (i) surface brightness in the R and I bands (I_R, I_I) at distance of 5000 AU (12″ at D=400pc) from a star and (ii) sizes of the shell projection which gives half of the total flux at 3, 10, 25 and 100 μm (R_3, R_{10}, R_{25} and R_{100}).

We varied the model parameters, keeping τ_V constant since it is relatively well constrained by observational data, and found that:

- $I_{R,I}$ and R_λ for $\lambda > 10\mu m$ strongly depend on α;
- $I_{R,I}$ and R_λ for $\lambda < 8\mu m$ change with a_{min} and q;
- τ_V affects only $I_{R,I}$ $(I \sim \tau\Lambda)$;
- R_{in} is important only for R_λ with $\lambda < 15\mu m$;
- neither $I_{R,I}$ nor $R_{3,10,25,100}$ depend on R_{out} and a_{max}.

Thus, from the intensity maps in the R and I bands one may expect to constrain α (and possibly a_{min} and q); while from the far IR scans/maps and from the near IR observations with very high angular resolution - α and R_{in} (and a_{min}, q), respectively.

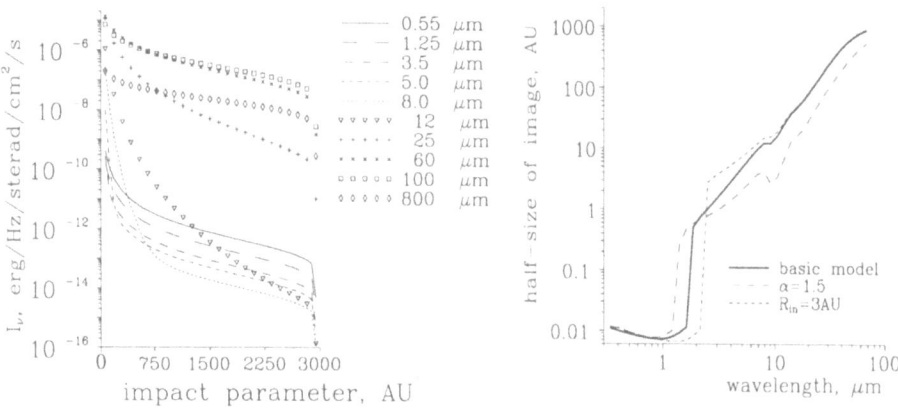

Fig. 1. The distribution of specific intensity vs. impact parameter. The model used —
$T_{eff} = 8000K$, $L = 25L_\odot$, $D = 250pc$, $R_{in} = 1AU$, $R_{out} = 3000AU$, $\alpha = 1.0$, $\tau_V = 0.9$, $a_{min} = 0.007\mu m$, $a_{max} = 0.25\mu m$, $q = 4.0$

Fig. 2. The size of image area producing half of the total flux at given wavelength. For basic model parameters see Fig. 1.

4   The Example of AB Aur

The shells around UX Ori-like stars practically were not mapped so far. High resolution observations have been made only for brighter HAEBE stars without Algol-like minima, for example, AB Aur. As an illustration, we compare available observational images of this star with theoretical maps. For AB Aur we adopt Sorrell's (1990) model which gives a good fit to the observed IR spectrum and extinction curve (for $\lambda > 0.16\mu m$): $\tau_V = 0.05$, $\alpha = -1.0$, $R_{in} = 0.2AU$, $R_{out} = 20000AU$, $q = -3.5$, $n_{si}/n_{gr} = 0.6$, $a_{min} = 0.005\mu m$, $a_{max} = 1.0\mu m$.

Four works presenting observations of the vicinity of AB Aur with a high angular resolution were found in the literature:
- scans of AB Aur environment at λ 100 μm made by Di Francesco et al. (1994) have shown FWHM to be < 14'' (or <2000AU);
- at λ 11.7 and 17.9 μm the shell of AB Aur was observed at the 5-m telescope by Marsh et al. (1994) with FWHM of about 40 and 80AU, respectively;
- speckle-interferometry in the K band with resolution of 0.03'' (5 AU) made by Leinert et al. (1994) gave no way to resolve the shell;
- coronographic images in the R and I bands obtained by Nakajima & Golimowski (1995) have shown a non-spherical matter distribution with the brightness of about 20 mag/□'' at distances 10-20'' from the star.

Our calculations show that Sorrell's model gives $R_{100} = 5000AU$, $R_{20} = 70AU$, $R_{11} = 20AU$, $R_{2.2} < 0.01AU$, and $I_{I,R} \sim 13$ mag/□'' at a distance $\sim15''$.

The difference between the observed and theoretical values would be much smaller if we selected steeper density distribution ($\alpha = -1.5$). However, we do not consider this model in more detail because its inner radius is certainly wrong.

At the distance of 0.2AU from the star grains would have temperature higher than 2500K, i.e. they cannot exist there.

The HAEBE stars with Algol-like minima for which our model was developed are at least 2.5 times more distant than AB Aur (D=160pc for the latter). At the moment it seems impossible to resolve their shells at 10 or 100 μm, but faint ($\sim$ 21 mag/$\Box''$) reflection nebulae could be detected by coronographic techniques.

5 Conclusions

We have computed intensity maps for simple shell models giving the observed spectral energy distributions and colour–magnitude diagrams for HAEBE stars with Algol-like minima. We have considered the main features of the shell images and their dependence on the model parameters. We found clear relationships between surface brightness distributions and some of the parameters. Thus, the shell images, providing important information on geometry of the shells, could give new constraints on their physical parameters (but only when being considered together with other observational data!).

As a next step we are going to model existing polarization data as well, using a spheroidal model of the shells and a Monte-Carlo code (Fischer et al. 1994). This must add one model parameter, the shell semiaxes ratio, but would allow one to involve more data into consideration: total polarization of the objects at different wavelengths and polarization maps of scattered light. We should mention that with the polarization data we shall be abble to involve into consideration all the data available for our stars.

Acknowledgments. We thank Dr. B. Stecklum for comments. V. I. acknowledges the support from the Alexander von Humboldt Foundation. N. K. thanks the Government of St. Petersburg for the financial support (Young Scientists Program grant 138-2-2.2).

References

Di Francesco, J., Evans II, N. J., Harvey, P. M., Mundy, L. G., Butner, H. M. (1994): ApJ **432**, 710–719
Fischer, O., Henning, Th., Yorke, H. W. (1994): A&A **284**, 187–209
Krivova, N. A., Il'in, V. B., (1995): This volume
Leinert, Ch., Richichi, A., Weitzel, N., Haas, M. (1994): eds. Thé, P. S., Perez, M. R., van den Heuvel, E. P. J. "The nature and evol. status of HAEBE stars", p. 155
Marsh, K. A., van Cleve, J. E., Mahoney, M. J., Hayward, T. L., Houck, J. R. (1994): BAAS **26**, 931
Men'shchikov, A. B., Henning, Th. (1995a): This volume
Men'shchikov, A. B., Henning, Th. (1995b): A&A (submitted)
Nakajima, T., Golimowski, D. A. (1995): AJ **109**, 1181–1198
Prusti, T., Natta, A., Palla, F. (1994): A&A **292**, 593–598
Thamm, E., Steinacker, J., Henning, Th. (1994): A&A **287**, 493–502
Weintraub, D. A., Kastner, J. H., Zuckerman, B., Gatley, I. (1992): ApJ **391**, 784–804

Multiwavelength Study of HAEBE Stars with Algol-Like Minima

Natalia Krivova[1] and Vladimir Il'in[2]

[1] Astronomical Institute, St. Petersburg University, 198904 St. Petersburg, Russia
[2] Max Planck Society, Research Unit "Dust in Star-Forming Regions", Schillergäßchen 3, D-07745 Jena, Germany

Abstract. The properties of circumstellar dust shells around several HAEBE stars with Algol-like minima are discussed. A numerical radiative transfer modelling technique for a spherical shell is used to fit the observed IR energy distribution, the circumstellar extinction curves (UV) and the colour-magnitude diagrams in minima. Such a simultaneous multiwavelength study enables some of the model parameters (such as optical depth, size and density distribution law of the envelope) to be found with some degree of certainty.

1 Introduction

The existence of dust shells around Herbig Ae/Be (HAEBE) stars shows itself in intense infrared (IR) emission and extinction of stellar radiation. Hence for the properties of circumstellar dust to be studied, it is reasonable to construct infrared spectra of the stars (Hartmann et al. 1993; Hillenbrand et al. 1992; Natta et al. 1993) as well as circumstellar extinction curves in ultraviolet (Voshchinnikov et al. 1995) and to compare them with the observational data. However, simultaneous modelling the far-ultraviolet (UV) through far-IR observations for dust embedded stars provides a better chance to strengthen constraints on dust grain properties (see e.g. Sorrell 1990).

For the group of HAEBE stars with non-periodic Algol-like brightness minima (UX Ori, WW Vul, etc), the observations of the brightness variations may tell something about the circumstellar (CS) dust environments as well (Voshchinnikov 1989).

Here we report the results of a multiwavelength modelling for a few UX Ori-like stars taking into account their colour-indexes behaviour in minima. Note that in previous papers there were no attempts to fit IR spectrum, colour-magnitude diagrams, and CS extinction curve simultaneously.

2 Model

To model radiative transfer in a spherical dust envelope with a power-law dust density distribution the numerical code described by Chini et. al. (1986) with some modifications is used. Inner boundary of the shell is determined by evaporation of dust particles. It is assumed that non-periodic Algol-like minima are caused by circumstellar dust clouds, that move around the star and screen it from time to time inducing the "blueing" effect (Grinin, 1988). The diffuse dust in the envelope responsible for this effect is held to be as in MRN-mixture of graphite and silicate grains.

In such a manner we have modeled, for each star, the spectral energy distribution (SED) in IR region, the variation of the colour-indexes (U-B, B-V, V-R, V-I, V-J, V-H, and V-K) with the stellar brightness, and the circumstellar extinction curves with the following parameters — dust shell: inner R_i and outer R_o radii, exponent of the dust density distribution law α, optical depth τ; dust grains: minimum a_{min} and maximum a_{max} sizes, index of the size power-law distribution q, and graphite and silicate particles ratio n_C/n_{Si}. For more details about the approach and modelling technique see Krivova (1995).

3   Stars

We considered five HAEBE stars with non-periodic minima of the brightness — UX Ori, WW Vul, BF Ori, CQ Tau, and NX Pup. For each of them, we have collected from the literature the observed fluxes from UV to IR and the brightness variations with time in the U, B, V, R, I bands. The fluxes are reduced for interstellar extinction using the available distances and colour excesses E_{B-V}. For NX Pup the observations in the UV from the IUE and ANS satellites were also used (de Boer 1977; Tjin A Djie et al. 1989).

The main observational characteristics of the stars are listed in Table 1 (top part).

Table 1. Observational characteristics of the stars and parameters of the models

Star	UX Ori	WW Vul	BF Ori	CQ Tau	NX Pup
Maximum visual brightness, m_V	9.61	10.25	9.55	9.48	9.84
Max. brightness variation, Δm_V	2.73	2.37	2.88	1.77	1.32
Spectral type	A1-A4	A0-A2	A0-A2	A2-A8	A0-A2
Luminosity, $L/L_\odot$	55	80	50	25	70
Distance, pc	450	550	400	200	450
Colour excess E_{B-V}, total	0.15	0.35	0.11-0.16	0.6	0.5-0.6
$\quad E_{B-V}$, interstellar	0.06	$\gtrsim 0.2$	--	0.3	0
Density distribution, α	-1.4	-1.5	-1.5	-1.0	-1.3
Optical thickness, τ_V	0.3	0.3	0.2	0.8-0.9	1.0-1.2
Inner boundary R_i, AU	1.5	2	2.5	1	2
Outer boundary R_o, AU	$\sim 4 \times 10^4$	$\sim 2 \times 10^4$	$\sim 5 \times 10^4$	$\sim 3 \times 10^3$	$\sim 3 \times 10^4$

4   Results

In an attempt to understand the parameter dependences of the results, numerous calculations have been made. We have drawn the following conclusions:

- fitting the IR SEDs has a potential for yielding information about the density distribution (α), optical depth (τ), and inner radius (R_i) of the shell;
- fitting the colour-magnitude diagrams enables to strengthen constraints on the optical depth and give an insight into albedo of grains;
- to clear up a point of the dust mixture one has also to take a look at the modelling of CS extinction curves.

Figure 1 presents two models of the dust envelope of NX Pup. The models differ by the grain mixture used. The parameters of the shell are identical to those listed in Table 1.

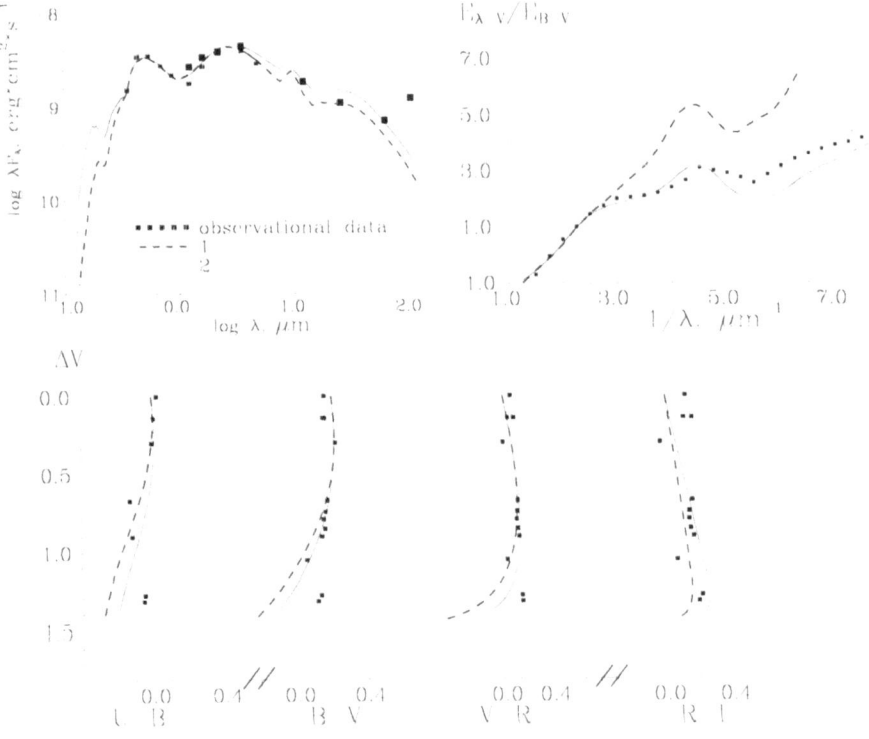

Fig. 1. IR spectra, CS curves, and colour-magnitude diagrams for shell models of NX Pup. Shell parameters: see Table 1. Dust mixture parameters: model 1 — $a_{min} = 0.06\mu m, a_{max} = 0.24\mu m, q = 5.0$ (obtained by Voshchinnikov and Grinin, 1991 for WW Vul); model 2 — $a_{min} = 0.0075\mu m, a_{max} = 0.96\mu m, q = 1.0$, without grains with $a = 0.03 - 0.2\mu m$

The results of the modelling for other stars are also presented in Table 1. In the same manner as in the case of NX Pup, the shell parameters were obtained using more that one grain mixture. It should be noticed that our objects fell into two groups — dust density in the deeper shells seems to have slower decrease to the outside.

Unfortunately, lack of reliable UV data as well as of information about the interstellar extinction in front of the stars makes it difficult to discuss the grain mixture properties.

5 Acknowledgments

We are grateful to Dr. E. Krügel for kind permission to use his radiative transfer code. N. K. acknowledges the support by the KCFE grant 138-2-2.2 for young scientists. V. I. thanks the Alexander von Humboldt Foundation for the financial support.

References

de Boer, K. S. (1977): Far-UV Observations of T-Tau Like Stars. A&A **61**, 605–608

Chini, R., Krügel, E., Kreysa, E. (1986): Dust Emission Spectra from Star-Forming Regions. A&A **167**, 315–324

Grinin, V. P. (1988): On the Blue Emission Visible During Deep Minima of Young Irregular Variables. SvAL **14**, 27–28

Hartmann, L., Kenyon, S. J., Calvet, N. (1993): The Excess Infrared Emission of Herbig Ae/Be Stars — Disks or Envelopes? ApJ **407**, 219–231

Hillenbrand, L. A., Strom, S. E. (1992): Herbig Ae/Be stars — Intermediate-Mass Stars Surrounded by Massive Circumstellar Accretion Disks. ApJ **397**, 613–643

Krivova, N. A. (1995): The Herbig Ae Stars with Algol-like Minima: the Modelling of IR Spectra and Colours Behaviuor During Minima. SvAL (submitted)

Natta, A., Prusti, T., Krügel, E. (1993): Very Small Dust Grains in the Circumstellar Environment of Herbig Ae/Be Stars. A&A **275**, 527–533

Sorrell, W. H. (1990): Constraints on Astronomical Silicate Dust. ApJ **361**, 150–154

Tjin A Djie, H. R. E., Remijn, L., Thé, P. S. (1984): A Study of the Herbig Ae-type Stars UX Ori and CD-44° 3318 Based on IUE Spectra, and on Visual and Infrared Photometry. A&A **134**, 273–283

Voshchinnikov, N. V. (1989): Dust Around Young Stars. Model of Algol-like Minima of UX Orionis Stars. Afz **30**, 309–317

Voshchinnikov, N. V., Grinin, V. P. (1991): Dust Around Young Stars. Model of Envelope of the Ae Herbig Star WW Vul. Afz **34**, 84–94

Voshchinnikov, N. V., Molster, F. J., Thé, P. S. (1995): Circumstellar Dust and Circumstellar Extinction Curves. Ap&SS **224**, 223–226

Investigating the Circumstellar Morphology of Herbig Ae/Be Stars

D. Lorenzetti[1], B. Nisini[1], S. Pezzuto[2], F. Strafella[2], and F. Berrilli[3]

[1] Istituto di Fisica Spazio Interplanetario CNR - Frascati (Italy)
[2] Dipartimento di Fisica - Università di Lecce - Lecce (Italy)
[3] Dipartimento di Fisica - Università di Roma Tor Vergata - Roma (Italy)

Abstract. We present preliminary results obtained by consistently fitting, from UV to radio wavelengths, the observed spectral energy distributions (SEDs) of a selected sample of Herbig Ae/Be stars (HAEBE). An emission model is used to compute the continuum emission emerging from star, circumstellar gas and dust arranged in a spherical geometry. Comparing the numerical results with the observations we explore the sensitivity of the model to different size distributions for dust. Spherical envelopes with rather shallow density gradients and constituted by a dust grain mixture typical of circumstellar environments, seem a suitable model to account for the observed SEDs.

1 Introduction

In the field of the Pre-Main Sequence (PMS) evolution, despite many observational and modeling efforts to understand the morphology of the circumstellar (CS) matter distribution around HAEBE stars no conclusive evidences have been reached. The situation is in fact still controverted although the accountance for the HAEBE composite SEDs, the origin of their variability and the correct interpretation of the observed line profiles and widths, are all facts which critically depend on whether the matter is organized in flat or spherical envelopes. Given the potential capability of the mm-continuum emission to further constrain the modeling of the emitted continuum and then possibly to discriminate among different morphologies of the CS matter, we have defined a sample of HAEBE stars observed from UV to mm-wavelengths. Starting with the compilation of 108 HAEBE stars and candidates published by Thè et al. (1994 *AAS* 104,315), we selected 33 objects detected so far in the mm-continuum. The observed flux densities between 350 μm and 1.3mm are taken from the following papers: Mannings (1994 *MNRAS* 271,587); Osterloh & Beckwith (1994 *ApJ* 439,288); Hillenbrand et al. (1992 *ApJ* 397,613); Henning et al. (1993 *AA* 276,129),(1994 *AA* 291,546); Reipurth et al. (1993 *AA* 273,221); Casey & Harper (1990 *ApJ* 362,663); Giannini et al. (1995 *this issue*). Then we searched the literature for the photometric data (UBVRI, JHKLM, NQ, narrowband "silicate" filters, IRAS, and cm-wavelength observations) obtaining a final sample of 32 HAEBE stars.

2 Model Calculations

To generate a grid of synthetic spectra we use our spherically symmetric model whose details are given in a previous paper (Berrilli et al. 1992 *ApJ* 398,254). Gas and dust, after the interaction with the stellar photons emit typical SEDs which are obtained by considering different emission processes (free-free, free-bound, electron scattering and dust emission). The absorption coefficient k_ν is computed as a function of the radial distance r once the density law $\rho(r) = \rho_0 (r_0/r)^\alpha$ and the temperature law $T_d(r) = T_d(r_d/r)^n$ are given.

To check the sensitivity of our model to different dust mixtures we used two different blends, whose properties are given in Table 1, that have been prepared according to the prescriptions given by Mathis, Rumpl & Nordsieck (1977 *ApJ* 217,425) and using the optical constants by Draine & Lee (1984 *ApJ* 285,89). As one can see the mixture labeled # 2, due to the larger grain sizes, is more appropriate for CS environments while the # 1 is more representative of the interstellar medium.

Table 1. Dust mixtures

#	Composition (in mass)		Grain size (μm) a_{min} a_{max}		A_V/E(B-V)	N_H/E(B-V) $(cm^{-2}mag^{-1})$
1(IS)	C	0.66%	0.005	0.1	3.1	$6 \cdot 10^{21}$
	Si	0.33%	0.04	0.4		
2(CS)	C	0.75%	0.044	0.5	6.7	$9 \cdot 10^{21}$
	Si	0.25%	0.035	0.6		

3   Results

A sample of fits obtained by using the CS mixture (# 2) are reported in Fig. 1. The selection of the best models has been performed with the usual method of the weighted χ^2, where the weights are estimated according to the photometric accuracy.

A quite satisfactory fit to the data points is obtained over a wide range of wavelengths (more than five orders of magnitude), for a quite large number of objects. In total more than 50% (17 out of 32) sources of our initial sample are well fitted by a spherical envelope with a dust mixture typical of CS regions.

The selection of the best models is accomplished without constraining any free parameter, thus we can evaluate the global consistency of our procedure by comparing the parameters derived from the model selection with the values derived from the literature. In Table 2 the model parameters which refer to the 17 well fitted HAEBE are reported. We can see that the sources belong both to Group I and II (1st column), indicating this distinction (Hillenbrand et al. 1992) could not be so compelling. The dust radii (R_d) are similar to those estimated by Hartmann, Kenyon & Calvet (1993 *ApJ* 407,219). Quite flat density gradients (α

is always less than unity) have been already determined for some HAEBE (Natta et al. 1993 ApJ 406,674), and according to the theory of star formation are typical of regions not participating to the central collapse and thus compatible with the PMS nature of the HAEBE. However, a consequence of shallow density gradients is that the millimetric flux, produced by low temperature dust grains, is critically dependent on the choice of a shell radius, implying a careful examination of beam size effects. The models selection has proven to be able to match the observed spectral types (SpT) and visual extinction [A_V, given in the last column as $A_V(CS)/A_V(IS)$]. The correlations between the model and the literature data are shown in Figure 2a and 2b. A global consistency can be recognized although the model tends to slightly overestimate the extinction.

On the contrary, by inspecting the results and the χ^2 selection, we find that when an IS mixture of dust (# 1) is used, very few HAEBE are fitted with marginally consistent values of the model parameters.

Therefore we conclude that the observed SEDs of the majority of HAEBE stars require grains larger than the interstellar ones but not necessarily the presence of CS (accretion) disks.

Table 2. Model Parameters

Source	G	R_d $(R_\star)$	α	n	ρ_0 (cm^{-3})	M_{env} $(M_\odot)$	SpT	A_V (mag)
MacC H12	I	27	0.8	0.5	$3.2\ 10^8$	0.06	A3	5.4 (2.4/3.0)
LKHα 198	II	46	0.8	0.5	$6.3\ 10^8$	0.50	B8	6.9 (6.9/0.0)
V376 Cas	II	152	0.8	0.6	$3.2\ 10^8$	1.58	B2	8.3 (6.3/2.0)
AB Aur	I	41	1.0	0.5	$6.3\ 10^8$	0.03	B9	2.4 (0.9/1.5)
V380 Ori	I	60	0.8	0.6	$3.2\ 10^8$	0.40	B6	3.9 (3.9/0.0)
Haro 13a	II	152	1.0	0.5	$1.0\ 10^8$	0.04	B2	20.3 (3.0/20)
VY Mon	I	35	0.8	0.6	$6.3\ 10^8$	0.25	A0	9.0 (6.0/3.0)
HD 259431	I	69	0.9	0.5	$3.2\ 10^8$	0.12	B5	2.0 (1.5/0.5)
Z CMa	II	35	0.8	0.6	$6.3\ 10^8$	0.25	A0	6.5 (6.0/0.5)
ESO 313-10	I	345	0.5	0.5	$3.2\ 10^5$	0.20	B0	1.8 (0.3/1.5)
He 3-741	I	27	1.0	0.5	$1.0\ 10^9$	0.02	A3	1.7 (1.2/0.5)
IRAS 12496	II	393	0.6	0.6	$1.0\ 10^6$	0.20	O9	7.8 (0.3/7.5)
KK Oph	I	345	1.0	0.6	$3.2\ 10^8$	0.25	B0	3.4 (0.9/2.5)
MWC 297	I	463	0.8	0.6	$1.0\ 10^7$	0.16	O8	8.8 (0.3/8.5)
PV Cep	-	35	0.8	0.5	$6.3\ 10^8$	0.25	A0	12.5 (6.0/6.5)
LKHα 234	I	41	0.3	0.5	$1.0\ 10^6$	1.00	B9	7.2 (7.2/0.0)
MWC 1080	I	69	0.9	0.6	$6.2\ 10^8$	0.25	B5	6.5 (3.5/3.0)

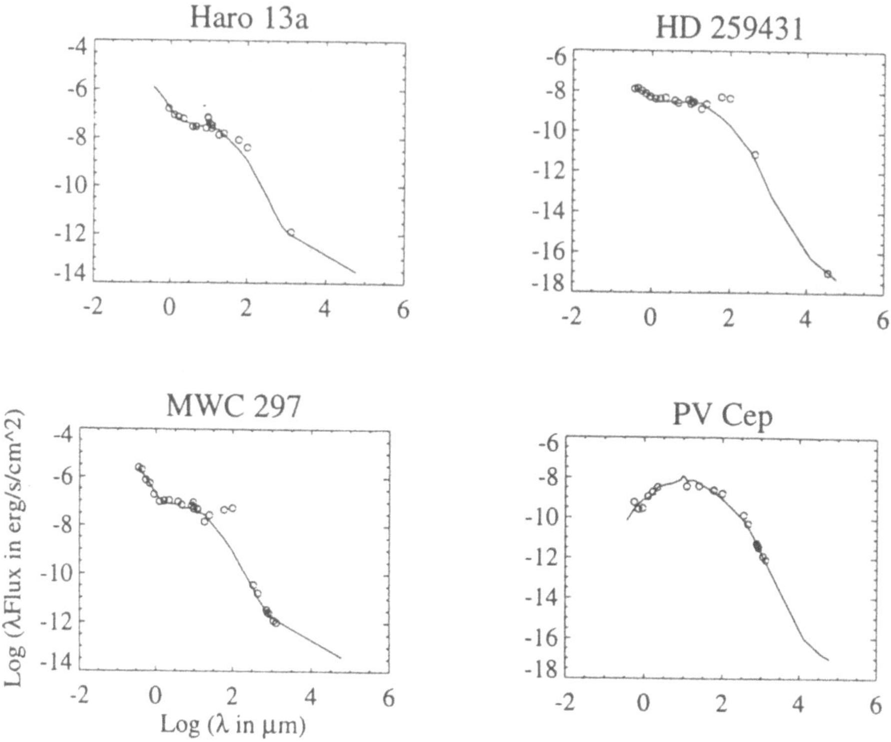

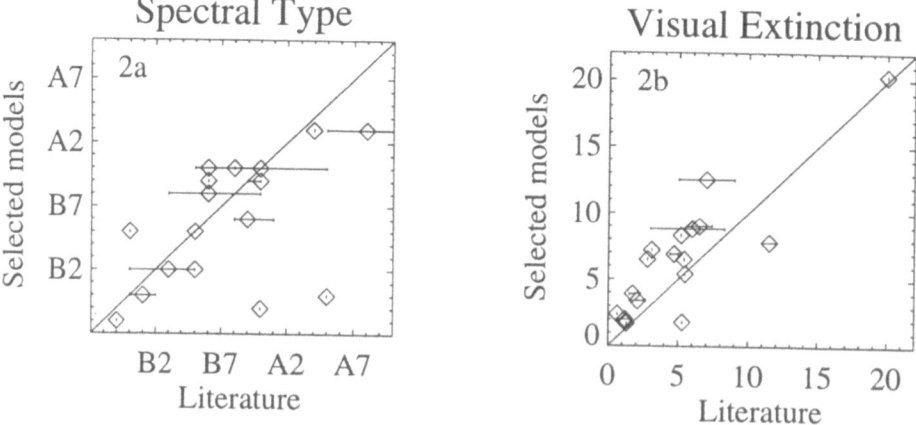

Fig. 1. Some examples of best fits to the observed SEDs

Fig. 2. Correlations between the model parameters and the literature data. **2a** - Spectral types; **2b** - visual extinction

Coronographic Search for Dust Around Main Sequence Stars

David Mouillet[1], Anne-Marie Lagrange[1] and Jean-Luc Beuzit[1]

Laboratoire d'Astrophysique de Grenoble, Université J. Fourier, BP53X,
F-38041 Grenoble Cedex, France

Abstract. The dust around young stellar objects is supposed to be blown out or processed very quickly compared to the stellar lifetime. Nevertheless, large amounts of dust have been detected around β Pictoris and many other main sequence (MS) stars, according to their IR excess. The study of the evolutionary status, origin and lifetime of such environment is related to scenarii of star and planetary system formation, but it is still very poorly observationally constrained. Recent improvements in observational techniques, with the ESO adaptive optics system and a coronograph, allow further investigations of the β Pictoris disk and the characterization of the dust around other MS stars.

1 Circumstellar Dust Around MS Stars

IRAS data evidence the presence of circumstellar dust around a large proportion of MS stars. This phenomenon is observed for stars of all spectral types and ages (see Artymowicz, this volume ; Backman & Paresce 1993). When inverting the IRAS data to derive the distribution and properties of the grains orbiting the stars, many different combinations of the amount, distributions and optical properties of dust appear to reproduce the far IR excesses (Gillet 1986, Backman et al 1992). Theoretical expectations on grain lifetimes can further constrain the possible nature of these grains but this implies some knowledge on the general status of the disk, its age, origin and evolution. In the case of β Pictoris, various observations suggest a very complex system with different components from gas to large solid bodies, including the possible detection of short lived species.

Finally, to be able to describe and understand such circumstellar media, one should take advantage of comparing various kinds of observational data. Among them, spatial information obtained through imaging observations is critically important to discriminate between different components and limit the number of free parameters to interprete the data.

2 How to Image Such Circumstellar Dust?

In the case of β Pictoris, imaging of both the thermal emission (Lagage & Pantin 1994) and the scattered light (Kalas & Jewitt 1995, and references therein) has been performed. We focus here on the observation of the scattered light. This signal is expected to be extended and very faint compared to the central star.

Let us define here what should be the observational capabilities to be able to get valuable information on this signal :

– *very high rejection capability* of the stellar light in order to detect the faint disk with a good S/N.

– to explore the *very close environment of the star* so that to detect some disks around stars more distant than β Pictoris, having thus a smaller angular size. Also in the case of β Pictoris, the close environment is crucial since this is the place where the nature and density of grains may change, with important implications on the origin of the disk.

– *a good angular resolution* in the whole field of view to give a precise description of the disk where it is detected. A resolution smaller than the spatial scales of the disk blurs the information and lead to puzzling inversion problem to derive the actual density distribution of matter (Artymowicz et al 1989).

– also, the observation at *different wavelengths* to constrain the scattering properties of grains.

3   A Coronograph with Adaptive Optics

3.1   Description

The instrumental combination of a coronograph with an adaptive optics (AO) system is very efficient for the previously described purposes. Indeed, the AO provide a very high angular resolution over the whole field, but also a point spread function (PSF) very stable in position and shape. This allows the use of a coronograph with high rejection rates, even with masks substantially smaller than without AO.

In this perspective, we developed a coronograph dedicated to the ESO adaptive optics system (ADONIS) on the 3.6-meter telescope at La Silla. Diffraction limited images in K are performed, which means an angular resolution of 0.12" (Gendron et al 1991). The available occuting masks of the coronograph are small, with projected diameters ranging from 0.3" to 2". The coronograph is composed exclusively of reflecting surfaces to avoid chromatic effects when switching from one wavelength to another. The possible wavelength range is then determined by the camera : J, H and K bands with the SHARP II camera up to now and also L and M bands with the forthcoming COMIC camera. Note that these ranges are complementary with the previous B to I bands observations of β Pictoris.

3.2   Performance

We developed an observational and reduction procedure fitted to this kind of instrument (Beuzit et al 1995). It is based on the comparison with a reference star observed in the same conditions in order to remove the residual diffracted light of the occulted star. Our procedure allows to evaluate the uncertainty level step by step. On Dec 1994 observations, we measured a detection capability of

10^{-5} fainter than the star at 2". Another illustrative example is the detection of a faint companion ($\Delta m = 9$) of HR 4796 situated 4.7", which could have been detected down to 2".

4   First Results and Perspectives

4.1 β Pictoris

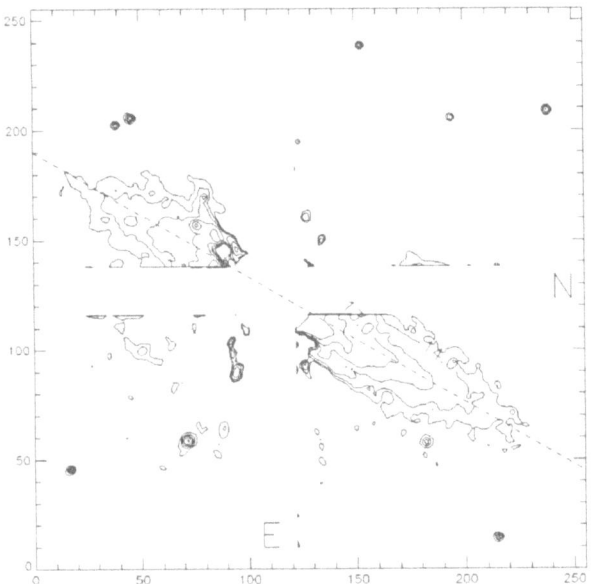

Fig. 1. Observation of the close environment of β Pictoris, December 22-23 1994. This image results from a 6 minute integration time in the K' band. The field-of-view is 13" × 13" with a sampling of 0.05" per pixel, smoothed to 0.15". Quantitative measurements are possible down to 1.5" from the star whereas uncorrected images provide information only down to 6" and tip–tilt correction or anti-blooming detectors provide quantitative information down to 2.4". Regularly spaced isophotes are given from 12 to 14.5 mag arcsec^{-2}

We observed the β Pictoris disk with this instrumental configuration in December 94 and March 95 (Figure 1). The disk is detected down to 1.5" (24 AU) from the star, to be compared with the previous 2.4" best limit (Mouillet et al 1995). Conversely to what was observed further away than 50 AU, the radial brightness distribution does not follow a power law closer to the star but gets flatter. A better angular resolution is also achieved. Since the luminosity gradients of the disks are sharp, the high resolution is very valuable and give more precise indications of the actual vertical and radial distribution of the disk.

In particular, the disk here appears vertically much thinner than previously reported. Further on, we will compare these data to previous ones since they are complementary. This requires to fully take into account the reduction and angular resolution effects. Then, the consistency of the various data will be tested and the constraints on the astrophysical medium derived. Next January, the disk will be observed with a better S/N and in other bands to complete this study.

4.2 Disks around other stars

By now, no other disk has been detected around other IRAS excess MS stars. Disks similar to the β Pictoris one are expected to appear angularly smaller around more distant stars and also fainter in case of fainter IRAS excesses, suggesting that they are less massive. These reasons can explain the so far non detections. The improvement realized with this instrumental configuration in terms of rejecting capability and vicinity to the star justified a new survey, started in August 95. If the dust around other stars has the same structure than the β Pictoris disk, it should be detected for the best candidates. In the case of HR 4796, one of the MS stars with the strongest relative IR luminosity , we detect no disk which allows to exclude a structure similar to the one of the β Pictoris disk. One possibility is that the circumstellar matter is gravitationally confined very close to the star because of its binary status. A sample of about 20 best candidates is to be observed in order to conclude whether the β Pictoris disk is a unique phenomenon or not.

References

Artymowicz, P, Burrows, C, Paresce, F. (1989): The structure of the beta Pictoris circumstellar disk from combined IRAS and coronographic observations. ApJ. **337**, 494.

Backman, D.E. Paresce, F. (1993): *Protostars and Planets III*, 1253–1304

Backman, D.E., Gillet, F.C., Witteborn, F.C. (1992): Infrared observations and thermal models of the β Pictoris disk. ApJ. **395**, 680

Beuzit, J.-L., Mouillet, D., Lagrange, A.-M. (1995): A stellar coronograph for the COME-ON-PLUS adaptive optics system : I. Description and performance A.& A. accepted

Gillet, F.C. (1986): *Light on Dark Matter* (ed. F.P. Israel, D. Reidel Publishing Company), 61–69

Gendron, E. Cuby, J.-G., Rigaut, F. et al. (1991) *Active and Adaptive Optics* (ed. M.A. Ealey, Proc. SPIE, 1542, 297)

Kalas, P. , Jewitt, D. (1995): Asymmetries in the beta Pictoris dust disk. A.J. in press

Lagage, P.O., Pantin, E. (1994): Dust depletion in the inner disk of β Pictoris as a possible indicator of planets. Nature **369**, 628–630

Mouillet, D., Lagrange, A.-M., Beuzit, J.-L., Renaud, N. (1995): Stellar coronograph on the COME-ON-PLUS adaptive optics system : II. First astronomical results. A.& A. , submitted

First Results of a Spectropolarimetric Survey of Herbig Ae/Be Candidates

René D. Oudmaijer[1] and Janet E. Drew[1]

Imperial College of Science, Medicine and Technology, Blackett Laboratory, Prince Consort Road, London, SW7 2BZ, U.K.

1 Introduction

We report the first results of our spectropolarimetric survey of Herbig Ae/Be stars. The circumstellar geometry of these supposedly intermediate mass, pre-main sequence objects is currently controversial. The preferred model of some, based on the interpretation of the strong IR excesses typical of the class, is that these stars are still in an accreting phase, encircled by an active disc (e.g. Hillenbrand et al. 1992). On the basis of the same data, it has also been argued that a more nearly spherical dusty envelope is the dominant factor, with the implication that continuing high levels of accretion are not required (e.g. Hartmann et al. 1993).

Spectropolarimetry is a powerful tool to help distinguish between these extremes. In particular, intermediate resolution spectropolarimetry across the Hα emission line allows one to discriminate between the limiting cases of scattering of continuum light in a circumstellar disc, which would result in depolarization across the line, and scattering of Hα photons within a dusty circumstellar envelope which can give increased polarization instead.

2 Observations

The data were taken during a run with the spectropolarimeter mounted on the 3.9 meter Anglo-Australian Telescope during the nights of 11-13 January 1995. The observations that will be discussed here were taken using the 1200V grating which provides a wavelength coverage from 6335 - 6882 Å, the resulting spectral resolution is ~ 1 Å.

The instrumental set-up was normal; a half-wave plate rotator was set at angles of 0, 45, 22.5 and 67.5° to obtain the Stokes parameters, while a Calcite block was used to separate the light leaving the retarder into perpendicularly polarized light waves, respectively the O(rdinary) and E(xtra-ordinary) rays.

3 Results

In Figs. 1-3 we show the spectra of three objects. Each figure consists of three panels, which show respectively the total spectrum (lower panel), the polarization in % (middle panel), and the position angle Θ in degrees (upper panel).

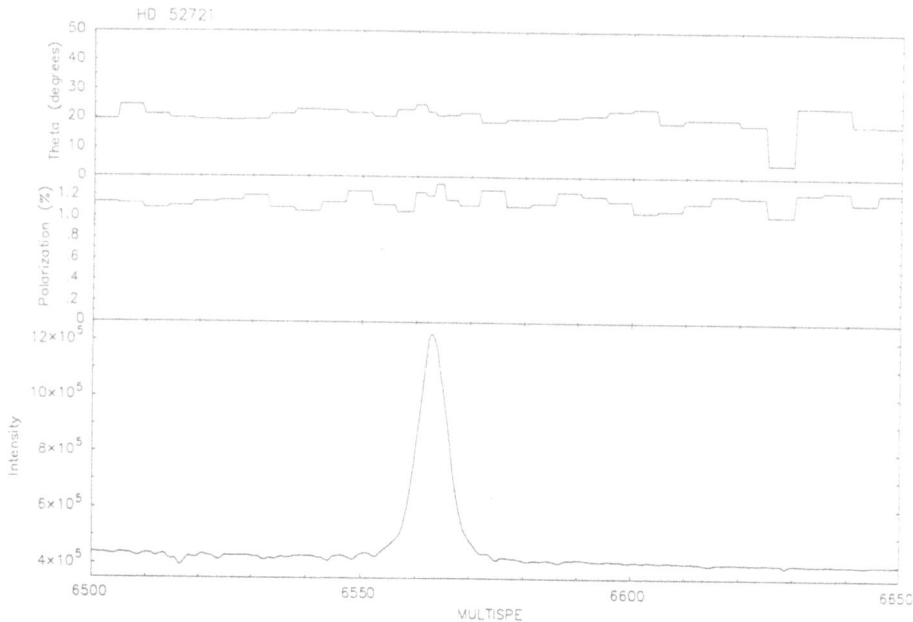

Fig. 1. The polarization spectrum of HD 52721. The lower panel shows the total spectrum, the middle panel presents the polarization in %, while the upper panel shows the polarization angle Θ.

The spectra have been re-binned into variable sizes to a constant error in the polarization of 0.075 % per bin. Note that the spectra have not been corrected for interstellar polarization. Based on these data, the Hα polarization can be distinguished into three classes;

1. No obvious difference in polarization between the continuum and the line (cf. HD 52721, Fig, 1). In this case the polarizing material is not strongly deviant from a spherical geometry. On the other hand, based on the relatively small polarization which could be interstellar, it is possible that the circumstellar extinction is too small to significantly contribute to the polarization.

2. Depolarization in Hα (cf. HD 50138, Fig, 2). HD 50138 shows a strong double peaked emission line. As in the case of Be stars, the emission line is depolarized with respect to the continuum, which is due to the attenuation of the radiation within the scattering medium. Interestingly, the (fainter) blue-shifted emission peak is not depolarized. This implies that the blue emission originates from a smaller volume than the most red-shifted emission. The smaller depolarization provides strong constraints on the geometry of the ionized volume around the star.

3. Enhanced polarization in Hα (cf. HD 45677, Fig. 3). An interesting result is the larger polarization of the Hα line of HD 45677 with respect to the continuum, confirming the result of Schulte-Ladbeck et al. (1992). These authors speculate that the line forming region is located closer to the scat-

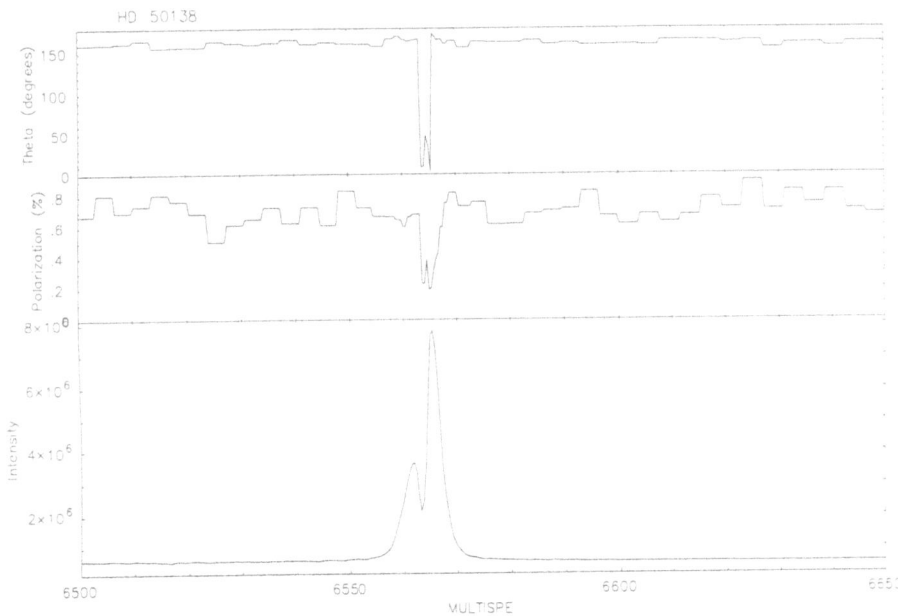

Fig. 2. As the previous figure, now for HD 50138

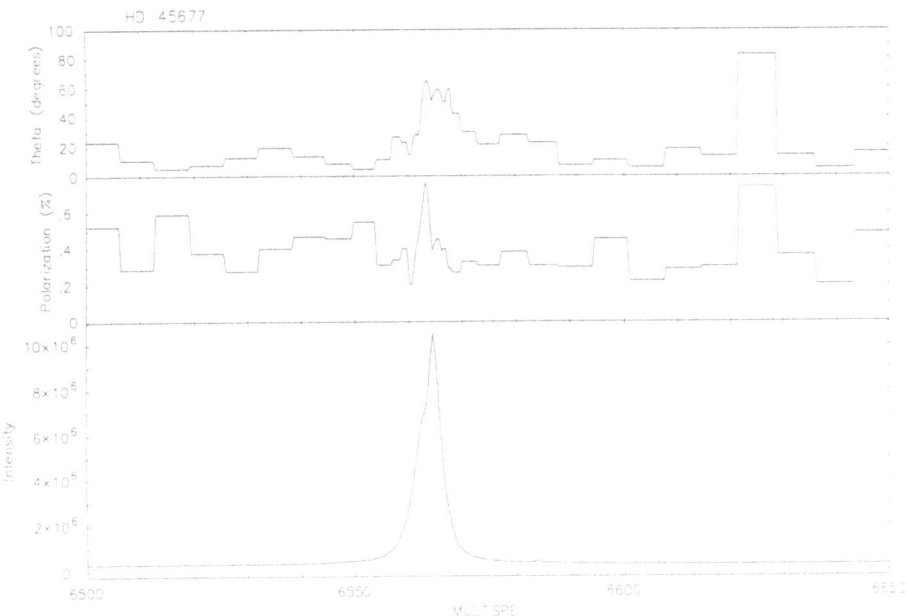

Fig. 3. As before, now for HD 45677

tering medium than the continuum radiator. A new result is the rotation of Θ by about 40-50 degrees with respect to the continuum. It is hard to interpret this new result. Schulte-Ladbeck et al. (1992) conclude from their their broad-band spectropolarimetric observations that the circumstellar material around HD 45677 has a bi-polar geometry. The red stellar light that penetrates the edge-on dusty disk in the line of sight has a position angle of 90° less than the blue light which is scattered off the bi-polar lobes. Possibly, the Hα line emission is being scattered off bi-polar lobes in much the same way as the blue stellar continuum. The rotation by only 40-50° in contrast to the 90° in the blue light however remains as yet unclear.

Acknowledgements This work was entirely funded by the Particle Physics and Astronomy Research Council of the United Kingdom.

References

Hartmann, L., Kenyon, S. J., Calvet, N., 1993, ApJ, 407, 219

Hillenbrand L.A., Strom S.E., Vrba F.J., Keene J. 1992, ApJ 397, 613

Schulte-Ladbeck R.E., Shepherd D.S., Nordsieck K.H., Code A.D., Anderson C.M., Babler B.L., Bjorkman K.S., Clayton G.C., Magalhaes A.M., Meade M.R., Taylor M., Whitney B.A. 1992, ApJ 401, L105

Forbidden Emission Lines in the Winds of Classical T Tauri Stars

Gernot Paatz and Max Camenzind

Landessternwarte Königstuhl, D-69117 Heidelberg, Germany

Abstract. We present a model for the origin of the forbidden emission lines (FELs), especially [OI]$\lambda\lambda$ 6300, 6363 and [SII]$\lambda\lambda$ 6716, 6731, which are observed in the winds of classical T Tauri stars (CTTSs). This model is based on numerical calculations of the wind structure so that for the first time a physically motivated scenario of the FEL formation is given. Our calculations suggest that the low–velocity component (LVC) and the high–velocity component (HVC) of the double–peaked FELs are formed in spatially different regions of the young stellar outflow with the LVCs being formed near the star. The differences between their profiles can be explained by their different critical densities and the rotation properties of the wind in the formation region. The HVC of the lines is formed in a shock which occurs in the collimation region probably due to intersecting flow characteristics. Furthermore, the steepening of the field lines in that region leads to a deceleration of the wind with increasing distance from the star, as also indicated by observations.

1 Introduction

Forbidden emission lines (FELs), e.g. the [OI]$\lambda\lambda$ 6300, 6363 Å and the [SII]$\lambda\lambda$ 6716, 6731 Å lines are a very important indicator for outflow activity associated with young stellar objects such as classical T Tauri stars (CTTS) (Edwards *et al.* 1993, Hirth *et al.* 1994). TTSs in many cases show a double–peaked line profile in the FELs consisting of a high–velocity component (HVC) of about -150 km/s and a low–velocity component (LVC) blueshifted by a few 10 km/s (e.g. Appenzeller *et al.* 1984, Edwards *et al.* 1987; Hirth *et al.* 1994). The blueshifts of the line components can be attributed to the existence of a circumstellar disk obscuring an anisotropic bipolar outflow (Appenzeller *et al.* 1984).

To explain the two velocity components several purely *kinematic* models have been proposed (for details see Paatz & Camenzind 1996) and in all of them the double peak profile results from special projection effects, e.g. a hollow cone geometry as in Edwards *et al.* (1987). Temperatures of 10^4 K necessary for the formation of the FELs can be achieved by shock excitation (Oyed & Pudritz 1993) or ambipolar diffusion heating (Königl 1989, 1994). However, these models cannot account for some observed properties of the FELs, such as the different excitations and formation regions of the two line components (Solf 1989; Solf & Böhm 1993; Hirth *et al.* 1994). Furthermore these authors observe a broad red wing in the LVC of the [OI]λ 6300 line in a number of objects which indicates that this component is formed in a rotating region.

Different to the above–mentioned models, the kinematic model by Kwan & Tademaru (1988) assumes the components of the FELs to be formed in two physically and spatially different regions, namely in a slow disk wind and a fast stellar jet.

In this contribution we introduce a physically motivated scenario of the FEL formation which does not require special projection effects nor the *ad hoc* assumption of two independent flow components with distinct velocities. This scenario is based on numerical modelling of the structure of rotating magnetospheres of TTSs (for details see Paatz & Camenzind 1995) and is supported by recent observations.

2   Magnetically Driven Winds

The observed characteristics of the FELs are closely related to the properties of the TTS winds. The source of these winds may be the star itself as well as the surrounding accretion disk. Presumably they are driven by the magnetic properties of the underlying protostellar system, i.e. Poynting flux which is converted into kinetic energy of the wind (Camenzind 1990). The effenciency of this process depends on the form of the magnetic flux tubes. Both observations (André *et al.* 1992) and theoretical arguments (Camenzind 1990) indicate that the magnetospheres of CTTSs are dominated by a strong stellar magnetic field. The precise form of the flux tubes is however not known, so that we have to rely on approximative treatments (see Paatz & Camenzind 1995). The wind is treated as a stationary, polytropic magnetohydrodynamic flow in an axisymmetric geometry. Initial conditions appropriate for protostellar systems lead to outflow velocities of several hundred kilometers per second both for disk winds and for stellar winds, in good agreement with the observations (Mundt 1993). Due to corotation with the magnetic field the rotation velocity of the plasma (emanating from the disk or the star) reaches values in excess of 200 km/s and begins to decrease around the Alfvén point, where the plasma decouples from the field lines. Both temperature and density decrease with increasing radius and plasma velocity, their quantitative behaviour considerably depending on the polytropic index (Paatz & Camenzind 1995).

3   Theoretical Line Profiles

The above–summarized results indicate the following scenario of the FEL formation: The LVC and the HVC of the FELs form in spatially distinct regions of a stellar or/and a disk wind. The LVC is excited in the stellar vicinity (i.e. at a distance of several AU), where the density in the wind has fallen below the critical density of the line. The emission of this component ceases when the wind temperature becomes too small. As the critical densities of the [OI] lines lie two orders of magnitude above those of the [SII] lines, [OI] forms closer to the star (or the disk), where the rotation velocity of the plasma reaches high values due to corotation with the magnetic field lines. This leads to the formation of the broad red wing in the [OI] LVC (Fig. 1, left), as observed in a number of

objects (Hirth *et al.* 1994) (an analogous blue wing is not seen as it is covered by
the HVC). Due to their lower critical densities the [SII] lines are formed further
out in the wind, where the plasma rotates considerably slower as the corotation
has broken down. For this reason, the LVC of the [SII] lines does not show a
corresponding red wing (Fig. 1, right). The HVC is formed in the collimation
region of the wind where shocks are required to heat the emitting gas. In this
region intersections of the flow characteristics due to the curvature (steepening)
of the magnetic field lines could lead to the formation of shocks and thus to
the excitation of the HVC. Another possible mechanism of shock excitation is
a variable wind mass flux. Evidence for wind velocity variations on time scales
of months to years comes from the observed variations of FEL profiles (Hirth
et al. 1994). Furthermore the steepening of the field lines in the collimation region corresponds to an increase of the flux–tube function which decelerates the
plasma and leads to the observed velocity decrease of the HVC along the flow
direction.

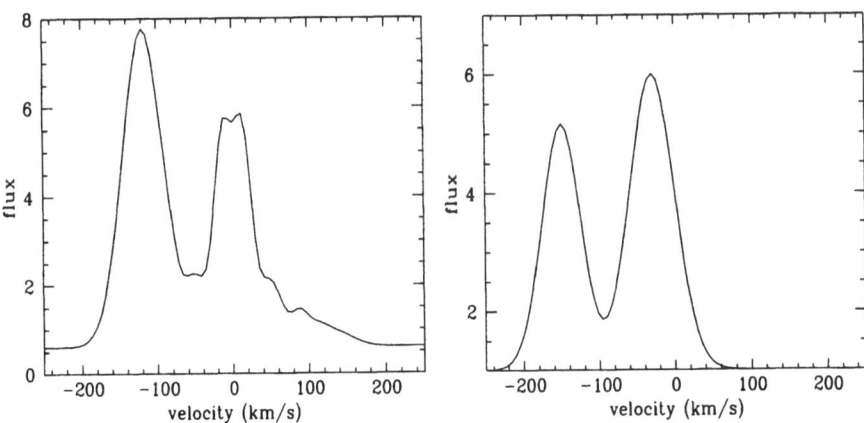

Fig. 1. Calculated profile of the [OI]λ 6300 Å (left) and the [SII]λ 6731 Å (right)
lines. The LVC of the [OI] line shows a broad red wing ranging to about 200 km/s, as
observed in a number of objects. This effect is due to the high rotation velocity of the
plasma in the formation region. Contrary, the LVC of the [SII] line does not show a
red wing. Due to its lower critical density this line is formed further out in the wind
where the rotation velocity of the plasma has considerably decreased. The blueshift of
the HVC of both lines lies at -100 to -150 km/s, depending on the wind velocity and
the viewing angle. The intensity is normalized to the continuum.

The resulting profiles of the FELs can be calculated with the escape probability method (Sobolev 1960) with the calculated wind parameters as input. The
line shapes are reproduced correctly, if the angle i between the jet axis and the
line of sight lies between $i = 10°$ and $25°$. Too small values of i make the red
wing of the [OI] LVC vanish. If, on the other hand, the inclination is too large,
the LVC splits into two components and, furthermore, the blueshift of the HVC

strongly decreases. In the chosen parameter range, the LVC of the [OI] line shows the characteristic red wing which is missing in the [SII] line. The blueshifts of the HVC lie around -150 km s^{-1} (Fig. 1). The width of the HVC results from the velocity profile of the wind in the collimation region, which, on the other hand, depends on the form of the (only poorly known) flux–tube function. The line fluxes lie in the range of 10^{28} erg s^{-1} if $M_w \simeq 10^{-9} M_\odot$ yr^{-1}.

We emphasize that the *spectral* properties of the FELs can in principle be produced in a stellar *or* a disk wind *alone*. This idea has to be checked further in the light of the following remarks concerning the *spatial* properties of the FELs: From the observation of a HVC in [OI] and $H\alpha$ which spatially overlaps with the LVC (both being formed at about 10 AU) in the case of DG Tau, Solf & Böhm (1993) concluded that this HVC is formed in a stellar wind whereas the LVC arises in a disk wind. However, this observation can as well be explained by projection effects (Paatz & Camenzind 1996). Mundt (1984) argues from the observation of broad blueshifted NaD absorption profiles in several objects that the stellar wind must be accelerated to 300 to 400 km/s within two stellar radii which means that the stellar wind cannot contribute to the LVC (which forms at several AU). This observation poses a problem for the wind theory in general.

Acknowledgement: This work is supported by the Deutsche Forschungsgemeinschaft.

References

André, P., Phillips, R.B., Lestrade, J.-F., Klein, K.L. 1991, ApJ 376, 630

Appenzeller, I., Jankovics, I., Östreicher, R. 1984, A&A 141, 108

Appenzeller, I., Mundt, R. 1989, A&A Review 1, 291

Camenzind, M., 1990 in *Rev. Mod. Astron.* 3, Ed. G. Klare, Springer (Heidelberg), 234

Edwards, S., Cabrit, S., Strom, S.E., Heyer, I., Strom, K.M., Anderson, E. 1987, ApJ 321, 473

Edwards, S., Ray, T., Mundt, R. 1993, in *Protostars and Planets III*, Eds. E.H. Levy, J.I. Lumine, University of Arizona Press

Hirth, G.A., Mundt, R., Solf, J. 1994, A&A 285, 929

Königl, A. 1989, ApJ 342, 208

Königl, A., 1994 in *Theory of Accretion Disks – 2*, Eds. W.J. Duschl, J. Frank, F. Meyer, E. Meyer-Hofmeister, W. Tscharnuter, Kluwer Academic Publishers, Dordrecht, 53

Kwan, J., Tademaru, E. 1988, ApJ 332, L41

Mundt, R. 1993 in *Stellar Jets and Bipolar Outflows*, Eds. L. Errico, A. Vittone, Kluwer, Dordrecht, 91

Mundt, R., 1984, ApJ 280, 749

Oyed, R., Pudritz, R.E. 1993, ApJ 419, 255

Paatz, G., Camenzind, M. 1995, A&A, in press

Paatz, G., Camenzind, M. 1996, A&A, in prep.

Sobolev, V.V., 1960, *Moving Envelopes of Stars*, Harvard University Press, Cambridge

Solf, J., 1989 in: ESO Conf. Proc. 33, *Low Mass Star Formation and Pre-Main Sequence Objects*, Ed. B. Reipurth (Garching: ESO), 399

Solf, J., Böhm, K.-H. 1993, ApJ 410, L31

Composition and Distribution of Dust in the β-Pictoris Disk [*]

Eric Pantin[1] and Pierre-Olivier Lagage[1]

CEA/DSM/DAPNIA Service d'Astrophysique F-91191 Gif-sur Yvette Cedex, France

Abstract. In 1984, the mission of the infrared satellite IRAS led to the discover that many sequence stars are surrounded by a tenuous dust disk (by opposition to the thick disks around YSO's). The most famous example is the disk around β Pictoris may be because of a combination of favourable factors. Since then, this disk was observed intensively, but the inner part was remaining unreachable because of the use of coronographical techniques. Using a 10 μm camera we were able to probe the inner part and thanks to a model of thermal emission of the dust, we could deduce the radial density. On the other hand, one of the major unknown lies in the grain composition and size distribution. Trying to take in account some observables of the disk (photometry, spectrum, neutral scattering), we show that we can exclude many of the possibilies in terms of composition and sizes.

1 Introduction

In the Southern constellation of the Pictor, the second star in brightness (β Pictoris) is in the light of the projectors since the IRAS infrared satellite detected that as well as hundreds of other main-sequence stars, β Pictoris shows an unexpected infrared excess (Aumann, 85) attributed to a circumstellar dust disk, maybe planetary. The β Pictoris dust disk remains unique, maybe because of a combination of favourable elements : inclination (edge-on, Smith and Terrile, 84) and remoteness (at 16.6 parsecs). The use of coronographic techniques to remove the starlight prevents to have access to the innermost regions : the disk has been imaged down to 4 arcsec (≈65 AU from the star) in the R-band (Kalas and Jewitt, 95), 2.5 arcsec (≈40 AU) in the K-band using adaptive optics (Mouillet, this conference). We have obtained new 10 μm images of the disk in order to probe the inner regions and get the radial surface density of the dust.

2 Observations

2.1 Why to observe at 10 μm

Contrarily to the visible or near-infrared range, the mid-infrared observations are not critically dominated by the starlight. In addition, the seeing is less limitative in the mid-infrared range. On the other hand, one needs a model of thermal emission of the dust to be able to interpret the data. We have obtained new images

---

[*] Based on observations made at the European Southern Observatory, La Silla

of the disk in November 1994 of the disk using the mid-infrared camera TIMMI placed on the 3.6m ESO telescope (Chili). We used the 10.5-13.3 μm band-pass filter. The air-mass was always below 1.3 and we took very often images of the reference star α Carinae. After image reduction (including removal of the star contribution), we got an image of the dust disk with a flux of 1.1 Jy. Because of a strong blurring of the image by diffraction and seeing, we deconvolved the images using α Car images as an experimental PSF. We compared the resuls of different methods of deconvolution : Maximum Entropy (Skilling and Gull, 84), Multiscale Maximum Entropy (Pantin and Starck, 95), Richardson-Lucy with wavelets regularization (Starck and Murthag, 94); each one gave similar results. The resulting image is shown on Fig.1. The disk is extended and detected up to a distance of 100 AU from the star showing an asymmetry in favour of the S-W extension. We detect a wedge disk with an opening angle of about 15°.

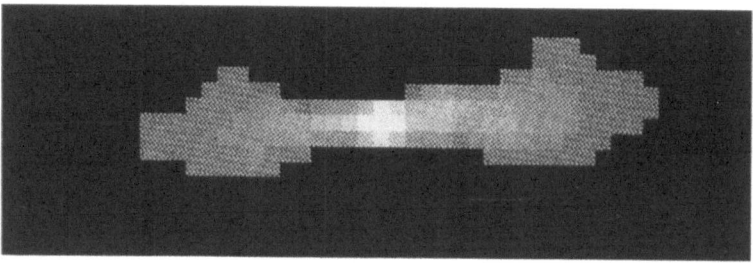

Fig. 1. The β Pictoris dust disk at 10 μm . The pixel field of view is 0.3"/pixel so that each pixel represents a distance of $\approx$ 6 AU in β Pictoris . The image has been deconvolved. The maximum corresponds to a flux of 2.8 Jy/arcsec^2.

3   Modeling of the Thermal Emission of the Dust

Formerly, to study the β Pictoris dust disk, we had used the optical constants of Draine and Lee astronomical silicates (Lagage and Pantin, 94, LP94 hereafter).

Since these constants are not realistic (built artificially to fit the interstellar medium observations) and have a mean albedo below 0.1, we have tried to choose a better material, taking in account the observations of the disk : an albedo between 0.66 and 0.76 (Artymowicz et al., 89) and a 10 μm spectrum showing a broad silicate feature with superimposed crystalline peaks (9.5 and 11.3 μm). We have therefore taken a composition based on a matrix of basaltic glass, with inclusions of crystalline olivine, and calculated absorption coefficients of porous grains (porosity=0.2) using Maxwell-Garnett effective medium theory combined with Mie calculations (Bohren and Huffamn, 83). Using the β Pictoris star spectrum, we could derive the temperature of the grains.

As the radiation pressure effect is very selective on the particles sizes in the β Pictoris system (L=6.5 L$_\odot$), submicronic grains are rapidly expelled from the system (Artymowicz, 88). Moreover, observational results like neutral scattering of the starlight (Paresce et al. 87) or polarization (Gledhill et al., 91) suggest that submicronic grains cannot be dominant in the β Pictoris system. On the other hand, other observational evidences infer the presence of submicronic grains : too large grains would produce a too flat 10 μm silicate feature uncompatible with the 10 μm spectrum (Knacke et al., 93), observations with an array of 3x3 bolometers at 10 and 20 μm (Telesco et al. 88). Small grains could be replenished by mechanism such as: collisions of larger ones, dust supply by cometary sublimation (Lagrange-Henry et al., 88, Beust et al., 90) in competition with submicronic blow-out. We have therefore chosen a size distribution following a power-law, with a lower steep for the small grains: $n(a)da \propto a^{-2}$ if $a \leq 2\mu m$, $n(a)da \propto a^{-4.2}$ otherwise and cut-off in sizes at 0.1 and 1000 μm.

4   Getting the Radial Density

We consider the equation giving the projected flux F_{iobs} as a function of the radial profile F(r). Assuming azimuthal invariance on one extension (N-E or S-W) of the density, one has :

$$F_{iobs}(y) = 2 \int_{r=y}^{r=rmax} F(r) \frac{r}{\sqrt{r^2 - y^2}} dr \; , \tag{1}$$

where the axis y is perpendicular to the line of sight. In reality, the experimental data F_{iobs} are corrupted with noise, leading to an ill-posed problem, making the use of the Abel transform inefficient. We have proceeded to the inversion of the equation (4) by regularizing the problem by the cross-entropy (Shannon, 48, Jaynes , 57), in order to obtain the simpliest solution compatible with the data. Once the radial flux obtained, we derived the surface density by comparing the deprojected flux to the flux emitted by a constant density. The resulting surface density is shown in Fig.2. As in LP94, the density has an inner clearing zone ($r \leq 40$-50 AU). Moreover, as seen on Fig.2, the existence of the "inner void" doesn't depend on the chosen material, and therefore, cannot be an artifact created by the processing chain to get the density.

5   Conclusion

Using the 10 μm images of the β Pictoris dust disk to deduce the density, we have constructed a thermal model of the infrared emission of the particles to be able to interpret the data. We have tried to find a grain material that fullfills the observational constraints as close as possible, however, there might be other good candidates (choosing for instance a different matrix material). The following uncertainty on the density at distances larger than 50 AU could be fixed thanks to the coronographic observations giving another measure of the density (at 65 AU from the star in the visible), or taking images of the disk at 20 μm

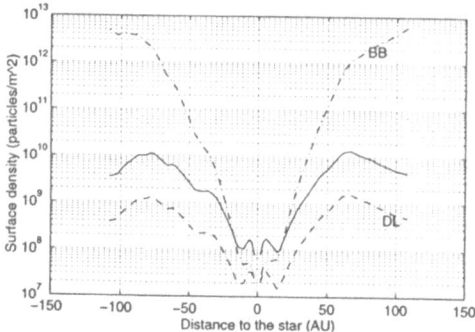

Fig. 2. The surface density of the particles. The plain line is the density obtained using glassy basalt grains with crystalline olivine inclusions. The other ones (DL, blackbody) show the effects of the grains optical properties on the density (considering Draine and Lee silicates or blackbodies).

sensitive to colder grains (at larger distances) and giving an model-independent measure of the temperature of the grains.

Acknoledgments : We wish to thank Pawel Artymowicz for valuable discussions and suggestions.

References

Artymowicz, P., Burrows, C. & Paresce, F. 1989, Ap.J., **337**, 494

Artymowicz, P. 1988, Ap.J.Lett, **335**, L79

Aumann, H. H. 1985, Pub. Astr. Soc. Pacif., **97**, 885

Backman, D. E., Gillet, F. C., & Witteborn, F. C. 1992, Ap. J, **385**, 670

Beust H., Lagrange-Henri, A. M., Vidal-Madjar, A. & Ferlet , R. 1990, A&A. **236**, 202

Bohren, C.F. and Huffmann, D.R. 1983 *Absorption and scattering of light by small particles*, Wiley, New York.

Draine, B.T. 1985, Ap.J.Supp., **57**, 587

Jaynes, E.T. 1957, Phys.Rev., **106**, 620

Kalas, P. & Jewitt, D. 1995, Ap.J., **110**, 794

Knacke, R. F., Fajardo-Acosta S. B., Telesco C. M., Hackwell J. A., Lynch D. K. & Russell, R. W. 1993, Ap. J., **418**, 440

Lagage, P.O. & Pantin, E. 1994, Nature, **369**, 628

Lagrange-Henri, A. M., Vidal-Madjar, A. & Ferlet, R. 1988, A&A, **190**, 275

Lucy, L. B. 1974, Ap. J., **79**, 745

Mouillet, D., ESO proceedings of the conference *Stardust*

Pantin, E. & Starck, J.L., submitted to A&A

Richardson, B. H. 1972, J. Opt. Soc. Am., **62**, 55

Shannon, C.E., 1948, Bell System Tech. J., **27**, 379

Skilling, J. 1989 in *Maximum Entropy and Bayesian methods, Cambridge, England, 1988* (ed. J.Skilling), 45

Starck, J. L. & Murtagh, F. 1994 A&A, **288**, 342

Telesco, C. M. & Knacke, R. F. 1991 Ap. J. Lett., **372**, L29

Telesco, C. M., Becklin, E. E., Wolstencroft, R. D. & Decher, R. 1988, Nature, **335**, 51

Observations of Binaries and Envelopes in Herbig Ae/Be Stars

Andrea Richichi[1] and Christoph Leinert[2]

[1] Osservatorio Astrofisico di Arcetri,
   Largo E. Fermi, 5 – 50125 Firenze, Italy
[2] Max–Planck–Institut für Astronomie,
   Königstuhl 17, 69117 Heidelberg, Germany

Abstract. We have conducted a survey of Herbig Ae/Be stars by near-IR speckle interferometry, similar to a recent one on T Tauri stars (Leinert et al. 1993). Although the sample is numerically smaller, there is some indication that the occurrence of binaries and circumstellar envelopes might be higher among the Herbig Ae/Be stars than in their lower mass counterparts. As an illustration of the variety of observational evidence encountered in this survey, we discuss – also in the light of new additional measurements – three examples: V380 Ori is a binary star whose components have comparable spectral energy distributions, while in the case of HK Ori, it appears that the companion is much redder than the (visual) primary. Finally, relatively extended structure is resolved around Elias 1, and we discuss its geometry and brightness ratio with varying wavelength.

1 Introduction

High angular resolution observations of Herbig Ae/Be stars are important, in order to detect the presence of circumstellar material, to determine its geometry, to identify the physical processes responsible for the emission and ultimately to constrain the properties of the particles. Also, it is becoming increasingly evident that at least in the lower–mass counterparts, i.e. T Tauri stars, binarity is the rule rather than the exception (Ghez et al. 1993, Leinert et al. 1993, Richichi et al. 1994, Simon et al. 1994). It is then interesting to establish whether this applies also to Herbig Ae/Be stars, and if possible to understand the implications for what concerns the formation and initial evolution of massive stars. Also intriguing is the question of the occurrence of circumstellar envelopes, their mass and their shape. We have investigated such characteristics by observing at several wavelengths and, when possible, by using speckle polarimetry.

Partly because of their larger distance and relative scarcity, and partly because HAe/Be stars have begun attracting major interest only in recent years, observations of these sources by high angular resolution techniques, and by speckle interferometry in particular, have generally been less extensive than for T Tauri stars. In this contribution, we present our database of observations collected in the course of a program started already several years ago, and we present results for three case studies: V380 Ori, HK Ori and Elias 1. This is part of a more extensive work that is about to be submitted for publication.

2   Observations and Data Reduction

The observations were carried out mostly at the 3.5m telescope of the Calar Alto Observatory, using the locally available instrumentation for IR speckle interferometry: the slit–scanning specklegraph, the $1 - 5\mu m$ IR camera, and more recently the new 256x256 MAGIC camera. Some additional observations were also collected at La Silla, using the 3.6m telescope with the 1-D IR specklegraph and the NTT with the SHARP instrument.

Data reduction is performed using software packages developed in the group for high angular resolution at the MPIA. These include the standard Fourier processing for the computation of the visibilities, and methods such as Knox-Thompson and bi-spectral analysis for the reconstruction of the object phase. We have developed also fitting programs that allow us to derive the parameters of best models to the Fourier components by binary and/or envelope models. We also use image space sharpening algorithms such as Shift–and-Add or Lucy's deconvolution.

3   Results

Our database of speckle observations covers $\approx$30 objects. The details can be found in Leinert et al. 1994. The 30 objects in this list have quite an inhomogeneous observational coverage, from a few individual observations to more extensive studies, some of which have been published in individual papers already (Leinert et al. 1991, Haas et al. 1992, Leinert et al. 1993). Work is currently in progress, to analyze the bulk of the data and publish a detailed list of results. In this contribution, we limit ourselves to three case studies, which provide a good illustration of the diverse morphological features encountered on HAe/Be stars.

3.1   HK Ori

The star has a companion at 0.33″ (see Fig. 1). Judging from the spectral energy distributions that we derive, it seems that the the brightest star at $\lambda > 1\mu m$ is probably only a faint companion at shorter wavelengths, a situation often encountered also in T Tauri stars. Infrared speckle interferometry is the ideal technique for revealing such companions.

3.2   V380 Ori

Also this star has a nearby companion at 0.165″ (Fig. 2), but here the two components appear to have the same spectral energy distribution (as confirmed also by data taken at 0.9μm and not shown here). It is interesting to note that the visibility at H is better fitted by using a model with an additional halo component. If confirmed, this is most certainly scattered emission since it appears to vanish at the longer wavelengths.

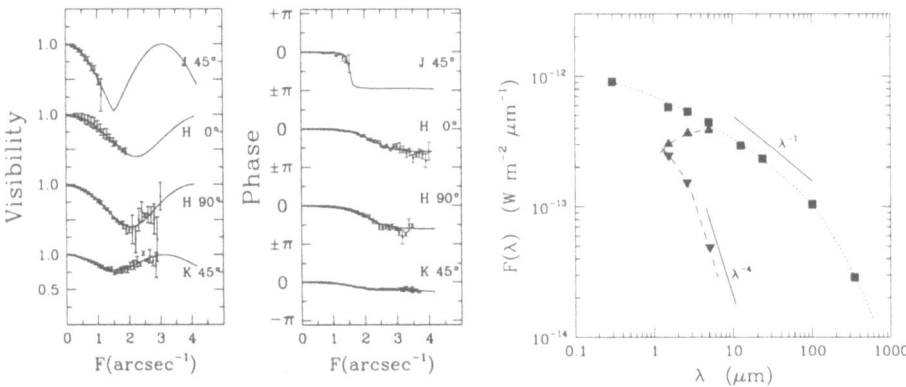

Fig. 1. *Left:* 1-D visibility and phases for HK Ori, at various PAs and wavelengths. The companion has $d = 0.33 \pm 0.01''$ at PA=48 ± 1°. *Right:* Spectral energy distribution of HK Ori as a whole (squares), and separated into the two components (triangles).

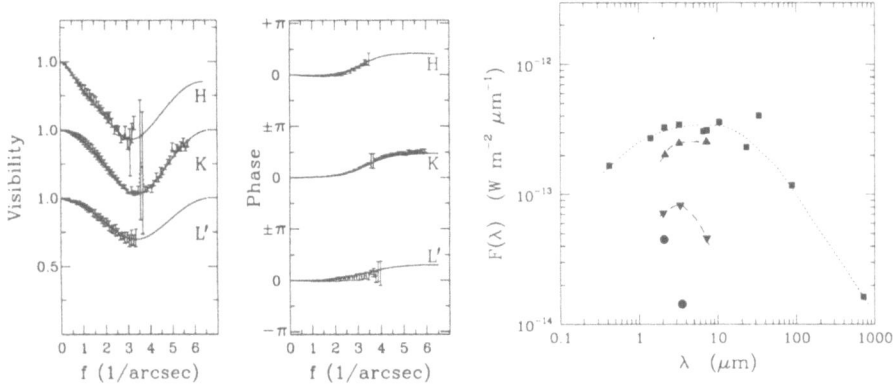

Fig. 2. *Left:* 1-D visibility and phases for V380 Ori, at various PAs and wavelengths. The companion has $d = 0.065 \pm 0.005''$ and PA=202 ± 2°. *Right:* Spectral energy distribution of HK Ori as a whole (squares), and separated into the two components (triangles).

3.3   Elias 1

This star shows a third and different scenario. There is a relatively distant companion, at about 4.1″ to the NE, well detected in our shift–and–add images. The companion has Δmag≈ 4 with about the same colors as the primary. Elias 1-B is found to be a VLA source. Furthermore, speckle interferometry shows the presence of a scattering halo around the primary. In Fig. 3 it is shown how the size differs greatly in the NS and EW direction (particularly at J and H), and how the intensity rapidly decreases with wavelength. Evidence points at a scattering halo with a flattened geometry.

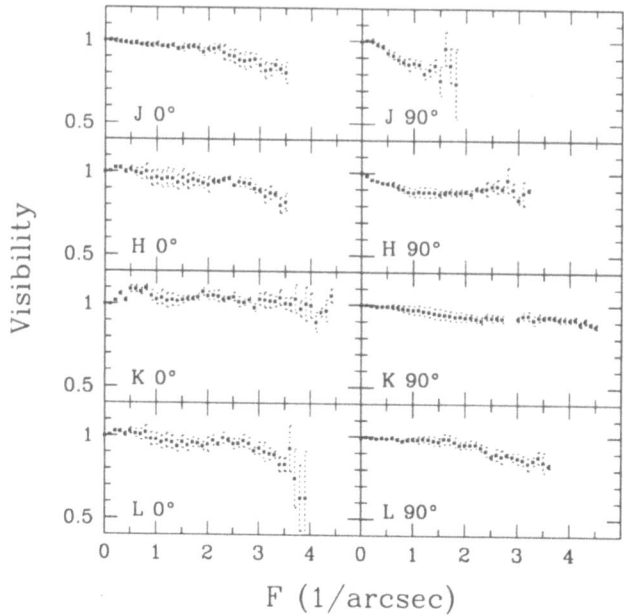

Fig. 3. 1-D visibilities at different PAs and wavelengths for Elias 1, indicative of the presence of a scattering halo with a flattened geometry.

As a final comment, we note that the 7 stars out of our sample of 30 sources are binary (KK Oph, Z CMa, HR 5999, MWC 1080 in addition   to the three above). Very tentatively, and being aware of the smallness of the sample, it appears that the binary frequency among HAe/Be is probably higher than those of solar-like stars in our neighbourhood, as is already the case for the T Tauri stars at least in Taurus and Ophiuchus.

References

Ghez, A., Neugebauer, G., Matthews, K. 1993, AJ, 106, 2005
Leinert, Ch., Zinnecker, H., Weitzel, N., Christou, J., Ridgway, S.T., Jameson, R., Haas, M., Lenzen, R. 1993, A&A, 278, 129
Richichi, A., Leinert, Ch., Jameson, R., Zinnecker H. 1994, A&A, 287, 145.
Simon, M., Ghez, A.M., Leinert, Ch., Allen, D.A., Chen, W.P., Howell, R.R., Jameson, R., Matthews, K., Neugebauer, G., Richichi, A. 1994, ApJ, 443, 625
Leinert, Richichi, Weitzel, Haas 1994, *Proceedings of the 1st Conference on the Nature and Evolution of HAEBE Stars*, Thé, Pérez and van den Heuvel, eds.
Leinert, Ch., Haas, M., Lenzen, R. 1991, A&A, **246**, 180
Haas, M., Leinert, Ch., Lenzen, R. 1992, A&A, **261**, 130
Leinert, Ch., Haas, M., Weitzel, N. 1993, A&A, **2716**, 535

Dust Shells Around Herbig Ae/Be Stars

Nikolai V. Voshchinnikov

Astronomical Institute, St. Petersburg University, St. Petersburg, 198904 Russia

Abstract. The properties of the dust grains and dust shells of Herbig Ae/Be stars obtained from the modelling of circumstellar extinction. Polarization curves are discussed. The minimum size of circumstellar grains appears to increase with the stellar luminosity L_*. It is also found that the mass of dust is proportional to L_*.

1 Dust Grains in Circumstellar Shells

1.1 Stars

Dust shells are the attributes of the young (evolving) and old (evolved) stars. The first group consists of the cocoon stars, T Tauri stars, and Herbig Ae/Be (HAeBe) stars. Their dust shells are the remnants of the clouds from which the stars were originated. The second group includes red giants and supergiants, Miras, RCB stars, novae, etc. Dust grains are formed in the atmospheres and outflows of these stars. Note that the observational effects of the circumstellar (CS) dust grains as well as the light scattering processes operating in dust shells are the same in both groups of the stars.

1.2 Observational Indications

Dust grains located in the CS space produce the following observational effects.

1. *Infrared (IR) excess* appears at wavelengths $\gtrsim 1\mu m$ as an extra radiation above the star flux calculated from the models of stellar atmospheres.
2. *Anomalous extinction* is observed as the deviations of the extinction curves from the standard interstellar extinction law. The peculiarities of these curves for hot stars are well seen in the ultraviolet (UV) (e.g. Sitko et al. 1981).
3. *The blueing effect* is connected with the special behaviour of tracks on the colour-magnitude diagrams when a star fades. At first stages, the stellar radiation becomes redder but after some turning point it alters to bluer. This effect is well known for HAeBe stars (see Thé 1994 for discussion) and was also observed for T Tauri stars (Grinin et al. 1995). For late-type stars, the blueing effect is difficult to distinguish due to the overlapping molecular lines.
4. *The polarization* of stars with dust shells is variable. The ratio of polarization degree to star colour excess is usually larger than that for the interstellar medium.

1.3   Light Scattering

The effects mentioned are the consequences of the absorption and scattering of star radiation by CS grains. During the modelling, the interaction between the radiation and dust can be taken into account in the following form.

1. *The line-of-sight extinction* of star light in the shell.
2. *The scattering* of star light.
3. *The absorption and re-emission* of radiation by dust grains.

For the major part of stars, we cannot resolve the CS shells in a telescope. Then, the observed flux from a star with dust shell consists of three components

$$F_{obs}(\lambda) = F_*(\lambda) \exp\left[-\tau^{ext}(\lambda)\right] + F_{sca}(\lambda) + F_{IR}(\lambda) , \qquad (1)$$

where first term depends on the optical path of star light in the shell only. Other two components are determined by the dust optical properties and the dust distribution around the star.

Note also that the observed IR excesses arise due to the absorption and re-emission of the radiation. Other three observational effects mentioned in Section 1.2 appear as a result of the combined influence of the line-of-sight extinction and the scattered radiation.

1.4   Dust Components

So-called 'big' grains with the sizes $a \gtrsim 0.01\,\mu m$ which optical properties are described by Mie theory are usually used to model the light scattering in dust shells from the far-UV to far-IR.

In the UV, the polycyclic aromatic hydrocarbons (PAHs) may also influence the line-of-sight extinction (Joblin et al. 1992). The contribution of PAHs to the near and mid-IR emission, apparently, is rather important (see, e.g. Natta et al. 1993). At the same time, due to their very low albedo, the radiation scattered by PAHs can be neglected.

1.5   Location of Dust in the Shells

All dust grains located on the ray between the star and observer are responsible for the extinction described by first term of Eq. (1).

Because of the dilution, the scattered light becomes unimportant at large distances from the star. Its intensity drops with the distance as $\propto r^{-2}$.

The contribution of different shell areas to the IR fluxes is determined by dust temperature. At the near-IR, hot grains located near the star are the most bright source of radiation. The cold dust is responsible for the major part of the far-IR radiation coming from the outer layers.

2   Modelling

2.1   Types of Models

Many models treat the dust shells as spheres. They are mainly focused on the interpretation of the IR spectra and can be used to get some information about big grains and PAHs located at the different distances around the stars. Note that in the frame of the spherical models the consideration of the CS extinction and the blueing effect is also possible (e.g. Krivova and Il'in 1995).

To explain the polarimetric observations, some asymmetry of the scatterers in the form of non-spherical grains or/and non-spherical shells must be introduced. Monte Carlo simulation is used most often in order to calculate the multiple scattering of polarized radiation in non-spherical geometries. However, this technique is ineffective in the case of calculations of grain temperature and dust emission. Thus, Monte Carlo approach may be useful for the modelling of the CS extinction, polarization and the blueing effect. As a result, we estimate the properties of big grains located in the inner parts of dust shells where the scattered radiation is still strong.

In any case, it should be borne in mind that all results obtained from the modelling are *model-dependent*.

2.2   Used Model

Below we discuss some properties of dust shells around HAeBe stars obtained from the modelling of the curves of CS extinction and polarization in the visual-UV part of spectrum. Our model is based on the special modification of Monte Carlo method (see Voshchinnikov and Karjukin 1994 for details).

We considered oblate spheroidal shells with various aspect ratios (A/B), viewing angles (i), and the optical thicknesses along the main semiaxis of the shell $[\tau_{90}^{ext}(\lambda_0)]$. In general, the power-law density distribution of dust grains was assumed

$$n_d(r) = \begin{cases} 0, & r < r_0, \\ n_d(r_0)\left(\frac{r}{r_0}\right)^{-s}, & r_0 \leq r < r_{out} . \end{cases} \tag{2}$$

Here, r_0 is the inner radius of the dust shell, in which the number density of grains is equal to $n_d(r_0)$ and r_{out} its outer radius.

The properties of CS dust were modeled by changing the parameters of the silicate-graphite mixtures of grains (MRN mixture; Mathis et al. 1977): the lower (a_-) and upper (a_+) size cutoffs, the index of the power-law size distribution (q), and the ratio of the number densities of silicate and graphite grains (n_{Si}/n_C).

3   Results and Discussion

3.1   Program Stars

Table 1 summarizes the dust model parameters found for seven HAeBe stars. The stellar luminosities were determined from the spectral types of stars given by Thé et al. (1994). The dust model parameters were obtained in the result of the interpretation of the observations of the extinction in the UV and visual, and the visual polarization. For UX Ori and WW Vul, we included the results of Voshchinnikov et al. (1988), Voshchinnikov and Grinin (1992) where single scattering model was applied to explain the behaviour of the colour-magnitude diagrams and the wavelength dependence of polarization (for WW Vul only).

Table 1. Characteristics of dust shells around HAeBe stars obtained from modelling

Star	$L_\star$ $(L_\odot)$	Dust grains				Dust shell					A (AU)	M_{dust} $(M_\odot)$
		a_- (μm)	a_+ (μm)	q	$\frac{n_{Si}}{n_C}$	$\frac{A}{B}$	i $(°)$	$\tau_{90}^{ext}(V)$	$\frac{r_0}{B}$	s		
AB Aur	41.0	0.020	0.25	3.5	2.0	2.0	30	0.88	0.1	1.0	270	$1.0\,10^{-6}$
UX Ori	40.4	0.04	0.15	3.5	1.07	4.0	90	0.3	0.0	0.0	190	$2.8\,10^{-8}$
HD 45677	510	0.05	0.25	5.0	1.07	2.0	90	1.06	0.1	0.0	1020	$5.1\,10^{-6}$
HD 259431	204	0.015	0.22	3.2	1.07	2.5	90	1.2	0.1	0.5	590	$5.2\,10^{-6}$
HR 5999	36.0	0.01	0.25	3.6	0.75	1.5	90	0.6	0.1	0.0	150	$8.9\,10^{-8}$
WW Vul	21.4	0.055	0.25	5.0	0.25	3.0	90	0.4	0.05	0.0	220	$3.8\,10^{-8}$
HD 200775	650	0.05	0.25	5.0	0.5	5.0	65	2.0	0.01	0.0	1160	$4.2\,10^{-6}$

The CS extinction curves were calculated in the following form

$$A^{cs}(\lambda) = -2.5 \log \left\{ \exp\left[-\tau_i^{ext}(\lambda)\right] + F_{sca}(\lambda)/F_\star(\lambda) \right\} , \qquad (3)$$

where $\tau_i^{ext}(\lambda)$ is the optical path of the stellar radiation inside the shell in the direction of an observer. Equation (3) gives the usual interstellar extinction if the scattered radiation is neglected.

The details of the modelling can be found in the papers of Voshchinnikov et al. (1995a) for HD 45677, Voshchinnikov et al. (1995b) for AB Aur, HD 259431 and HD 200775, and Thé et al. (1995) for HR 5999.

3.2 Grain Size

Table 1 shows that the parameters of CS dust grains are generally distinguished from those of standard MRN mixture ($a_- = 0.005\ \mu m$, $a_+ = 0.25\ \mu m$, $q = 3.5$, $n_{Si}/n_C = 1.07$). The importance of grain parameters was discussed by Voshchinnikov et al. (1995b) where it is shown that the minimum size of dust mixture determines the behaviour of the extinction curves in the UV, while the variations of a_+ are the lest important.

As follows from Fig. 1, the minimum cutoff of dust size distribution, apparently, grows with the stellar luminosity. This is quite natural because the smallest grains are uncharged and are quickly swept out by the radiation pressure (see

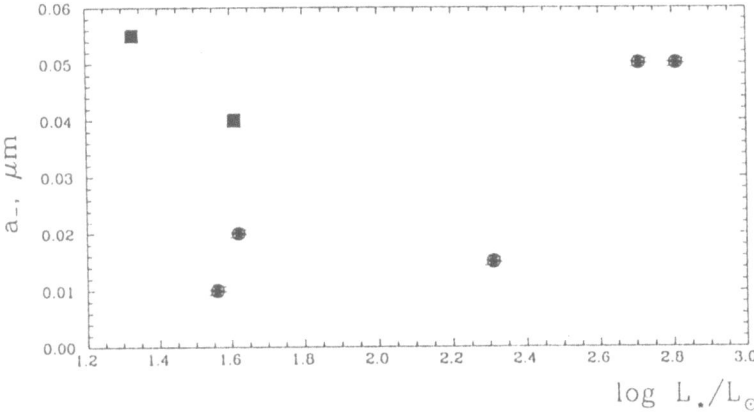

Fig. 1. Dependence of the minimum size of dust mixture on stellar luminosity. The data for UX Ori and WW Vul are shown by squares

Il'in and Voshchinnikov 1993 for discussion). Note that the extinction of UX Ori and WW Vul in the UV, where it is the most sensitive to the changes of a_-, did not modeled.

3.3 Shell Shape

The visible shape of dust shells can be calculated from the model parameters A/B and i. It mainly determines the level of the stellar polarization in the model used that is seen from Fig. 2 where the intrinsic polarization of HAeBe stars near their brightness maxima is shown.

The data presented indicate the distribution of dust grains around the stars considered is far from to form disk-like structures. This conclusion is made from the modelling with smooth radial dust distributions. Apparently, the dust clumps coming close to the star are also able to contribute to the polarization.

3.4 Scattered Radiation

The scattered radiation presented in dust shells can be estimated quantitatively if we introduce the ratio of the CS ("scattered") optical thickness to the "extinction" one $\tau_i^{cs}(\lambda)/\tau_i^{ext}(\lambda)$

$$\exp\left[-\tau_i^{cs}(\lambda)\right] = \exp\left[-\tau_i^{ext}(\lambda)\right] + F_{sca}(l)/F_*(\lambda) \ . \tag{4}$$

If $F_{sca}(\lambda) = 0$, then $\tau_i^{cs}(\lambda)/\tau_i^{ext}(\lambda) = 1$, whereas $\tau_i^{cs}(\lambda)/\tau_i^{ext}(\lambda) < 1$ in the presence of scattered radiation. This ratio in visual is plotted in Fig. 3 for program stars. Figure 3 shows that the influence of the scattered radiation is moderate for all stars excluding AB Aur. In the shell of AB Aur, the scattered radiation is more important because it has a smaller viewing angle and an excess of silicate grains with high albedo in the visual.

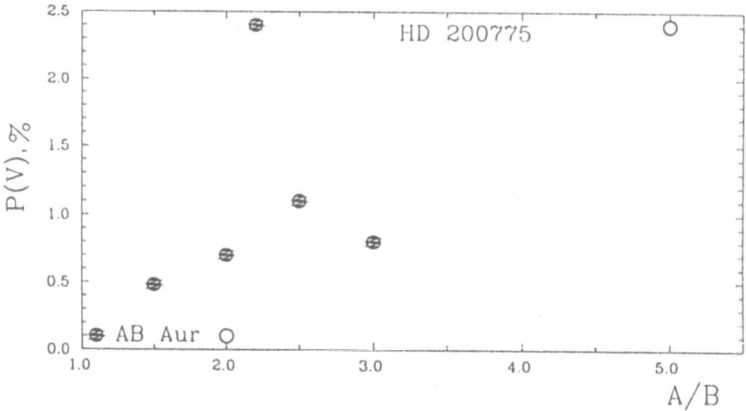

Fig. 2. Dependence of the observed polarization in the V band (for HR 5999 in the B band) on the "visible" aspect ratio of dust shell obtained from the modelling. For AB Aur and HD 200775 the model values of A/B are also shown

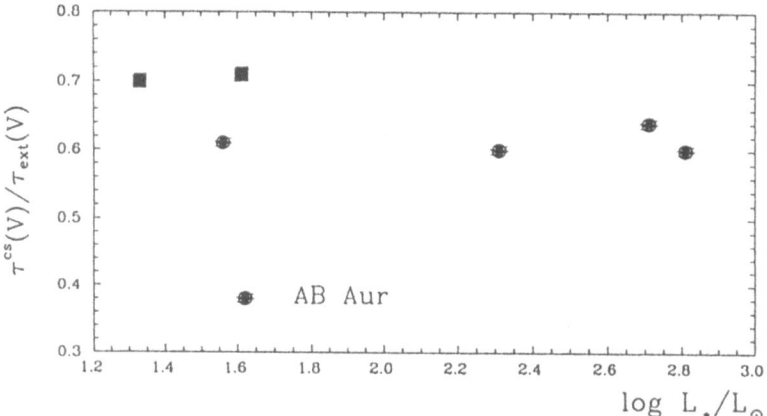

Fig. 3. Ratio of the calculated CS optical thickness to the model "extinction" one in dependence on stellar luminosity

3.5   Dust Mass

We also found the mass of the CS dust responsible for the extinction and polarization. The calculated dust masses are plotted in Fig. 4 by filled symbols. Because our model is independent of the absolute values of the inner and outer shell sizes, we estimated the distance of the outermost boundary assuming that there the intensity of the scattered radiation is 10^3 times smaller than at the inner boundary r_0, i.e. $A \approx \sqrt{1000}r_0$. The values of r_0 were scaled according to the relation

$$r_{0_\star} = r_{0\,WW\,Vul} \left(\frac{T_\star}{T_{WW\,Vul}} \right)^2 , \qquad (5)$$

where $r_{0WW\,Vul} = 7$ AU (Il'in and Voshchinnikov 1993). The effective stellar temperatures were found from the spectral types of stars.

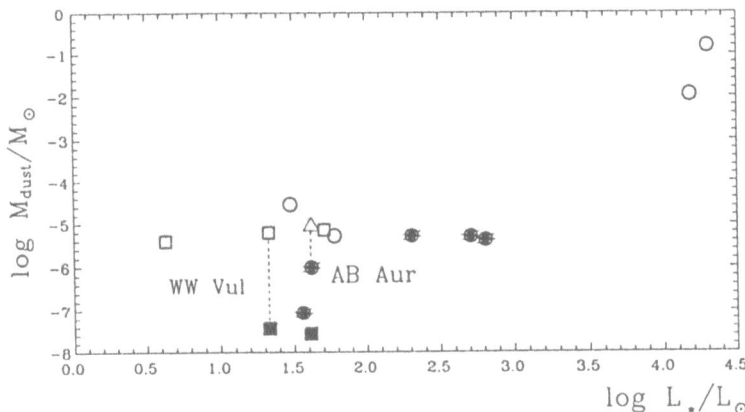

Fig. 4. Mass of CS dust in dependence on stellar luminosity. The data for "scattering" dust are taken from Table 1 and are shown by filled symbols. The mass of "emitting" dust is represented by open symbols ($\bigcirc$ – Henning et al. 1994; $\square$ – Friedemann et al. 1994; $\triangle$ – Sorrel 1990)

In Fig. 4, open symbols show the mass of CS dust for several HAeBe stars as given from the modelling of IR fluxes in spherical shells. These data are related to the "emitting" dust which occupies the large CS environments. The gas masses from Henning et al. (1994) were recalculated using the gas to dust ratio $M_{gas}/M_{dust} = 10^3$. Note also that Friedemann et al. (1994) took the outer radius of the shell for WW Vul as large as 4700 AU. After the scaling to our value (220 AU), we obtain the mass $M_{dust} \approx 5.5\,10^{-8}\,M_\odot$, i.e., rather close to that of the "scattering" dust for this star (see Table 1).

The general behaviour of data plotted in Fig. 4 may be considered as some specific mass – luminosity relation. It shows that the masses of the star and the CS dust shell are tightly connected that tells about their common origin.

4   Conclusions

Our results can be summarized as follows.

1. The modelling of the circumstellar extinction and polarization curves may be used to conclude about the properties and spatial distribution of big grains located in the inner parts of dust shells where the contribution of the scattered radiation is important.

2. For Herbig Ae/Be stars, the minimum grain size and the mass of circumstellar dust, apparently, increase with the growth of the stellar luminosity.

3. The spatial distribution of circumstellar dust grains responsible for the scattered radiation is far from being disk-like.

References

Friedemann, C., Reimann, H.-G., Gürtler, J. (1994): Cloudy circumstellar dust shells around young variable stars. Ap&SS **212**, 221–229

Grinin, V.P., Kolotilov, E.A., Rostopchina, A. (1995): Dust around young stars. Photopolarimetric observations of the T Tauri star BM And. A&AS **112**, 457–473

Henning, Th., Launhardt, R., Steinacker, J., Thamm, E. (1994): Cold dust around southern Herbig Ae/Be stars. A&A **291**, 546–556

Il'in, V.B., Voshchinnikov, N.V. (1993): Dynamics of dust grains in the shells of Ae Herbig stars. Astron. Rep. **37**, 362–367

Joblin, C., Léger, A., Martin, P. (1992): Contribution of polycyclic aromatic hydrocarbon molecules to the interstellar extinction curve. ApJ **393**, L79–L82

Krivova, N., Il'in, V. (1995): Multiwavelength study of HAEBE stars with Algol-like minima. This volume

Mathis, M.L., Savage, B.D., Meade, M.R. (1977): The size distribution of interstellar grains. ApJ **217**, 425–433

Natta, A., Prusti, T., Krügel, E. (1993): Very small grains in the circumstellar environment of Herbig AeBe stars. A&A **275**, 527–533

Sitko, M.L., Savage, B.D., Meade, M.R. (1981): Ultraviolet observations of hot stars with circumstellar dust shells. ApJ **246**, 161–183

Sorrell, W.H. (1990): Constraints on astronomical silicate dust. ApJ **361**, 150–154

Thé, P.S. (1994): The photometric behaviour of Herbig Ae/Be stars and its interpretation. in: The Nature and Evolutionary Status of Herbig Ae/Be Stars, eds. P.S. Thé et al., ASP Conference Series **62**, 23–29

Thé, P.S., de Winter, D., Pérez, M.R. (1994): A new catalogue of members and candidate members of the Herbig Ae/Be (HAEBE) stellar group. A&AS **104**, 315–339

Thé, P.S., Pérez, M.R., Voshchinnikov, N.V., van den Ancker, M.E. (1995): The variable Herbig Ae star HR 5999: XII. Its circumstellar extinction law. A&A, in preparation

Voshchinnikov, N.V., Grinin, V.P. (1992): Dust around young stars. Model of the shell Ae-Herbig star WW Vul. Afz **34**, 84–95

Voshchinnikov, N.V., Grinin, V.P., Kiselev, N.N., Minikulov, N.Kh. (1988): Dust around young stars. Polarization observations of UX Ori in deep minima. Afz **28**, 182–193

Voshchinnikov, N.V., Karjukin, V.V. (1994): Multiple scattering of polarized radiation in circumstellar dust shells. A&A **288**, 883–896

Voshchinnikov, N.V., Molster, F.J., Thé, P.S. (1995a): Circumstellar extinction and circumstellar extinction curves. A&SS **224**, 223–226

Voshchinnikov, N.V., Molster, F.J., Thé, P.S. (1995b): Circumstellar extinction of pre-main-sequence stars. A&A, in press

Silicate Dust Around β-Pic-Like Stars

H. J. Walker[1], H. M. Butner[2], D. Wooden[3], and F. Witteborn[3]

[1] CLRC Rutherford Appleton Laboratory, Chilton, Didcot, Oxon, OX11 0QX, UK
[2] Dept of Terrestrial Magnetism, Carnegie Institution of Washington, 5241 Broad Branch Road N. W., Washington, DC 20015, USA
[3] NASA Ames Research Center, MS:245-6, Moffett Field, CA 94035, USA

Abstract. Silicate dust features at 10μm have been observed in a few β-Pic-like stars. Recent data from HIFOGS (the High Resolution Faint Object Grating Spectrograph) are presented here, and show a wide range of different types of silicate feature. The data are compared with those of other authors, spectra from Young Stellar Objects, comets and Interplanetary Dust Particles (IDPs). Our spectra are found to resemble those due to Young Stars and IDPs.

1 Introduction

Silicate emission was first detected from the disc of β Pic by Telesco and Knacke (1991), using spectrophotometry. They compared the feature they found to that produced by comets like Halley. More recent spectra of β Pic were compared in detail, by Knacke et al. (1993), with comet spectra and found to produce a good match. They also compared the silicate feature from β Pic with that from several other sources, and showed that data from Draine and Lee (1984), a Young Star (GW Ori) and M stars did not fit very well. They concluded that the silicate dust around β Pic had undergone some heating.

The data for GW Ori came from a paper by Cohen and Witteborn (1985), using FOGS data. They found the silicate feature (at 10μm) in young stars had many different shapes. GW Ori had a sharper feature compared to many in the paper. Hanner et al. (1994a) observed Elias 1 and found a feature at 11.2μm in addition to the main silicate emission feature. This feature was attributed to PAHs (Polycyclic Aromatic Hydrocarbons). They also noted a feature around 11.06μm and possibly others at 11.6 and 11.76 μm. Apart from these extra features, Hanner et al. judged that the silicate feature in Elias 1 arose from normal astronomical silicates.

Earlier work by Hanner et al. (1990) on comets, identified a feature in Comet Bradfield around 11.3μm, which was attributed to olivine. Hanner et al. (1994b) showed that comets had a variety of silicate emission features (at 10μm). Bradley et al. (1992) worked on the 10μm feature in Interplanetary Dust Particles (IDPs), and showed that these too had a wide variety of shapes, dependent on the precise composition of the IDP examined. They showed two samples, labelled glass-rich (with silicate crystals) which gave good fits to data from comets Halley and Bradfield, supporting the suggestion of the feature in comets arising from crystalline silicates.

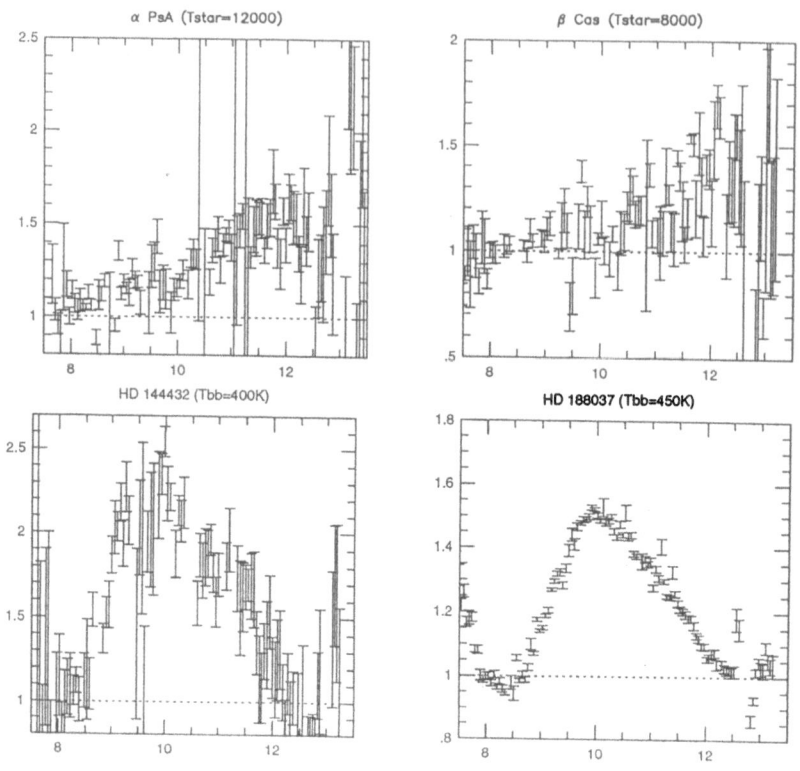

Fig. 1. 10μm data for sources observed in 1994 (a) α PsA (b) β Cas (c) HD 144432 (d) HD 188037. Wavelength in μm is plotted against the ratio of total to blackbody flux.

2   Observations

Only β Pic among the four prototype Vega-like stars (Gillett, 1986) has an excess at 12μm according to the IRAS data. Vega is still used as a calibration source at 10μm. Observations by us of α PsA (see fig. 1), show that the excess emission due to the thermal dust continuum is just starting to appear around 10μm. Sources from the Walker and Wolstencroft (1988) lists of Vega-like candidates have been observed by Sylvester et al. (1995), and they found three objects with silicate emission and three objects with emission features they attribute to PAHs.

HIFOGS (the High Resolution Faint Object Grating Spectrograph) works in the region 8 – 13 μm, see Witteborn et al. (1991) for a full description. There are 120 detectors covering this wavelength range, giving a resolution of 0.05μm. Our data come from two observing runs on NASA IRTF, with excellent weather on the most recent run. The data in fig. 1 and fig. 2 have the atmospheric transmission removed, however a little noise remains around 10μm due to the subtraction of the atmospheric ozone feature. An underlying blackbody continuum is also removed from the data, to show the silicate feature clearly. The exact temperature of the blackbody chosen can influence the shape of the silicate feature.

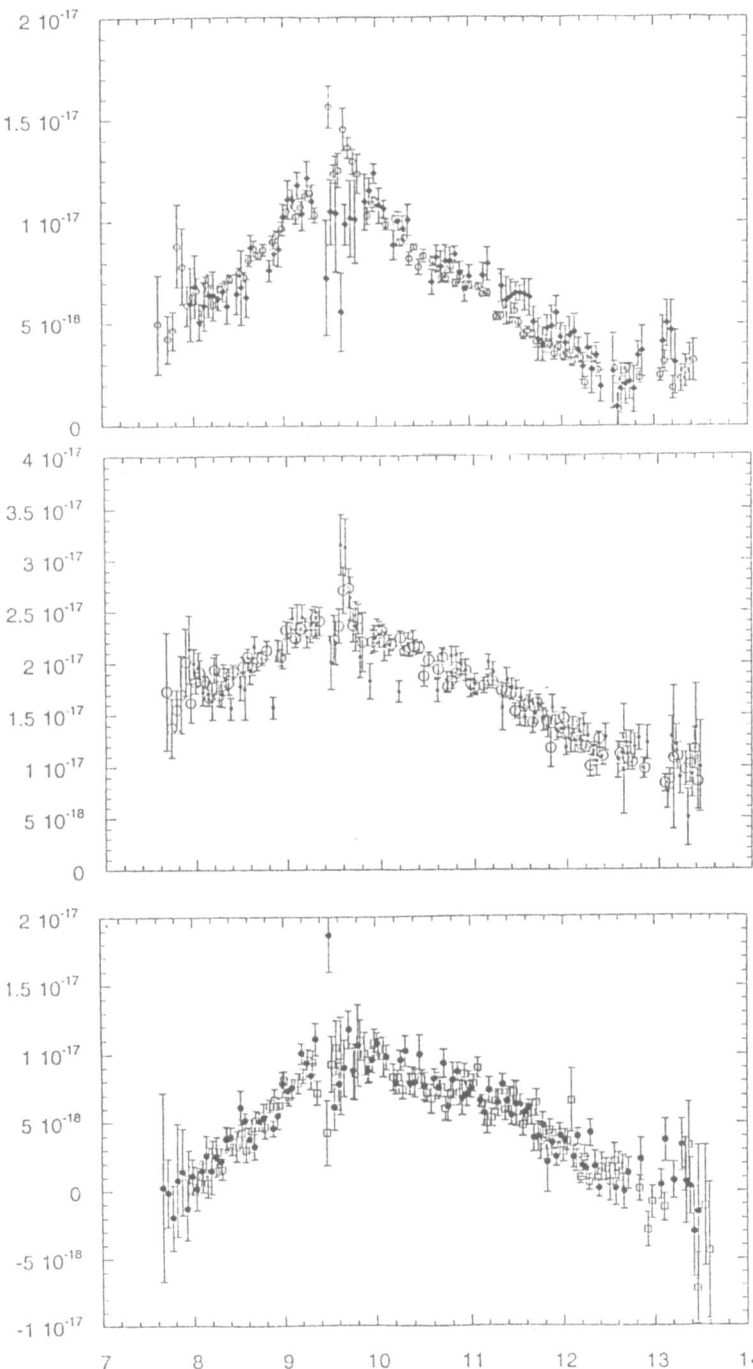

Fig. 2. 10μm data for sources observed in 1995 (a) HD 144432 (b) HD 142666 (c) 51 Oph. Wavelength in μm is plotted against relative flux in w.cm^{-2}.μm^{-1}.

3   Discussion

HD 188037 (SAO 87856) has been reported to have an M star companion. Fig. 1 confirms this, in that the silicate feature found here resembles that due to normal astronomical silicates. The feature shown for HD 144432 (SAO 184124) appears sharper (fig. 1 and fig. 2), and is reminiscent of that due to GW Ori. HD 142666 (SAO 183956) has a broad silicate feature (fig. 2), which resembles that found for HD 98800 (SAO 179815) by Sylvester et al. (1995). Both α PsA and β Cas (fig. 1) show the start of a thermal dust continuum in the 10μm region. Our data show no PAH-like emission features in any of the sources we observed. Sylvester et al. report the features in three sources HD 169142 (SAO 186777), HD 135344 (SAO 206462) and HD 141569 (SAO 140789).

We have observed 51 Oph, which is a Be star with a dust disc, reported by Fajardo-Acosta et al. (1993) to have silicate emission similar to β Pic and comets. They note that the dust is warmer, and conclude that the grains will be larger than those around β Pic. Sylvester et al. (1995) have also observed it, but make no comment on the emission, apart from noting that it is silicate emission. Our own data have been carefully reduced (fig. 2), with the atmospheric transmission effects removed, and a 550K blackbody subtracted. The feature appears unusual due to the broad shoulder in the emission beyond 10μm, and one of the glass-rich IDPs from Bradley et al. (1992) looks very similar in shape to that which we find in 51 Oph. This was not so evident in our earlier data (Walker and Butner, 1995).

The work is not yet at a stage where conclusions can be reached, but the data we (and others) have collected suggest that there may be a variety of environments giving rise to a β-Pic-like dust disc.

References

Bradley, J. P., Humecki, H. J. and Germani, M. S. (1992): ApJ **394**, 643

Cohen, M., and Witteborn, F. (1985): ApJ **294**, 345

Draine, B. T., and Lee, H. M. (1984): ApJ **285**, 89

Fajardo-Acosta, S. B., Telesco, C. M. and Knacke, R. F. (1993): ApJ **417**, L33

Gillett, F. C. (1986): *Light on Dark Matter* ed. F. Israel, (Reidel, Dordrecht), 61

Hanner, M. S., Newburn, R. L., Gehrz, R. D., Harrison, T., Ney, E. P., and Hayward, T. L. (1990): ApJ **348**, 312

Hanner, M. S., Brooke, T. Y., and Tokunaga, A. T. (1994a): ApJ **433**, L97

Hanner, M. S., Lynch, D. K. and Russell, R. W. (1994b): ApJ **425**, 274

Knacke, R. F., Fajardo-Acosta, S. B., Telesco, C. M., Hackwell, J. A., Lynch, D. K. and Russell, R. W. (1993): ApJ **418**, 440

Sylvester, R. J., Barlow, M. J. and Skinner, C. J. (1995): ApSS **224**, 405

Telesco, C. M. and Knacke, R. F. (1991): ApJ **372**, L29

Walker, H. J. and Butner, H. M. (1995): ApSS **224**, 389

Walker, H. J. and Wolstencroft, R. D. (1988) PASP **100**, 1509

Witteborn, F. C., Bregman, J. D., Rank, D. M. and Cohen, M. (1991): *Proc. of the 1991 North American Workshop on Infrared Spectroscopy,* ed. R. E. Stencel (Boulder: Colerado) 29

Part V

Properties of Dust Around Young Stellar Objects

Grain Properties in Different Phases
of the Interstellar Medium

G. Barbaro[1], A. Di Bartolomeo[1], P. Patriarchi[2], and M. Perinotto[3]

[1] Dipartimento di Astronomia,Università di Padova, Padova, Italy
[2] GNA/CNR, Osservatorio Astrofisico di Arcetri, Firenze, Italy
[3] Dipartimento di Astronomia e Scienza dello Spazio, Università di Firenze, Firenze, Italy

1 Introduction

While a variety of data is available on the extinction properties in several lines of sight, attempts to correlate them with the characteristics of the interstellar medium have been barely successful. Detailed analyses exist in the literature but they are limited to few interesting lines of sight (e. g. the regions of Orion and ρ Ophiuchi). This limitation does not allow to outline the general properties of the dust grains in the different phases of the interstellar medium.

On the other hand approaches involving a larger number of lines of sight are restricted only to the examination of the shape of the extinction curves or, when attempting a correlation with the characteristics of the involved interstellar medium, did not actually succeed in clearly separating regions with different dust properties, also due to the difficulty to have a sufficiently extended sample of homogeneous data.

A determination of the dust properties in the *pure* phases of the interstellar medium seems fundamental to the study of the grains and of their evolution. By *pure* phases we mean physically well separated regions as molecular clouds, HI regions and HII regions can considered to be. In the following we are indeed restricting our attention to these three main phases of the interstellar medium.

We have faced the task of searching for relations between two important extinction parameters of the dust, i.e. the reddening E(B–V) and the total-to-selective extinction $R_V = A_V / E(B-V)$, and the status of the associated gaseous component whether ionized, neutral or molecular, as it is represented by the respective column densities.

Our data base consists of about seventy lines of sight, where the required quantities have been collected as follows:

1. The HI and H_2 column densities come from the sample of Savage *et al.* (1977); the data for other four stars have been taken from Shull & Van Steenberg (1985) and Bohlin *et al.* (1983);
2. E(B–V) and the spectral type are taken from the above sources;
3. R_V has been derived either by a direct method (Cardelli, Clayton & Mathis, 1988) using infrared photometric observations of Thé *et al.* (1986), Johnson

(1966) and Gezari *et al.* (1987) or by the UV extinction curves of Savage *et al.* (1985), Aiello *et al.* (1988) and Papaj, Wagner & Krelowski (1992) according to the relations of Cardelli, Clayton & Mathis (1989).

4. The HII column density has been estimated from a simple Stroemgren sphere model with an assumed density of 10^3 cm^{-3}. This essentially because of lack of proper observational data. Also a density of 10^2 cm^{-3} has been used to test the influence of this choice.

2   Data Analysis

First of all we have searched for the relation between E(B–V) and the column densities. Assuming that each line of sight successively crosses the three *pure* regions, the solution of the equation of the radiative transfer gives:

$$A_\lambda = 1.086[k_\lambda(HI)N_d(HI) + k_\lambda(H_2)N_d(H_2) + k_\lambda(HII)N_d(HII)] \quad (1)$$

k and N_d being respectively the extinction coefficients and the dust column densities. By using the dust–to–gas ratios $N_d(HI)/ N(HI)$, $N_d(H_2)/ N(H_2)$ and $N_d(HII)/ N(HII)$, N(HI), N(H$_2$) and N(HII) being the gas column densities, one obtains:

$$E(B - V) = a(HI)N(HI) + a(H_2)N(H_2) + a(HII)N(HII) \quad (2)$$

Using this expression as fitting function, the least square method applied to our sample gives the following values for the coefficients:

$$a(HI) = 7.77 \cdot 10^{-23} \qquad (7.76 \cdot 10^{-23}) \qquad (3)$$

$$a(H_2) = 6.17 \cdot 10^{-22} \qquad (6.19 \cdot 10^{-22}) \qquad (4)$$

$$a(HII) = 5.02 \cdot 10^{-23} \qquad (1.06 \cdot 10^{-22}) \qquad (5)$$

In parentheses are the results for a density of n=10^2 cm^{-3} in the HII regions. These coefficients can be interpreted as the $E(B - V)/N$ ratios in the *pure* regions, where N equals N(HI), N(H$_2$) or N(HII), respectively.

In a N(H$_2$)/N(HI) vs. N(HII)/N(HI) diagram, the zone where the extinction is dominated by the dust associated with the HI regions can be singled out by the condition:

$$a(HI)N(HI) > a(H_2)N(H_2) + a(HII)N(HII) \quad (6)$$

while analogous conditions define the H$_2$–dominated and HII–dominated zones. In this way three categories are introduced and a line of sight belonging to one of them approximates the behaviour of the dust of a *pure* region.

We mention that various lines of sight HI–dominated resulted to have R_V values considerably larger than the value generally attributed to the diffuse medium.

These larger values are considered to be rather typical of regions like Orion where the HII regions are dominant. A possibility is that for these objects the HII column density might have been underestimated, as further HII regions, excited by neighbouring stars, could be present in the foreground. Were their contribution added, these line of sigths would move into the the HII–dominated category. This problem deserves clearly further analysis, because in principle the mentioned possibility cannot be ruled out also for the H_2–dominated lines of sight, although only two of them present large R_V values.

	(1)	(2)
HI–dominated	3.61	3.64
H_2–dominated	3.16	2.34
HII–dominated	3.99	7.69

Table 1. R_V values: 1) mean values obtained for the three categories; 2) values derived from the least square fit

The correlation of the total–to–selective extinction R_V with the column densities can be looked for by the least square method as well.

It is also easy to derive the following expression:

$$R_V = \frac{a + bN(H_2)/N(HI) + cN(HII)/N(HI)}{d + eN(H_2)/N(HI) + fN(HII)/N(HI)} \qquad (7)$$

With this non–linear function a least square fitting of all our data provided the coefficients a, b, c, d, e and f. From them we obtained values of R_V representative of the *pure* phases. They are shown in Table 1 together with the mean values of the observed R_V in the three categories: HI–dominated, H_2–dominated and HII–dominated.

The R_V values derived for the pure regions by the non–linear interpolation are affected by a considerable uncertainty due to the large number of coefficients involved. This can explain the large difference existing between this evaluation and the average R_V values derived from the categories selected from the sample.

3 Conclusions

Analyzing a large sample of lines of sigth, we have determined on a relatively good statistical basis the behaviour of important parameters of dust and gas separately for regions where the gas is in molecular, neutral or ionized status:

1. Dust grains in the ionized medium tend to raise the value of R_V and depress the FUV extinction compared with the "mean" interstellar medium, a well

known effect in Orion and other similar regions. On the contrary, dust residing in regions of molecular gas tends to reduce R_V and correspondingly to
raise the FUV extinction. This behaviour had been observed in some lines
of sight (e.g. HD 62542 and HD 204827).

2. The $E(B-V)/N(gas)$ ratio is much higher in the molecular dominated regions
   than in the neutral dominated ones. This suggests an higher dust-to-gas
   ratio in the molecular regions.
3. The interpolation formulae we have derived fix some constraints on the extinction cross section averaged over the distribution of grain sizes and the
   composition of the dust in the *pure* regions.
4. In the literature the evaluation of the H column densities has been usually
   done by ignoring the contribution of the ionized gas. Clearly this is a bad
   approximation at least for the HII-dominated lines of sight.

	(1)	(2)	(3)
HI-dominated	$7.77 \cdot 10^{-23}$	$1.18 \cdot 10^{-22}$	$6 \cdot 10^{-23} - 3 \cdot 10^{-22}$
H_2-dominated	$6.17 \cdot 10^{-22}$	$1.65 \cdot 10^{-22}$	$1 \cdot 10^{-22} - 3 \cdot 10^{-22}$
HII-dominated	$5.03 \cdot 10^{-23}$	$5.37 \cdot 10^{-23}$	$5.03 \cdot 10^{-23} - 8 \cdot 10^{-23}$

Table 2. $E(B-V)/N$: 1) coefficients of the least squares fit of $E(B-V)$; 2) mean values
obtained for the three categories; 3) maximum/minimum values

References

Aiello, S., Barsella, B., Chlewicki, G., Greenberg, J.M., Patriarchi, P., & Perinotto, M.,
    1988, A&AS, 73, 195.
Bohlin, R.C., Hill, J.K., Jenkins, E.B., Savage, B.D., Snow, T.P., Spitzer, L. jr. &
    York,D.G., 1983, ApJS, 51, 277.
Cardelli, J.A., Clayton, G.C. & Mathis, J.S., 1988, ApJ, 329, L33.
Cardelli, J.A., Clayton, G.C. & Mathis, J.S., 1989, ApJ, 345, 245.
Gezari, D.Y., Schmitz, M. & Mead, J.M., 1987, NASA Ref. Pub. 1196.
Johnson, H.L., 1966, ARA&A, 4, 193.
Papaj, J., Wagner, W. & Krelowski, J., 1991., MNRAS, 252, 403.
Savage, D., Bohlin, R.C., Drake, J.F. & Budich, W., 1977, ApJ, 216, 291.
Savage, D., Massa, D., Meade, M. & Wesselius P.R., 1985, ApJS, 59, 397.
Shull, J.M. & Van Steenberg, M.E., 1983, ApJ, 294, 599.
Thé, P.S., Wesselius, P.R. & Janssen, I.M.H.H., 1986, A&AS, 66, 63.

A Molecular Conglomerate Model of Small Interstellar Dust

Frank O. Clark[1], R. F. Shipman[1], R. Assendorp[2], D. Kester[3], and M.P. Egan[1]

[1] Phillips Laboratory, 29 Randolph Road, Hanscom AFB, MA 01731-3010, USA
[2] Astrophysikalisches Institut Potsdam, an der Sternwarte 16, D-14482 Potsdam-Babelsberg
   Germany
[3] Laboratory For Space Research, Postbus 800, 9700 AV Groningen, The Netherlands

Abstract. We analyze emission of the interstellar dust from the vicinities of three heating sources. We find that as the radiation sources are approached, relative emission in both IRAS short wavelength bands expressed as $I(12)/I(100)$ and $I(25)/I(100)$ decline together. This result runs counter to existing models. We propose a molecular conglomerate model of the small interstellar dust grains to explain these data.

1 Introduction

We have used the InfraRed Astronomical Satellite (IRAS) measurements of emission from the interstellar dust as radiation density increases near two H II regions, the Rosette nebula ($6^h 29.5^m + 5°04'$) and λ Orionis ($5^h 34^m + 9°55'$), and the Pleiades cluster ($3^h 40^m + 24°$) with earliest stars respectively of O4, O8 III, and B6 III. We have modeled the data using a composite dust grain model of silicates, graphite, and amorphous carbon, taking into account temperature fluctuations of small grains (Draine and Lee 1984, Draine and Anderson 1985, Mathis and Whiffen 1989). In all cases large fields of view have been used to anchor the data to the undisturbed diffuse interstellar medium.

2 Data

We used both the IPAC data described by Wheelock et. al. (1994), and the Space Research of the Netherlands IRAS Software Telescope (Assendorp et al. 1995)

3 Discussion

We analyzed ratios of the IRAS short wavelength bands to 100 μm, $I(12)/I(100)$ and $I(25)/I(100)$ to minimize effects of column density. As the three sources are approached from afar, the data in all cases reveal a declining $I(12)/I(100)$, a result already demonstrated by Cox, Deharving, and Leene (1990) and Boulanger et al. (1990). Our data additionally reveal that $I(25)/I(100)$ declines at the same location as $I(12)/I(100)$, beginning well outside of any ionized regime.

3.1   Implication

A decline in emission at 12 or 25 μm could result from a reduction in excitation or abundance of the respective carriers. We estimate a characteristic size and number of atoms for the carriers at each wavelength using the amorphous carbon model of Mathis and Whiffen, for the carriers at 12 and 25 μm respectively of 40 and 1000 C atoms.

3.2   Excitation

It is possible that the drop in emission at the two IRAS bands is caused by a change in excitation. Boulanger et al. (1990) rule out excitation for the case of simple ultraviolet excitation of particles with constant physical characteristics.

3.3   Thermal Spiking

Mayo Greenberg (private communication) points out that thermal spiking will be reduced as the average grain temperature rises. We modeled this effect, but find that it cannot reproduce the data.

3.4   Thermal Conduction

As the radiation field increases the grains may undergo a thermal conduction change, transitioning from cold to warm from very poor thermal conductors to good thermal conductors. Poor thermal conductors could restrict photon excitation and thermal spiking to a small localized hot spot. A transition to a conducting state would cause each absorbed photon to heat the entire grain, reducing the effective temperature. Although physically possible, we were unable to identify a realistic material with appropriate properties.

3.5   Emissivity

These grains may go through an exotic phase change from insulator to semiconductor, like the coal model of Guillois et al. (1994) in which an initially low infrared emissivity transitions to a more efficient emitter. This again lowers the effective temperature. A carbon compound which would undergo such a change is physically reasonable, but again, we could not identify such a compound.

3.6   Abundance

The abundance of both carriers may be reduced as the radiation field increases. There are three possible abundance models which might explain the observed declines in emission as radiation field is increased: 1) molecular bond breaking, 2) runaway sublimation (Guhathakurta and Draine), and 3) conversion of a conglomerate 25 μm emitter into 12 μm emitters. Molecular bond breaking would require destroying larger molecular emitters more rapidly than smaller ones, which is incompatible with experiment. Runaway sublimation would destroy smaller solids faster than larger ones, which is counter to the observations. A conglomerate model may be possible for molecular conglomerates. Solid state conglomerates may not disintegrate with the small changes in radiation field over the regimes observed.

4   A Conglomerate Model of the Small Dust Grains

The observed results may be reproduced with a conglomerate model in which 25 μm carriers are weakly bound conglomerates, 25% of which are converted to a size that will emit at 12 μm. The conglomerates could either be solid state or molecular, although a solid state conglomerate model is difficult to dissociate with the very small increases in radiation field observed because of the large bonding energy.

4.1   Molecular Conglomerates

A molecular conglomerate will work if 12 μm carriers consist of hexagonal cyclic carbon structures, and 25 μm carriers consist of weakly bound associations of these same structures, such that 25% of the 25 μm carriers are converted to those emitting at 12 μm.

A conglomerate model of this type must be able to replenish the 12 μm carriers from the 25 μm carriers at their rate of destruction. This yields a simple relation between the carriers emitting at 12 μm and 25 μm:

$$dN(12)/dt = (bsu + 1)eff \ dN(25)/dt. \tag{1}$$

where bsu represents the number of basic structural units contained in the 25 μm carriers, and eff is the efficiency of release of basic structural units which then emit at 12 μm.

Carriers of 40 and 1000 carbon atoms respectively and a perfect efficiency of production, require a rate of destruction of the 12 μm carriers that is 26 times that of the 25 μm carriers to reproduce the observations. This result suggests that the stability to radiative dissociation between particle size and radiation density is linear, in perfect agreement with experiment (Jochims et al.1994).

5   Summary

In summary we propose that the majority of carriers emitting in the IRAS 12 μm band are hexagonal cyclic carbon structures containing 40 C atoms, and the majority of carriers emitting in the IRAS 25 μm band are weakly bound conglomerates of these cyclic structures of approximately 1000 carbon atoms.

References

Boulanger, F., Falgarone, E., Puget, J., Helou, G. (1990): Variations in the Abundance of Transiently Heated Particles in Nearby Molecular Clouds. Ap.J. **364**, 135–145.

Buchta, C., D'Alessio, A. D'Anna, A., Gambi,G., Minutolo, P., Russo, S. (1995): The Optical Characterization of High Molecular Mass Carbonaceous Structures Produced in Premixed Laminar Flames Across the Soot Threshold Limit. PASS, in press.

Draine, B. and Anderson, N. (1985): Temperature Fluctuations and Infrared Emission from Interstellar Grains. Ap.J. **292**, 494–499.

Draine, B. and Lee, H.M. (1984): Optical Properties of the Interstellar Graphite and Silicate Grains. Ap.J. **285**, 89–108.

Guhathakurta, P. and Draine, B. (1989): Temperature Fluctuations in Interstellar Grains. I. Computational Method and Sublimation of Small Grains. Ap.J. **345**, 230–244.

Jochims, H.W., Rühl, E., Baumgärtel, H.,Tobita, S., and Leach, S. (1994): Size Effects on Dissociation Rates of PolycyclicAromatic Hydrocarbon Cations: Laboratory Studies and Astrophysical Implications. Ap.J. **420**, 307–317.

Mathis, J., and Whiffen, G., (1989): Composite Interstellar Grains. Ap.J., **341**, 808–822.

Wheelock, S.L. et. al., (1994): IRAS Sky Survey Atlas Explanatory Supplement. JPL Publication 94-11 (Pasadena: JPL).

On the Stability of Dust Aggregates in Collisions

C. Dominik[1] and A.G.G.M. Tielens[2]

[1] Sterrewacht, P.O. Box 9513, 2300 RA Leiden, The Netherlands
[2] NASA Ames Research Center, Mail-Stop 245-3, Moffett Field, CA 94043, USA

1 Abstract

Coagulation is suspected of being an important process changing dust properties in star forming regions. This can be so only if the aggregates actually can form and survive in these environments. We have investigated the mechanical properties of dust aggregates, showing that aggregates a fragile objects that can be restructured or destroyed in collisions with other grains or aggregates at velocities less than 0.1 km/s.

2 Introduction

Coagulation of dust grains is one of the processes considered to be important for altering the dust properties in star forming environments. There is some evidence that the size distribution in these regions is different from the one found in the interstellar medium (Mathis, Rumpl, Nordsiek, 1977). An increasing amount of model calculations involving dust aggregation in star forming environment has appeared recently in the literature (c.f. Ossenkopf 1993, Weidenschilling and Ruzmaikina 1994, and also Sablotny and Henning, this volume). However, these models are mostly still on somewhat shaky grounds as the basic physics involved in particle coagulation is not known with certainty. This, however, may change in the near future. Not only have experimental results come within reach (Blum 1995), but theoretical studies are advancing, too. Chokshi et al. (1993) studied head-on collisions between elastic spheres attracted by van der Waals forces and found a critical velocity for sticking. Dominik and Tielens (1995a,b) studied the effects of tangential forces on contacts between grains, which might lead to sliding, rolling or even break-up of the contact. Detailed results will be published in (Dominik and Tielens 1996).

In this poster we represent model calculation results of collisions between dust aggregates and single grains. The results will clearly indicate that these aggregates are quite fragile and coagulation during star formation will be limited by the damage these processes might do.

3 Basic Physics

Our model is based on Johnson's solution for the contact problem of two elastic spheres with adhesion (Johnson, 1976). Such a contact has a total of six different

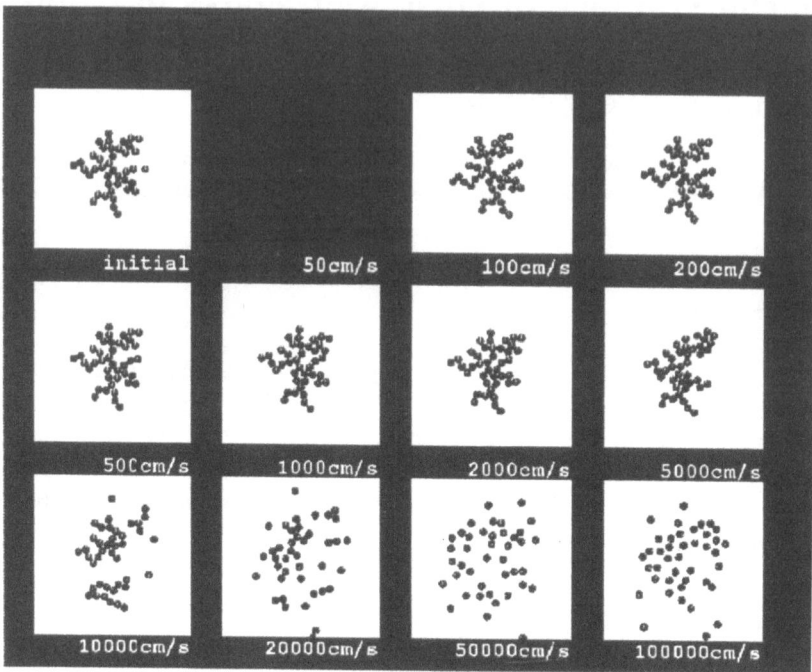

Fig. 1. Result of a collision between an aggregate made of 0.1 μm ice particles with a single grain of the same size. The different pictures show the initial state before the collision, and "end products" after the collision for different collision velocities (which are indicated below the pictures). One can easily see that velocities below 5000 cm/s lead to compression of the aggregates, while higher velocities have the ejection of at least some particles as a consequence. Above 50000 cm/s, complete destruction is the case.

degrees of freedom (one pull-off, two sliding, two rolling, one spinning). In a collision between aggregates, forces will strain the material and try to move the contact around, or break it.

The critical force that leads to breakup is given by the pull-off force in John-sons theory which is given by $F_c = 3\pi\gamma R$. γ is the surface energy of each surface and R is the reduced radius of the two spheres in contact. Tangential forces first strain the material at the contact. The contact slips at a critical force which is dependent on material parameters. Processes contributing to this force are wear-less friction on atomic scales, and surface roughness (Dominik and Tielens 1995b). The critical force to initiate sliding is usually greater than the pull-off force. Therefore, an attempt to initiate sliding is likely to break the aggregate. The critical force to initiate rolling, on the other hand, is much smaller (Dominik and Tielens 1995a). Thus, rolling will be the major process for restructuring of aggregates.

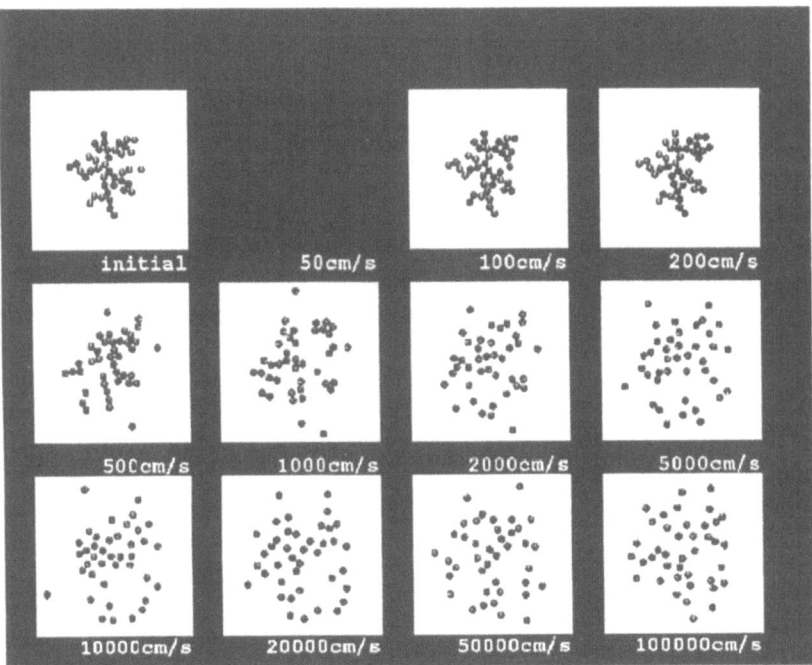

Fig. 2. Same as Fig. 5, but with silicate particles. Since silicate is much less sticky that ice is, destruction sets on already at about 500cm/s.

4   Results

Figures 1 and 2 show results of collision simulations. The code used is a two-dimensional N-body code which models the motion of the individual particles as influenced by the different contact forces. In both cases we study collisions between an aggregate composed of 40 particles with a diameter of 10^{-5}cm each. In figure 1, the particles are made of ice while in figure 2 we use silicates.

We can see that for very low relative velocities the incoming grain simply sticks to the aggregate. At intermediate velocities, some restructuring occurs near the impact point. Higher velocities lead to a partial destruction of the aggregates. A collision velocity of approximately 50000 cm/s is sufficient to dissolve the aggregates into individual particles.

This effect is even stronger for silicate particles which are much less sticky (Chokshi et al. 1993) than ice is. Here complete destruction already happens at a collision velocity of 500 cm/s.

Other simulations we have done show that smaller aggregates of smaller particles are less easily destructed. However, at velocities typical for the turbulent motions in star-forming regions (a few km/s) destruction or at least considerable compression and/or restructuring of aggregates has to be expected.

5   Conclusions

Our results clearly show that collisional aggregates are destroyed by collisions with other particles at relatively low velocities. The velocities are smaller than typical turbulent velocities in dark clouds and, therefore, coagulation in these environments will be limited.

Stability of aggregates increases as the individual particles get smaller. Also, particles accreted onto a big grains are relatively protected since the impact of a small grain onto a large one does transfer little energy. Another stabilizing processes might be the growth of ice mantles after coagulation. Coagulation in star forming regions might, therefore, be possible. However, our results indicate that it is not likely to be an one-way process. Turbulence and on-going star formation will stir up the medium and induce grain-grain collisions. Destruction and restructuring of aggregates should be considered when dust models for these environments are developed.

Acknowledgment: This work was made possible by an NRC postdoc grant for CD at NASA Ames. Theoretical studies of interstellar dust at NASA Ames is supported under task 399-20-01-30 through NASA's Theory Program.

References

Blum, J. 1995. Adv. Space Res. Vol. 15, Nr. 10.
Chokshi,A., Tielens, A.G.G.M., Hollenbach, D., 1993. APJ 407, 806.
Mathis, J.S., Rumpl,W., Nordsieck, K.H., 1977. APJ 217, 425.
Dominik, C., Tielens, A.G.G.M., 1995a. Phil. Mag. A, in press.
Dominik, C., Tielens A.G.G.M. 1995b. Phil. Mag. A, submitted.
Dominik, C., Tielens A.G.G.M. 1996. ApJ in preparation.
Ossenkopf, V., 1993. A&A 280, 617.
Sablotny, R., Henning, T. 1995. This volume.
Weidenschilling, S.J., Ruzmaikina, T.V. 1994. APJ 430, 713.

Dust in the Environment of YSOs

A. Evans

Department of Physics, Keele University, Keele, Staffordshire, ST5 5BG, U.K.

Abstract. The properties of interstellar dust and of the dust around young stellar objects are discussed. The discussion includes their physical, and optical properties, and the effects of dust processing in the YSO environment.

1 Introduction

As is well known, dust particles are produced in the environments of a variety of evolved stars (Gehrz 1989), ejected into the interstellar medium (ISM), subsumed into molecular clouds where they catalyze the formation of new stars, and the cycle stars all over again. During the recycling process however the grains are subject to a variety of modifying processes – both in the ISM (e.g. Jones et al. 1994) and in the environments of Young Stellar Objects (YSOs) – so that, in the case of some Solar System solids, all that remains to tell us of their origins is the isotopic evidence (e.g. Tang et al. 1989).

While we have indirect evidence about the nature of grains, for example from depletions and the nature of dust-producing stars, the only observationally accessible property of grains – whether in the ISM or around YSOs – is the way in which they interact with radiation, by extinction, scattering and emission. Every other property, such as grain size and shape, chemical composition, physical nature etc. is ultimately deduced from the optical properties of the grains.

We begin by summarizing briefly the properties of interstellar (IS) dust, before looking at the properties of dust in the YSO environments.

2 Interstellar Dust

The ISM is a complex and dynamic medium. Even the simplest model (McKee & Ostriker 1977) involves a three-phase medium with components having different temperatures, states of ionization and grain contents. Thus whatever property of ISM dust we observe is some ill-defined average over substantial path lengths along a very inhomogeneous (physically, chemically) line of sight. It is not at all surprising therefore that there are small but significant differences in the extinction laws in different directions (e.g. Witt et al. 1984; Fitzpatrick & Massa 1988). Thus while it may be helpful to construct a 'standard' IS extinction law (see e.g. Mathis 1990 and references therein) there is in reality no such thing as a 'standard' IS dust grain population.

For ISM dust one can measure the extinction properties over a broad range of wavelengths, the emissivity in the infrared (IR) and millimetre, and the scattering properties in the ultraviolet (UV), optical and IR (see e.g. Mathis 1990 for a review). Abundance and spectroscopic considerations imply that ISM grains are very likely composed of a mixture of silicate and carbonaceous components, although how these components are actually mixed together is still to some extent unclear (e.g. Mathis 1990; Cardelli & Clayton 1991).

2.1 Extinction

Several 'mean' extinction curves ISM dust have been published over the years. Recently it has been shown (Cardelli et al. 1989) that the IS extinction law may be parametrised by the ratio of total-to-selective extinction R_V to values as high as $R_V \simeq 6$ for a variety of environments. Extinction is also responsible for the phenomenon of IS polarization, via differential extinction by elongated grains aligned by the IS magnetic field. Both extinction and polarization imply particles of dimensions $\sim 0.01 \ldots 0.1 \mu m$.

There are several features in the extinction curve that provide clues to the composition of ISM dust. The $10 \mu m$ silicate feature is seen in absorption, e.g. towards the Galactic Centre (e.g. Roche 1988). The feature profile indicates that the silicate is amorphous, and indeed is very similar to that seen in the circumstellar (CS) shells of evolved (dust-producing) objects, such as μ Cep (Roche & Aitken 1984). Laboratory experiments on the thermal annealing of amorphous silicates (Thompson et al. 1996) indicate that heating to 900° C results in crystallization, suggesting that ISM and evolved star silicates have not been heated to these temperatures. The '2175Å' bump is a ubiquitous feature of the ISM extinction curve. It is widely believed, on the grounds of abundance and optical properties, to arise in some form of carbonaceous material (such as graphitic carbon) although silicates are also a possibility. The 'interstellar diffuse bands' – which still defy identification – seem to be just that: their presence in the CS shells seems to be extremely rare.

2.2 Scattering

The scattering properties of IS dust are deduced by observation of the diffuse galactic light and of reflection nebulae. Since the scattering cross section of dust grains depends on direction as well as on wavelength and grain size, knowledge of scattering geometry is essential to model the scattered light. Uncertainty in the relative locations of nebulosity and illuminating star(s) is nearly always a complication, but observations indicate that IS grains have high albedo and are strongly forward-scattering in the optical. However the albedo falls significantly around 2175Å, suggesting that this is predominantly an absorption feature.

2.3 Emissivity

The emissivity of ISM dust comes largely from IRAS, COBE and ground-based millimetre measurements. As is well-known (e.g. Puget & Léger 1989) the far-IR data are well-fitted by grains of 'standard' size. However there is a substantial excess shortward of the peak of emission, due to thermal 'spiking' of very small grains. Similar grains are present in YSO environments, where they give rise to the so-called 'UIR' features. The $3.28\mu m$ UIR feature is also seen in the ISM, where it is correlated with the far IR emission from the ISM grain population (Giard et al. 1989).

Millimetre observations indicate a β–index for ISM dust in the range $1\ldots2$ (Wright et al. 1991; Fischer et al. 1995 and references therein), over the wavelength range $\sim 100\ldots3000\mu m$, with some evidence for hotter (cooler) grains with steeper (flatter) emissivity laws at low (high) Galactic latitudes. Values of $\beta \simeq 2$ are sometimes taken as evidence for crystallinity, $\beta \simeq 1$ indicating amorphous grains. However uncertainties in values of the ISM β–index (typically $\pm \sim 0.4$) are such that it is probably premature to draw any firm conclusions about the nature of ISM grains on this basis.

2.4 Optical properties

The interpretation of the observations requires knowledge of the variation of the (complex) refractive indices of likely grain materials with wavelength. These may be based on laboratory measurements of bulk solids, or on astronomical data (such as spectral features). Draine & Lee (1984; hereafter DL) compiled a set of optical constants for 'astronomical silicate' and carbon which reproduced well the mean ISM extinction. The DL constants are widely used in the literature, although they are often used to model situations (e.g. YSO and evolved star dust) for which they are not ideally suited nor intended.

2.5 Size distribution

For many years the 'MRN' size distribution (Mathis et al. 1977), in which both carbon and silicate grains are present with a size distribution $n(a)\,da \propto a^{-3.5}\,da$ in a way that is consistent with cosmic abundances, has been the standard grain size distribution for ISM dust. A size distribution of this form has attractions because, to a first approximation, it results from grain shattering, as might happen in the environments in which the grains form. However Kim & Martin (1994, 1995) have recently used extinction and polarization data to show that the MRN distribution is only an approximation to a distribution that falls off approximately exponentially at the large particle end.

3 Dust Around YSOs

In the case of YSO dust grains we have the same difficulty as with ISM dust: the integrated columns may be comparable in magnitude with – or even greater than – those in the ISM but it is likely that the material through which we are looking is just as non-uniform. For example the sight-line to a YSO might include an accretion disc, the greater CS environment and the surrounding molecular cloud (as well as the foreground ISM). Modelling observations may therefore be difficult unless one has data at sufficiently high spatial and spectral resolution.

3.1 Extinction

In order to deduce the extinction of material around a YSO we must know the flux distribution of the YSO itself and the foreground cloud and IS extinction; in general, none of these may be known with any certainty. Also, care must be exercised in comparing the extinction laws in the ISM and in YSO environments because we might expect a greater proportion of scattered light from the CS shell of a YSO than from a sight-line in the ISM. Hence the measured extinction per column would effectively be less for the former and we would get different extinction laws for the two cases even if the grain populations were identical.

It is well-known (e.g. Evans et al. 1982; Cardelli & Clayton 1991) that the ratio of total-to-selective extinction R_V is larger in molecular clouds and in YSO environments than it is in the ISM. Since R_V is a grain size indicator it seems that substantial grain growth has occurred in these environments. Furthermore, the parameterization of the extinction law in terms of R_V suggests that any processing the grains have suffered is essentially independent of grain size.

For YSO environments (such as dense clouds), the *larger* the value of R_V, the *less pronounced* is the '2175 Å' bump; indeed the latter is generally weak in regions populated by YSOs (Sorrell 1990). Cardelli & Clayton (1991) suggest that the small grains responsible for the '2175' feature are swept up by larger grains and removed from dark cloud environments, accounting for the anticorrelation between bump area per H-column density and R_V. Further, the *larger* the value of R_V, the *less steep* the rise in extinction to the far UV. This correlation between R_V and UV extinction (Cardelli et al. 1989) suggests that, in environments where R_V is high, UV may penetrate further into the YSO dust distribution than earlier 'mean' extinction laws might indicate. This has implications for the processing not only of the YSO grains themselves but also (e.g. by ionization) of the environment in which they are located.

3.2 Scattering

As in the IS case, a limiting factor is the lack of knowledge of the scattering geometry. However for some reflection nebulae near YSOs this may deduced or guessed, for example if the star is seen through a disc.

Many attempts to model the observed scattering properties of dust grains, both in the ISM and in YSO environments, use the MRN grain size distribution. For example, Monin et al. (1989) have concluded that near-IR images of the T Tauri star HL Tau are consistent with scattering by 'MRN' grains in a slightly-inclined disc. Pendleton et al. (1990) have modelled reflection nebulae around luminous IR sources in the OMC with ice-covered grains. They find that the intensity is equally well fitted by either MRN or 'large grain' models – the latter being essentially MRN displaced to larger sizes – but that the polarization data are consistent only with the larger grains. Kenyon et al. (1993) have modelled near-IR colours, images and flux distributions of Class I PMS stars in the Taurus-Auriga molecular cloud and find that the data are best fitted by the MRN size distribution and DL optical constants.

3.3   Emissivity

Infrared. It is well-known that the UIR emission bands are associated with (i) carbonaceous material of some form (most likely PAH molecules) and (ii) sources of UV radiation. These sources range from YSOs to highly evolved objects (see Allamandola et al. 1989 for a review). In contrast to the ISM case, there is evidence that the silicate feature in YSO environments shows some structure, suggesting a degree of crystallinity; certainly silicates in the Solar System (e.g. cometary and interplanetary dust particle [IDP] silicates) show evidence of crystallinity, which must have been established when the Sun was a YSO.

Millimetre. Gear et al. (1988) suggested that the β−index for grains in YSO environments changes from $\lesssim 1$ to $\gtrsim 1$ between sub-millimetre and millimetre wavelengths. More recently observations of T Tauri (Beckwith & Sargent 1991; Mannings & Emerson 1994) and Herbig Ae/Be stars (Mannings 1994) have indicated that $\beta \lesssim 1$ for the former, while $\beta \simeq 1$ for the latter. On the other hand the β−index of the dust around the proto-typical 'Class 0' object VLA 1623, whose age is probably $\lesssim 10^4$ yr (André et al. 1993), is $\simeq 1.5$; similar values have been found for other extremely young objects. While the T Tauri star β−indices seem less than the IS value, we should again stress that it is perhaps premature to make comparisons and to draw any conclusions about the respective millimetre emissivity laws of IS and YSO dust.

As discussed by Mannings & Emerson (1994) there are several ways of interpreting the low β−indices for the YSO dust. For example, the grains around YSOs may be more amorphous than they are in the ISM, or they may be fractal. Mannings & Emerson, however, conclude that the most likely interpretation is that the low−β grains are substantially larger than those in the ISM, which would be consistent with other grain size indicators in YSO environments.

3.4   Optical properties

As already noted, several workers start with DL optical constants and the MRN size distribution in modelling YSO dust. However the processing that grains suffer (e.g. growth, annealing) in YSO environments means that this starting point is almost certainly incorrect. There are also other unknowns that follow from this, such as the way in which fundamental solid state properties (such as degree of crystallinity, importance of defects, etc., and their effect on the optical properties) depend on grain temperature. Correct choice of optical constants is of vital importance not only for fitting the spectral energy distributions (SED) of YSOs but also for the determination of CS dust mass.

The effects of coagulation and the growth of ice mantles on the IR and mm emissivity and extinction of both carbon and silicate grains have been calculated in detail by Krügel & Siebenmorgen (1994). Their calculations include a MRN-type grain size distribution (with adjusted lower and upper bounds to account for the removal of small grains by larger) and the effects of grain porosity. Even for the MRN-type size distribution, there are large effects on the extinction coefficient, which is an order of magnitude higher for fluffy, ice-coated grains than for the ISM MRN grain population.

Similar calculations have been made for 'dirty' ice mantles on carbon grains by Preibisch et al. (1993; this paper contains a tabulation of complex refractive indices of amorphous carbon and 'dirty' ice for the wavelength range $0.1 \ldots 800\mu m$). These authors have used their results to calculate the SED of protostellar objects and they conclude that uncertainties in the optical properties of the dust particles can lead to uncertainties of a factor ~ 5 in the CS dust mass.

4 Processing of Dust in YSO Environments

4.1 Grain growth

Coagulation. As already noted, there is much evidence (R_V, β–index) to suggest that dust grains in dense clouds and around YSOs are significantly larger than those in the ISM, although not all dark clouds display large values of R_V (Cardelli & Clayton 1991). It seems unlikely that substantial particle growth can occur via accretion from the gas phase and the most likely mechanism is coagulation of smaller grains. Although there is ample evidence that such coagulation occurs (e.g. in IDPs) the physics of coagulation remains poorly understood.

This problem has been addressed by Chokshi et al. (1993) and by Ossenkopf (1993) and, in the context of cometary grains, by Bailey (1987). The physics is complicated by the fact that real grains will be irregular, electrically (probably negatively) charged, covered with icy mantles etc. Chokshi et al. show that, for two colliding (ideal) grains there is a critical velocity below which coagulation is likely to occur. They conclude that coagulation in dense star-forming cores requires that the grains be carried by turbulence (rather than, for example, Brownian motion) in regions where the gas densities $\gtrsim 10^4\,\mathrm{cm}^{-3}$; even then a grain's mass is doubled only within a dynamical evolution time scale. However small ($\ll 0.1\mu m$) grains are very efficiently swept up by larger grains, and are removed from the grain size distribution. The observational consequences of this are two-fold: (i) the extinction law for YSO dust is (see above) expected to be very different from that in the ISM, particularly in the UV where small particles dominate, and (ii) there will be significant differences in the IR emission, which in the ISM is caused by temperature spiking of small ($\lesssim 100\,\text{Å}$) grains.

Ossenkopf (1993) has carried out extensive numerical simulations of the evolution of dust in dense cores, including the effects of grain charge, ice accretion etc. He concludes, as do Chokshi et al., that turbulence drives the coagulation for gas densities $\lesssim 10^8$ H atoms cm^{-3}, with Brownian motion dominating at higher densities. He has also examined the evolution of the size distribution of an initially-MRN grain size distribution. The results make sobering reading for anyone tempted to apply the MRN size distribution to the theoretical modelling of these environments.

Mantle growth. There is much evidence for the existence of ice mantles on grain surfaces in molecular clouds (e.g. Whittet 1992 and references therein), and even on T Tauri star grains (e.g. Cohen 1975). During mantle formation ice grows in crystalline form if the gas-phase species can find the most favourable site on the grain surface for binding, and is amorphous otherwise. The latter requires that the effective flux of H_2O from the gas phase onto the solid exceeds a critical value (e.g. Gail & Sedlmayr, 1984; Kouchi et al. 1994) and in molecular clouds ice will grow in amorphous form. Further, for conditions in molecular clouds, the crystallization time-scale for water ice far exceeds cloud lifetimes (Kouchi et al. 1994) so that ice will remain amorphous. For icy grains in YSO environments there are two cases, namely (i) preservation of (amorphous) molecular cloud ice and (ii) the recondensation of ices on YSO dust. Kouchi et al. find that, for a pre-Solar-type YSO, the critical temperature for crystallization is $\sim 110\,\mathrm{K}$, so close to the sublimation temperature of ice that amorphous ice is preserved. On the other hand, *recondensation* of ice on refractory grains is likely to result in crystalline ice. These results have obvious implications not only for YSO dust but also for cometary ices in the Solar System.

4.2   Radiation processing in the YSO environment

As already noted, ISM silicates seem to be amorphous. However there is evidence that the silicate in the YSO environments is significantly different; for example, in the case of the Trapezium the feature is considerably broader than that in the diffuse ISM, while spectropolarimetry of the compact molecular cloud source AFGL 2591 by (Aitken et al. 1988) suggests a polarization component arising from annealed silicate. There is also evidence that Solar System silicates are crystalline, and there seem to be broad similarities between the $10\mu m$ profile of cometary, IDP and molecular cloud silicates. The work of Thompson et al. (1996) suggests that where silicates do show evidence of crystallinity the processing is likely to have occurred in the YSO environment.

A significant source of processing in dense clouds and in YSO environments is irradiation by cosmic ray H and He. The possibility that the GEMS (Glassy grains with Embedded Metal and Sulphides; Bradley 1994) found in IDPs are a remnant of the ISM grain population has been discussed by Martin (1995). Laboratory evidence suggests that they have suffered irradiation by H and He, which could either have occurred in the ISM (in supernova shocks) or in the pre-Solar nebula. Their $10\mu m$ silicate feature seems to resemble those of cometary and molecular cloud rather than ISM silicates and Flynn (1994) argues for irradiation in the pre-Solar nebula (when the Sun was a YSO).

Several star-forming regions have been surveyed at x-ray wavelengths. Most recently Casanova et al. (1995) have carried out observations of the ρ Oph cloud with ROSAT and they find a large fraction of x-ray sources associated with embedded (Class I) sources; indeed a large fraction of the IR sources in the ρ Oph cloud are x-ray emitters. The CS environments of these x-ray/IR sources must be subject to complex ionization and heating effects by the x-radiation. There are implications here for the irradiation of grains by virtue of their proximity to intense sources of x-radiation (a situation that does not apply to ISM grains). These include (i) the formation of carbonaceous materials in ice mantles,

(ii) photon-stimulated desorption, in which PAH-like units are detached from the grain surface and which would offset the sweeping-up process discussed in §4.1, (iii) the production of structural defects in grains, such colour centres.

5   Concluding Remarks

As discussed above it is common, when interpreting observations of YSOs, to assume grain parameters which are unlikely to be appropriate for the YSO situation, for example the DL optical constants and the MRN grain size distribution. In view of the known differences between ISM and YSO dust, *using ISM dust model parameters to model dust around YSOs is unlikely to lead to reliable conclusions.* What is surprising (even worrying) is that they often give a reasonable description of the data.

Likewise, modelling only one observational aspect of a system can lead to misleading conclusions and it is important to note that *grain properties are well constrained only by simultaneously fitting two or more independent data sets.* These might include, for example, polarization, IR SED, spatial distribution of reflection nebula intensity.

On an observational front some effort might be devoted to seeing whether the β-index of the dust in YSO environments is correlated with grain size indicators, such as R_V, or with the evolutionary state of YSOs. This would be a direct check on whether the variations that appear to occur in the β-index are due to particle size effects or to (for example) environmental processing.

There is a need for laboratory investigations of (i) the effects on the physical properties (e.g. optical constants, crystalline structure) of likely astronomical solids of irradiation by UV- and x-radiation, H and He, such as is likely to be found in a YSO environment; (ii) the temperature dependence of these physical properties over relevant temperature ranges.

Acknowledgements. I thank the Royal Society for financial support.

References

Aitken, D. K., Roche, P. F., Smith, C. H., James, S. D., Hough, J. H., (1988): MNRAS, **230**, 629–638

Allamandola, L. J., Tielens, A. G. G. M., Barker, J. R., (1989): ApJS, **71**, 733–775

André, P., Ward-Thompson, D., Barsony, M., (1993): ApJ, **406**, 122–141

Bailey, M. E., (1987): Icarus, **69**, 70

Beckwith, S. V. W., Sargent, A. I., (1991): ApJ, **381**, 250–258

Bradley, J. P., (1994): Science, **265**, 925–929

Cardelli, J. A., Clayton, G. C., (1991): AJ, **101**, 1021–1032

Cardelli, J. A., Clayton, G. C. Mathis, J. S., (1989): ApJ, **345**, 245–256

Casanova, S., Montmerle, Th., Fiegelson, E. D., André, P., (1995): ApJ, **439**, 752–770

Chokshi, A., Tielens, A. G. G. M., Hollenbach, D. J., (1993): ApJ, **407**, 806–819

Cohen, M. H., (1975): MNRAS, **173**, 279–293

Draine, B., Lee, H. M., (1984): ApJ, **285**, 89–108

Evans, A., Bode, M. F., Whittet, D. C. B., Davies, J. K., Kilkenny, D., Baines, D. W. T., (1982): MNRAS, **199**, 37P–43P

Fischer, M. L., Clapp, A., Devlin, M., Gundersen, J. O., Lange, A. E., Lubin, P. M., Meinhold, P. R., Richards, P. L., Smoot, G., (1995): ApJ, **444**, 226–230

Fitzpatrick, E. L., Massa, D., (1988): ApJ, **328**, 734

Flynn, G. J., (1994): Nature, **371**, 287–288

Gail, H.-P., Sedlmayr, E., (1984): A&A, **132**, 163–167

Gear, W. K., Robson, E. I., Griffin, M. W., (1988): MNRAS, **231**, 55P–62P

Gehrz, R. D., (1989): In *Interstellar Dust*, proceedings of IAU Symposium 135, Eds L. J. Allamandola, A. G. G. M. Tielens, Kluwer Academic Publishers, 445–452

Giard, M., Pajot, F., Lamarre, J. M., Serra, G., Caux, E., (1989): A&A, **215**, 92–100

Jones, A. P., Tielens, A. G. G. M., Hollenbach, D. J., McKee, C. F., (1994): ApJ, **433**, 797–810

Kenyon, S. J., Whitney, B. A., Gomez, M., Hartmann, L., (1993): ApJ, **414**, 773–792

Kim, S.-H., Martin, P. G., Hendry, P. D., (1994): ApJ, **422**, 164–175

Kim, S.-H., Martin, P. G., (1994): ApJ, **431**, 783–796

Kouchi, A., Yamamoto, T., Kozasa, T., Kuroda, T., Greenberg, J. M., (1994): A&A, **290**, 1009-1018

Krügel, E., Siebenmorgen, R., (1994): A&A, **288**, 929–941

Mannings, V., (1994): MNRAS, **271**, 587–600

Mannings, V., Emerson, J. P., (1994): MNRAS, **267**, 361–378

Martin, P. G., (1995): ApJ, **445**, L63–L66

Mathis, J. S., (1990): ARAA, **28**, 37–70

Mathis, J. S., Rumpl, W., Nordsieck, K. H., (1977): ApJ, **217**, 425–433

McKee, C. F., Ostriker, J. P. (1977): ApJ, **218**, 148–169

Monin, J.-L., Pudritz, R. E., Rouan, D., Lacome, F., (1989): A&A, **215**, L1–L4

Ossenkopf, V., (1993): A&A, **280**, 617–646

Pendleton, Y., J., Tielens, A. G. G. M., Werner, M. W., (1990): ApJ, **349**, 107–119

Preibisch, Th., Ossenkopf, V., Yorke, H. W., Henning, Th., (1993): A&A, **279**, 577–588

Puget, J. L., Léger, A., (1989): ARAA, **27**, 161–198

Roche, P. F., (1988): In *Dust in the Universe*, Eds Bailey, M. E., Williams, D. A., Cambridge University Press, 415–433

Roche, P. F., Aitken, D. K., (1984): MNRAS, **208**, 481–492

Sorrell, W. H., (1990): ApJ, **361**, 150–154

Tang, M., Anders, E., Hoppe, P., Zinner, E., (1989): Nature, **339**, 351–354

Thompson, S. P., Evans, A., Jones, A. P., (1996): A&A, in press

Whittet, D. C. B., (1992): *Dust in the Galactic Environment*, IOP Publishing, Bristol, New York

Witt, A. N., Bohlin, R. C., Stecher, T. P., (1984): ApJ, **279**, 698–704

Wright, E. L. et al., (1991): ApJ, **381**, 200–209

Dust Opacities for Molecular Cloud Cores and Protoplanetary Accretion Disks

Thomas Henning

Max Planck Society, Research Unit "Dust in Star-forming Regions", Schillergäßchen 3, D–07745 Jena, Germany

Abstract. In this review, the properties of dust populations typical for molecular cloud cores and protoplanetary accretion disks will be discussed. The paper especially deals with the optical properties of the particles.

1 Introduction

Infrared and submm/mm continuum radiation coming from tiny dust grains located in star-forming regions, protostellar and protoplanetary disks, and the tori/envelopes surrounding active galactic nuclei contains important information on the geometrical structure and the orientation of the objects. However, without accurate knowledge of the optical properties of the grains – either from experiments or basic theory – radiative transfer models have only limited predictive power (for a more detailed discussion see, e.g., Butner et al. 1990, 1991, Henning et al. 1992, Wolfire & Churchwell 1994, Stenholm 1994, Efstathiou & Rowan-Robinson 1995).

Dust grains do not only play a role as tracers for protostars and disk-like structures, they are also a very active factor in the star formation process. They influence the thermal structure and the chemistry of molecular cloud cores, the coupling of the magnetic field to the gas, and the formation of massive stars by the strong radiation pressure acting on the grains. Large opacity gradients may trigger instabilities during the protostellar collapse phase or the evolution of protoplanetary accretion disks (Lin & Papaloizou 1985, Wuchterl 1990, Balluch 1991ab; see also Duschl 1993, Kürschner 1994, Yorke & Henning 1994).

All this together means that the knowledge of the dust properties is of decisive importance for both the interpretation of astronomical data and the description of the dynamical evolution during star formation. Here, one has to realize that the dust opacity depends not only on the local conditions, i.e. the chemical composition and the two state variables density and temperature, but also on the overall evolutionary history of the system. This is particularly true for dust opacities in the stellar/protostellar environment (Lenzuni et al. 1995, Duschl et al. 1995). In any case, it is important to note that, if present, the dust opacity exceeds the molecular or atomic contributions to the total opacity by several orders of magnitude, at least in a broad wavelength range or in case of the Rosseland mean opacity in a certain temperature interval (see, e.g., Alexander & Ferguson 1994). To demonstrate the importance of the dust opacity, Fig.1

shows the Rosseland mean dust opacity κ_R for protostellar cores taking into account sublimation and sputtering of refractory grains in comparison with the gas opacity.

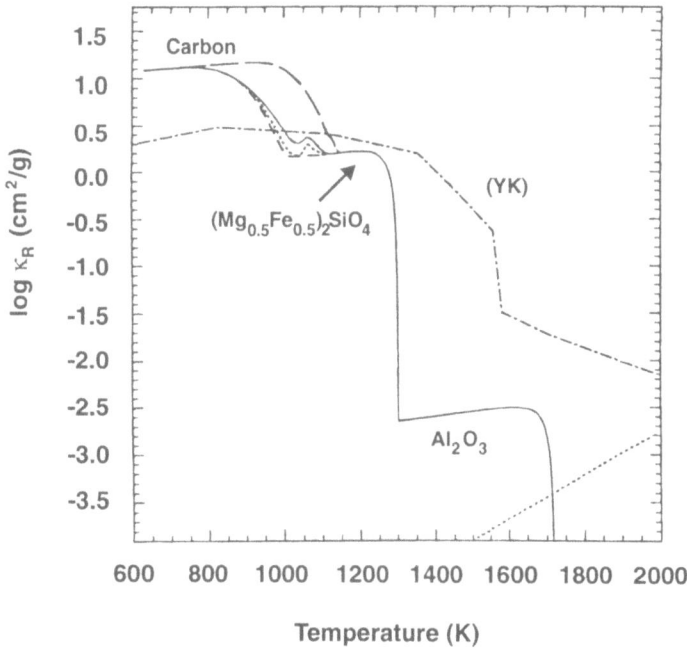

Fig. 1. Evolution of the Rosseland mean opacity in protostellar cores. Labels indicate materials which provide the dominant contribution in specific temperature regimes. The different lines in case of the carbon destruction refer to different sputtering mechanisms/yields. The gas opacity is indicated by the dotted line. The results of the model by Yorke & Krügel (1977; dashed-dotted line) are also shown. After Lenzuni et al. 1995

In this review, the evolution and optical properties of grains in cold molecular cloud cores and protoplanetary accretion disks will be discussed. In the next section, the different grain populations are shortly characterized. In section 3, the main steps in constructing a grain model and the relevant basic optical data for cosmic dust analogues will be summarized. In section 4, the results of recent models will be presented.

2   Dust Populations and Metamorphosis

Four basic types of cosmic dust populations can be distinguished, which are:

1. Stellar outflow dust (stardust)
2. Dust in the diffuse ISM (interstellar dust)

3. Dust in dense cool clouds (molecular cloud dust)
4. Dust in protostellar and protoplanetary accretion disks (disk dust).

For completeness a fifth type, interplanetary dust, should be included. There is evidence for subdivisions of the above main populations as well as for further dust populations (for an extensive review, see Dorschner & Henning 1995). For example, dust grains in photodissociation regions and the "hot cores" of massive star formation have other properties (e.g., evaporation and photolysis of ice mantles and destruction of particles by shocks) than those in the cold and very dense pre-protostellar and protostellar cores. The dust populations are multi-component systems containing particles of different chemical composition, solid-state structure, and size/shape. The conditions of grain formation and the subsequent metamorphosis by grain growth and destruction, chemical evolution, and structural transformation due to heating and irradiation strongly modifies the grain properties and, therefore, their behaviour in the interaction with light.

Due to the relatively high densities in molecular cloud cores ($n_H \gtrsim 10^4$ cm^{-3}), the adsorption of gaseous species proceeds at a faster rate than in the diffuse ISM, grain surface chemistry becomes more important, and the grains develop ice mantles. One can expect an ice mixture consisting of H_2O, CH_3OH, CO, and some other simple molecules (Whittet 1993), the existence of which is proven by infrared spectroscopy. Due to additional accretion of very small particles (e.g., PAHs, small amorphous carbon grains), the mantle material may be considered to be a "dirty" ice (Preibisch et al. 1993). The second important process expected in dense quiescent cores is the coagulation of particles (Ossenkopf 1993, Weidenschilling & Ruzmaikina 1994), which is based on the observational fact that small grains are deficient in the outer regions of molecular clouds (Mathis & Whiffen 1989, Kim et al. 1994). Accretion of atoms/molecules is always fast compared to the coagulation process. Therefore, one can assume that the grains are ice-covered before coagulation starts. For grains larger than a few nanometers, the thickness of the ice layers   should be independent of the grain size.

Based on efficient grain destruction by shocks in the diffuse ISM deduced from recent models (Jones et al. 1994), a considerable grain growth in dense clouds is expected because the dust destruction rate derived from the calculations exceeds the dust formation rate in the envelopes around AGB stars. However, no large increase of grain size should occur in molecular clouds even if all volatile molecules would be frozen out onto the grains. In this case, which may be appropriate for very dense and cold regions, one obtains an ice mass relative to the mass of refractory components of about 1.4. The coagulation simulations for the physical conditions of cold molecular cloud cores by Ossenkopf (1993) have shown that dust aggregates with radii of some micrometers but not very large dust particles are produced.

The grain evolution in protoplanetary disks (working definition: disk mass $\ll$ stellar mass; disk evolves in the gravitational field of the central protostellar core) is much more complex due to the complicated velocity, thermal and density (and even magnetic) structure of the disks and the non-linear re-coupling of any opacity change to the dynamical evolution of the disk (see, e.g., Duschl 1993, Henning et al. 1995a). In addition, the physical and chemical properties of the initial dust population modified by passing the accretion shock front are poorly known (Lunine et al. 1991, Prinn 1993). Grain growth, destruction by evaporation, chemical reactions and shocks as well as re-condensation of material can be expected (Völk et al. 1980, Mizuno et al. 1988, Mizuno 1989, Sterzik & Morfill 1994, Duschl et al. 1995, Henning et al. 1995a).

3   How to Construct a Dust Model?

Typical steps in constructing a dust model are:

1. Assume (or attempt to calculate ab initio) chemical composition, shape, and size of each grain type.
2. Use dielectric properties of cosmic dust analogues measured in the laboratory or based on theoretical calculations.
3. Calculate the optical properties (absorption and scattering efficiencies) for each grain type.
4. Construct appropriate mean values and work with the model.

Table 1. Optical constants of selected cosmic dust analogues

Material	Wavelength region [μm]	Reference
Amorphous silicate (cosmic composition)	0.2 – 500	Jäger et al. 1994
Amorphous silicate (varying Fe/Mg ratio)	0.2 – 500	Dorschner et al. 1995
Iron/magnesium oxides (varying Fe/Mg ratio)	0.2 – 500	Henning et al. 1995c
Aluminium oxide	0.2 – 500	Koike et al. 1995, Begemann et al. 1995
Iron	0.6 – 285	Ordal et al. 1988
Iron/magnesium sulfides	10 – 500	Begemann et al. 1994
Amorphous carbon	0.1 – 800	Preibisch et al. 1993
Ices	2.5 – 200	Hudgins et al. 1993

Step 1 for dust present in molecular cloud cores and protoplanetary accretion disks will be discussed in somewhat more detail in the next section. Factors which can strongly influence the optical behaviour of the grains are: (a) size of the particles (e.g., "Rayleigh" particles vs. grains large compared to the wavelength); (b) shape of the grains (e.g., spherical vs. elongated grains); (c) chemical structure and formation of additional dust components (e.g., amorphous carbon vs.

graphitic particles, formation of Fe/FeO/FeS in protoplanetary accretion disks); (d) presence of ice mantles; (e) agglomeration of particles; (f) temperature. For a more comprehensive discussion of these factors, we refer to the review by Henning et al. (1995b) and references therein.

Optical constants for cosmic dust analogues were compiled by Pollack et al. (1994) for wavelengths between 0.1 μm and 100 mm. Recently, more laboratory data for such analogue materials became available, which are summarized in Tab. 1. The next step is the calculation of absorption and scattering efficiencies. Exact solutions for the scattering problem are only available for simple shapes/configurations like homogeneous spheres, multilayered spheres, prolate and oblate spheroids, infinite cylinders, and simple multi-sphere systems. However, there is considerable progress in the theoretical description of the optical properties of inhomogeneous and irregularly shaped particles (for a review, see Henning et al. 1995b and for a detailed discussion Stognienko et al. 1995).

4   Dust Models for Molecular Cloud Cores and Protoplanetary Accretion Disks

As already mentioned in the previous section, a grain model for a special phase of the ISM should start with the determination of the size, structure, and composition of the particles. Based on the simulations of grain growth and subsequent aggregate formation in dense and cold molecular cloud cores by Ossenkopf (1993), a self-consistent calculation of the dust opacities and their temporal evolution at different densities was performed by Ossenkopf & Henning (1994). These authors considered amorphous carbon, silicates, and dirty ice mantles as the most important grain species and explicitly treated the fluffy structure of the aggregates produced during the coagulation process; the opacities are tabulated for wavelengths between 1 μm and 1.3 mm. In addition, an extensive discussion of the uncertainties and the factors influencing the opacities can be found in this paper (see also Krügel & Siebenmorgen 1994). The effect of dirty ice mantles was studied in detail by Preibisch et al. (1993). Recommended dust opacities based on these models are summarized in Tab. 2. Here, one should caution the reader against uncritically using observationally-based opacities. The derivation of the frequency dependence of the mass absorption coefficient κ_m from the observations, often assumed to be a power law, is not without assumptions and problems (see, e.g., Beckwith & Sargent 1991). In addition, one should keep in mind that we have a multi-component system and only an ensemble-average can be derived from the observations.

The most comprehensive dust model for protoplanetary accretion disks was worked out by Pollack et al. (1994). Based on the composition of primitive material in the solar system, equilibrium calculations for the determination of the evaporation temperatures, and additional theoretical considerations, this model includes olivine and orthopyroxene, volatile and refractory organics (major condensed carbonaceous species in this model), water ice, troilite (FeS), and metallic iron as the main dust components. Henning & Stognienko (1995) extended this

Table 2. Recommended mass absorption coeffcients κ_m per gram dust at $\lambda=1.3$ mm

Dust component	Model	κ_m $[cm^2 g^{-1}]$
Diffuse ISM	Draine & Lee (1984): bare graphite + silicate grains	0.24
Cloud envelopes	Preibisch et al. (1993): amorphous carbon + silicates with dirty ice mantles	0.50
Dense and cold cloud cores $(n_H \geq 10^7 \ cm^{-3})$	Ossenkopf & Henning (1994) : coagulated grains; (amorphous) carbon + silicates, dirty ice	1.10

model by explicitely studying the influence of a fluffy structure expected for these particles (see Fig.2). A major result of this study is the fact that the iron distribution in the different dust species plays a crucial role for the optical properties of the protoplanetary dust population. Evaporation of grain material may not be the only destruction process. In contrast to the evaporation, Duschl et al. (1995) investigated the possibility that carbon may be destroyed by a slow "combustion" due to oxidation reactions with OH radicals.

Several authors discussed the influence of a further grain growth on the opacity. It is a well-known fact that the absorptivity of compact particles first starts to increase and then decreases if the particle size becomes comparable to/larger than the wavelength. This means that for large compact particles (mm sizes and larger), the far-infrared opacity decreases and the temperature gradient of the Rosseland mean opacities changes (see, e.g., Henning et al. 1995a). This could have dramatic effects on the further disk evolution. However, one should keep in mind that large agglomerates produced by cluster-cluster aggregation (CCA) do not show an optical behaviour different from that of small CCA clusters as long as there is no large contribution by metallic particles (Ossenkopf & Henning 1994). An opacity change caused by the formation of very large aggregates may, therefore, only amount to a few percent. If compaction sets in, one returns to the behaviour of compact particles. For very large compact particles ($\gtrsim 10$ mm), a grey behaviour of the monochromatic opacity is expected. This typical behaviour may open a way for the identification of considerable particle growth in disks.

5   Further Perspectives

Further studies should concentrate on a self-consistent treatment of dust evolution (formation, growth, destruction) and the change of the dust opacities. These studies should rely on basic laboratory data for cosmic dust analogues. The coupling between the evolution of the dust properties and the dynamical

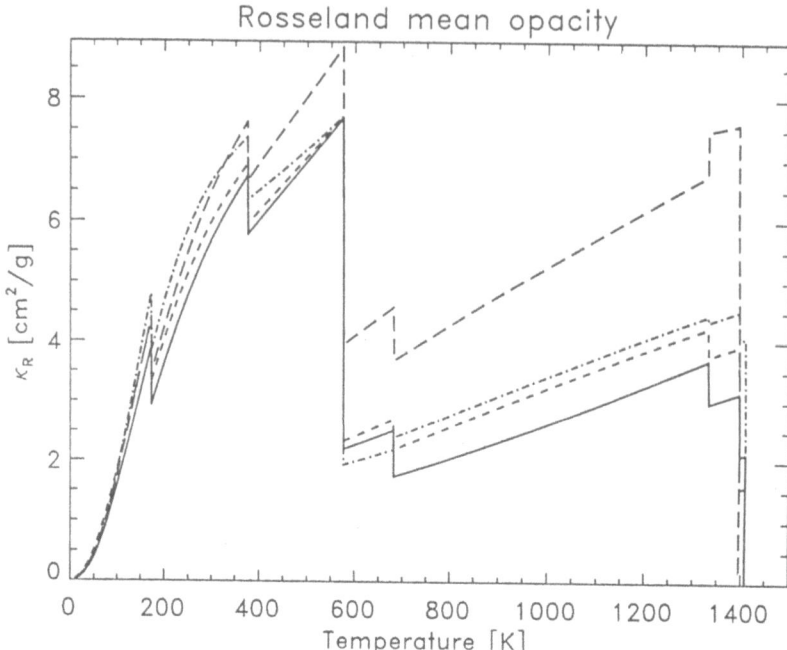

Fig. 2. Rosseland mean opacity including aggregation of composite grains with iron-poor (long-dashed line) and iron-rich (dashed-dotted line) silicates. The solid line represents the values given by Pollack et al. (1994) and the dashed line corresponds to the Pollack et al. model including aggregation of composite grains. After Henning & Stognienko 1995 (see also for monochromatic opacities)

evolution of the systems is an additional step which is especially important for protostellar and protoplanetary accretion disks.

References

Alexander, D.R., Ferguson, J.W. (1994): ApJ **437**, 879.
Balluch, M. (1991a): A&A **243**, 168.
Balluch, M. (1991b): A&A **243**, 187.
Beckwith, S.V.W., Sargent, A.I. (1991): ApJ **381**, 250.
Begemann, B., Dorschner, J., Henning, Th., Mutschke, H., Thamm, E. (1994): ApJ **423**, L71.
Begemann, B., Henning, Th., Mutschke, H., Dorschner, J., Nass, R. (1995), in preparation.
Butner, H.M. et al. (1990): ApJ **364**, 164.
Butner, H.M. et al. (1991): ApJ **376**, 636.
Dorschner, J., Henning, Th. (1995): A&AR, in press.
Dorschner, J., Begemann, B., Henning, Th., Jäger, C., Mutschke H. (1995): A&A **300**, 503.
Draine, B.T., Lee, H.M. (1984): ApJ **285**, 89.

Duschl, W.J. (1993): In: J.M. Greenberg, C.X. Mendoza-Gómez, V. Pironello (Eds.), The Chemistry of Life's Origins (Kluwer, Dordrecht), p.55

Duschl, W.J., Gail, H.-P., Tscharnuter, W. (1995): A&A, submitted

Efstathiou, A., Rowan-Robinson, M. (1995): MNRAS **273**, 649.

Henning, Th., Stognienko, R. (1995): A&A, submitted.

Henning, Th., Chini, R., Pfau, W. (1992): A&A **263**, 285.

Henning, Th., Schmitt, W., Klahr, H., Mucha, R. (1995a): In B.Å.S. Gustafson, M.S. Hanner (Eds.), Physics, Chemistry, and Dynamics of Interplanetary Dust (IAU Coll. No. 150), in press

Henning, Th., Michel, B., Stognienko, R. (1995b): Planet. & Space Sci., in press

Henning, Th., Begemann, B., Mutschke, H., Dorschner, J. (1995c): A&AS **112**, 143.

Hudgins, D.M., Sandford, S.A., Allamandola, L.J., Tielens, A.G.G.M. (1993):ApJS **86**, 713.

Jäger, C., Mutschke, H., Begemann, B., Dorschner, J., Henning, Th. (1994): A&A **292**, 641.

Jones, A.P., Tielens, A.G.G.M., Hollenbach, D.J., McKee, C.F. (1994): ApJ **433**, 797.

Kim, S.-H., Martin, P.G., Hendry, P.D. (1994): ApJ **422**, 164.

Koike, C. et al. 1995: Icarus **114**, 203.

Krügel, E., Siebenmorgen, R. (1994): A&A **288**, 929.

Kürschner, R. (1994): A&A **285**, 897.

Lenzuni, P., Gail, H.-P., Henning, Th. (1995): ApJ **447**, 848.

Lin, D.N.C., Papaloizou, J. (1985): In: D.C. Black, M.S. Matthews (Eds.), Protostars and Planets II (Univ. of Arizona Press, Tucson) p.981.

Lunine, J.I., Engel, St., Rizk, B., Horanyi, M. (1991): Icarus **94**, 333.

Mathis, J.S., Whiffen, G. (1989): ApJ **341**, 808.

Mizuno, H., Markiewicz, W.J., Völk, H.J. (1988): A&A **195**, 183.

Mizuno, H. (1989): Icarus **80**, 189.

Ordal, M.A., Bell, R.J., Alexander Jr., R.W., Newquist, L.A., Querry, M.R. (1988): Applied Optics **27**, 1203.

Ossenkopf, V. (1993): A&A **280**, 617.

Ossenkopf, V., Henning, Th. (1994): A&A **291**, 943.

Pollack, J.B. et al. (1994): ApJ **421**, 615.

Preibisch, Th., Ossenkopf, V., Yorke, H.W., Henning, Th. (1993): A&A **279**, 577.

Prinn, R.G. (1993): In: E.H. Levy, J.I. Lunine (Eds.), Protostars and Planets III (Univ. of Arizona Press, Tucson & London) p.1005

Stenholm, L. (1994): A&A **290**, 393.

Sterzik, M.F., Morfill, G.E. (1994): Icarus **111**, 536.

Stognienko, R., Henning, Th., Ossenkopf, V. (1995): A&A **296**, 797.

Völk, H.J., Jones, F.C., Morfill, G.E., Röser, S. (1980): A&A **85**, 316.

Weidenschilling, S.J., Ruzmaikina, T.V. (1994): ApJ **430**, 713.

Whittet, D.C.B. (1993): In: T.J. Millar, D.A. Williams (Eds.), Dust and Chemistry in Astronomy (Inst. of Physics Publ, Bristol, Philadelphia) p.9

Wolfire, M.G., Churchwell, E. (1994): ApJ **427**, 889.

Wuchterl, G. (1990): A&A, **238** 83.

Yorke, H.W., Henning, Th. (1994): In: U.G. Jorgensen (Ed.), Molecules in the Stellar Environment (Springer, Berlin, Heidelberg) p.186.

Yorke, H.W., Krügel, E. (1977): A&A **54**, 183.

On the Absence of Diffuse Band Carriers in Opaque Circumstellar Disks

Jacek Krełowski[1] and Walter Wegner[2]

[1] Institute of Astronomy, Nicholas Copernicus University, Chopina 12/18,
   Pl-87-100 Toruń, Poland
[2] Institute of Mathematics, Pedagogical University, Chodkiewicz 30,
   Pl-85-064 Bydgoszcz, Poland

Abstract. The Be stars, characterized by the emission in the Fraunhofer D3 HeI line (5876Å) and the specific profiles of the hydrogen Balmer lines, originating in part in rotating circumstellar disks, are apparently immersed in opaque circumstellar matter. Such circumstellar shells are capable of producing quite a substantial reddening but they do not contain the DIB carriers.

1 Introduction

Diffuse Interstellar Bands (DIBs) remain unidentified since their discovery by Heger in 1922 despite the fact that all conceivable forms of matter have been proposed as the possible carriers, ranging from dust grains, to free molecules of very different sizes and structures, to even the hydrogen negative ion. It has been demonstrated during the last decade that more than one carrier must be responsible for the observed DIBs (see Herbig, 1995 for a review). The relations between E_{B-V} and DIB strengths have never been found as very tight (Herbig, 1975, 1995). This is most probably due to the varying intensity ratios of the DIBs – the phenomenon being apparently the result of varying mutual abundance ratios of the different DIB carriers in individual clouds. It seems thus interesting to investigate the DIBs in different media which can make possible an identification of the conditions which facilitate the formation or destruction of the carriers of different DIBs.

Recently we found that in spectra of Be stars the DIBs are weak in relation to E_{B-V}. Stars are classified as Be objects if any of the strong stellar lines (usually H_α) is observed in emission. Some of them show the P Cygni profiles, some others – the broad emission components with some absorption in their centers (Fig.1): evidently originating in circumstellar disks. The latter objects are considered in this paper. The disk, rotating much slower than the star itself, can be assumed to be the remnant of the star parent cloud in contrast to the circumstellar matter producing a P Cygni profiles which reveal a fast expansion of the matter from the star's atmosphere. We can assume that such media differ considerably in their histories and the current physical conditions.

The circumstellar disks can also produce some continuous extinction. The recent atlas of extinction curves (Papaj, Wegner and Krełowski, 1991) shows several "peculiar" extinction curves derived from spectra of Be stars. Such curves have been described for the first time by Sitko et al. (1981). The paper of Krełowski et al. (1992) clearly suggests that DIB intensity ratios are strongly related to the shapes of extinction curves.

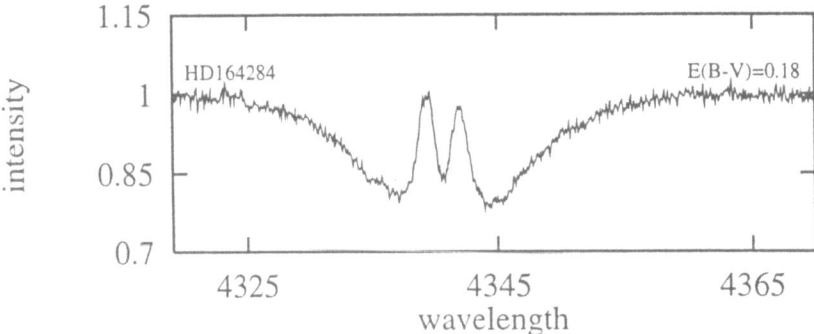

Fig. 1. The H_γ profile in the spectrum of the Be star. Observed at Pic du Midi in July 1995.

2 Diffuse Bands in Spectra of Be and Normal Stars

The high resolution, high S/N spectra acquired in 1993 at the McDonald Observatory and in 1995 at the Pic du Midi Observatory allow to observe many of the DIBs together with the stellar lines of hydrogen and helium. We have selected three objects to illustrate the strengths of DIBs in spectra of "normal" and different Be stars. One of them: HD170740 is the "normal" object, without any emissions in its spectrum. The star HD2905 shows only the hydrogen H_α emission; in the spectrum of HD148184 we can see hydrogen as well as helium emission lines (Fig. 2,3).

The comparison of the strengths of the strong DIBs: 5780 and 5797 (Fig. 4) is quite interesting: the bands are apparently strong in both: HD170740 and HD2905 despite the H_α emission clearly seen in the latter. The situation in HD148184 is quite different. We observe here the emissions in HeI as well as in Balmer lines, the profiles clearly suggesting the observations through a pretty opaque disk. In this case the intensities of DIBs are much lower in relation to E_{B-V}.

3 Discussion

Is the above described phenomenon common for all Be stars? Fig. 5 demonstrates the relation between the equivalent width of the 5797 DIB and E_{B-V} for about ~100 spectra of our McDonald and Pic du Midi sample. The result is quite clear:

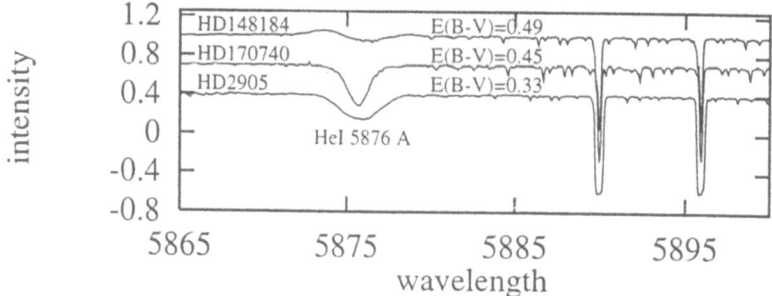

Fig. 2. The interstellar sodium dublet and stellar HeI lines in the McDonald spectra of our examples

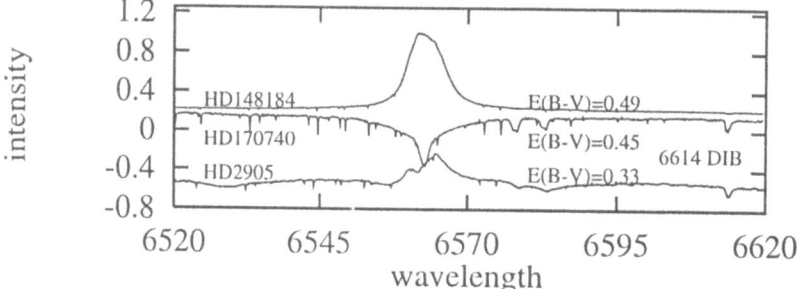

Fig. 3. The H_α profiles and the interstellar 6614 feature (note its weakness in HD148184!) in the spectra of our examples.

DIBs are much weaker in relation to E_{B-V} in all the spectra in which we can observe emissions in HeI lines. The emission in H_α only does not influence the observed DIB intensities.

Are the observed reddenings of Be stars caused by a circumstellar extinction of grains being quite different from those in the diffuse interstellar clouds or their colors are altered by some local free–free scattering? The recent review of Krełowski and Papaj (1993) suggests that the intrinsic flux distributions of both B and Be stars are identical, however altered by different extinction laws. The extinction curves observed towards Be stars may suggest that circumstellar disks are populated by bigger grains than diffuse interstellar clouds. Perhaps the DIB carriers are incorporated into these big grains in the course of the cloud evolution leading to the star formation. The stars with HeI emission lines are apparently obscured by circumstellar disks containing dust grains quite different from "typical" and not containig DIB carriers at all which also allows to understand the scatter of the relations between DIB strengths and E_{B-V}.

Acknowledgements The authors thank the McDonald and Pic du Midi Observatories for allocation of the observing time and to the Kosciuszko Foundation and the French–Polish project PICS for the financial assistance.

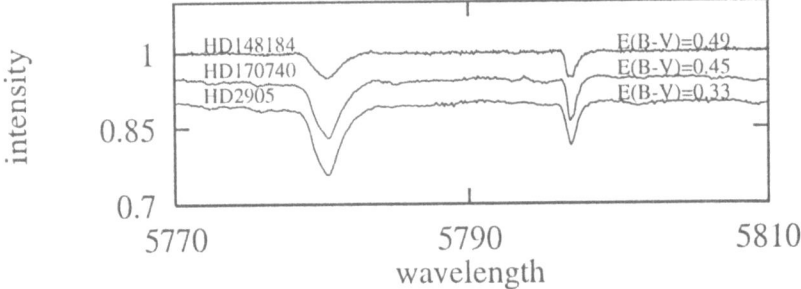

Fig. 4. Two major DIBs in the spectra of our examples. In HD148184 the bands are weak in relation to E_{B-V}.

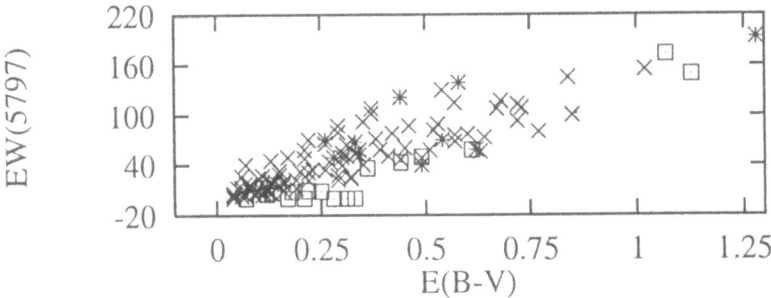

Fig. 5. The equivalent widths of the 5797 DIB plotted vs. E_{B-V}. Open circles – normal B stars, open squares – Be stars with H_α emission only, filled squares – Be stars with emissions in both: hydrogen and helium lines.

References

Herbig, G. H. (1975): The diffuse interstellar bands. IV. The region 4400–6850 Å. ApJ **196**, 129–160

Herbig, G. H. (1995): The Diffuse Interstellar Bands: a review. ARA&A **33** (in press)

Papaj, J., Wegner, W., and Krełowski, J. (1991): Atlas of extinction curves derived from ultraviolet spectra of the TD-1 staellite. MNRAS **252**, 403–407

Krełowski, J., Snow, T. P., Seab, C. G., Papaj, J. (1992): Spatial correlation between CH, CN and the diffuse interstellar band carriers. MNRAS **258**, 693–696

Krełowski, J., Papaj, J. (1993): The Interstellar extinction Curve. PASP **105**, 1209–1221

Sitko, M. L., Savage, B. D., Meade, M. R. (1981): Ultraviolet observations of hot stars with circumstellar dust shells. ApJ **246**, 161–183

Dust in Protostellar Cores

Endrik Krügel[1] and Ralf Siebenmorgen[2]

[1] MPIfR, Bonn, Auf dem Hügel 69, D–53121 Bonn, Germany
[2] ESA – ISO Science Operations, Vilafranca del Catillo, P.O. Box 50727, E–28080 MADRID

Abstract. Inside cold and dense clouds dust grains are modified in three manners: They acquire ice mantles, they coagulate and thus become bigger, and they obtain during coagulation a fluffy structure. We calculate how these modifications change the wavelength dependent extinction coefficient of the dust. The result should be considered as an improved guess for the dust cross section in protostellar clouds.

1 Observational Evidence

Stars form out of an environment that is initially dense and cold. We know that the dust there is different from that in the diffuse interstellar medium, which is sometimes referred to as the standard dust.

First, the interstellar extinction curve of the standard dust has a uniform shape over all directions in the sky and a value $R = A_V/E_{B-V} = 3.1$. In molecular clouds, on the other hand, the shape is variable, the far UV extinction is reduced, and R–values up to 5 are observed. Extinction studies can, however, only scratch the cloud edge and are limited to $A_V \leq 10$ mag.

Second, there are broad IR absorption bands at various wavelengths (3.1μm, 4.7μm, ...) which are absent in the diffuse medium. In these features one can penetrate much deeper into a cloud, up to $A_V \leq 100$ mag.

Third, as a last piece of evidence on a list that is not complete, protostellar dust emits at submm wavelengths much more efficiently than the dust in the diffuse medium. This fact is less known and there are practically no limitations by obscuration. The most astounding example is the protostellar candidate HH24 MMS (Chini et al. 1993, Krügel & Chini 1994, Ward–Thompson et al. 1995). Consider the derivation of its gas mass M_{gas}. We may obtain it from the 1.3mm dust emission or from the column density N of $C^{18}O$, a molecule with optically thin transitions. Usually, these two estimates roughly agree: $M_{gas}[1.3\text{mm dust}] \approx M_{gas}[N(C^{18}O)]$, where in square brackets we indicate the observational method. However, in the case of HH24 MMS, $M_{gas}[N(C^{18}O)]$ is 250 times smaller. Now one might surmise this is because $C^{18}O$ has frozen out, which indeed is likely. But that does not fully explain the discrepancy. Because if we use a purely kinematic procedure and derive the gas mass from the width Δv of an optically thin $C^{18}O$ line, we are independent of the amount of $C^{18}O$ gas, but the estimate is still ten times smaller: $M_{gas}[1.3\text{mm dust}] \approx 10 \cdot M_{gas}[\Delta v(C^{18}O)]$. Evidently, to derive $M_{gas}[1.3\text{mm dust}]$ we must not use the 1.3mm absorption coefficient of the standard dust, but a significantly larger value.

These observationally established peculiarities of dust in a dense cloud are best interpreted by grain growth, deposits of ice and a fluffy composition. Indeed, as small grains are responsible for the far UV extinction, its reduction in molecular clouds implies that they have disappeared and the average grain size has increased. Indirect, but convincing arguments lead to the idea that grains coagulate in dense clouds making them bigger and giving them a fluffy structure. Finally, the IR bands are explained by the excitation of vibrations in functional groups, like OH, of molecules in the solid phase.

2   How to Calculate Cross Sections

The next goal is to quantify this picture. We may envisage a dust grain to consist of refractory subparticles, probably each with its own ice coating, loosely bound together with spaces of vacuum in between. To evaluate the cross section of such a composite grain, we need to know the overall geometric configuration; the chemical composition of the subparticles and the ice; and their microscopic solid structure, namely whether they are amorphous or crystalline and how much they are contaminated by impurities. As any such aggregate is much too complicated to be computed exactly and as there is an infinite variety of such aggregates, a radical simplification is needed.

There are basically two ways of approximating a composite grain: either by an effective medium theory (EMT) or by the discrete dipole (or multipole) approximation (DDA). In EMTs, one derives from the various optical constants of the components an average value representing the whole grain and assumes an overall simple shape, usually a sphere. In the DDA, the grain is replaced by a few hundred or thousand of dipoles and one can then calculate their excitation by the incoming electromagnetic wave and their subsequent emission. Eventually, one needs a numerical algorithm, which is Mie theory in the case of an EMT and something more complicated for the DDA. These algorithms are practically public, so with a little bit of guidance anyone can calculate cross sections of dust in a protostellar cloud.

In our approach, we assume the grains to consist of four distinct chemical components: astronomical silicate, amorphous carbon (aC), dirty ice and vacuum. Their optical constants (n, k) are for silicate from Draine (1985), for aC from Rouleau & Martin (1991), and for dirty ice from Preibisch et al. (1993); vacuum has $(n, k) = (1, 0)$. The volume fractions follow from cosmic abundances, roughly $f_{Si} : f_{aC} : f_{ice} = 1 : 1 : 5$. The degree of porosity is unknown, but $f_{vac} \approx 0.5$ may not be a bad guess. We then apply the Bruggeman theory, which is probably the most appropriate EMT for this configuration. If one feels one has more realistic (n, k)-values for any of the chemical components, one may readily update the cross sections.

It is an appealing and at the same time a weak aspect of our approach that we need only the few parameters (n_i, k_i) and f_i to describe the physics of the grain structure. If one wishes to do better, one has to tackle the formidably complex processes accompanying coagulation and frosting. Attempts in this direction have been made with basic studies by Chokshi et al. (1993) and Ossenkopf (1993). But before falling into euphoria about the accuracy of cross sections, one should bear in mind that EMT and DDA are not at all strict theories, instead they give at times contradictory results.

3   How Size, Ice and Porosity Change K

In Fig.1 we illustrate the effect of grain size, fluffiness and ice coating on the absorption coefficient K. To be specific, we consider 1g of silicate grains at a wavelength λ=1mm, where $(n, k) = (3.5, 0.05)$. We use such units that this 1g of grains has a total absorption coefficient of one.

Now first, we deposit on each grain an ice mantle with a mass about the same as that of the core, as is suggested by cosmic abundances when the freeze out is complete. The absorption coefficient then rises. The relevant parameter is the imaginary part of the optical constant of ice. Estimates for k_{ice} at this wavelength are around 0.01, so we expect that an ice mantle alone enhances K by a factor of three or so (Fig.1 left).

Second, the effect of particle radius a is best seen when K is plotted against a/λ. While increasing the size of the grains, we keep their total mass of 1g fixed. As expected, K is constant when the grains are small and falls like a^{-1} when they are big. Significant amplification occurs when $a/\lambda \approx 1$ (Fig.1 center).

Third, porosity also has the effect to make K larger. For f_{vac}=0, the silicate grains are compact; moving on the abscissa to the right, they become fluffy. For reasonable estimates of the porosity $(0.4 \leq f_{vac} \leq 0.8)$, K is roughly doubled (Fig.1 right).

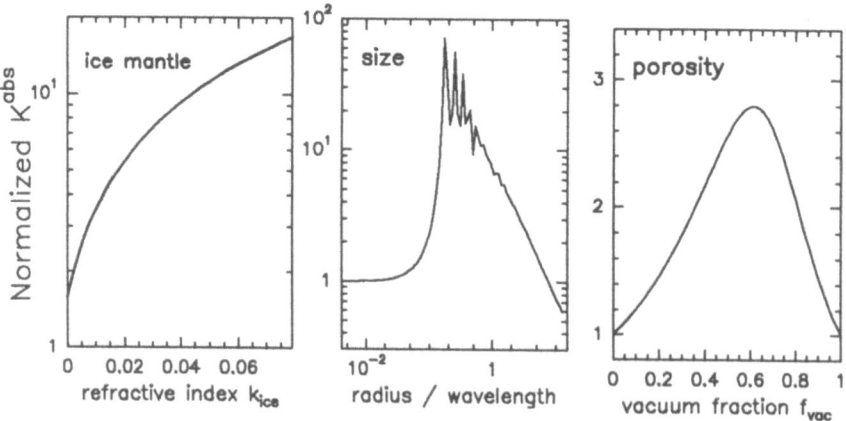

Fig. 1. The influence of an ice mantle, grain growth and porosity on the mass absorption coefficient, here for silicates at λ=1mm.

4   The Mass Extinction Coefficient of Protostellar Dust

Our main result is the thick solid curve in Fig.2 (Krügel & Siebenmorgen 1994). It gives the extinction coefficient per gram of interstellar matter over the entire wavelength range for grains in a protostellar cloud. They are fluffy aggregates consisting of dirty ice, astronomical silicate and amorphous carbon. We now have to specify the grain size distribution $n(a)$. Out of sheer ignorance, we assume the usual power law $n(a) \propto a^{-3.5}$, without any physical argument to support it. As a lower size limit we take $a_- = 0.03\mu$m, for the upper $a_+ = 0.3\mu$m. But note, even if we shift a_+ up to 3μm, the curve stays the same at $\lambda > 10\mu$m, i.e. at the observationally relevant wavelengths for protostellar clouds.

Comparing with the standard dust of the diffuse medium, displayed dotted and labeled MRN, we see that the dust extinction coefficient in a dense cloud is greater everywhere and there is a constant enhancement by a factor of eight in the FIR and submm region. Likewise, the decline with wavelength there has the same ν^2–dependence. The dashed curve depicts what happens when a_+ is truly large and equal to 30μm.

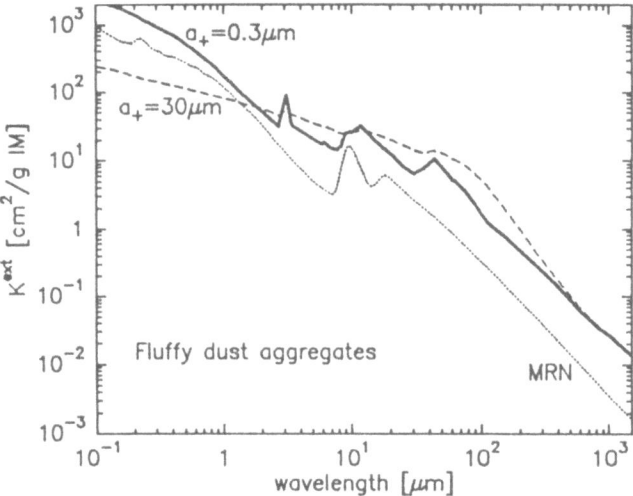

Fig. 2. The extinction coefficient of dust in protostellar clouds per gram of interstellar matter. Grains are icy and fluffy aggregates (*solid*). For comparison, standard dust (*dotted*) of diffuse clouds and very large aggregates (*dashed*).

5   Consequences for Star Formation

Masses. In the study of protostellar evolution, the gas mass is a fundamental parameter, be it of the initial gravitational unstable clump or the accreting disk or the envelope. If we derive it from the dust emission with flux S_ν according to the formula

$$M_{gas} = \frac{S_\nu D^2}{K_\nu B_\nu(T_d)} \qquad (1)$$

we have to use for K_ν the solid curve in Fig.2, and not the dotted one. Thus, at the most frequently employed wavelength of 1.3mm, the absorption coefficient equals 0.02 cm^2 per gram of interstellar matter. Compared to the MRN curve, this brings the mass down by a factor of eight. Reassuringly, for HH24 MMS, where we mentioned in Section 1 the large discrepancy, we now get a mass estimate in good agreement with that from the width of a C^{18}O line: $M_{gas}[\Delta v(C^{18}O)]$ $\approx M_{gas}[1.3mm$ dust]. In fact, considering the indirect way in which dust cross sections in dense clouds have to be derived, this agreement is even a corroboration that our value for $K_{1.3mm}$ is not far off the truth.

Temperatures. An almost equally important quantity is the dust temperature. At the densities prevailing in protostellar clouds, it is also the temperature of the gas because of the strong coupling between the two. Somewhat surprising, there are hardly any changes involved for T_d when using the thick curve in Fig.2 instead of the dotted. The grains acquire their temperature in a radiation field J_ν from the balance between absorption and emission

$$\int K_\nu J_\nu \, d\nu = \int K_\nu B_\nu(T_d) \, d\nu \qquad (2)$$

and as in a protostellar clump J_ν peaks in the far IR, K_ν is raised by the same amount on both sides of the equation, so T_d stays fixed. The same holds for the color temperature which, in contrast to the dust temperature, is a directly measurable quantity that follows from a flux ratio at two wavelengths.

6   The Protostellar Candidate HH24 MMS

For illustration, we have calculated the radiative transfer for HH24 MMS using the dust appropriate for protostars (solid line in Fig.2). The spectral energy distribution (SED) of the model is shown by the solid line of Fig.3.

Note that the broad band photometric observations alone do not give any hint of the anomalous dust properties. Taking the standard dust and the same gas mass, we obtain the lower dotted curve of Fig.3, which yields the same bolometric luminosity and color temperature in the submm region, but does not fit. Of course, one can increase M_{gas} and the data can then be fit (upper dotted curve of Fig.3). However, this violates other constraints on the gas mass, as we have from $\Delta v(C^{18}O)$.

Because HH24 MMS has an embedded VLA cm–nebulosity (Bontemps et al. 1995), we incorporate in the model an internal energy source. Fits are then only

The Composition and Distribution of Dust in Galactic H II Regions

Russell F. Shipman and Frank O. Clark

Geophysics Directorate, Phillips Laboratory, 29 Randolph Rd. Hanscom AFB, MA 01731, USA

Abstract. We have modeled the far infrared emission of two resolved H II regions, the Rosette Nebula (NGC 2237-46) and the λ Orionis H II region (S264). Visible extinction maps and radio continuum observations are combined with infrared data from the Infrared Astronomical Satellite (IRAS) to determine the composition and distribution of dust associated with these nebulae.

Various grain models were used to explain both the 100 μm emission and the visible extinction. The models that best match the data require that the grains producing the visual extinction and the 100 μm emission are larger than average interstellar grains. Since current theories link grain size to the total-to-selective extinction, R_V, our results imply that within and near H II regions R_V is larger than the average ISM value of 3.1.

Furthermore, in order to explain the observed $I_\nu(60)/I_\nu(100)$ ratio, a grain component separate from the grains emitting at 100 μm is required to supply the excess 60 μm emission. Some reasonable possibilities for these emitters are either small (50 to 200 Å) grains that experience strong temperature fluctuations or "astronomical" iron grains that attain high equilibrium temperatures.

1 Introduction

Resolved H II regions provide a natural laboratory for exploring the nature of interstellar dust grains. The high spatial resolution allows for a better determination of the three-dimensional geometry of the region while the high luminosity implies a strong infrared source. Furthermore, many if not all of the H II regions observed by IRAS have been well observed and studied at other wavelengths. This study exploits the fact that the two H II regions, NGC 2237-46 and S264, have been well observed at wavelengths from the ultraviolet to the radio. These previous observations provide many physical properties (e.g., total luminosity, gas and dust distribution) which are necessary for detailed infrared emission models.

Current theories on dust grains rely on the grain size distributions to explain the observed variations in the interstellar extinction curve (e.g., Mathis, Rumpl and Nordseik 1977). Overall, different sizes of grains contribute to different features of the extinction curve. The infrared emission is also dependent on the grain size (Draine and Lee 1984). This study combines infrared emission and visual extinction to constrain the sizes of dust grains associated with H II regions.

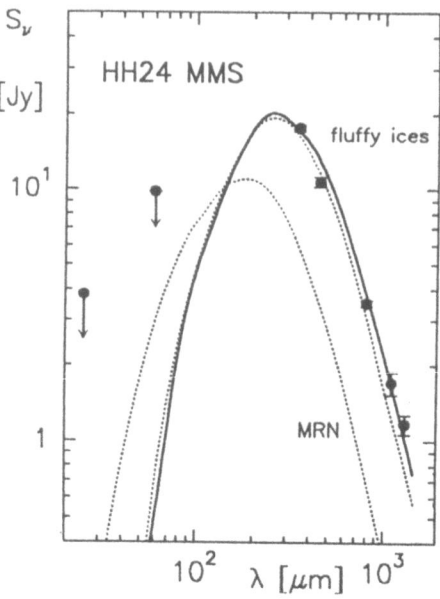

Fig. 3. A radiative transfer model (*solid*) calculated with the dust of Fig.2 (thick line) for the extreme Class 0 object HH24 MMS.

possible as long as its luminosity stays below $3L_\odot$, otherwise the SED of the model overshoots at $350\mu m$. As the gas mass of HH24 MMS is about $4M_\odot$, a premain sequence luminosity lower than $3L_\odot$ implies that the accretion shock has not yet fully developed. This is a strong argument for the youth of this object. It is supported by the SED which peaks at $250\mu m$, so the dust in HH24 MMS is very cold (< 15K), and almost half of the bolometric luminosity is emitted at submm wavelength. Therefore, HH24 MMS must be considered as an extreme Class 0 object.

References

Bontemps S., André P., Ward–Thompson D., 1995, A&A 297, 98

Chini R., Krügel E., Haslam C.G.T., Kreysa E., Lemke R., Reipurth B., Sievers A., Ward–Thompson D., 1993, A&A 272, L5

Chokshi A., Tielens A.G.G.M., Hollenbach D., 1993, ApJ 407, 806

Draine B., 1985, ApJ Suppl 57, 587

Krügel E., Siebenmorgen R., 1994 A&A 288, 929

Krügel E., Chini R., 1994 A&A 288, 929

Ossenkopf V., 1993, A&A 180, 617

Preibisch Th., Ossenkopf V., Yorke H.W., Henning Th., 1993, A&A 279, 577

Rouleau F., Martin PG., 1991, ApJ 377, 526

Ward–Thompson D., Chini R., Krügel E., André P., Bontemps S., 1995, MNRAS 274,1219

2   Methods

A complete infrared emission model requires estimates of the visible to UV radiation field, the gas and dust distribution, and a grain model. Each of these is briefly discussed below.

For both H II regions, the infrared emission data were obtained from the IRAS Sky Survey Atlas (ISSA Explanatory Supplement). The IRAS images were smoothed to the same resolution as the extinction maps and the infrared intensities were obtained from the same positions as the extinction. For S264, the extinction was obtained from star counts by Coulson et al. (1978). For NGC 2237-46, the extinction was obtained from a comparison of the measured Hα emission and the predicted Hα emission (Celnik 1986).

2.1   Radiation Field

The three components to the radiation field are stellar photons, nebular photons, and the interstellar radiation field. The stellar component is obtained from the spectral classification of the stars in the central cluster. Radio continuum observations also measure the rate of emission of ionizing photons from these stars. The nebular radiation field is found by modeling ionization structure of the H II region while the interstellar radiation field was taken to be that of the local neighborhood (Mezger, Mathis, and Panagia 1982).

2.2   Gas and Dust Distribution

Modeling the distribution of material is also essential for modeling the infrared emission. Both H II regions are symmetric and are therefore assumed to be spherical. The radial variation of gas and dust is obtained from radio continuum observations, extinction maps, and H I observations.

2.3   Grain Models

The grain models are composed of bare silicate and graphite grains (Draine and Lee 1984) in a power law size distribution (Mathis, Rumpl, and Nordsieck 1977). To produce various extinction curves, the minimum and maximum grain sizes were varied according to the method described by Maccioni and Perinotto (1994). The different grain models are identified by the total-to-selective extinction they would produce. However, as Maccioni and Perinotto point out, the size distributions are not unique. Therefore, We consider a range of grain models where the ratio of the mass in silicates to mass in graphite is varied.

3   Results

Since, similar results were found for the Rosette Nebula as for S264, we only present the results for S264. Figure 1 shows the $I_\nu(100)/A_B$ for the S264 along

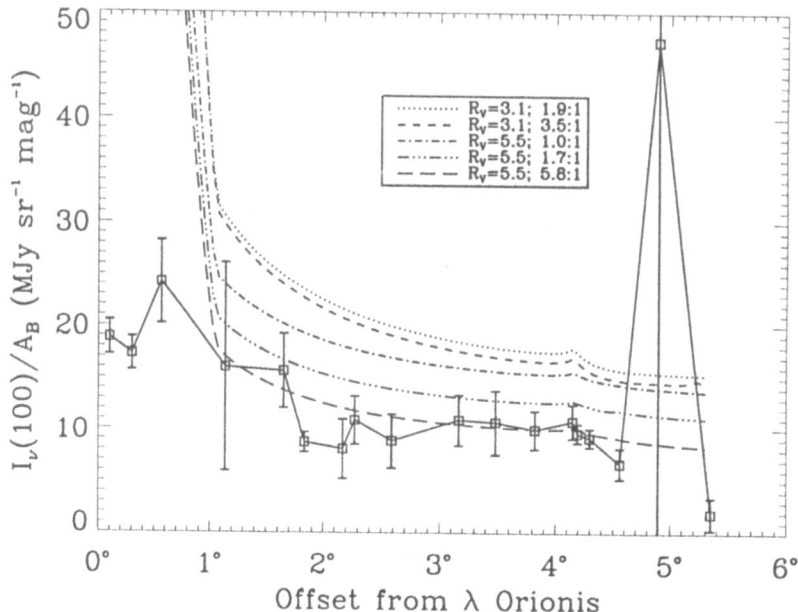

Fig. 1. The $I_\nu(100)/A_B$ ratio as a function of angular distance for the northeast quadrant of S264. Squares show data with errors, while dashed and dotted lines indicate models. The models are labeled by the total-to-selective extinction, R_V, and the silicate-to-graphite mass ratio.

with the model calculations. The best match with the data occurs for a grain model with larger than average grains ($R_V = 5.5$).

The core of S264 is not well modeled by a spherical distribution around the central star. A possible explanation for this problem is that the star λ Ori has wandered from the center of the cloud.

Figure 2 shows the radial variation of the ratio $I_\nu(60)/I_\nu(100)$. The $R_V = 3.1$ grains are excluded based on the emission-to-extinction ratio (**Figure 1**). This results in an underestimation of the 60 μm emission throughout the cloud. Based on the radio continuum data, the strength of nebular lines is unable to account for the strength of the 60 μm emission. The 60 μm excess can only be explained by another grain component that is not part of the size distribution used in the grain models presented here. There are two possibilities presented in the literature: very small (50–200 Å) temperature fluctuating grains (Desert, Boulanger, and Puget 1990) or 'astromonical' iron grains (Chlewicki and Larueijs 1988).

4   Conclusions

In both H II regions, the 100 μm emission and visual extinction indicate that the grains are better represented by larger than average grains. The grain model that

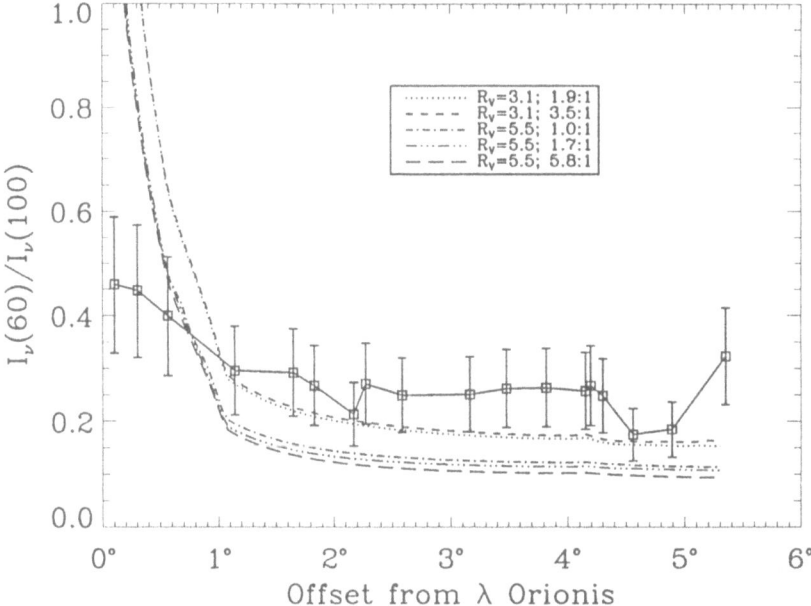

Fig. 2. Here is shown the $I_\nu(60)/I_\nu(100)$ ratio as a function of angular distance for the northeast quadrant of S264. The squares indicate data with errors, while the dashed and dotted lines indicate models. The models are labeled as in **Figure 1**.

best matches the data would produce an $R_V = 5.5$. However this grain model underestimates the 60 μm emission which suggests that another grain component is necessary to explain the strength of the short wavelength emission.

References

Celnik, W. E. (1986): A&A **160**, 287–294

Chlewicki, G. and Laureijs, R.J., (1988): A&A **207**, L11

Coulson, I. M., Murdin, P. G., MacGillivray, H.T., and Zealey, W. J. (1978): MNRAS **184**, 171–180

Desert, F.X., Boulanger, F. and Puget, J.L. (1990): A&A **231**, 215

Draine, B. T., Lee, H. M. (1984): ApJ **285**, 89–108

Maccioni, A. and Perinotto, M. (1994): A&A **284**, 241–247

Mathis, J. S., Rumpl, W., and Nordsieck, K. H. (1977): ApJ **217**, 245

Mezger, P.G., Mathis, J.S., Panagia, N. (1982): A&A **105**, 372

Wheelock, S. L. et al. (1994): *IRAS Sky Survey Atlas Explanatory Supplement* (JPL Publication 94-11, Pasadena: JPL)

The Effects of Clumping on Calculated Grain Properties

Russell F. Shipman

Geophysics Directorate, Phillips Laboratory, 29 Randolph Rd. Hanscom AFB, MA 01731, USA

Abstract. We apply current theories of radiative transfer through a clumpy medium to explore the effects of clumping on visible to far ultraviolet extinction and mid to far infrared emission. In general, clumping can easily mimic the observed environmental variations in extinction. These variations are commonly attributed to changes in the dust grains themselves. Specifically, larger grains are thought to be responsible for high values of total-to-selective extinction, R_V. Such a change in the grains would produce lower emission at far infrared wavelengths.

In this paper, we explore the possibility that clumping could mimic trends in both the visible extinction and the infrared emission. Using a grain model appropriate for the diffuse ISM ($R_V = 3.1$), we determine the clumping parameters (filling factor, clump size, and relative clump density) that would produce the same far infrared emission as a grain model of larger grains appropriate for high values of R_V.

1 Introduction

Line-of-sight variations in the interstellar extinction curve are commonly attributed to physical changes in the dust grains themselves. For example, a high value of the total-to-selective extinction, R_V, implies larger than average grains (e.g., Cardelli, Clayton, and Mathis 1989).

Any physical changes in the dust grain will also be reflected in the grain emission. Specifically, larger grains will reach a lower temperature in a given radiation field. For the same visual extinction, a larger dust grain will emit less than a smaller grain.

Natta and Panagia, (1984), showed that significant variations in the extinction curve can be obtained if the distribution of obscuring material is inhomogeneous. Therefore, for extended objects, clumping of foreground material can mimic large values of the total-to-selective extinction (**Figure 1**). The purpose of this study is to determine whether a clumpy or inhomogeneous cloud can mimic the expected changes in dust emission as well.

2 Radiation Field

The radiative transfer within a clumpy medium is calculated following Boissé (1990). The cloud opacity is defined by a two-phase (clump/interclump) Markov stochastic process. The mean intensity within each phase can be related to the mean intensity of an equivalent homogeneous cloud. The radiation transfer is solved numerically for the homogeneous cloud using the code CSDUST3 (Egan, Leung and Spagna 1988).

Five parameters must be specified to describe the cloud: the optical depth, albedo of the grains, ratio of clump density to interclump density, volume filling factor and the scale length of a clump. These parameters can be varied to find a set that would produce an $R_V = 5.5$ extinction curve (**Figure 1**).

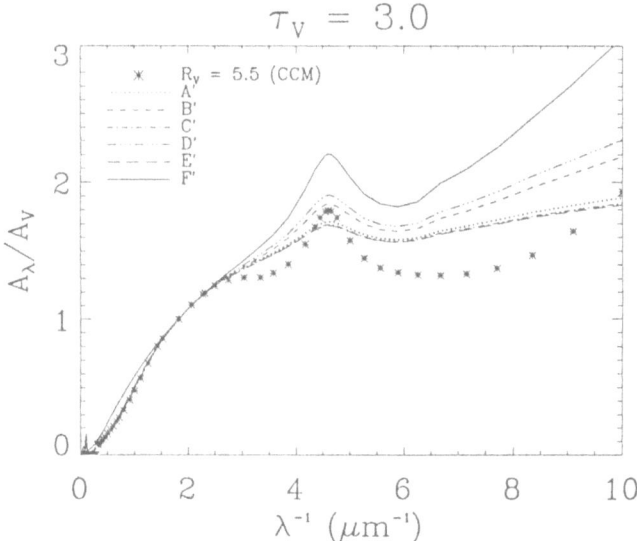

Fig. 1. The extinction curves for various clumping models are compared with an average $R_V = 5.5$ extinction curve (Cardelli, Clayton, and Mathis (1989)). The grain model is the standard bare silicate and graphite grain model (DL and MRN) where $R_V = 3.1$. Each clumping model ($A\prime - F\prime$) was chosen to produce $R_V = 5.5$ for a cloud with an optical depth of 3.0. All models except $F\prime$ agree well with the $R_V = 5.5$ extinction curve (stars).

3   Grain Model

To represent average interstellar grains, the standard bare silicates and graphite grain models were used. The optical properties were taken from Draine and Lee (DL 1984) and the size distribution is that of Mathis, Rumpl and Nordsieck (MRN 1977).

All the grains used in the clumping models follow the MRN size distribution appropriate for the diffuse ISM ($R_V = 3.1$). There is no difference between the grains within the clumps and the grains between the clumps.

To calculate the expected emission from homogeneous clouds with different total-to-selective extinctions, the standard grain model was altered. The minimum and maximum sizes of the silicates and graphite grains were varied according to Maccioni and Perinotto (1994) in order to produce an $R_V = 5.5$ extinction curve. The expected emission from these grains is shown in **Figure 2**.

4   Conclusions

The expected changes in dust emission with large R_V are reproduced within a clumpy medium (**Figure 2**). This implies that observations of visual extinction and infrared emission of extended objects are not sufficient to distinguish between changing grain properties (grain growth) or an intrinsically clumpy medium. This offers an alternate interpretation for the observations of two galactic H II regions (Shipman and Clark 1995) The results presented here are robust since a wide range of clumping parameters can produce the low infrared emission expected for grains with $R_V = 5.5$.

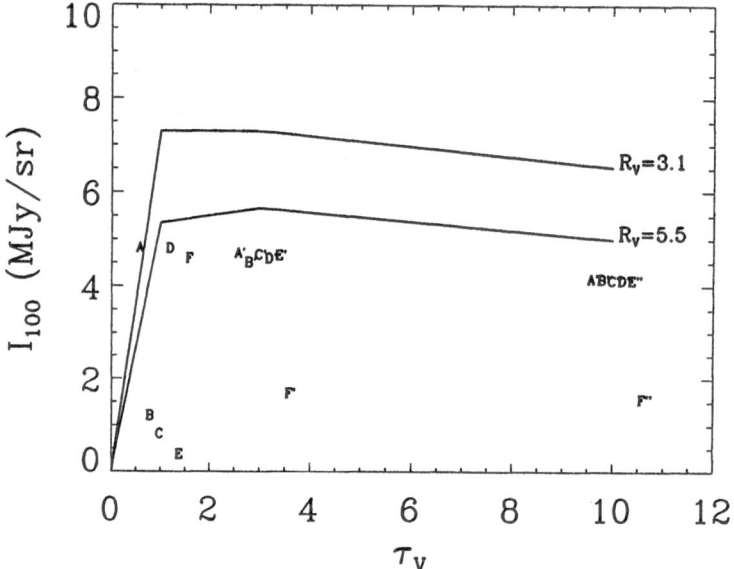

Fig. 2. This figure shows the $I_\nu(100)$ surface brightness as a function of optical depth for a 10 pc slab heated on one side by the interstellar radiation field. The two solid lines show the expected emission from a homogeneous slab for two different grain models. The $R_V = 3.1$ grain model is the standard bare silicate and graphite model (DL and MRN). The $R_V = 5.5$ grain model uses the same grains with different minimum and maximum grain sizes (Maccioni and Perinotto 1994). The points labeled by letters show the predictions of the clumping models. Note that all of the clumping models use the standard $R_V = 3.1$ grain model and yet they easily produce the emission expected from an $R_V = 5.5$ grain model.

References

Boissé, P. (1990): A&A **228**, 483–502
Cardelli, J. A., Clayton, G. C., and Mathis, J. S. (1988): ApJ **329**, L33–L37
Cardelli, J. A., Clayton, G. C., and Mathis, J. S. (1989): ApJ **345**, 245–256

Draine, B. T., and Lee, H. M. (1984): ApJ **285**, 89–108

Egan, M.P., Leung, C. M., Spagna, G. F. Jr. (1988): Computer Physics Communications **48**, 271–292

Maccioni, A. and Perinotto, M. (1994): A&A **284**, 241–247

Mathis, J. S., Rumpl, W., and Nordsieck, K. H. (1977): ApJ **217**, 245

Natta, A. and Panagia, N. (1984): ApJ **287**, 228–237

Shipman, R.F and Clark, F.O. (1995): *these proceedings*

Signature of the Dust Composition Towards Herbig–Haro Energy Sources

Ralf Siebenmorgen[1], Roland Gredel[2], and Jean–Luc Starck[3]

[1] ESA – ISO Science Operations, Vilafranca , P.O. Box 50727, E–28080 MADRID
[2] ESO, Casilla 19001, Santiago 19, Chile
[3] CEA/DSM/DAPNIA, CE-SACLAY, F-91191 Gif sur Yvette Cedex

Abstract. We have observed a sample of Herbig–Haro energy sources with the long slit spectrograph IRSPEC ($R \approx 1500$). The spectra are analyzed by using the *à trous* wavelet transform algorithm. An emission feature near $\approx 2850 cm^{-1}$, with a width of $\Delta\nu \approx 25 cm^{-1}$, is seen towards the HH 100, VMR IRS 8/2 and HH 52–54 energy sources, which we assign to aliphatic CH_2 vibrations. Frozen methanol, near $\approx 2820 cm^{-1}$, is detected towards the HH 100 and HH 65 energy sources. Unidentified bands occur at $2795 cm^{-1}$, $2766 cm^{-1}$ and $2719 cm^{-1}$.

1 Introduction

The optical emission line nebulae known as Herbig-Haro objects are believed to arise from shock waves that are generated from young stars (see e.g. review by Reipurth & Cernicharo 1995). The young stars, or Herbig-Haro energy sources (HHES), are in general deeply embedded into their parent molecular clouds, and the question arises as to what extent the energetic environment causes a re-processing of the surrounding dust and its *icc* mantles in particular. It is well established that the formation and the constitution of grain ice mantles depends sensitively on the physical conditions of the environment. The 3.08 μm O-H stretch is detected in dense molecular clouds towards lines of sight with visual extinction $A_V > 3$ mag, and the optical depth of the absorption, $\tau_{3.1}$ correlates with A_V (Smith et al., 1993). A shallow and relatively broad absorption feature has been detected at wavelengths between 3.4–3.5 μm, superimposed to the red wing of the ice band. There are contributions from interstellar *Methanol* ices, with narrow absorptions near 3.54 μm (Grim et al. 1991, and from hydrogenated, sp^3 bonded carbon or *diamonds*, with absorptions near 3.47 μm (Allamandola et al. 1992). A study of the H_2O feature towards HHES at low spectral resolution, did not show any sign of those L-band features (Graham & Chen, 1991).

2 Observations

A sample of HHES were observed on La Silla using the ESO near-infrared spectrograph IRSPEC on the 3.5m NTT. Details of the instrument are given elsewhere (Gredel & Moorwood 1994). The resolution was set to R = $\lambda/\Delta\lambda \approx 1500$. The length of individual spectra is only ≈ 0.06 μm, and interleaved measurements, with 50% overlap between adjacent spectra, were employed to construct the 3.2–3.8 μm spectra. For each grating setting, explicit sky observations were carried out by moving the source 40″ along the slit. Each full scan was followed by observations of a standard star, and then repeated to accumulate S/N. This technique proved very helpful to judge the quality of individual integrations on the object, and to reduce uncertainties in the transformation to absolute fluxes.

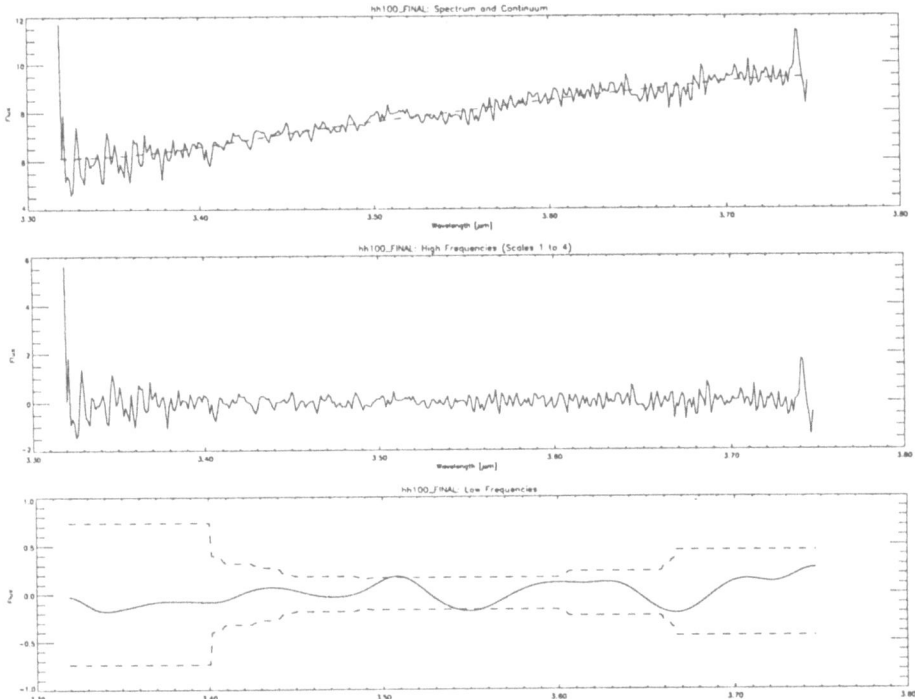

Fig. 1. Results from HH 100. From top to bottom: a) Spectrum and continuum (dashed). b) High frequencies. c) Low frequencies and 3 σ detection limits (dashed). d) Low resolution signal and continuum (dashed). e) Optical depth with standard deviation (dashed).

3   The Wavelet Transform

The wavelet transform (WT) provide a powerful addition to conventional Fourier analysis. Here we apply the *à trous* algorithm (Holdschneider et al., 1989). The wavelet transform of a signal by this algorithm produces at each scale j, a set $\{w_j\}$ which we will call a wavelet scale. The original signal c_0 can be expressed as the sum of all the wavelet planes and the smoothed array c_p:

$$c_0(\lambda) = c_p(\lambda) + \sum_{j=1}^{p} w_j(\lambda) \tag{1}$$

λ μm	ν cm^{-1}	Object	SNR	Opticaldepth τ	Identification
3.508	2850	HH100	3	emission	$cyclo - (CH_2)_6$
3.508	2850	VMR IRS 8/2	3	emission	$cyclo - (CH_2)_6$
3.515	2845	HH52	3	emission	$aliphatic - (CH_2)$
3.516	2844	IRC10011	3	0.042	$aliphatic - (CH_2)$
3.535	2829	V380	3	emission	?
3.535	2829	V883	< 3	emission	?
3.545	2820	HH65	3	0.019	Methanol $CH_3 - OH$
3.545	2821	HH100	3	0.025	Methanol $CH_3 - OH$
3.578	2795	VMR IRS 8/2	3	emission	?
3.578	2795	HH52	< 3	0.013	?
3.615	2766	V380	5	0.018	?
3.678	2719	HH52	< 3	0.029	?

Table 1. Observational results

The smoothed array is a good estimation of the continuum. It contains all the information at a very low resolution and so provides a baseline fit. For large band detection, we are particularly interested in the low frequencies. So we decompose the signal in three parts:

$$\text{High frequencies} : H(\lambda) = \sum_{j=1}^{b} w_j(\lambda) \tag{2}$$

$$\text{Low frequencies} : L(\lambda) = \sum_{j=b+1}^{p-1} w_j(\lambda) \tag{3}$$

$$\text{Continuum} : C(\lambda) = c_p(\lambda) \tag{4}$$

H and L have zero mean. We use for our data $p = 7$, $b = 4$. We speak of detection after inspection of the wavelet space. For this we compare the wavelet coefficients $c_j(\lambda)$ to the standard deviation of the noise σ_j at the same wavelet scale j. We estimate σ_j from the noise in the data. A band is detected at wavelength λ and scale j if $c_j(\lambda) > 3 \cdot \sigma_j$. The optical depth is computed by using the advantage of the WT method to separate continuum, signal and noise from the original spectrum, viz:

$$\tau(\lambda) = -ln \left(\frac{L(\lambda) + C(\lambda)}{C(\lambda)} \right) \tag{5}$$

4 Results

The results obtained towards HHES are summarized in Table 1. A typical spectrum is shown for the source HH 100 (Figure 1). The wavelet analysis resulted in the detection of an emission feature at $2850cm^{-1}$ towards the energy sources of HH 100, VMR IRS 8/2 and HH 52–54. An absorption feature, at $2820cm^{-1}$, is seen towards HH 65 and HH 100. Comparing these features with laboratory measurements (Hudgins et al., 1993), we assign the 2850 cm^{-1} band to $aliphatic - CH_2$ vibrations, such as $cyclo - (CH_2)_6$, and the $2820cm^{-1}$ absorption to frozen methanol. The spectra contain unidentified bands as well: a band near $2795cm^{-1}$ occurs in emission towards VMR IRS 8/2 and in absorption towards HH 52–54. Other absorption features occur at $2719cm^{-1}$ towards HH 52–54. The present assignment of the observed features should be confirmed by observations of vibrational transitions of the same species that are expected at longer wavelengths, as e.g. discussed by d'Hendecourt & Muizon (1989). Available IRAS LRS spectra have to low S/N ratios to confirm the identifications proposed here.

References

Allamandola, L.J., Sandford, S.A., Tielens, A.G.G.M., Herbst, T.M. (1992): ApJ 399, 134

Graham, J.A., Chen, W.P. (1991): AJ 102, 1405

Gredel, R., Moorwood, A.F.M. (1994): IRSPEC Operating Manual, electronic version URL: http: //lw10.ls.eso.org/manuals/irspec/irspec94.html)

Grim, R.J.A., Baas, F., Geballe, T.R., Greenberg, J.M., Schutte, W. (1991): A&A 243, 473

d'Hendecourt, L.B., and de Muizon, M.J. (1989): AA 223, L5

Holdschneider, M., Kronland-Martinet, R., Morlet, J., Tchamitchian, Ph. (1989): "A real-time algorithm for signal analysis with the help of the wavelet transform", in *Wavelets* (J.M. Combes, A. Grossmann and Ph. Tchamitchian, Eds.), pp. 286–297, Springer-Verlag, Berlin

Hudgins, D.M., Sandford, S.A, Allamandola, L.J., Tielens, A.G.G.M. (1993): ApJS 86, 713

Reipurth, B. Cernicharo, J., 1995, in *Circumstellar disks, outflows, and star formation*, Rev. Mex. de Astron. y Astrophis., series de conferencias, 1, 43

Smith, R.G., Sellgren, K., Brooke, T.Y.(1993): MNRAS 263, 749

Starck J.L., and Murtagh F. (1994): AA 288, 342

Williams, D.A., Hartquist, T.W., Whittet, D.C.B. (1992): MNRAS 258, 599

The Composition of Interstellar Dust

A.G.G.M. Tielens

Space Sciences Division, NASA Ames Research Center, Moffett Field
CA 94035-1000, USA

Abstract. Various lines of evidence point towards the presence of Mg– and MgFe–silicates, iron, graphite, hydrocarbons, hydrogenated amorphous carbon, and PAHs in the diffuse interstellar medium. A major fraction of the solid C is presently unidentified. It may be in the form of large graphite, amorphous carbon, or diamond grains. Various stardust components, isolated from meteorites, are discussed and related to interstellar dust components.

1 Introduction

There are various pieces of evidence on the composition of interstellar dust. These include direct observations of spectral structure in the interstellar extinction curve, elemental gas phase depletions, studies of stardust stellar birth sites, and stardust recovered from meteorites and Interplanetary Dust Particles (IDPs). In general, much of this evidence is indirect and/or incomplete and requires creative interpretation. Small wonder that the composition of interstellar dust is highly controversial. Models for interstellar dust can be separated into two broad classes: 1) Stardust models in which dust is formed in the ejecta from stars. Interstellar dust is then thought to consist of silicates and graphite (cf., Mathis et al. 1977). 2) Interstellar dust models in which injected stardust is supplemented by dust formed in the interstellar medium. In that case, besides silicates and graphitic stardust, various hydrocarbon grains play an important role (Greenberg 1979; Jones et al. 1990). In this paper, the composition of interstellar dust is summarized from astronomical observations (§2) and from analysis of stardust isolated from meteorites (§3), emphasizing underlying links to (uncertain) interpretations. Finally, some of the pressing problems facing the various proposed models of interstellar dust are pointed out in §4.

2 The Composition of Interstellar Dust

Figure 1 summarizes my, highly subjective, views on the composition of interstellar dust (cf., Tielens et al. 1996), loosely subdivided into mineral and carbonaceous components. The total dust volume required to explain the interstellar extinction curve, derived from a Kramers–Kronig analysis, is 7×10^{-27} cm^3/(H–atom) (Purcell 1969). The observed strength of the 10 μm feature implies that almost all of the elemental Si is in the form of silicates, in good agreement with interstellar gas phase depletion studies. Silicon is injected by oxygen–rich red

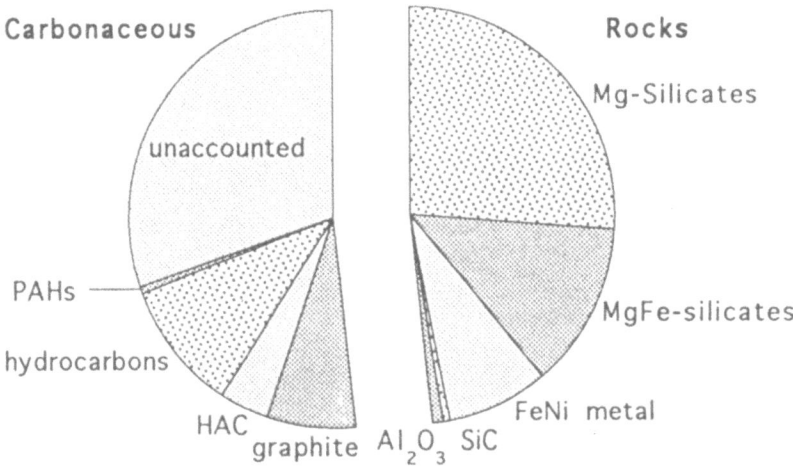

Fig. 1. One man's guide to the dusty universe. Righthand side: mineral components. Lefthand side: carbonaceous components. See text for details.

giants and type II supernovae. In Figure 1, it is assumed that Si injected by red giants is in the form of iron–magnesium silicates. The actual materials condensed in SN ejecta depend on the amount of mixing of the ejecta before dust formation. If there is little mixing, supernovae form Mg–silicates, iron, and C–grains. With mixing, only MgFe–silicates would result. It is difficult to establish observationally whether Fe is locked in silicates or in metals (cf., Tielens 1990a). Depletion studies in the ISM do give some indication that Fe, Mg, and Si are not well mixed in interstellar dust (Sofia et al. 1994). There is some evidence for SN-produced graphite grains in meteorites. Also, Mg–silicates and iron grains have been detected during the Halley missions. Hence, Figure 1 assumes that microscopic mixing during the SN explosion is unimportant and derives   the relative volumes of FeMg–silicates, Mg–silicates, and metalic iron from the estimated contribution of these different stellar birthsites, assuming that all their Si, Mg, and Fe condenses in the form of dust (Tielens 1994). With these assumptions, these mineral components comprise $\simeq 47\%$ of the interstellar dust volume. Aluminumoxide and SiC grains – identified as stardust from their isotopic composition – have been isolated from meteorites. They have also been observed in stellar outflows, but not in the ISM, yet. Figure 1 corresponds to the observational upper limit in the ISM ($< 4 \times 10^{-29}$ cm^3/(H–atom).

The remainder of the interstellar dust has to be carbonaceous in nature. The 2175 Å bump is often attributed to small graphite grains (cf., Draine 1989). Although that assignment has its problems – observations place strict constraints on the shape, size, and coatings of graphite grains (Draine and Malhotra 1993; Mathis 1994) – likely the carrier is carbonaceous in nature. For reasonable oscillator strength, the carrier requires $\simeq 10\%$ of the elemental C. Figure 1 corresponds to the required value for the leading candidate, graphite. Absorption features at

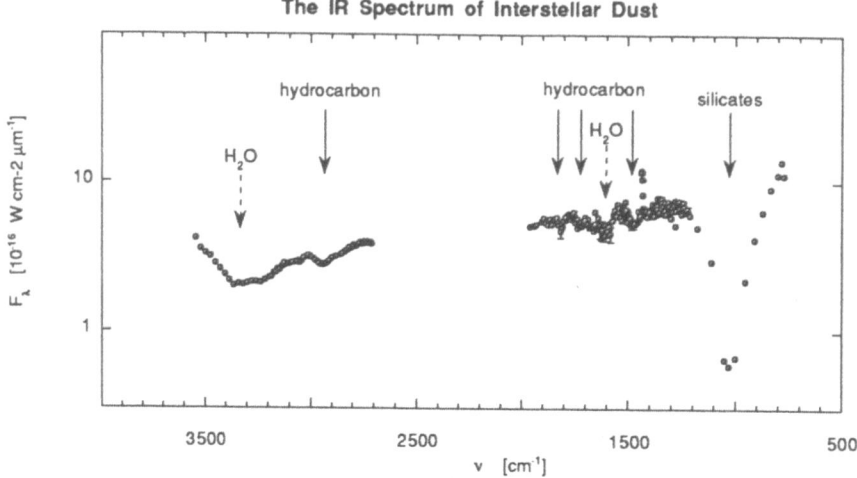

Fig. 2. The IR spectrum of interstellar dust measured towards Sgr A (Tielens et al. 1996). Arrows identify absorption bands due to various dust components. The H_2O features are local to the galactic center. Except for the 3μm H_2O band, the integrated strength per unit dust volume of the various bands are comparable. Hence, equal areas of absorption indicate equal dust volumes. The *observed* interstellar dust volume is dominated by silicates and much of the carbon-dust volume must be hidden from view.

3.4 and 6.8 μm reveal the presence of hydrocarbon grains in the diffuse ISM (Adamson et al. 1990; Pendleton et al. 1994; Tielens et al. 1996). This dust component contains about 8% of the elemental C and very little oxygen (0.5% of the elemental O). Likely, this component has a structure similar to (slightly oxidized) hydrogenated amorphous carbon (HAC). If so, then it is also the carrier of the Extended Red Emission (ERE). Otherwise, another $\simeq$8% of the C is needed for HAC grains carrying the ERE. Finally, about 1% of the C is locked up in interstellar PAHs with a typical size of 50 C-atoms. These PAH molecules are the carriers of the IR emission features at 3.3, 6.2, 7.7, and 11.3 μm. Observed spectra also show broad plateaus underneath these narrow features. These are thought to be carried by larger PAH–clusters with typically $\simeq 10^3$ C-atoms, locking up another 1% of the elemental C (Allamandola et al. 1989).

About half of the carbon–dust volume has presently not been identified (Fig. 1). This is illustrated in a slightly different way in Figure 2, which shows the IR spectrum of the interstellar dust as measured towards the galactic center (Tielens et al. 1996). The *observed* interstellar dust volume is dominated by far by silicates. Clearly, the missing carbon–dust material can have only weak IR vibrational modes. This clue points towards diamond, amorphous carbon, or large graphite grains. Diamond is a perfectly monovalent crystal and has no IR active modes. Disorder will introduce some IR activity and amorphous diamonds show weak and broad absorption near the peak of the phonon density of states ($\simeq 5$ μm), which will be difficult to detect. Surface functional groups will show

IR activity but for large grains their number will be small[1]. The onset of the UV band edge is also difficult to detect (Lewis et al. 1989). Graphite is a semi metal and its IR absorption is dominated by interband electronic transitions. In contrast to small graphite grains, large grains also show no discernable features in the UV (Draine and Lee 1984). The IR spectrum of amorphous carbon or highly dehydrogenated HAC shows only weak C–C stretching modes in the 7-8μm range and these could have escaped detection (Tielens et al. 1996).

3   Interstellar Dust in Meteorites

SiC, graphite, diamond, TiC, and Al_2O_3 grains have been isolated from meteorites with an isotopic composition characteristic for the nucleosynthetic processes of specific stardust birth sites. Hence, these are genuine stardust which survived the formation of the solar system. Very sophisticated techniques have been developed to analyze these grain; the larger ones ($\gtrsim 1\mu$m) individually, the smaller ones in agglomerates. Most of those studies have concentrated on the detailed nucleosynthetic processes taking place deep inside stellar interiors and captured in these grains. However, these grains also provide a direct window on the formation and evolution of stardust in the ISM (Tang and Anders 1988; Zinner et al. 1995). For example, isotopic analysis shows that each individual SiC and graphite grain originated in a different star. The sun probably formed in giant molecular cloud whose gas and dust ($\simeq 10^5 M_\odot$) must have originated from at least 10^5 stars. Unlike the live ^{26}Al in the solar nebula, this stardust was not a late ($\lesssim 10^6$ yr) addition injected by a single, nearby stellar source.

This stardust may have been extensively processed prior to and during the formation of solar system bodies and any record has to be carefully evaluated. Comets retained 20% of the elemental C in solid form, while IDPs and carbonaceous meteorites have only 10 and 2% , respectively. This should be contrasted with the ISM, where 50% of the C is in solid form. Hence, this processing was probably more extensive in the inner solar system than at the outskirts, but even comets do not store pristine stardust records. As to cometary grains, 22% of the grains analyzed by the comet Halley fly–bys was C–based (Fomenkova et al. 1994). Most strikingly, the CHON particles appear to be mixtures of various carbons and organic components at the submicron level, possibly related to meteoritic kerogen material. Only 19% of these are pure elemental C particles. Most of it is hydrocarbon grains with various amounts of substitutional groups consisting of O, N, and H, perhaps akind to interstellar hydrocarbon grains.

Much of the C ($\simeq 75\%$) in carbonaceous meteorites is in the form of a fine-grained, amorphous, macromolecular kerogen–like material with an enhanced D abundance, betraying it's or its precursor's interstellar roots (Kerridge 1989). The C stardust is at the parts per million level; 600 ppm for diamonds and $\simeq 10$ ppm for graphite and SiC (Anders and Zinner 1993). Among the other C–phases

---

[1] A 3.46 μm feature in protostellar spectra has been ascribed to H bonded to diamond surfaces (Allamandola et al. 1992), but its carrier may be limited to dense molecular clouds and may not be present in the diffuse ISM (Brooke et al. 1996).

are carbonates, revealing the importance of interaction with fluid water on the meteoritic parent body. The isotopic anomalies in SiC, graphite and diamond each betray a different stellar mix. The SiC comes mainly from AGB stars – very much in line with the astronomical evidence – with a minor contribution (1%) from type II SN or WC stars (Anders and Zinner 1993; Zinner 1995).

Graphite stardust is of particular interest in view of §2. Graphite grains can be separated into spherulitic aggregates, "blocky" grains, and round grains. The latter can be further subdivided by morphology into onion– and cauliflower–type grains and these are the grains which carry the noble gas anomalies indicative of a stellar origin (Zinner et al. 1995). The onions consist of layers of well crystalized graphite (ordered on 150–250 Å sizescale) that frequently mantle an amorphous carbon core. They carry the C–isotope signature of He burning and probably originated in type II SN or WC stars. Cauliflowers on the other hand consist of concentric layers of small scale ($\lesssim$ 150Å), poorly graphitized C and originated in AGB outflows. Both show evidence for the presence of organic material (Zinner et al. 1995). In particular, PAHs with an (anomalous) C–isotopic signature consistent with the parent grain's stellar birthsite have been extracted from some graphite grains (Clemett et al. 1994). Neither of these graphite stardust classes shows evidence for SiC cores and, thus, heterogeneous nucleation played no role during their formation. The spherulitic graphite aggregates are built up from $\simeq$ 0.2μm particles. Except for a modest ^{15}N enrichment, there are no isotope anomalies associated with these grains and they seem to sample the main reservoir of interstellar carbon. Since the C isotopes are injected by different stellar sources, this suggest that these grains were formed in the ISM from the major reservoir of carbon. They do show a high concentration of trace elements but whether the subgrains have a core–mantle structure as envisioned by some interstellar models remains to be seen. In that respect, note that no silicate core–carbon mantle structures have been observed in interplanetary dust particles. Various origins for the tiny ($\simeq$10Å) diamond grains recovered from meteorites have been proposed (cf., Anders and Zinner 1993). Since their C–isotopes are rather mundane, no single stellar source is implied. In contrast, direct diamond formation from the gas by CVD methods requires very specific conditions, including a substrate, which are unlikely to be characteristic of all stellar outflows. Moreover, there is likely only one C-condensate per type of outflow and graphite stardust originating from red giants and from SN/WC stars have already been identified. Furthermore, diamonds carry only the Xe–isotopic signature (but not C) of type II SNe, not of AGB stars. These points argue again for an interstellar rather than a circumstellar process. Likely they are formed by grain–grain collisions in interstellar shock waves (Tielens et al. 1987). Since there is no graphitic precursor with the Xe–HL anomaly, this signature likely reflects implanatation into a small subset of grains near the SN which were directly exposed to the ejecta. The spherulitic graphite and the diamonds differ slightly in their C and N isotopic composition. Since most of the C–dust was processed in the ISM and in the solar nebula, the former might still be (related to) the parent of the latter.

4   Problems

In evaluating the different models for interstellar dust, we should take two additional points into consideration. First, theoretical studies have shown that interstellar dust is rapidly destroyed in interstellar shock waves (Jones et al. 1994). This is in line with elemental abundance studies in high velocity (and hence recently shocked) gas which invariable show much less depletion than the general ISM (Jenkins 1989). These theoretical studies yield dust life times in the ISM of $\simeq 5 \times 10^8$ yr, much less than the stardust injection timescale ($\simeq 2 \times 10^9$ yr). With these timescales, the average fraction of an element locked up in dust is equal to 0.2 of the initial fraction injected in the form of dust by stars. If SN form dust efficiently, this initial fraction is ~ 1 for Si. For carbon, however, this initial fraction is estimated to be only 0.2 with the remainder in gaseous CO and atomic C (Tielens 1990b). Thus, any interstellar dust model based solely upon dust injection by stellar sources, such as MRN, faces serious problems explaining the large fraction of various elements locked up in dust.

Second, one might circumvent this problem by efficiently growing dust through accretion in the ISM (Greenberg 1979). At the low temperatures of the ISM, formation of simple ices is expected. Interstellar ice is indeed an abundant dust component in the protected environment of dense molecular clouds. However, such simple ice mantles are very volatile and readily destroyed in the diffuse ISM by shocks in $\simeq 10^6$ yr (Jones et al. 1994). More refractory material (ie., organic residue) could be produced by FUV photolysis (producing radicals) and warm-up (leading to radical diffusion and reaction) of these ices. Laboratory studies support this general scenario but the measured efficiency of this process is fairly low ($\lesssim 1\%$ conversion efficiency of ice into organic residue) and many cycles of accretion, photolysis, and warm-up are required during one sojourn into a molecular cloud (cf., Tielens and Allamandola 1987).

The presence of hydrocarbon grains in the diffuse ISM has often been claimed as "evidence" for the importance of this process (Butchard et al. 1986; Sandford et al. 1991). However, the 3.4 μm feature, the tell-tale signature of hydrocarbon grains, has only been observed in the diffuse ISM and is notably absent in dense cloud environments where this material is supposed to be formed (Allamandola et al. 1993). Indeed, there is very little observational evidence for photolyzed residues in molecular cloud material. Moreover, while laboratory residues provide reasonable fits to the detailed shape of the 3.4 μm feature, they have strong features in the 5-8 μm window due to carbonyl (C=O) bonds, in contrast to the observations (Fig. 2). Interstellar hydrocarbon dust consists just of hydrogen and carbon with a titbit of oxygen. Hence, if the carrier of these interstellar IR bands is formed by photolysis from ice material, the resulting residues must have been almost completely carbonized in molecular clouds, perhaps by prolonged photolysis. There is some support from lab studies for this point of view (Jenniskens et al. 1993), but neither this process nor its importance in the ISM has been clearly established.

Almost complete carbonization might solve the lifetime crisis and produce a grain material with few observational signatures [a theorists's dream !]. However, this would then also imply that the interstellar hydrocarbon absorption features (ie., the 3.4μm band) are not direct evidence for the photolysis model. The absence of the hydrocarbon features in molecular cloud material suggests that its carrier is formed in the diffuse ISM itself. In that case, it is likely a transformation process of a preexisting grain material rather than a grain condensation process. Furton and Witt (1993) have shown that amorphous carbon will transform into HAC under exposure to H, particularly in the presence of strong UV fields. Alternatively, proton bombardment of graphite, amorphous carbon, or even diamond will lead to the formation of a 10-20Å layer of HAC in interstellar shocks (Tielens et al. 1994). Thus, summarizing, we may be driven to a synthesis of the existing dust models: Interstellar dust consist largely of silicates and graphite/amorphous carbon grains. However, much of the carbon-based dust is formed mainly in dense clouds in the ISM through accretion and prolonged photolysis. Processing by shocks or H-atoms in the diffuse medium leads then to the transformation of a thin surface layer into HAC.

My criteria to evaluate astronomical models is: Two ifs give one but and with two buts you sit down. From that perspective, no model for interstellar dust is left standing and the one outlined above forms no exception. To a large extent, this reflects the lack of data to test theories against. This may however be changing. The upcoming launch of ISO, the arrival of 10m–class, optical/IR, ground–based telescopes, and rapid progress in analytical techniques for micron–sized stardust and IDPs promises a shining future for studies of interstellar dust. Among the observational questions that I would single out at this workshop is the spectral differences of dust in the diffuse ISM as compared to dust in molecular clouds. Or in more generally: How does the spectrum of interstellar dust vary from one environment to another ? This includes spectral variations at molecular cloud edges, within clouds, from one starforming region to another, etc. The title of this workshop could also be reversed to "how does star formation influence dust evolution". Interstellar dust formation starts with ice formation and observations indicate that their evolution is very much influenced by nearby, newly formed stars. Quantifying this ice evolution is another aspect of this same question.

At present, the IR spectrum of dust in the diffuse ISM has been characterized from 3–20 μm only towards Sgr A and over a very limited wavelength region (3-4 μm) towards ~10 objects. The SWS on ISO can be expected to give good spectra from 3-45 μm for perhaps a dozen stars dominated by diffuse interstellar dust. Because of its light gathering powers and its longer lifetime, the VLT equiped with a medium resolution grating spectrometer can address these questions for a much larger sample over more limited wavelength regions. One serious problem is the small IR optical depth expected for typical sources ($\lesssim 0.01A_v$ with A_v a few). A successful program will require routinely high S/N ($>> 100$) and much attention will have to be payed to continuum baselines, "true" stellar standards, matching of spectral types, careful removing of atmospheric and photospheric features, etc. Of equal importance is the characterization of the physical conditions along the line of sight with high spectral resolution ($R \gtrsim 10^5$) spectrometers in IR, rovibrational, molecular lines on the VLT.

References

Adamson, A.J., Whittet, D.C.B., Duley, W.W., 1990, MNRAS, 243, 400

Allamandola, L.J., Tielens, A.G.G.M., Barker, J.R.,1989, ApJS, 71, 733

Allamandola, L.J., Sandford, S., Tielens, A.G.G.M., Herbst, T., 1992, ApJ, 399, 134

Allamandola, L.J., Sandford, S.A., Tielens, A.G.G.M., Herbst, T.M., 1993, Science, 260, 64

Anders, E. Zinner, E., 1993, Meteoritics, 28, 490

Brooke, T.Y, Sellgren, K. Smith, R.G., 1996, ApJ, in press

Butchart, I., McFadzean, A.D., Whittet, D.C.B., Geballe, T.R., Greenberg, J.M., 1986, A & A, 154, L5

Clemett, S.J., et al., 1994, Meteoritics, 29, 457

Draine, B.T.,1989, in *Interstellar Dust*, eds. L.J. Allamandola & A.G.G.M. Tielens, (Kluwer, Dordrecht), 313

Draine, B.T., and Lee, H.M., 1984, ApJ, 285, 89

Draine, B.T., Malhotra, S., 1993, ApJ, 414, 632

Fomenkova, M., Chang, S. Mukhin, L., 1994, Geochim Cosmochim Acta, 58, 4503

Furton, D.G. Witt, A.N., 1993, ApJL, 415, L51

Greenberg, J.M., 1979, in *Stars and Starsystems*, ed. B.E. Westerlund, (Reidel, Dordrecht), 173

Jenkins, E., 1989 in *Interstellar Dust*, eds. L.J. Allamandola, A.G.G.M. Tielens, (Kluwer, Dordrecht), 23

Jenniskens, P., Baratta, G.A., Kouchi, A., de Groot, M.S., Greenberg, J.M. Strazzulla, G., 1993, A & A, 273, 583

Jones, A.P., Duley, W.W. Williams, D.A., 1990, QJRAS, 31, 567

Jones, A.P., Tielens, A.G.G.M., Hollenbach, D.J. McKee, C.F., 1994, ApJ, 433, 797

Lewis, R.S., Anders, E. Draine, B.T., 1989, Nature, 339, 117

Mathis, J.S., Rumpl, W., Nordsieck, K.H., 1977, Ap.J., 217, 425

Mathis, J.S., 1994, ApJ, 422,176

Pendleton, Y. J., Sandford, S.A. Allamandola, L.J., Tielens, A.G.G.M., Sellgren, K., 1994, ApJ, 437, 683

Purcell, E.M., 1969, ApJ, 158, 433

Sandford, S.A., Allamandola, L.J., Tielens, A.G.G.M., Sellgren, K., Tapia, M. & Pendleton, Y., 1991, ApJ, 371, 607

Sofia, U.J., Cardelli, J.A., Savage, B.D., 1994, ApJ, 430, 650

Tang, M. Anders, E., 1988, ApJ, 335, L31

Tielens, A.G.G.M., 1990a, in *From Miras to Planetary Nebulae*, eds. M. Mennessier & A. Omont, (Press Frontiere, Montpellier), 186

Tielens, A.G.G.M., 1990b, in *Carbon in the Galaxy*, eds. J. Tarter et al., NASA CP 3061, 59

Tielens, A.G.G.M., 1994, in *Airborne Astronomy Symposium*, eds. M.R. Haas et al (ASP, San Francisco), 3

Tielens, A.G.G.M., Allamandola, L.J., 1987, in *Interstellar Processes*, eds. D. Hollenbach & H. Thronson, (Kluwer, Dordrecht), 397

Tielens, A.G.G.M., Seab, C.G. Hollenbach, D.J., McKee, C.F., 1987, ApJ, 319, L109

Tielens, A.G.G.M., McKee, C.F., Seab, C.G. Hollenbach, D.J., 1994, ApJ, 431, 321

Tielens, A.G.G.M., Wooden, D., Allamandola, L.J., Bregman, J., Wittebron, F.C., 1996, ApJ, in press

Zinner, E., 1995, in *Nuclei in the Cosmos*, eds. M. Busso, et al., (AIP, New York) 567

Zinner, E., Amari, S., Wopenka, Lewis, R.S., 1995, Meteoritics, 30, 209

The Shape of the Extinction Curve of Opaque Circumstellar Matter

Walter Wegner[1] and Jacek Krełowski[2]

[1] Pedagogical University, Institute of Mathematics, Chodkiewicza 30, Pl-85-064 Bydgoszcz, Poland
[2] Institute of Astronomy, N. Copernicus University, Chopina 12/18, Pl-87-100 Toruń, Poland

Abstract. The extinction curves observed in the spectra of Be stars, characterized by stellar emission of both hydrogen and helium lines, show several peculiarities. The extinction bump near 2200 Å is apparently absent in the extinction law of the circumstellar dust grains. The total-to-selective extinction ratio strongly suggests that the circumstellar grains are hot and, thus, they strongly radiate in the infrared spectral range. The intrisic stellar parameters of normal B and Be stars are seemingly identical.

1 Introduction

Surveys show a great variety of extinction curves differing from object to object which, however, can be grouped into "families" (apparently originating in similar clouds) – Fig.1 and Fig.2. Very peculiar extinction curves have been found

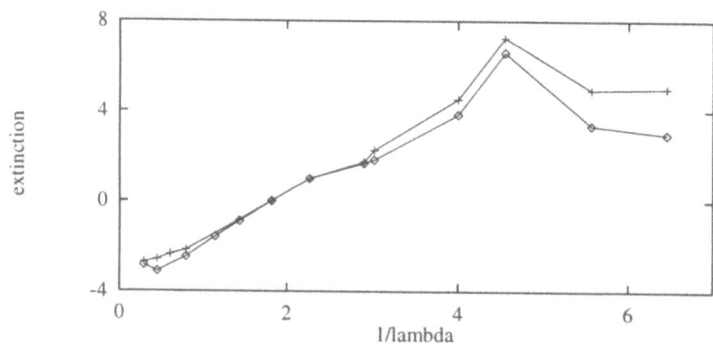

Fig. 1. Two extinction curves derived from photometric data for the objects representative for the σ (HD144470 – diamonds) and ζ (HD179406 – crosses) types. Note the evident difference in the vacuum–UV range

by Sitko *et al.* (1981) as well as by Papaj *et al.* (1991). Only a few actual extinction curves resemble the mean curve determined by Savage & Mathis (1979). It is clear that high reddenings are usually "composite cases" – the extinction originates in several clouds and so the extinction curves deviate much less from what is known as the average Galactic extinction curve. The extinction curves do not give us absolute extinction values, but only relative ones. To turn the

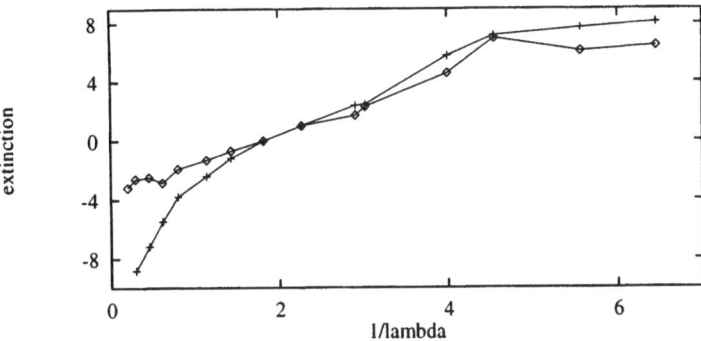

Fig. 2. The extinction curves derived from photometric data for two Be stars: HD2905 – diamonds (only H_α in emission) and HD202904 – crosses (also HeI lines in emission)

curve into absolute extinctions, it is necessary to determine at least one value of the absolute extinction, *e.g.*, in the V band. A widely adopted parameter is the total-to-selective extinction ratio $R = A_V/E_{B-V}$, where A_V represents the total V extinction. The value of R is represented by the point in which the curve intersects the ordinate axis (*i.e.* for $1/\lambda = 0$). The extinction for infinite wavelength should be zero by definition. However, in the infrared a serious difficulty is to be faced. In the radiative transfer equation in the ultraviolet or visual wavelength ranges we may neglect the re-emission term due to the low temperature of interstellar grains. In the infrared it can easily be a much too simplistic approximation. Thus, even while the applied standards are absolutely correct, the resultant extinction may be seriously contaminated by emission originating in dust grains due to the energy from all other spectral ranges absorbed by these grains. It is to be noted that in cases of Be stars the "abnormal" value of this important constant is most probably due to infrared emissions originating in circumstellar matter.

2   Results

We present several extinction curves derived from spectra of Be stars. It is strongly suggested that the extinction originating in opaque, circumstellar disks (which also produce emission helium lines) differs seriously from that originating in diffuse interstellar clouds. We present also an attempt to separate the emission and absorption in the infrared range in Fig.3 and Fig.4.

The interstellar extinction curves originating in opaque circumstellar disks differ substantially from those created in diffuse interstellar clouds. The especially striking is the lack of the 2200 Å extinction bump. This phenomenon may be due to the lack of fine carbon particles which can be incorporated into larger grains in the course of protostellar cloud collapse. Moreover the extinction curves suggest a very high value of the total–to–selective extinction ratio. The latter is most probbaly due to the infrared emissions caused by relatively hot circumstellar grains irradiated by the hot, central stars. The fact that the relations

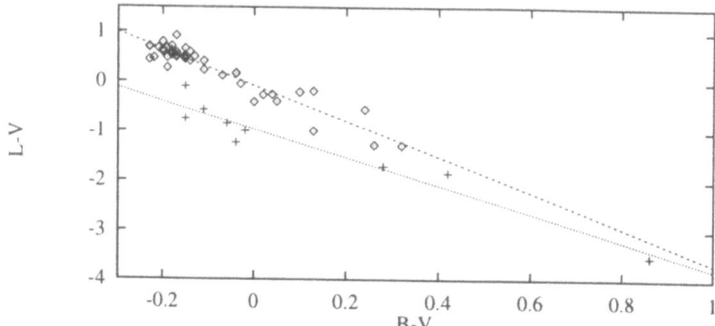

Fig. 3. The two–colour diagram relating IR and B-V colour indices of B0V – B3V stars. Normal B stars: diamonds, Be stars with HeI emission lines – crosses. The slopes of the straight lines represent the extinction law

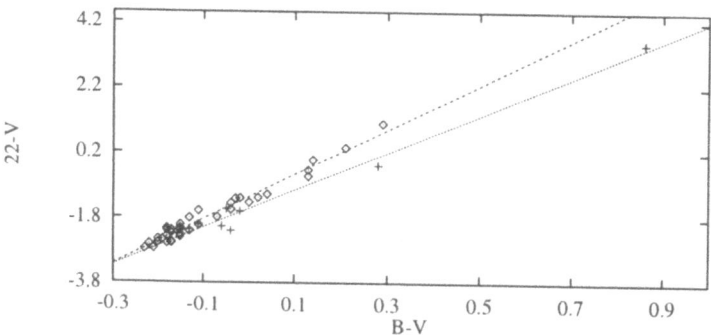

Fig. 4. The same as in Fig.3 but for the vacuum–UV band measuring the strength of the 2200 Å extinction band. The latter is apparently much weaker in all Be stars with HeI emissions

for normal B and Be stars, shown in Figs. 3 and 4, intersect each other close to $E_{B-V}=0$ proves that the parameters of the stellar photospheres of normal B and Be stars are the same, but the extinction law observed in both cases is quite different.

References

Sitko,M.L., Savage, B.D., Meade, M.R. (1981): Ultraviolet observations of hot stars with circumstellar dust shells. ApJ **246**, 161–183

Papaj, J., Wegner, W., Krełowski, J. (1991): Intrinsic energy distributions in spectra of early-type stars. MNRAS **252**, 408–414

Savage, B.D., Mathis, J.S. (1979): Observed properties of interstellar dust. ARA&A **17**, 75–111

Modelling of Interstellar Extinction in Single Clouds Using the Method of Regularisation

Victor G. Zubko[1,2], Jacek Krełowski[1] and Walter Wegner[3]

[1] Institute of Astronomy, N. Copernicus University, Chopina 12/18, Pl–87–100 Toruń, Poland
[2] Main Astronomical Observatory, National Academy of Sciences, Golosiiv, Kyiv–650, 252127, Ukraine
[3] Pedagogical University, Institute of Mathematics, Chodkiewicza 30, Pl–85–064 Bydgoszcz, Poland

Abstract. The method of regularisation for resolution of the inverse problems has been applied for an analysis of interstellar extinction originating in apparently single clouds seen towards the stars HD147165, HD179406 and HD202904. The multicomponent size distributions for various bare spherical dust grain mixtures containing graphite, silicates, amorphous carbon and SiC have been deduced. The results could be interpreted in terms of an evolution of interstellar clouds.

1 Interstellar Extinction

The important role of interstellar dust in various physical and chemical processes in interstellar clouds is well-known. For a long time one of the most powerful and plausible tools to explore the properties of dust is analysis of the extinction law of starlight as a function of wavelength. Interstellar extinction may be explained as a combined effect of absorption and scattering by small dust grains and, possibly, other molecular species (e.g., PAH). However, practically all efforts of investigators have been so far directed to construction and modelling of various *average* extinction curves (see, e.g., Bless and Savage 1972, Massa and Savage 1989, Kim et al. 1994). The parametrization of extinction with R_V (total-to-selective extinction ratio) found by Cardelli et al. (1989) applies to some average extinction law too. Evidently, any such extinction curve has no direct physical interpretation because it is a result of averaging over, as a rule, many clouds of different properties of both gas and dust. Therefore, an investigation of extinction of nearby stars seen through an apparently single cloud is of special importance. Krełowski and Wegner (1989) have shown that the extinction curves derived from the spectra of some nearby bright stars may be classified into, at least, three families. Moreover, this division is well supplied by the behaviour of diffuse interstellar bands and interstellar molecular lines (Krełowski and Sneden 1994, Snow and Krełowski 1994). So, the purpose of this contribution is to present some results of our detail modelling of the typical extinction curves of these families within some plausible dust grain models and using the state-of-the-art mathematical technique. As the output data we obtain the size distributions of dust grains in the most general form. We plan to describe more details and results in our forthcoming paper.

2   Model

We have involved into our analysis the data for the stars: HD147165, HD179406 and HD202904. The extinction curves were derived from spectra of these stars in all possible ranges of wavelengths from the vacuum–UV down to the IR using the method of artificial standards (Papaj et al. 1993, Wegner 1994). This is especially important to obtain a reliable size distribution of dust grains. We have taken the UV data from Wesselius et al. (1982), the UBV ones from Hoffleit and Jaschek (1982) and the IR from Gezari et al. (1984). Some fragments of the spectra of the analysed stars are displayed in Fig. 1 – the very narrow lines of sodium D_1 and D_2 give a strong evidence that we observe the analysed stars through single clouds.

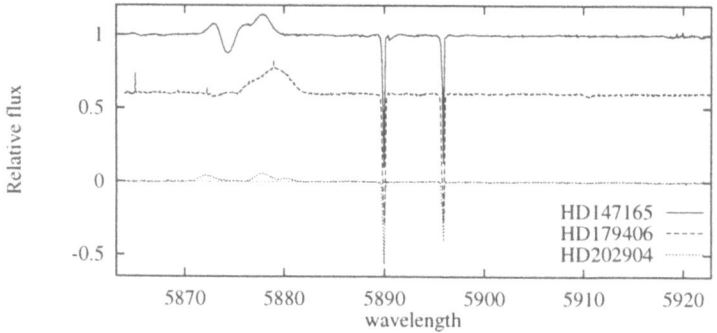

Fig. 1. Sodium lines D_1 and D_2 in the spectra of HD147165, HD179406 and HD202904 obtained by JK in May 1993 at McDonald Observatory

We have stated a problem to extract the size distributions of dust grains involving various mixtures of the bare spherical grains made up of graphite, "astronomical" silicate, SiC and amorphous carbon, in order to obtain the best fit of any extinction curve to within the errors of observations. Mathematically, this is a typical inverse problem – resolution of a Fredholm integral equation of the first order. We have developed the algorithm and the programme to resolve our problem on the base of the method of regularization using the technique of the conjugated gradient projections to minimize the main functional. (Tikhonov et al. 1990). It should be noted that the method of regularization has some advantages in comparison with the maximum enthropy method already used by Kim et al. (1994) to extract the size distribution of interstellar dust grains. In particular, there is no need to specify any default solution since the method of regularization generates a solution for very small and very big grains automatically. We have performed the calculations without and with the abundance constraints (typically, C/H=3.0 × 10⁻⁴ and Si/H=3.2 × 10⁻⁵). We have used the standard Mie theory for the calculation of the extinction efficiencies and the optical constants of graphite, "astronomical" silicate and SiC from Laor and Draine (1993) and of amorphous carbon (ACAR sample) from Zubko et al. (1995).

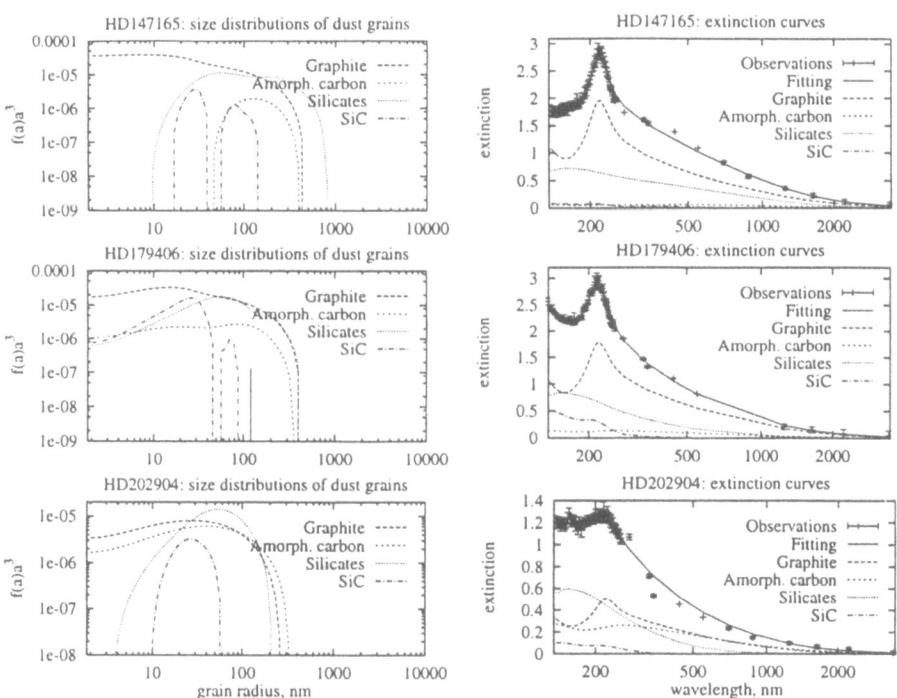

Fig. 2. Size distributions of dust grains (left panels) for the 4-component mixture and the respective extinction curves (right panels) with the contributions of each constituent

3   Results

Some results of our modelling for the 4-component mixture of graphite, "astronomical" silicate, SiC and amorphous carbon grains are shown in Fig.2. We would like to draw attention on a few features. Firstly, practically all the size distributions are essentially different from the well-known Mathis–Rumpl–Nordsieck power–law one. Secondly, the extinction for HD147165 is explained mainly by graphite+"astron." silicate. However, the contributions in extinction from SiC and amorphous carbon become noticeable and even comparable with graphite and "astron." silicate for HD179406 and HD202904, especially in the

UV. Thirdly, the large grain edge of the size distributions is in average shifted into the smaller sizes from HD147165 to HD202904. Fourthly, the size distribution of SiC grains has a tendency to be multimodal. Do our results mean that the clouds under analysis represent the links of some evolutionary sequence? This is a question to our forthcoming investigations.

References

Bless, R.C., Savage, B.D. (1972): Ultraviolet photometry from the orbiting astronomical observatory. II. Interstellar extinction. ApJ **171**, 293–308

Cardelli, J.A., Clayton, G.C., Mathis, J.S. (1989): The relationship between infrared, optical and ultroviolet extinction. ApJ **345**, 245–256

Gezari, D.Y., Schitz, M., Mead, J.M. (1984): *Catalogue of Infrared Observations*, NASA Reference Publs. 1118.

Hoffleit, D., Jaschek, C. (1982): *The Bright Star Catalogue* (Yale University Observatory, New Haven)

Kim, S.-H., Martin, P.G., Hendry, P.D. (1994): The size distribution of interstellar dust particles as determined from extinction. ApJ **422**, 164–175

Krełowski, J., Wegner, W. (1989): Interstellar extinction curves originated in single clouds. Astron. Nachr. **310**, 281–287

Krełowski, J., Sneden, C. (1994): "Very high resolutio, high signal-to-noise spectroscopy of diffuse interstellar bands". in Proc. the First Symposium on the Infrared Cirrus and Diffuse Interstellar Clouds, ed. by R.M. Cutri and W.B. Latter (Astron. Society of the Pacific, San-Francisco), 12–23

Laor, A., Draine, B.T. (1993): Spectroscopic constraints on the properties of dust in active galactic nuclei. ApJ **402**, 441–468

Massa, D., Savage, B.D. (1989): "Measurements of interstellar extinction". in Proc. IAU Symposium 135: Interstellar Dust, ed. by L.J. Allamandola and A.G.G.M. Tielens (Reidel, Dordrecht), 3–21

Papaj, J., Krelowski, J., Wegner, W. (1993): Intrinsic UV colours of OB stars. A&A **273**, 575–582

Snow, T.P., Krełowski, J. (1994): "A comparison of the diffuse clouds in front of ζ Ophiuchi and σ Scorpii". in Proc. the First Symposium on the Infrared Cirrus and Diffuse Interstellar Clouds, ed. by R.M. Cutri and W.B. Latter (Astron. Society of the Pacific, San-Francisco), 64–73

Tikhonov, A.N., Goncharsky, A.V., Stepanov, V.V., Yagola A.G. (1990): *Numerical Methods for the Solution of Ill-Posed Problems* (Nauka, Moscow)

Wegner, W. (1994): Intrinsic colour indices of OB supergiants, giants and dwarfs in the UBVRIJHKLM system. MNRAS **270**, 229–234

Wesselius, P.R., van Duinen, R.J., de Jonge, A.R.W., Aalders, J.W.G., Luinge, W., Wildeman, K.J. (1982): ANS ultraviolet photometry, catalogue of point sources. A&AS **49**, 427–474

Zubko, V.G., Mennella, V., Colangeli, L., Bussoletti, E. (1995): Optical constants of amorphous carbon extracted from recent laboratory extinction measurements, this volume

Part VI

Ices and Laboratory Studies

Laboratory Studies of Electric Charging of Dust Particles

Ivo Čermák[1], Eberhard Grün[1], and Jiří Švestka[2]

[1] Max-Planck-Institut für Kernphysik, Postfach 103980, D-69029 Heidelberg, Germany
[2] Prague Observatory, Petřín 205, CZ-11846 Prague 1, Czech Republic

Abstract. In an electrodynamic quadrupole inside an ultra-high-vacuum chamber, glass particles of micron sizes were suspended and charged by electron and ion beams of energies up to 5 keV. The temporal dependencies of the particle Q/M–ratios, surface potentials, electric field strengths and charging currents were obtained. The sticking coefficients of the primary electrons and ions on the particle surface were estimated. We also calculated the energy spectrum of secondary electrons emitted from the particle surface. During charging by ions, a discharging effect probably due to a field emission of ions from the surface of the particle was found. Since we see such an emission at much lower field strengths than we expected, the equilibrium surface potentials of very small (or irregularly-shaped) particles may be lower than former theoretical estimates predict.

1 Experiment

Dust particles of micron sizes are released by means of vibrations from a metallic dust reservoir, trapped in an electrodynamic quadrupole and exposed to beams of electrons and ions of an energy up to 5 keV (Čermák et al., 1995). During the measurements, a trapped particle is irradiated by He–Ne laser light. The light scattered off the particle is collected by a simple optical system and the coordinates of the light spot amplified by an image intensifier are measured. The coordinate signals are used to obtain the frequency of the particle oscillations from which the Q/M–ratio of the dust particle is calculated.

A new dynamical method has been developed to evaluate the measured and recorded temporal dependency of the particle's Q/M–ratio. By this method, the Volt-Amp'ere characteristics (current-voltage dependency) of the particle charging can be reconstructed. The necessary calibration of the particle's size can be carried out by observing the discharge of the particle by a low energy ion beam and by comparing the corresponding Volt-Amp'ere characteristics with the mean ion energy (Čermák et al., 1995).

2 Results and Discussion

The measurements were performed on spherical glass particles of diameters between 1–5 μm. The particles were exposed to He–, Ar– and H_2–ions and to electrons. The Volt-Amp'ere characteristics of the corresponding charging processes were reconstructed and analyzed.

During charging the particle by ions and by electrons, the Volt-Amp'ere characteristics show different shapes at low and high field strengths on the particle surface. At lower surface voltages (several hundred Volts for μ-sized particles) we found that the Volt-Amp'ere characteristics are very linear. The theoretical cross section of the particle for the ion and electron capture decreases linearly with its surface voltage Φ according to the expression $S \cdot (1 - e_0\Phi/E_i)$, where S denotes the geometrical cross section of the particle ($S = \pi R^2$; R is the particle radius), e_0 is the electronic charge, and E_i, the energy of the monoenergetic ion beam. The charging current I onto the particle surface is proportional to the theoretical cross section of the particle and to the sticking coefficient α of the primary ions on the surface of the dust particle. From the derived Volt-Amp'ere characteristics, the dependencies of the I/M-ratios on the surface voltage Φ of the particle can be obtained. Dividing the I/M-ratio by $1 - e_0\Phi/E_i$, the quantity $I_0/M \cdot \alpha$ is obtained which is proportional to the sticking coefficient α of the ions. The constant I_0/M in this expression represents the I/M-ratio on an uncharged particle - $I/M(\Phi) = I_0/M \cdot (1 - e_0\Phi/E_i) \cdot \alpha(E_i - e_0\Phi)$. The sticking coefficient α can be obtained by this method only in arbitrary units. Nevertheless, its energy dependency $\alpha(E_i - e_0\Phi)$ is nearly constant (see Fig. 1), indicating that the probable value of the sticking coefficient is about unity, depending only slightly on the effective energy $E_i - e_0\Phi$ at which the ions reach the particle's surface, and on the field strength on the particle surface.

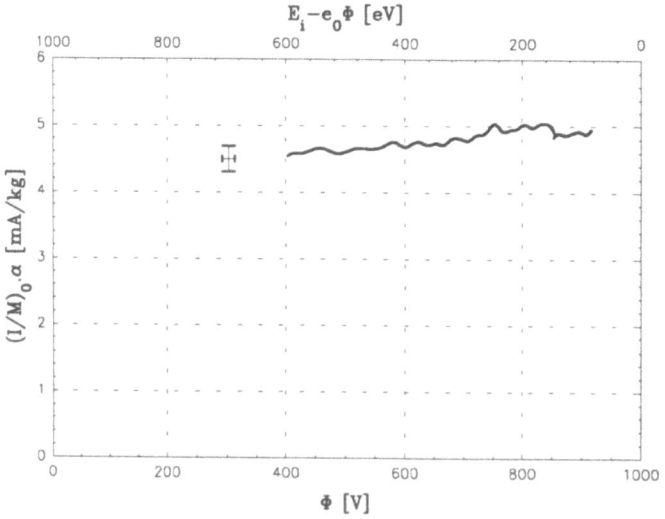

Fig. 1. Dependency of the sticking coefficient α of the 1 keV He-ions on the surface potential Φ of the particle.

During charging the particle by electrons, the Volt-Amp'ere characteristics differ significantly from the linear shape at very low positive particle surface potentials ($\Phi \lesssim 10\,\mathrm{V}$). The reason of this effect is the secondary electron emission from the dust particle. The positive current I_{SE} of the escaping secondary electrons lowers the effect of the charging current I_E of the primary electrons. At particle surface potentials of typically 5–10 V the currents compensate each other resulting in equilibrium state. From the obtained Volt-Amp'ere characteristics the current of the secondary electrons I_{SE} was determined by subtracting the charging current I_E of the primary electrons from the observed total charging current I. The yield of the secondary electron emission Y was calculated as the ratio between these currents $Y(\Phi) = I_{SE}(\Phi)/I_E(\Phi)$. The yield Y was plotted against the particle surface potential Φ (see Fig. 2, bold curve). The curve shows the energy spectrum of the secondary electrons emitted from the surface of the dust particle. This spectrum calculated from the measured data was fitted with theoretical energy spectra of the secondary electrons (fine curves 1–3 in Fig. 2). Curve 1 corresponds to a Maxwell energy spectrum with a mean energy of 3.7 eV, curve 2, to a Grard energy spectrum with a mean energy of 5.3 eV (Whipple, 1981) and curve 3, to the energy spectrum according to Draine and Salpeter (1979) with a mean energy of 3.5 eV. The yield of the secondary electron emission from an uncharged particle Y_0 resulted in all three cases in about 1.75. For comparison with the tabled data for glass, a Maxwell energy spectrum with a yield $Y_0 = 2.5$ and a mean energy of 2.7 eV is also depicted (curve 4). All characteristics are consistent with the mean energy of the secondary electrons of about 3–5 eV (see e.g. Goertz, 1989) and only curve 3 differs significantly from the measured data.

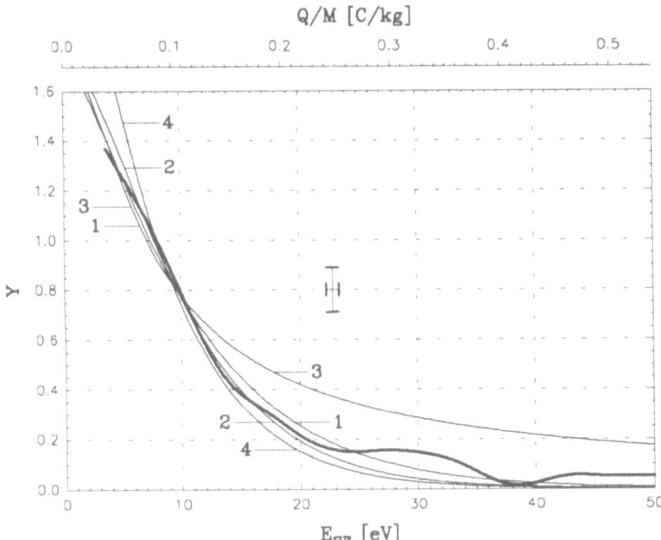

Fig. 2. Integral energy spectrum of the secondary electrons emitted from a glass particle with a diameter of about 1.8 μm during bombardment by 1 keV electrons

When charging the dust particles by ions, we found that the corresponding Volt-Amp'ere characteristics at high particle surface potentials are non-linear. The non-linearity is caused by a discharging process that we have further analyzed: The discharging process appears also after turning off the primary ion current so that it must be caused by the dust particle itself. Moreover, there is no pressure dependency of the discharging currents below 10^{-7} mbar. This shows that the dust particle doesn't interact with the rest gas in the experiment chamber and that a field ionization of the rest gas can be excluded as a possible explanation of the discharging process. The discharging currents depend on the particle's history: During cleaning of the particle surface the discharging currents decrease. Once a particle has been cleaned, the discharging currents remain low. This is consistent with the initial discharging by a field desorption of atoms and ions from the particle surface which becomes negligible after the surface has been cleaned. The discharging currents from a cleaned particle depend on the kind of ions which were used before to charge the particle. This indicates that a surface process is responsible for the discharging. The dependencies of the density of the discharging current j plotted in semi-logarithmic scale against the reciprocal value of the surface field strength F ($j = j(1/F)$) are nearly linear (Čermák et al., 1995). Such exponential dependencies are typical of a field emission (Elinson and Vasilev, 1958).

The threshold for the observed discharging amounts to about $5 \cdot 10^8$ V/m. This value is significantly lower than the threshold observed so far on conductive materials (e.g. Draine and Salpeter, 1979 assume a typical value of the field strength of about $3 \cdot 10^{10}$ V/m). The reason of this large difference of nearly two orders of magnitude are probably the different materials on which the measurements have been performed: For our measurements, we used dielectric spheres. In case of dielectric material the electric field penetrates into the particle and substantially alters the conditions on the particle surface. Moreover, the threshold found in our studies corresponds to the field strength needed for the pure electric break-down of the particle material (Franz, 1953). Nevertheless, an exact explanation of the observed discharging process is not yet available.

References

Čermák, I., Grün, E. and Švestka, J. (1995): New Results in Studies of Electric Charging of Dust Particles. Adv. Space Res. **15**, No. 10, 59–64

Draine, B. T. and Salpeter, E. E. (1979): On the Physics of Dust Grains in Hot Gas. Astrophys. J. **231**, 77–94

Elinson, I. and Vasilev, G. F. (1958): *Avtoelektronnaja Emissija* (Gosudarstvennoje Izdatelstvo Fyziko-Matematičeskoj Literatury, Moscow)

Franz, W. (1953): *Theorie des rein elektrischen Durchschlags fester Isolatoren* in: Ergebnisse d. exakt. Naturwiss., Band XXVII, 1–55

Goertz, C. K. (1989): Dusty Plasmas in the Solar System. Rev. Geophys. **27**, 271–292

Whipple, E. C. (1981): Potentials of Surfaces in Space. Rev. Prog. Phys. **44**, 1197–1250

Solid State Infrared Features: a Diagnostic Tool for Chemical Interactions Between Interstellar Gas and Grains

Louis d'Hendecourt[1] and Pascale Ehrenfreund[2]

[1] Institute Astrophysique Spatiale, Bat. 121, Campus d'Orsay, 91405 Orsay, France
[2] Leiden Observatory, P. O. Box 9513, 2300 RA Leiden, The Netherlands

Abstract. Mid-infrared spectroscopy is a unique tool to provide data on the chemical composition of interstellar and circumstellar dust grains as well as important constraints on some of their physical parameters. Processes such as condensation, surface reactions and bulk chemical evolution upon the occurrence of energetic processes (e.g. UV irradiation) can be directly simulated in laboratory experiments. Using the elegant tools of matrix isolation spectroscopy, out of equilibrium processes - common in interstellar solids - can be approached in a realistic way. Complex solid state effects on lineshapes, broadening and shifts can be understood and quantitatively evaluated. It is thus possible to determine accurately the abundances of simple solid state molecular species such as H_2O, CO, CO_2. Moreover, solid state effects (for example polarizability of solid state homonuclear molecules) may greatly help to detect species such as O_2, N_2 and H_2. A review of these effects will be presented together with recent experimental results. Reanalysis of IRAS-LRS data show valuable information about the presence of H_2O/CO_2 mixtures on grains in the lines of sight toward protostars. From such experiments, coupled to careful interpretation of observations, it is possible to evaluate the role of solid state molecules in interstellar chemistry. With high resolution and S/N data, and extensive spectral coverage (2.5 to 45 μm in particular), ISO is expected to revolutionize the field of grain chemistry. Besides molecules made of abundant elements (e.g. O, C, N), ISO data will help to understand the chemical distribution of more refractory and less abundant species.

1 Introduction

Icy grain mantles are dust components which are formed in dense interstellar clouds (Greenberg 1982, d'Hendecourt et al. 1985). These mantles cover small dust particles which are interspersed within the interstellar gas. They allow new pathways of molecule formation which are not possible in the gas phase. Many infrared observations towards obscured sources have been performed to determine the molecules residing on interstellar grains (e.g. Willner et al. 1982, Grim et al. 1991). These observations imply that the grain mantle constituents are either directly accreted from the gas phase or are produced by surface reactions. Energetic ultraviolet photoprocessing of icy grain mantles enables the formation of new molecules and radicals (d'Hendecourt et al. 1986).

Grain surface reactions play an active role in interstellar chemistry and strongly influence gas phase abundances. Desorption processes return molecules, that are different than the originally accreted species, back into the gas phase (Schutte & Greenberg 1991, Hasegawa & Herbst 1993). The infrared properties of astrophysically relevant ice mixtures have been studied extensively in the laboratory (see Tielens 1991 for a review). Comparison of these laboratory data with interstellar observations have led to the detection of many interstellar solid state molecules (see reviews by Whittet 1993, Schmitt 1994, Schutte 1996).

At low densities, atomic H is an important constituent in the gas phase and dominates in grain surface reactions forming mostly H_2O-rich ices (including CO, NH_3 and CH_4). At high densities, where heavy species are much more abundant than atomic H, grains might accrete a mantle of non-polar ices dominated by CO, CO_2, N_2 and O_2. Non-polar ices are much more volatile than solid H_2O and will be evaporated around luminous protostars. Thus the line of sight towards an embedded object may be dominated by grains with different mantle components: H_2O-rich ices close to the star and non-polar ices far away (Tielens et al. 1991).

1.1 Laboratory simulations

IR spectroscopy represents a powerful method to study solid interstellar particles and their laboratory analogues. To produce interstellar analogues in the laboratory a pure gas or gas mixtures are condensed in a high vacuum chamber on the surface of a cooled substrate (CsI window usually cooled to 10 K). Detailed comparison between laboratory and interstellar absorption profiles of ice components not only enables identification, but also gives information on the nature of the ice matrix surrounding the molecule. Radiation processing of the sample can be carried out with UV lamps or ion sources. The evaporation characteristics of ices can be studied by measuring spectra during warm-up.

An important aspect of laboratory simulations is the possibility to derive accurate abundances of individual components in interstellar ices. When the integrated absorbance A for a given species is measured in an appropriate matrix in the laboratory the column density N can be calculated using the equation

$$N = \frac{\tau \Delta \nu}{A} \tag{1}$$

where τ is the optical depth, $\Delta \nu$ the band width and A the integrated absorbance (in cm molecule $^{-1}$).

1.2 IR spectroscopy of ice absorption features

The primary astronomical targets for infrared spectroscopy of ices are protostars which contain circumstellar and foreground material. Energetic events (shocks, winds, UV radiation) might influence the interstellar environment around protostars. Observations of field stars can show ices remote from embedded sources and allow to determine the extent of grain processing in the protostellar environment.

The dominant ice features observed in the spectra toward many protostars are located at 3.08, 4.67, 6.0 and 6.85 μm. The strong band at 3.08 μm is due to the OH stretching vibration of H_2O ice and at 6 μm a weaker absorption is associated with the bending mode. The 4.67 μm band is due to solid CO and the band of 6.85 μm might be due to CH deformation in hydrocarbons and alcohols, but lacks exact identification.

Water ice. Water ice is the most important component in astrophysical ices and characterized by a strong absorption at 3 μm (Smith et al. 1989). Calculations indicate that only surface reactions produce H_2O ice mantles in sufficient abundances (Whittet 1993). Water ice formation also requires enhanced densities and UV shielding. Studies of the 3 μm absorption profile indicate the influence of different physical environments on the structure and purity of water ice (Hagen et al. 1981). A shoulder between 3.2 and 3.6 μm is observed in dense molecular clouds. This absorption is due to other molecules contained within the grain mantles, which are not yet revealed. Water ice may also be present as a minor constituent in non-polar ices (Ehrenfreund et al. 1995).

Methanol CH_3OH, Formaldehyde H_2CO, Ethanol CH_3CH_2OH. Radio observations of hot dense cores show high gas phase abundances of methanol. Solid methanol exhibits a number of spectral features, e.g. the CH stretch located at 3.53 μm which is detected in several sources (Grim et al. 1991, Allamandola et al. 1992) and the CO stretch at 9.7 μm, detected in GL2136 (Skinner et al. 1992). A strong band at 6.85 μm is observed towards many protostars which coincides with the deformation mode of methanol (Tielens & Allamandola 1987). However, the CH_3OH abundance derived from this feature is typically a factor 5-10 higher than obtained from the 3.53 and 9.7 μm bands (Allamandola et al. 1992, Schutte et al. 1996), indicating that CH_3OH is only a minor contributor to the 6.85 μm absorption.

The abundance of H_2CO in the gas phase in high latitude clouds can not be explained by gas phase chemistry (van Dishoeck & Black 1988). H_2CO is readily produced by UV and energetic particles in the laboratory in ices containing CO and H_2O. H_2CO ice displays CH stretches around 3.5 μm, and a C=O stretch at 5.81 μm. The detection of the H_2CO ice towards GL 2136 has recently been reported with an estimated abundance of 6% compared to H_2O (Schutte et al., 1995).

The detection of a large amount of gas phase ethanol (in the sub-mm) in the molecular cloud associated with the ultra-compact HII region G34.4+0.15 has been reported by Millar et al. (1995). This high abundance of ethanol cannot arise in gas phase reactions and it is very likely that ethanol is synthesized on the grain surface and thereafter released into the gas phase.

All these three important molecules do display features between 5 and 8 μm, a region not observable from the ground. This wavelength range probes in general CH, NH and OH bending and deformation modes and C=O stretches. ISO observations of this region will allow in the future to constrain the presence and abundance of these important molecules, which show large abundances in the gas phase.

CO and CO_2. CO is the most abundant ice component after water in the interstellar medium and should be the main constituent of non-polar ices. CO has been observed towards many sources (Tielens et al. 1991) and in particular in the Serpens and Taurus Dark Cloud systems (Chiar et al. 1994, 1995). Laboratory studies of the shape and peak position of the solid CO band in astrophysically relevant ice mixtures show a two-component structure of solid CO at 4.67 μm. A narrow CO band is observed in mixtures dominated by non-polar molecules (CO, CO_2, O_2, N_2), whereas a broad component originates in polar mixtures, such as H_2O ice. Observations show that many lines of sight contain (at least) two independent grain mantle components (Tielens et al. 1991).

The molecule CO_2 is efficiently produced by UV photolysis of grain mantles containing CO (d'Hendecourt et al. 1986). CO_2 is not observable from the ground. The detection of solid CO_2 through its weak ν_2 bending mode at 15.2 μm was reported by d'Hendecourt & de Muizon (1989) from IRAS-LRS spectra in 3 sources. The abundance of solid CO_2 toward AFGL 961 was estimated to be 10 % of H_2O. A search of this band towards a number of other embedded sources yielded CO_2 upper limits of 1-3 % relative to H_2O (Whittet & Walker 1991).

Recently a sample of protostars was selected in order to search for the H_2O libration mode at 13.3 μm and the CO_2 bending mode at 15.2 μm. Special care has been taken in the data reduction to detect these specific lines: first a fit of a continuum by using a single temperature between 7-25 μm was applied to the LRS spectra. The silicate absorption at 9.7 μm was then removed by using true interstellar silicates in the form of an IRAS-LRS spectrum of a M star. Thereafter the data were fitted with laboratory spectra. In Fig. 1 we show the reduced LRS spectrum of IRAS18316-0602 in comparison with laboratory data. More than 30 targets do display a significant CO_2 feature, which indicates that CO_2 may be an important constituent of interstellar ices.

Molecules containing N, S, P. Nitrogen is an abundant element and likely present in interstellar ices. N_2 might be continuously recycled between dust and gas phase by thermal evaporation like CO and might also be a dominant component in non-polar ices. The presence and abundance of solid NH_3 ice is still an unsolved problem. NH_3 should display a band at 2.96 μm, (which is not observed in the interstellar medium) and a broad feature extending between 3.2 and 3.6 μm, which is attributed to formation of ammonium hydrate groups in NH_3-H_2O mixtures. This feature may contribute to the "red shoulder" superimposed on the 3 μm H_2O band observed in dense clouds.

Radiative processes may lead to the formation of molecules with CN bond which arise between 4.5 and 4.7 μm. A band at 4.62 μm is observed in several sources, e.g. W33A, and WL5 and likely due to a "XCN" molecule, probably OCN^- or CH_3NC (Grim & Greenberg 1987). Recent detection of this band

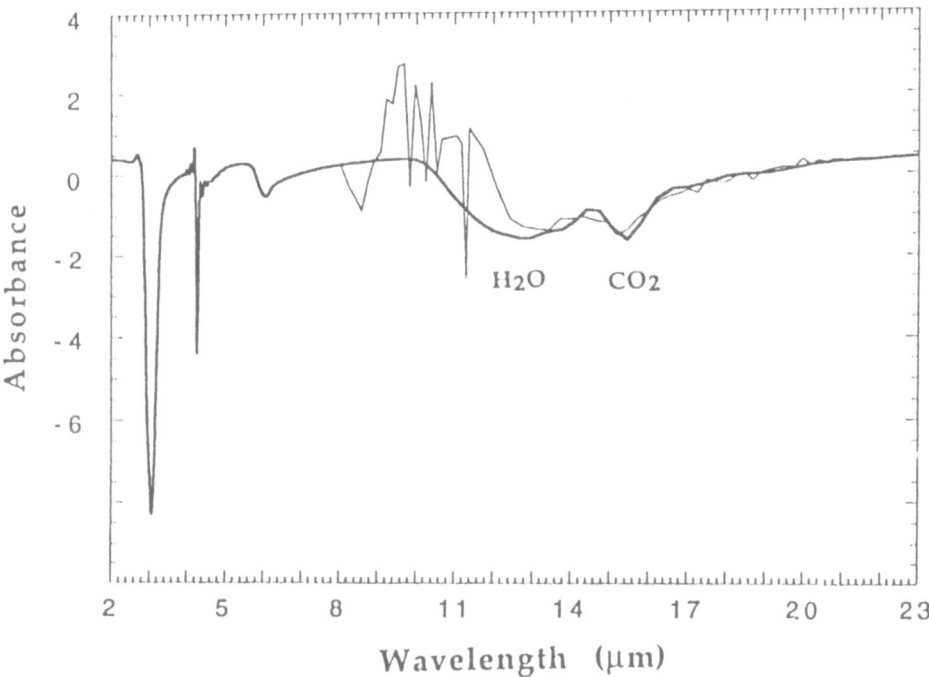

Fig. 1. Reduced LRS spectrum of IRAS18316-0602 (soft line) compared to a laboratory spectrum of H_2O-CO_2 ice (strong line). The H_2O libration mode is located at 13.3 μm and the CO_2 bending mode at 15.2 μm.

towards Elias 18 suggests photoprocessing of non-polar ices containing N_2 (Tegler et al. 1995).

H_2S is detected in the gas phase and might be also present on grains. So far the only Sulphur containing species observed is OCS (at 4.9 μm) with an abundance of 0.05 % in respect to H_2O towards W33A (Geballe et al 1985).

Phosphor could be depleted from the gas and be present in grain mantles as PH_3. PH_3 is rapidly destroyed upon return to the gas phase and further converted to P and PN (Charnley & Millar 1994).

Homonuclear diatomic molecules. H_2, O_2, N_2 as homonuclear diatomics are IR inactive and not observable at radio wavelengths. Interaction with neighbours can break the symmetry of the vibrations and modes may become weakly IR active.

 H_2 is the most abundant molecule in space formed by H recombination on dust grains in molecular clouds. The 2.41 μm Q1 pure vibrational transition of

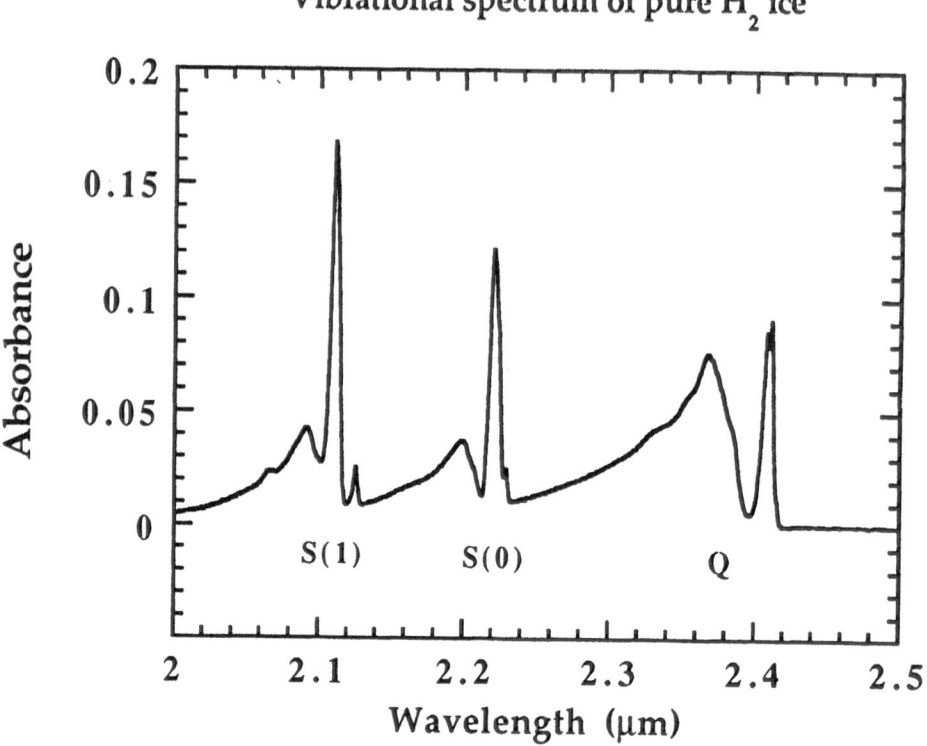

Fig. 2. The vibration rotation spectrum of H_2 ice with an apparition of phonon side-bands on the 2 - 2.5 μm range.

H_2 in H_2O rich ices has been studied in the laboratory (Sandford & Allamandola, 1993). It was thereafter detected in WL5 a protostar in the Oph cloud (Sandford et al. 1993). H_2 ice has a very rich IR spectrum in the 2-2.5 μm region. The binding energy of H_2 on H_2 is quite low (91 K) allowing no condensation above 2.1 K in interstellar conditions. However due to a higher binding energy with H_2O, H_2 may be trapped in interstellar grain mantles at T < 18 K. H_2 accretion is critically temperature dependent so that the detection of H_2 lines will constitute a powerful "grain-thermometer". The most promising lines of solid H_2 fall in the K band which is a window easily accessible to large telescopes (see Fig. 2).

The molecules O_2 and N_2 might be abundant grain mantle constituents in non-polar ices. O_2 is created by surface reactions as well as by accretion. The O=O fundamental vibration was detected in a CO_2 matrix at 6.4 μm (Ehrenfreund et al. 1992) and will be searched with ISO, the Infrared Space Observatory.

Irradiation of ices containing O_2 produce rapidly O_3, ozone which displays a strong band at 9.6 μm (Grim & d'Hendecourt 1986, Ehrenfreund et al. 1992).

N_2 shows its vibrational transition at 4.28 μm and will be therefore hard to detect due to the overlap with the CO_2 feature at 4.27 μm.

2   Conclusion

Comparison of laboratory absorption spectra with astronomical data indicates that ices are ubiquitous in dense clouds. The gas phase chemical composition in hot molecular cores in star-forming regions is determined by the evaporation of material from icy grain mantles. Solid state spectroscopy provides a unique tool to study the chemical interaction between interstellar gas and grains. Laboratory data of ices allow the identification of new interstellar solid components and derivation of their accurate abundances. Reanalysis of LRS spectra indicates that CO_2 is a widespread component in interstellar ices. Recent results show that H_2 can be efficiently accreted from the gas phase. Since H_2 accretion is only possible at temperatures below 18 K the search for solid H_2 in the 2 - 2.5 μm atmospheric window will help to detect very cold grain material in the line of sight of highly obscured infrared sources. Large amounts of O and N are apparently missing from the gas phase and could be depleted on grains likely in the form of N_2 and O_2. Future satellite observations with ISO in combination with laboratory data will lead to fundamental new discoveries regarding the composition of interstellar grains and their interaction with the gas phase environment.

References

Allamandola, L.J., Sandford, S.A., Tielens, A.G.G.M., Herbst, T. (1992): ApJ **399**, 134

Charnley, S.B., Millar T.J. (1994): MNRAS **270**, 570

Chiar, J.E., Adamson, A. J., Kerr, T. H., Whittet, D. C. B. (1994): ApJ **426**, 240

Chiar,J.E., Adamson, A. J., Kerr, T. H., Whittet, D. C. B. (1995): ApJ, in press

d'Hendecourt, L. B., Allamandola, L. J., Greenberg, J. M. (1985): A&A **152**, 130

d'Hendecourt, L. B., Allamandola, L. J., Grim, R., Greenberg, J. M. (1986): A&A **158**, 119

d'Hendecourt, L. B., Jourdain de Muizon, M. (1989): A&A **223**, L5

Ehrenfreund, P., Gerakines, P.A., Schutte, W.A., van Hemert, M., van Dishoeck, E.F. (1995): A&A, submitted

Ehrenfreund, P., Breukers, R., d'Hendecourt, L., Greenberg, J. M. (1992): A&A **260**, 431

Geballe, T.R., Baas, F., Greenberg, J.M., Schutte, W.A. (1985): A&A **146**, L6

Greenberg, J. M., (1982): *Comets*, ed. Wilkening, L.L. Tucson, University of Arizona Press, 131

Grim, R.J.A., d'Hendecourt, L. (1986): A&A **167**, 161

Grim, R.J.A., Greenberg, J.M. (1987): ApJ **321**, L91

Grim, R.J.A., Baas, F., Geballe, T.R., Greenberg, J.M., Schutte, W. (1991): A&A **243**, 473

Hagen, W., Tielens, A. G. G. M., Greenberg, J. M. (1981): Chem. Phys. **56**, 367

Hasegawa, T.I., Herbst, E. (1993): MNRAS **261**, 83

Millar, T.J., MacDonald, G.H., Habing, R.J. (1995): MNRAS **273**, 25

Sandford, S.A., Allamandola, L.J., Geballe, T.R. (1993): Science **262**, 400

Sandford, S.A., Allamandola, L.J. (1993): ApJ **409**, L65

Schmitt, B. (1994): *Molecules and grains in space*, ed. Nenner I. AIP press, New York, 735

Schutte, W.A., Greenberg, J. M. (1991): A&A **244**, 190

Schutte, W.A. (1996): *The cosmic dust connection* ed. Greenberg J. M., in press

Schutte, W.A., Gerakines, P.A., Geballe, T.R., van Dishoeck, E.F., Greenberg, J.M. (1995): A&A, in press

Skinner, C.J., Tielens, A.G.G.M, Barlow, M.J., Justtanont, K. (1992): ApJ **399**, L79

Smith, R.G., Sellgren, K., Tokunaga, A. T. (1989): ApJ **344**, 413

Tegler, S.C., Weintraub, D.A., Rettig, T.W., Pendleton, Y.J., Whittet, D.C.B., Kulesa C.A. (1995): ApJ **439**, 279

Tielens, A.G.G.M. (1991): *Solid state Astrophysics*, North Holland, eds. Bussoletti E., Strazulla G., 29

Tielens, A.G.G.M., Allamandola, L.J (1987): *Physical Processes in interstellar clouds*, eds. G.E. Morfill and M. Scholer, 333

Tielens, A.G.G.M., Tokunaga, A. T., Geballe, T. R., Baas, F. (1991): ApJ **381**, 181

van Dishoeck, E.F., Black, J.H. (1988): ApJ **334**, 771

Whittet, D.C.B., Walker, H.J. (1991): MNRAS **252**, 63

Whittet, D.C.B. (1993) *Dust and chemistry in astronomy*, eds. Millar T. J., Williams D. A., IOP Publ. Ltd. Bristol, 1

Willner, S.P. et al. (1982): ApJ **253**, 174

Infrared Properties of Isolated Water Ice

Pascale Ehrenfreund[1], Willem A. Schutte[1], and Perry Gerakines[1,2]

[1] Leiden Observatory, P.O. Box 9513, 2300 RA Leiden, The Netherlands
[2] Department of Physics, Rensselaer Polytechnic Institut, Troy, NY 12180-3590, USA

Abstract. Water ice is the most important component in astrophysical ices and is characterized by a strong broad absorption at 3 μm, observed in many interstellar spectra. In specific interstellar environments grain mantles might contain only a small amount of water, which is diluted in other volatiles dominant in these regions (e.g. CO, O_2, N_2, CO_2). Isolated water molecules display very sharp discrete bands between 2.5 and 2.9 μm, shortward of the 3 μm polymeric H_2O ice band. We have performed a detailed study of the infrared properties of H_2O diluted in various matrices and discuss the possible detection of isolated water on low temperature and ultraviolet shielded grain mantles covered with non-polar ices. Together with astronomical spectra taken by the ISO satellite these laboratory data will be extremely valuable for the determination of the grain mantle composition in dense clouds.

1 Introduction

Many infrared observations towards obscured sources have been performed to determine the molecules residing on interstellar grains (e.g. Willner et al. 1982, Grim et al. 1991). The infrared properties of astrophysically relevant ice mixtures have been studied extensively in the laboratory (see Tielens 1991 for a review). Comparison of these laboratory data with interstellar observations have led to the detection of many interstellar solid state molecules (see reviews by Whittet 1993, Schmitt 1994, Schutte 1996).

Water ice is the most abundant molecule in grain mantles. The strong broad absorption of water at 3 μm has been observed toward more than 100 sources. Recent work indicates layered structures on grain mantles and the existence of polar and non-polar ices (Tielens et al. 1991). Theoretical models demonstrate that in high density environments where grains accrete mainly non-polar gas phase molecules (CO, N_2, O_2), a non-negligible quantity of H_2O may still co-condense (d'Hendecourt et al. 1985). When H_2O molecules are isolated, sharp discrete bands due to monomeric H_2O (one single H_2O molecule surrounded by other molecules) and dimeric H_2O can be observed around 2.5 μm. The specific position and shape of these bands is determined by the interaction with adjacent molecules in the ice (Hagen & Tielens 1981).

The molecules O_2 and N_2 might be abundant grain mantle constituents in non-polar ices. O_2 and N_2 are homonuclear diatomic molecules, which are infrared inactive and unobservable at radio wavelengths. Interaction with adjacent molecules in the solid state can break the symmetry of the vibrations and modes become weakly IR active (Ehrenfreund et al. 1992). Indirect methods to infer the presence of homonuclear molecules are (i) to study the profile of the CO band and (ii) the band position of isolated water features (Ehrenfreund et al. 1995).

In a non-polar ice mixture on interstellar grains the water molecules will predominantly be surrounded by molecules such as CO, O_2, N_2 and CO_2. Therefore we studied the infrared properties of water ice diluted in such particular matrices.

2   Results

CO is the most abundant ice component after water in the interstellar medium and should be the main constituent of non-polar ices (Tielens et al. 1991).

Fig. 1 shows the spectrum of isolated water in matrices containing CO, O_2 and N_2. Comparison of the spectra shows that the CO molecule dominates the interaction within the matrix, even in the presence of N_2 or small amounts of O_2. However, the presence of equal amounts of CO and O_2 induces changes in band position indicative of a strong matrix perturbation. The band width increases strongly in multicomponent mixtures due to line blends and the band ratio of dimers and multimers to monomers is much higher in mixtures containing mainly CO and O_2.

We have also performed specific measurements in order to determine the integrated absorbance of monomeric and dimeric water ice in CO, O_2 and N_2 matrices, listed in Table 1. Our results indicate that water ice is detectable when present as a minor species on a 1 % level relative to CO in non-polar ices (yielding an equivalent width of a few %, corresponding to an optical depth $\tau \approx 0.01$).

Table 1. Infrared band intensities of isolated H_2O in various matrices. Symbols ν_3, ν_2 refer to the asymmetric OH stretching mode and the OH bending mode, respectively.

Ice matrix	Band	A
		cm molec^{-1}
H_2O:CO=1:100	ν_3 monomer	1.1×10^{-17}
	ν_3 dimer "eD"	5.8×10^{-17}
	ν_2 monomer	3.5×10^{-18}
	ν_2 dimer "eD"	1.0×10^{-17}
H_2O:O_2=1:100	ν_3 monomer	3.3×10^{-18}
	ν_3 dimer "eD"	1.5×10^{-17}
	ν_2 monomer	8.4×10^{-19}
	ν_2 dimer "eD"	9.0×10^{-18}
	ν_2 dimer "eA"	1.7×10^{-17}
H_2O:N_2=1:100	ν_3 monomer	7.8×10^{-18}
	ν_3 dimer "eD"	1.5×10^{-17}
	ν_2 monomer	6.1×10^{-18}
	ν_2 dimer "eD"	2.4×10^{-17}

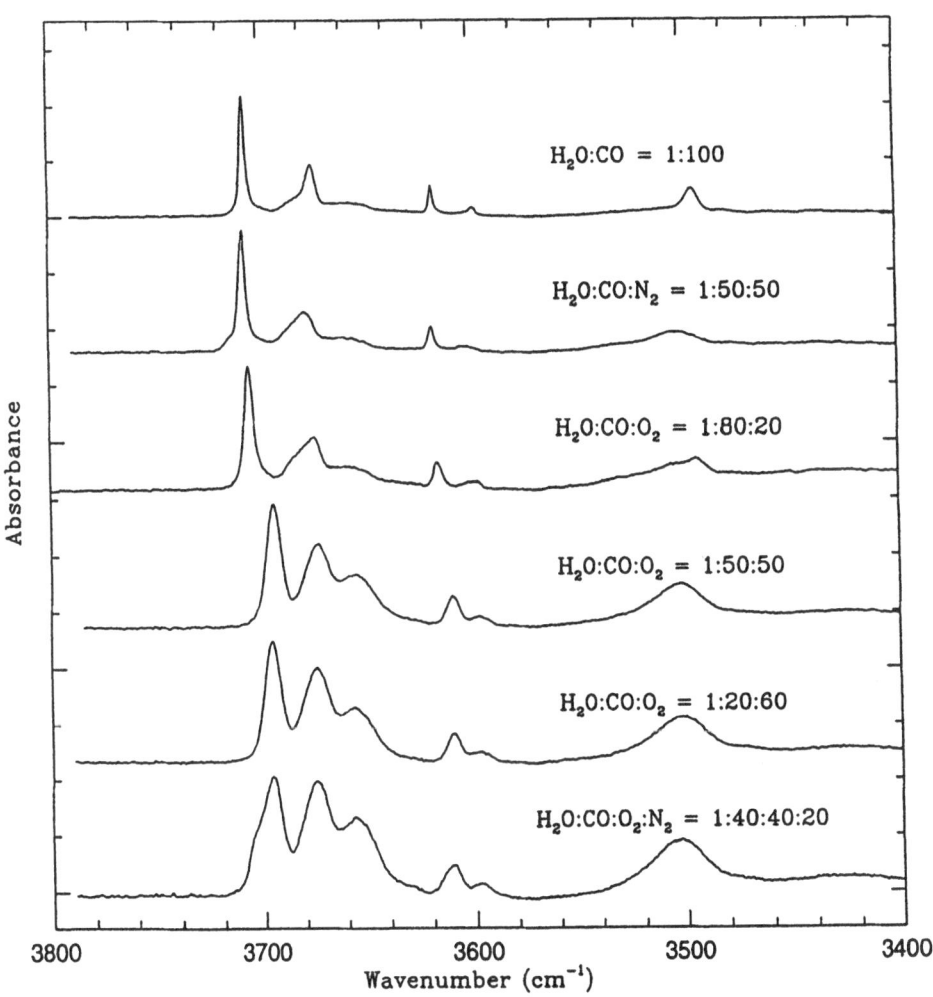

Fig. 1. Infrared absorption spectra of isolated H_2O in the stretching 3800-3000 cm^{-1} in various mixtures containing CO, N_2 and O_2. A strong shift of the monomer towards lower frequencies is observed in mixtures containing equal amounts of CO and O_2.

3   Discussion

We have measured the main features of isolated water, monomers and dimers, in different ice mixtures and concentrations in order to determine bands that can be detected by astronomical observations.

The derived integrated absorbance for the H_2O monomers in CO should allow a detection with ISO in lines of sight with a considerable non-polar ice column density ($\geq 5 \times 10^{17}$cm^{-2}). Since warm-up and UV processing efficiently destroy sites for isolated water molecules, the detection of water ice in this form requires cold and UV shielded environments. For this reason the best a priori targets for a search for these features are field stars obscured by a large column of dense cloud material. Furthermore the appearance of photolysis products can mask the signature of diluted water ice, e.g. CO_2 combination modes at 3708 and 3600 cm^{-1}. Only grains which contain little (a few %) CO_2 will allow us to measure the isolated water stretching modes in the 2.65-2.78 μm region. In case the abundance of O_2 equals the amount of solid CO on the grain mantle we can infer the presence of molecular oxygen, since the spectroscopic signature in this case is rather unique.

The detection of isolated water can provide important insights into interstellar chemistry and will help to confirm the presence of non-polar ices and constrain their O_2 content. Such information is essential to understand the chemical and physical conditions which gave rise to the non-polar component in dense cloud ices. The present results highlight that laboratory measurements using solid state spectroscopy are crucial for the identification of the grain mantle composition in dense clouds.

References

d'Hendecourt, L. B., Allamandola, L. J., Greenberg, J. M. (1985): A&A **152**, 130

Ehrenfreund, P., Gerakines, P.A., Schutte, W.A., van Hemert, M., van Dishoeck, E.F. (1995): A&A, submitted

Ehrenfreund, P., Breukers, R., d'Hendecourt, L., Greenberg, J. M. (1992): A&A **260**, 431

Grim, R.J.A., Baas, F., Geballe, T.R., Greenberg, J.M., Schutte, W. (1991): A&A **243**, 473

Hagen, W., Tielens, A. G. G. M. (1981): J. Chem. Phys. **75**, 4198

Schmitt, B. (1994): *Molecules and grains in space*, ed. Nenner I. AIP press, New York, 735

Schutte, W. A. (1996): *The cosmic dust connection*, ed. Greenberg J. M., in press

Tielens, A. G. G. M. (1991): *Solid state Astrophysics*, North Holland, eds. Bussoletti E., Strazulla G., 29

Tielens, A. G. G. M., Tokunaga, A. T., Geballe, T. R., Baas, F. (1991): ApJ **381**, 181

Whittet, D. C. B. (1993) *Dust and chemistry in astronomy*, eds. Millar T. J., Williams D. A., IOP Publ. Ltd. Bristol, 1

Willner, S. P.et al. (1982): ApJ **253**, 174

Nitrogen-Bearing Organic Molecules in Hot Cores

M.E. Kress[1,2] and S.B. Charnley[1,3]

[1] Mail Stop 245-3, Space Science Division, NASA Ames Research Center, Moffett Field, CA 94035-1000.
[2] Physics Department, Rensselaer Polytechnic Institute, Troy, NY 12180.
[3] Astronomy Department, University of California, Berkeley, CA 94720.

Abstract. We have modeled the gas-phase chemistry occurring in warm dense clumps of gas and dust near regions of massive star formation. Many large organic molecules have been previously observed in hot cores; their surprisingly high abundances have sometimes been attributed to grain surface chemistry. However, our models show that many complex organics can form in the warm gas upon injection of molecules believed to be icy mantle constituents (*e.g.* ammonia, methane, and simple alcohols) which will evaporate at hot core temperatures.

1 Introduction

The cold, dark environment of the dense interstellar medium hosts a wide variety of molecules, ranging from unsaturated carbon chain species such as C_4H, C_3N, HC_9N, etc., as well as simple saturated molecules which form icy mantles on dust grains. In regions of star formation, saturated molecules such as H_2O, CH_4, and NH_3, as well as more complex organic molecules such as methanol, ethanol, dimethyl ether, methyl formate, ketene, formaldehyde, acetylene, acetaldehyde, formic acid and several nitriles have been observed in the gas. (Blake et al. 1987; Sutton et al. 1995; Turner 1991).

Hot cores are dense, warm regions ($n_H > 10^6 cm^{-3}$, $T_{kin} > 100K$, $T_{dust} > 40K$) associated with young massive protostars. Icy grain mantles formed during the cold dark phase of the cloud's life are now released into the gas and participate in a dynamic chemistry. Models of hot core formation and evolution have provided a useful framework in which to understand the observed molecular inventory of star-forming regions (Charnley et al. 1992). By studying these regions, it is possible to distinguish those molecules which came off grain surfaces, formed either by surface reactions or in cold gas and subsequently accreted, from those formed in the hot gas. These models have shown that the observed CO, C_2H_2, H_2O, CH_3OH, C_2H_5OH, NH_3, CH_4, and H_2CO originate from grain surfaces and drive specific organic chemistry pathways in the gas. By extension of previous work, it is conjectured that evaporation of icy mantles ices containing CH_3OH, C_2H_5OH, NH_3, HCN, and CH_4 is the initiating step that leads to the specific mixture of organic molecules observed in star-forming cores. This theory provides viable gas-phase formation pathways to several molecules where none currently exist, and also suggests an observational test of a grain-surface origin for others.

2   Nitrogen-Bearing Organics

If some molecular mantles are differentiated between CH_4-rich and NH_3-rich cases, then, following evaporation, the specific chemistries driven by each can explain the large abundances of N-bearing species in cores adjacent to, but physically distinct from, cores containing large abundances of complex O-bearing molecules. One prediction of this theory is that high abundances of NH_3 and nitriles, specifically CH_3CN should coexist in such cores. Recent observations tend to confirm this picture (Wilner et al. 1994; Olmi et al. 1993) although, if anything, the correlation of CH_3CN and NH_3 is actually stronger than suggested by the models, where the CH_3CN formation time-scale is longer ($\sim 10^5$ years) than that for forming O-bearing organics from CH_3OH ($\sim$ a few $\times 10^4$ years). Obtaining CH_3CN from NH_3 relies on a slow requence of reactions to HCN, which is where the nitrile chemistry begins. Observations suggest that both HCN and DCN are present in mantles (Mangum et al. 1991), and so, if some observed HCN is injected from grains, the route to CH_3CN via CH_3^+ could be driven more efficiently.

There is some dispute in the literature regarding the precise location of specific molecules in the Hot Core and Compact Ridge sources of OMC-1. This tends to cast doubt on whether a clear distinction can really be made between purely N-rich and O-rich environments, as originally proposed by Blake et al. (1987). Indeed, the recent observations by Sutton et al. (1995) show that both these sources in fact contain the same complex molecules although the O-bearing ones have lower abundances in the Hot Core. It does appear, however, that only the Hot Core possesses a large ammonia abundance suggesting that perhaps the mantles in both sources had the *same* composition *except* for the amount of NH_3 present: if the Hot Core chemistry is more evolved than that of the Compact Ridge, the lower abundances of CH_3OH, C_2H_5OH, H_2CO, $(CH_3)_2O$ and $HCOOCH_3$ in the Hot Core could simply be an age effect. It is possible that there is some cross-over between alcohol chemistry and that of N-bearing molecules in hot cores (Charnley and Kress 1995, in preparation). We have extended this investigation to see if N-bearing complex molecules could be formed by reactions of protonated $CH_3OH_2^+$ and $C_2H_5OH_2^+$ with other mantle molecules such as NH_3 and HCN that are likely to be abundant on grains prior to evaporation.

Two organic molecules could be explained by gas-phase reactions of $CH_3OH_2^+$ and $C_2H_5OH_2^+$ with other abundant mantle molecules such as NH_3 and HCN. HCN is quite likely to be abundant on grains prior to evaporation, as witnessed by the presence of DCN in hot cores (Mangum et al. 1991). Two reactions of interest are

$$CH_3OH_2^+ \; + \; HCN \; \longrightarrow \; CH_3NCH^+ \; + \; H_2O \; . \tag{1}$$

$$C_2H_5OH_2^+ \; + \; HCN \; \longrightarrow \; C_2H_5NCH^+ \; + \; H_2O \; . \tag{2}$$

Both reactions have been studied in the laboratory and have measured rate coefficients of 1.8×10^{-11} cm^3 s^{-1} and 1.5×10^{-11} cm^3 s^{-1} respectively (Mautner and Karpas 1986).

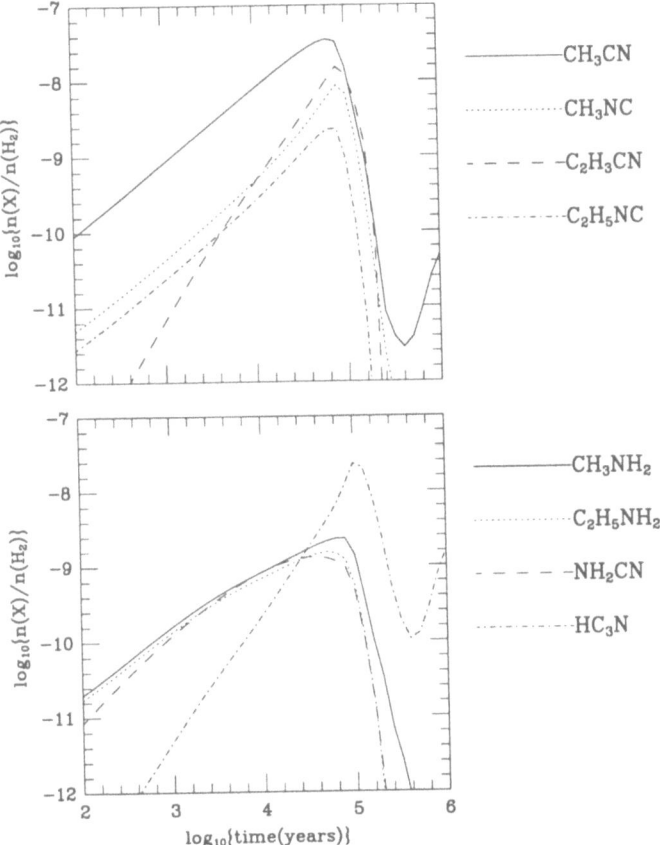

Fig. 1. Time evolution of several N-bearing organics under hot core conditions. In this model, NH_3, HCN, and other mantle ices are introduced via evaporation from grains at the onset of the calculation, $n_H = 2 \times 10^7 cm^{-3}$, $T_{kin} = 150K$, and photodestruction reactions are included.

The degree of rearrangement of the isocyanide ions produced in reactions (1) and (2) is uncertain, but if it occurs then these reactions would lead to CH_3CN and C_2H_5CN. There is no known gas phase route to the latter. We also considered formation of CH_3NH_2 (methylamine) and $C_2H_5NH_2$ (ethylamine) starting from methyl and ethyl cation transfer to ammonia. Neutral reactions involving CN are also potentially important, and we have studied the formation of HC_3N, CH_2CHCN, and NH_2CN by CN reactions with C_2H_2, C_2H_4, and NH_3.

The model we present here is similar to the one most appropriate to the Orion Hot Core as determined by Charnley et al. (1992), with the evaporated mantle composition used by Charnley et al. (1995) with HCN and NH_3 included, and $T_{kin} = 150K$. We have included cosmic-ray induced photoprocesses as well (Prasad and Tarafdar 1983).

3   Conclusions

The model results are shown in Figure 1. Comparison with observations of the Hot Core (see Sutton et al. 1995, Turner et al. 1991, and Blake et al. 1987) indicates that the abundances of the nitrogen-bearing organics can be explained via gas phase reactions. In many cases, our results peak at a value somewhat higher than the observed value; the model results come back down to observed values later in the core's evolution.

Our gas phase theory can account for the presence of many N-bearing organics in star forming regions. C_2H_5CN is an interesting case since no gas-phase formation route is currently known, but this molecule has been detected in the Hot Core. Perhaps the pathway to convert the isocyanides into the cyanides is rather efficient. If this is not the case, we predict methyl and ethyl isocyanides to be present in detectable quantities. We also predict $C_2H_5NH_2$ to be present as well. A more detailed investigation of these and other reactions is currently underway (Charnley and Kress, in preparation).

References

Blake, G.A., Sutton, E.C., Masson, C.R. and Phillips, T.G. 1987, *Ap.J.*, **315**, 621.

Brown, P.D., Charnley, S.B. and Millar, T.J. 1988, *M.N.R.A.S.*, **231**, 409.

Charnley, S.B. and Millar, T.J. 1994, *M.N.R.A.S.*, **270**, 570.

Charnley, S.B. 1995, *Ap.J.*, , in press.

Charnley, S.B., Kress, M.E., Tielens, A.G.G.M. and Millar, T.J. 1995, *Ap.J.*, **448**, 232.

Charnley, S.B., Tielens, A. and Millar, T.J. 1992, *Ap.J.(Letters)*, **399**, L71.

Karpas, Z. and Meot-Ner (Mautner) M. 1989, *J. Phys. Chem.* **93**, 1859.

Mangum, J., Plambeck, R.L. and Wootten, A. 1991, *Ap.J.*, **369**, 157.

Meot-Ner (Mautner) M. and Karpas, Z. 1986, *J. Phys. Chem.* **90**, 2206.

Olmi, L., Cesaroni, R. & Walmsley, C.M. 1993, *Astr.Ap.*, **276**, 489.

Sutton, E.C., et al. 1995, *Ap.J.Suppl.*, , in press.

Turner, B.E. 1991, *Ap.J.Suppl.*, **76**, 617.

UV Irradiation of Small Carbon Grains

V. Mennella[1], L. Colangeli[1], P. Palumbo[2],
A. Rotundi[3], W. Schutte[4], and E. Bussoletti[3]

[1] Osservatorio Astronomico di Capodimonte, via Moiariello, 16, I-80131 Napoli, Italy
[2] Dipartimento di Ingegneria Aereospaziale, Università Federico II, P.le Tecchio, 80 I-80125, Napoli, Italy
[3] Istituto di Fisica Sperimentale, Istituto Universitario Navale, via A. De Gasperi, 5, I-80133 Napoli, Italy
[4] Leiden University Observatory, Postbus 9513 NL-2300 RA Leiden, The Netherlands

Abstract. We present preliminary results of an experiment aimed at simulating the UV processing of hydrogenated amorphous carbon grains occurring in the interstellar medium. UV exposure of these grains induces significant changes in the UV - Vis spectrum and, in particular, activates a resonance at 215 ± 2 nm very close to the position of the interstellar extinction bump. The spectral variations depend on the UV dose deposited in the samples: as the dose increases, the band becomes more intense while its peak position remains stable.

We attribute the band to π - π^* electronic transitions in sp^2 ringed clusters forming the grains and interpret the spectral variations in terms of structural changes of the grains, as confirmed by the optical gap variations.

The results of the present experiment suggest that it is unlikely that hydrogenated amorphous carbon grains can be transformed into pure graphite grains by UV processing in a typical diffuse cloud timescale.

1 Introduction

The key role of cosmic dust grains in determining the thermal, dynamical and chemical structure of the interstellar medium is becoming more and more evident (see the review paper by Dorschner and Henning 1995). To get insight into the nature of cosmic grains, the processes responsible for their formation and evolution should be taken into account. Their influence on the grain properties during particle life cycle, such as the interchange between the molecular cloud phase and the diffuse medium, must be carefully considered.

In this work we consider one of the processes which characterize the evolution of dust particles in the diffuse medium: the exposure to UV radiation. In particular, we analyze in the laboratory the changes induced by energetic UV photons in the UV spectrum of hydrogenated amorphous carbon grains.

Among the possible carriers of the UV extinction bump at 217.5 nm small hydrogen free carbon particles have indeed been proposed. Hecht (1986) and Sorrell (1990) suggested that dehydrogenation and graphitization of small carbon grains take place when the hydrogenated carbon particles undergo annealing produced by exposure to ultraviolet radiation. Specific experimental results on this subject are lacking.

2   Experimental and Results

Hydrogenated amorphous carbon grains (ACH2) were produced by arc discharge in H_2 atmosphere (10 mbar). The samples are characterized by a chain-like structure of spherical aggregates composed of three - five spherical grains with an average diameter of 11 nm.

UV processing of ACH2 was carried out at room temperature in a chamber at a pressure less than 10^{-6} mbar. An hydrogen flow discharge lamp with a MgF_2 window, operating at a pressure of 2 mbar and 100 Volts, was used as a source of UV radiation. The energy flux at the sample position was $(2\pm1)\times10^{15}$ photons cm^{-2} s^{-1}, with an average energy per photon of 10 eV (Jenniskens et al. 1993). The doses deposited in ACH2 grains are listed in Table 1.

Table 1. Results of ACH2 exposure to UV radiation

UV Dose	$\lambda_p^{(a)}$	$E_g^{(b)}$
(eV cm^{-2})	(nm)	(eV)
0 $^{(c)}$	–	1.38±0.02
5×10^{20}	–	1.34±0.02
1×10^{22}	215±2	1.29±0.02
4×10^{22}	215±2	1.23±0.02

(a) UV peak position
(b) Optical gap
(c) ACH2 grains as produced

Spectrophotometric measurements were performed in the range 0.19 - 2.5 μm with a resolution of 2 nm by using a dispersive double beam spectrophotometer (Perkin Elmer mod. Lambda 9). Relevant spectra are reported in Fig. 1. No significant spectral variations were noted up to doses of 1×10^{21} eV cm^{-2}. On the other hand, a UV resonance at 215 nm is activated in ACH2 grains after an irradiation of 1×10^{22} eV cm^{-2}. The feature becomes more pronounced as the irradiation dose increases up to 4×10^{22} eV cm^{-2}, while the peak position remains stable.

Spectral changes take place at visible and near–IR wavelengths, where the spectral trend is characterized by a broad absorption edge which slowly changes due to the UV processing.

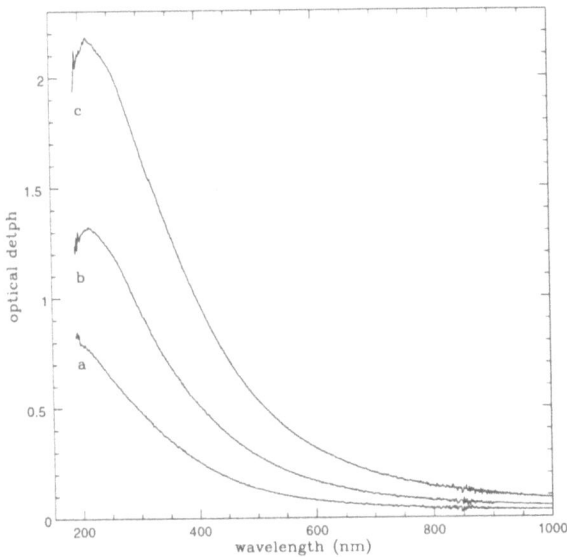

Fig. 1. UV - NIR optical depth of hydrogenated amorphous carbon grains not irradiated (a) and after an UV irradiation corresponding to doses of 1×10^{22} eV cm^{-2} (b) and 4×10^{22} eV cm^{-2} (c).

3   Discussion and Conclusions

The activation of the UV band is an evidence that modifications of the electronic structure in the grains take place due to energetic UV photons. According to the link between electronic and structural properties which exists in carbons (Mennella et al. 1995a,b), we attribute the band of irradiated ACH2 grains to π - π^* electronic transitions in sp^2 ringed clusters.

This conclusion is further supported by the behaviour of the optical gap, E_g, as a function of the irradiation dose (see Table 1). The gap decrease indicates an increase of the sp^2 clustering sites within the grains. According to Robertson (1991), $E_g \simeq 6 / \sqrt{M}$, where M is the number of rings forming the largest compact clusters, we find a M variation from 20 for ACH2 to 24 for the most irradiated sample.

Moreover, the increase of the UV optical depth with the irradiation can be interpreted in terms of growth in number of the smaller clusters which cannot be probed by the optical gap variations (Robertson 1986).

Finally, we note that a substantial hydrogenation degree of the grains is still present after a UV irradiation dose of 4×10^{22} eV cm^{-2}. According to the optical gap and hydrogen content values measured in previous experiments (Mennella et al. 1995a,b) we expect an atom ratio H/C $\simeq$ 0.3 for this last sample. H/C is 0.62 in ACH2 grains.

During the typical time of 3×10^7 yrs that grains spend in the diffuse cloud medium, the total dose of UV photons deposited in amorphous carbon grains is 3×10^{23} eV cm^{-2} (Jenniskens et al. 1993). This fluence is a factor of 7 larger than that obtained in the laboratory.

Extrapolating the logarithmic dependence of E_g on the dose (see Table 1 and Jenniskens et al. 1993) to the estimated fluence received by interstellar grains, we found $E_g = 1.19$ eV. Similar values of the optical gap (0.6 - 1.2 eV) are expected for the hydrogenated amorphous carbon produced from organic refractory grain mantles processed by UV radiation for a typical diffuse cloud medium (Jenniskens et al. 1993). These gaps correspond to graphitic clusters composed of 30–60 rings.

If our results are representative of the UV processing occurring in space, then it is unlikely that hydrogenated amorphous carbon grains can be transformed into pure graphite grains. These particles should form a disordered structure composed of sp^2 clusters and a sp^3 component whose relative amounts depend on residual hydrogenation. Moreover, the activation of a π plasmon in carbon requires a delocalization of the π bond which is inhibited by the presence of hydrogen. Therefore, a revision of the interstellar bump interpretation in terms of π plasmon in small carbon grains graphitized by UV radiation must be considered.

Acknowledgements
We thank S. Inarta, N. Staiano and E. Zona for their technical assistance during measurements. W. S. thanks M. de Groot for making his experimental set-up available for this research. This work has been supported by ASI, CNR and MURST research contracts.

References

Dorschner, J. and Henning, Th. (1995): Dust metamorphism in the Galaxy. A&A Rev., in press

Hecht, J. H. (1986): A physical model for the 2175 Å interstellar extinction feature. ApJ **305**, 817–822

Jenniskens, P., Baratta, G. A., Kouchi, A., de Groot, M. S., Greenberg, J. M., Strazzulla, G. (1993): Carbon dust formation on interstellar grains. A&A **273**, 583–600

Mennella, V., Colangeli, L., Blanco, A., Bussoletti, E., Fonti, S., Palumbo, P., Mertins, H. C. (1995a): A dehydrogenation study of cosmic analogue grains. ApJ **444**, 288–292

Mennella, V., Colangeli, L., Bussoletti, E., Monaco, G., Palumbo, P., Rotundi, A. (1995b): On the electronic structure of small carbon grains of astrophysical interest. ApJS **100**, 149–157

Robertson, J. (1986): Amorphous carbon. Adv. Phys. **35**, 317–374

Robertson, J. (1991): Hard amorphous (diamond-like) carbons. Prog. Solid State Chem., **21**, 199–333

Sorrell W. H. (1990): The λ 2175 Å feature from irradiated graphitic particles. MNRAS **243**, 570–587

Laboratory Experiments on CO and CO$_2$ Ices

M. Elisabetta Palumbo and Giovanni Strazzulla

Istituto di Astronomia Università di Catania and Osservatorio Astrofisico,
V.le A. Doria 6, I-95125 Catania, Italy

Abstract. We present results on the effects of ion irradiation, with 3 keV He$^+$ ions, on CO and CO$_2$ ices and mixtures with H$_2$O, CH$_3$OH and CH$_4$. We have studied changes in the profile of IR bands, formation of new species and their relative abundance in view of their importance to understand physical and chemical characteristics of interstellar and/or circumstellar icy grain mantles.

1 Backgrounds

Infrared spectra of several stellar objects show an absorption feature at about 2140 cm^{-1} (4.67 μm) which is attributed to solid (frozen) carbon monoxide (CO) in interstellar and/or circumstellar grain mantles (Lacy et al. 1984; Tielens et al. 1991; Kerr et al. 1993; Chiar et al. 1994, 1995). These objects are both young stars still embedded in their placental cloud and field stars located behind a dark cloud. The detailed shape of the 2140 cm^{-1} band varies from source to source and its origin has not yet been completely understood. In most cases it is possible to separate two independent absorption components, namely a narrow feature ($\Delta\tilde{\nu}$=3-9 cm^{-1}) at about 2140 cm^{-1} and a broader ($\Delta\tilde{\nu} \sim$10 cm^{-1}) and generally weaker feature at about 2136 cm^{-1}. Several laboratory experiments of astrophysically relevant mixtures, useful to the identification of the carrier(s) of these two components, have been performed (Sandford et al. 1988; Allamandola et al. 1988; Schmitt et al. 1988; Palumbo and Strazzulla 1992, 1993). Laboratory results have shown that the profile (shape, width and peak position) of the solid CO band strongly changes when CO is mixed in with other frozen gases.

Several models predict the presence of solid carbon dioxide (CO$_2$) in icy grain mantles. CO$_2$ in not predicted to have appreciable abundance in the gas phase in dense clouds so that its condensation on grains can be neglected while its presence may be due to surface processing such as UV and/or cosmic rays irradiation (e.g., Chiar et al. 1995). The profile of CO$_2$ bands in several ice mixtures has been studied by Sandford and Allamandola (1990). They showed that the exact profile of CO$_2$ bands depends on the matrix in which the molecule is frozen. Because of telluric absorption the strong CO$_2$ band at 2340 cm^{-1} (4.27 μm) can not be observed with groundbased or airborne instruments while the 650 cm^{-1} (15.3 μm) band can not be easily identified because of a strong and very broad water ice band at 750 cm^{-1} (13.3 μm; d'Hendecourt and Jourdain de Muizon 1989). This will change after the launch of the Infrared Space Observatory (ISO).

2 Experiments and Discussion

Ice samples accreted on a cold (10-30 K) substratum have been irradiated with 3 keV He^+ ions and IR spectra have been taken before, during and after irradiation. The penetration depth of 3 keV He^+ ions in frozen gases (mixtures), here studied, is only about 0.05 μm. Thus, gases have been irradiated during deposition in order to obtain samples thick enough to exhibit a good spectrum also for a study of weak features.

Pure CO ice at 10 K shows an absorption feature at about 2140 cm^{-1} with FWHM ~ 4 cm^{-1} on an optical depth scale. After ion irradiation of pure CO a band at 2340 cm^{-1} appears and the FWHM of the CO band increases. This latter is a general result: ion irradiation causes broadening of the bands. The newly formed band is easily attributed to CO_2. Laboratory experiments have shown that the CO band profile strongly depends on the mixture it is embedded in and it is modified after ion irradiation. Futhermore in mixtures with H_2O the CO peak position depends on the deposition rate of the sample. When pure methanol and $H_2O:CH_3OH$ mixtures are irradiated new bands appear thus testifying the formation of new species. Among these the most abundant and easily identifiable are CO and CO_2. However the CO profile in a $CH_3OH:CO$ mixture is different from that of CO produced after ion irradiation (Palumbo and Strazzulla 1993).

On the basis of these experimental results, comparisons between astronomical and laboratory spectra, which take into account the interaction of grain mantles with cosmic rays and the presence of methanol, have been proposed (Palumbo and Strazzulla 1993; Chiar et al. 1995). Fits relative to embedded stars were obtained using, for the narrow component, the 2140 cm^{-1} band of irradiated CO at 10 K (dose=12 eV/16amu) and, for the broad component, the 2136 cm^{-1} band of CO produced by irradiation of a $H_2O:CH_3OH=2:1$ mixture (dose=40 eV/16amu) warmed up to 67 K (Palumbo and Strazzulla 1993).

According to the model proposed, it is possible to estimate the abundance of CO_2 predicted to be present in icy grain mantles along the line of sight. Figure 1 shows, as an example, the profile of the expected CO_2 band for the embedded source AFGL 989. The CO_2 band is given by the sum of two contributions as it is for the CO band. CO_2 is produced after ion irradiation both of pure CO and a $H_2O:CH_3OH$ mixture.

Palumbo et al. (1995) have studied the profile of the CO_2 band and the CO/CO_2 ratio (Fig. 2) after ion irradiation of several ice mixtures (CH_3OH, $H_2O:CH_3OH$, $H_2O:CH_4$). It has been found that the profile of the 2340 cm^{-1} band formed after ion irradiation weakly depends on the relative abundance of species of the original mixture and dose. On the other hand the CO/CO_2 ratio strongly depends on the initial mixture and dose. In the case of $H_2O:CH_3OH$ mixtures, the CO/CO_2 ratio decreases as the dose increases and as the ratio H_2O/CH_3OH increases (Baratta et al. 1994). Moore and Nuth (1995) have studied the CO/CO_2 ratio after irradiation, with 700 MeV protons, of similar ice mixtures. They report CO/CO_2 values higher than those of Figure 2. This is probably due to the different thickness of the irradiated samples ($\sim \mu m$), suggesting that results obtained for thin samples can be extrapolated to the con-

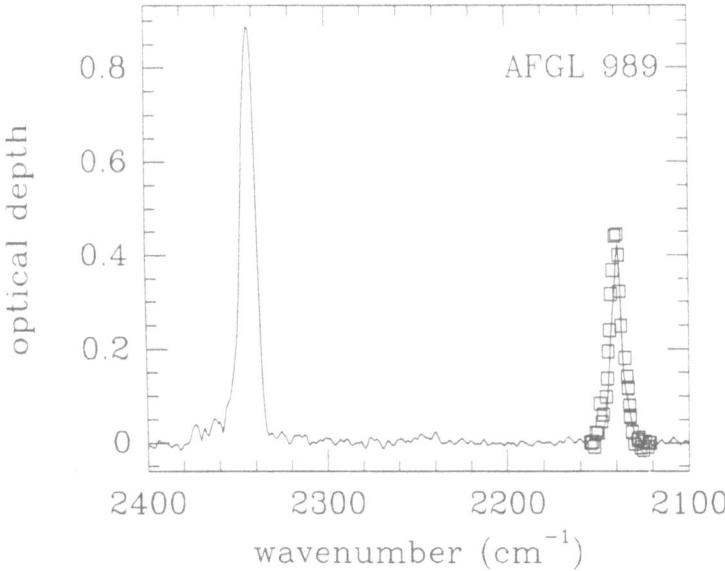

Fig. 1. Fit of the CO band observed toward AFGL 989 (Tielens et al. 1991), in the spectral range 2000-2500 cm^{-1}, giving the profile of the CO$_2$ band at 2340 cm^{-1} according to the model discussed in the text

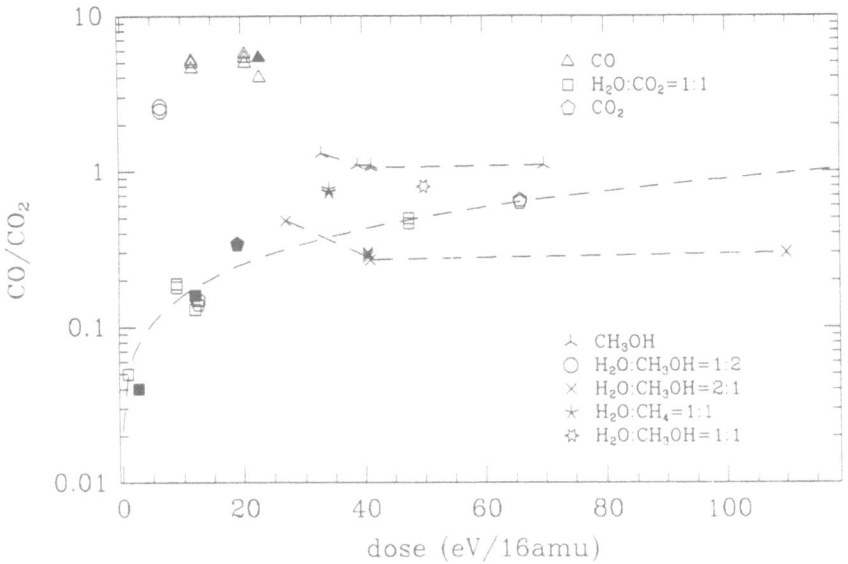

Fig. 2. CO/CO$_2$ ratio as a function of irradiation (3 keV He$^+$) dose in several ice mixtures. Dashed lines, which connect points relative to similar experiments, have been drawn to guide the eye. Solid points refer to results obtained after irradiation with 1.5 keV H$^+$ (Brucato et al. 1995)

ditions of the interstellar and/or circumstellar ice mantles while those obtained for thick sample to ices in the solar system. Allamandola et al. (1988) have estimated CO/CO_2 ratio after UV irradiation of $H_2O:CH_3OH=2:1$ mixtures. This values 1.25 which is very close to the value obtained after ion irradiation of pure CH_3OH. A possible explanation is that water is almost unaltered by UV photons and the products of water radiolysis, mainly H, OH, H_2, O_2 (e.g., Johnson 1990), do not partecipate in the UV-induced radiation chemistry. However the comparison of the results obtained after UV or ion irradiation is not straightforward because of the difficulties in estimating the amount of energy deposited (eV/molecule) during UV irradiation.

Observations in the 4.1-4.5 μm spectral region (available after the launch of ISO) will help to understand chemical and physical properties of icy grain mantles. It will be possible to fit CO and CO_2 bands simultaneously thus testing models proposed for the origin of both molecules.

References

Allamandola, L.J., Sandford, S.A., Valero, G.J. (1988): Icarus **76**, 225

Baratta, G.A., Castorina, A.C., Leto, G., Palumbo, M.E., Spinella, F., Strazzulla, G. (1994): Planet. Space Sci. **42**, 759

Brucato, J.R., Palumbo, M.E., Strazzulla, G. (1995): Icarus, submitted

Chiar, J.E., Adamson A.J., Kerr, T.H., Whittet D.C.B. (1994): ApJ, **426**, 240

Chiar, J.E., Adamson A.J., Kerr, T.H., Whittet D.C.B. (1995): ApJ, in press

d'Hendecourt, L.B., Jourdain de Muizon, M. (1989): A&A, **223**, L5

Johnson, R.E. (1990): *Energetic Charged Particle Interactions with Atmospheres and Surfaces* (Springer Verlag, Berlin)

Kerr, T.H., Adamson A.J., Whittet D.C.B. (1993): MNRAS, **262**, 1047

Lacy, J.H., Baas, F., Allamandola, L.J., Persson, S.E., McGregor, P.J., Lonsdale C.L., Geballe T.R., van de Bult, C.E.P. (1984): ApJ, **276**, 533

Moore, M.H., Nuth, J.A., (1995): Planet. Space Sci., submitted

Palumbo, M.E., Strazzulla, G. (1992): A&A, **259**, L12

Palumbo, M.E., Strazzulla, G. (1993): A&A, **269**, 568

Palumbo, M.E., Brucato, J.R., Castorina, A.C., Strazzulla, G. (1995), in preparation

Sandford, S.A., Allamandola, L.J. (1990): ApJ, **355**, 357

Sandford, S.A., Allamandola, L.J., Tielens, A.G.G.M., Valero, G.J. (1988): ApJ, **329**, 498

Schmitt, B., Greenberg, J.M., Grim, R.J.A. (1989): ApJ, **340** L33

Tielens, A.G.G.M., Tokunaga, A.T., Geballe T.R., Baas, F. (1991): ApJ, **381**, 181

Formaldehyde and Methanol Dominated Ices Toward GL 2136

W.A. Schutte[1], P.A. Gerakines[1,2],
T.R. Geballe[3], E.F. van Dishoeck[1], and J.M. Greenberg[1]

[1] Leiden Observatory, P.O. Box 9513, 2300 RA Leiden, the Netherlands
[2] Department of Physics, Rensselaer Polytechnic Institute, Troy, NY 12180-3590, USA
[3] Joint Astronomy Centre, 660 N. A'ohoku Pl., Hilo, HI 96720, USA

Abstract. Infrared spectroscopy towards the embedded massive protostellar object GL 2136 reveals 2 absorption features at 3.47 and 3.54 μm. Through comparison with spectra of laboratory ice mixtures, an identification with, respectively, the ν_5 feature of solid formaldehyde and a blend of the ν_1 formaldehyde and the ν_3 solid methanol mode is proposed. Detailed comparison with a variety of ice mixtures indicates that the methanol and formaldehyde ice on one hand and water ice on the other reside, for the greater part, in separate phases.

1 Introduction

Formaldehyde (H_2CO) is a well–known component of the gas phase in dense interstellar clouds. In warm star forming regions enhanced abundances of H_2CO and its reaction product $HCOOCH_3$ are observed. This has been interpreted in terms of evaporation from grain mantles (Caselli et al. 1993). An origin on grains is furthermore indicated by the very high deuterium fractionation of the formaldehyde (Turner 1990).

A direct detection of solid H_2CO would be of importance to clarify the role of grains in the chemistry of star forming regions. Furthermore, probing the contents of various phases of the icy grain mantle condensates provides a record of the chemical conditions occurring in the protostellar collapse. Formaldehyde has a number of infrared features. The strongest bands, at 5.80 and 6.69 μm (ν_2 and ν_3, respectively) can only be accessed by air-borne or space observatories. Weaker bands fall at 3.47 and 3.54 μm.

In this paper we report a search of the 3.47 and 3.54 μm H_2CO bands toward the high-mass protostar GL 2136 with the United Kingdom Infrared Telescope (UKIRT). This object was originally flagged as a promising candidate relating to low resolution airborne spectroscopy giving indications of the presence of the 5.80 and 6.69 μm features (Schutte 1988, and references therein).

2 Observations and Laboratory Spectra

Fig. 1 presents the spectroscopic data. This "optical depth" plot was obtained after subtracting a straight baseline (in the λ– $log(F_\lambda)$ plane) to eliminate the original strong red slope caused by the red shoulder on the 3 μm H_2O absorption and the emission spectrum of the source. Two narrow features can be seen centered at 3.474 and 3.536 μm. The 3.47 μm feature has not been observed

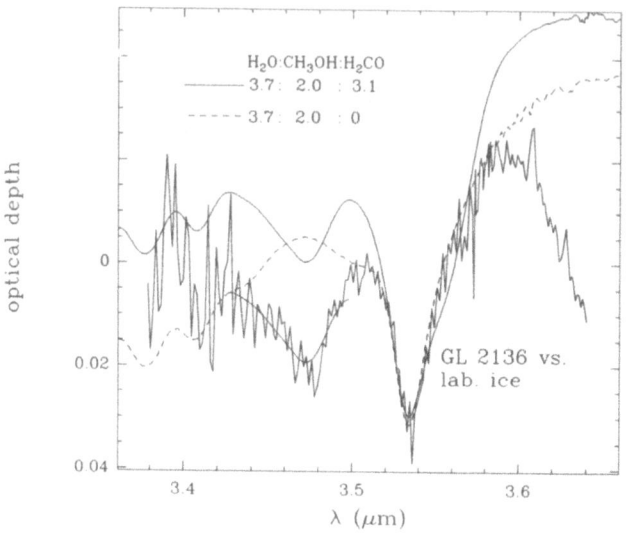

Fig. 1. GL 2136 "optical depth" plot compared to the ices $H_2O/CH_3OH/H_2CO$ = 3.7:2.0:3.1 and 3.7:2.0:0. The laboratory spectra is a temperature average 50K:70K:100K = 2:4:1.

previously towards embedded sources. It falls close to the ν_5 solid formaldehyde feature. Such an assignment would imply that part of the 3.54 μm band is produced by the H_2CO ν_1 mode. However, a very similar feature has been observed towards other high–mass protostars (Grim et al. 1991, Allamandola et al. 1992) which has been assigned to the ν_3 mode of solid CH_3OH. Thus the 3.54 μm band could be a blend of H_2CO and CH_3OH modes.

To further explore the proposed assignments we made a laboratory spectroscopic survey of ices containing CH_3OH and H_2CO as well as H_2O. Fig. 1 shows the best fit, obtained with the mixture $H_2O:CH_3OH:H_2CO = 3.7:2.0:3.1$. The laboratory spectrum is a temperature average 50K:70K:100K = 2:4:1. This ratio was choosen to account for the presence of relatively warm ice, as indicated by the position of the H_2O feature at 3.103 μm (Smith et al. 1989, Geballe 1986). The match corresponds to abundances of 6 $\pm$ 2 % for H_2CO and 4 $\pm$ 1 % for CH_3OH relative to solid H_2O.

The positions of the ν_5 H_2CO feature and the combined ν_1 H_2CO and ν_3 CH_3OH band show considerable shifts with matrix composition (Table 1). Matching the GL 2136 features leads to the requirement $[H_2CO]/[CH_3OH]$ ≈ 1.5 and $\frac{[H_2O]+[CH_3OH]}{[H_2O]} = 0.7 - 1.8$. The spectroscopically indicated H_2CO to CH_3OH ratio is very similar to the column densities, indicating that these species are well mixed. On the other hand, the admixed quantity of H_2O can only be a small fraction of the column density.

Table 1. Summary of the experimental results; Feat. 1 denotes the ν_5 H_2CO feature, Feat. 2 the combined ν_1 H_2CO and ν_3 CH_3OH feature

mixture			Feat. 1	Feat. 2
H_2O	CH_3OH	H_2CO	cm^{-1}	cm^{-1}
		1	2884.4	2822.1
	1			2828.5
	1.8	3.0	2881.0	2821.0
	2.3	3.0	2880.4	2825.0
3.6		3.0	2886.9	2828.4
66		3.0	2889.2	2825.6
3.7	2.0			2829.9
1.9	1.0	3.2	2882.7	2826.2
3.8	1.0	3.0	2882.4	2829.8
3.7	2.0	3.1	2879.8	2829.4
3.7	2.0	3.5	2880.8	2829.1
7.6	1.0	3.1	2883.5	2830.7
7.6	2.0	3.3	2879.9	2830.3
11	2.0	2.9	2871.0	2831.0
GL	2136		2878.7 ± 1.5	2828.6 ± 1.0

3   Discussion

Matching the 3.47 and 3.54 μm features with various ice mixtures of H_2CO, CH_3OH and H_2O indicates that, on one hand, H_2O ice, and, on the other, the H_2CO and CH_3OH dominated ice are separated. Henceforth we will denote these phases with type I and type II ice, respectively. They likely correspond to condensation at different gas phase chemical conditions. The alternate possibility, separation through a desorption mechanism discriminating components according to volatility, seems unlikely, due to the rather small difference in volatility between, in particular, CH_3OH and H_2O (Sandford and Allamandola 1993). Although the mechanism forming H_2CO and CH_3OH on grain surfaces is still unclear (Hiraoka et al. 1994), it seems probable that type II ice formed under conditions with abundant atomic hydrogen gas, producing such hydrogenated species as opposed to CO. Future space observations of the strong 5.8

and 6.7 μm H_2CO bands may reveal whether such conditions occur throughout dense clouds, or are typical to the protostellar environment. Finally, we note that the apparent isolation of solid H_2CO and CH_3OH in the protostellar environment may relate to the large variation in the abundances of these molecules relative to H_2O as observed towards comets (Bockelée-Morvan et al. 1995).

References

Allamandola, L.J., Sandford, S.A., Tielens, A.G.G.M., Herbst, T.M. (1992): Infrared spectroscopy of dense clouds in the C–H stretching region: Methanol and "diamonds", ApJ **399**, 134–146

Bockelée-Morvan, D., Brooke, T.Y., Crovisier, J. (1995): On the origin of the 3.2-3.6-μm emission feature in comets, Icarus **116**, 18–39

Caselli, P., Hasegawa, T.I., Herbst E. (1993): Chemical differentiation between star-forming regions: The Orion hot core and compact ridge, ApJ **408**, 548–558

Geballe, T.R. (1986): Some recent infrared spectroscopy of interstellar processes, *Summer School on Interstellar Processes: Abstracts of Contributed Papers*, Hollenbach, D.J., Thronson, H.A. (eds.). NASA, Moffett Field, p. 129–130

Grim, R.J.A., Baas, F., Geballe, T.R., Greenberg, J.M., Schutte, W. (1991): Detection of solid methanol toward W33A, A&A **243**, 473–477

Hiraoka, K., Ohashi, N., Kihare, Y., Yamamoto, K., Sato, T., Yamashita, A. (1994): Formation of formaldehyde and methanol from reactions of H atoms withh solid CO at 10–20 K, Chemical Physics Letters **229**, 408–414

Sandford, S.A., Allamandola, L.J. (1993): Condensation and vaporization studies of CH_3OH and NH_3: Major implications for astrochemistry, ApJ **417**, 815–825

Schutte, W.A. (1988): *The evolution of interstellar organic grain mantles*, PhD thesis, University of Leiden, Leiden

Smith, R.G., Sellgren, K., Tokunaga, A.T. (1989): Absorption features in the 3 micron spectra of protostars, ApJ **344**, 413–426

Turner, B.E. (1990): Detection of doubly deuterated interstellar formaldehyde: An idicator of active grain chemistry, ApJ **362**, L29–L33

Ice Cocktails in Molecular Cloud Cores

Teresa C. Teixeira[1,2], James P. Emerson[1], and Frank P. Pijpers[3]

[1] Physics Dept., Queen Mary & Westfield College, Mile End Rd., London E1 4NS, UK
[2] Centro de Astrofísica da Universidade do Porto, Porto, Portugal
[3] TAC, University of Aarhus, Aarhus, Denmark

Abstract. Spectra of the solid CO feature have been taken towards embedded Young Stellar Objects in nearby molecular clouds. The likely composition of the ices is analysed by fitting the observations with laboratory data. Taking a model for a protostar, the predicted and observed column densities of the ices are compared. Finally, the possibility is being explored of using the ice mantles as probes of protostellar activity in the embedded stage of the life of a star.

1 Ice Cocktails

An important goal in star formation studies is to understand the structure and composition of protostellar envelopes, in order to determine the properties of the embedded young stars and their disks. The main motivation for this work has been the need to constrain the parameters going into the current models of core collapse and star formation. One way of probing the physical conditions along the envelope is by comparing the observed column densities of silicates and ices with the results of protostellar models. Moreover, the profiles of those features may provide information on the composition, structure and thermal history of the grains (for a review see Whittet & Duley 1991).

Solid CO in grain mantles is identified by an absorption feature at $4.67\,\mu m$, corresponding to the fundamental stretching mode of the CO molecule. As shown by laboratory studies (Sandford et al. 1988, Schmitt et al. 1989), the peak position and width of the solid CO band depend on the chemical composition of the icy matrix in which CO is embedded, and on the extent of thermal processing of the ices. The observed features toward dark clouds generally show two components: a narrow dominant one at $\sim 4.674\,\mu m$, and a broader weaker feature at $\sim 4.681\,\mu m$ appearing as a long wavelength wing. The dominant component is due to CO in a nonpolar environment (CO, CO_2, O_2, N_2), and the broader one can be explained by CO trapped in polar mixtures (H_2O-dominated). These two types of ices present in the mantles may point to different stages in the grain mantle evolution, which lead to an "onion" skin structure of the ice mantle (Tielens et al. 1991).

2 The Observations

In this work, 4.5-$4.9\,\mu m$ spectra were taken with CGS4 at the UKIRT, toward embedded Young Stellar Objects in the Taurus, Perseus and L1641 molecular clouds. The solid CO band was detected in 6 of the 13 observed sources. An attempt has been made to determine the likely composition of the CO ices by comparison of the observed features with laboratory data from Sandford et al.(1988), Schmitt et al.(1989), and Tielens et al.(1991). The result of χ^2-minimization is shown in fig.1 for two of the sources.

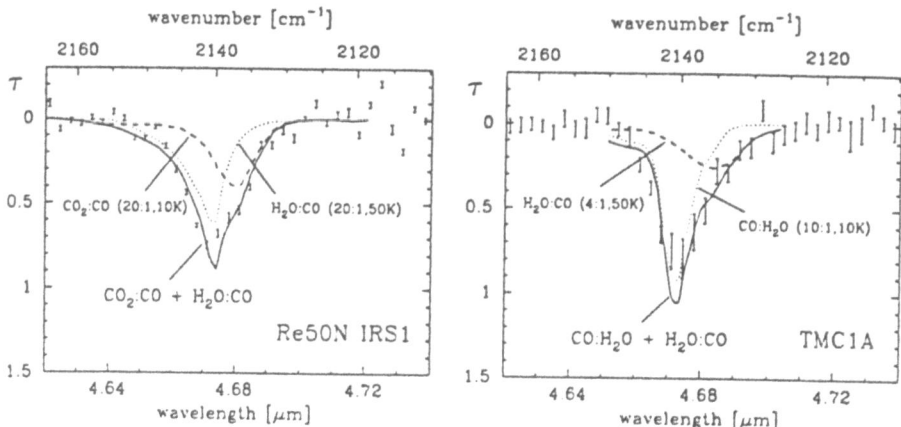

Fig. 1. Spectra of Re50N IRS1 and TMC1A in optical depth units, and best laboratory data fits (solid lines). The dotted and dashed lines correspond to the nonpolar and polar components of the ice mixtures, respectively.

3   Around a Protostar

A forming star heats up its surroundings, creating a temperature gradient along the protostellar envelope. That will cause the composition of the grain mantles in the envelope to vary with the distance to the centre (Tielens et al. 1991). Closest to the star the temperature is too high for any grains to survive. For radii where the temperature drops below $\sim 2300\,\mathrm{K}$, bare grains can exist but their icy mantles will have evaporated. Farther out, in regions of $T \leq 100\,\mathrm{K}$, the grains will have H_2O-rich mantles, with increasing trapped CO content as the temperature decreases. The very volatile nonpolar ices (CO-rich) will only survive in the coolest outer parts of the envelope, where $T < 20\,\mathrm{K}$.

The Adams, Lada & Shu(1987, hereafter ALS) model for a low mass protostar, produces a density and a temperature profile along the envelope. Integrating over the range of temperatures where CO can exist in the grain mantles, the total CO column density was estimated from the model, assuming 75% by mass of H_2 in the gas and $N(CO)/N(H_2) = 8 \times 10^{-5}$ (Frerking et al. 1982, van Dishoeck et al. 1993). Figure 2 plots the solid CO column density derived from the spectra, $N(CO)_s$, against the total (gas + solid) CO column density estimated from the model, $N(CO)_{ALS}$, for some YSOs in this study and from the literature. It can be seen that for some sources there is more *observed solid* CO than the *total* amount of CO *predicted* to be available. The reasons for this discrepancy might involve *a)* the CO to H_2 ratio which might be higher in star forming cores due perhaps to a different chemistry, *b)* the density profile along the envelope which might be shallower than that assumed by ALS, and *c)* the size of the core assumed in the model which might also be larger. It is, however, acknowledged that it is difficult to assess whether all the material in the line of sight belongs to the collapsing core or if a part of it belongs to foreground material in the quiescent molecular cloud.

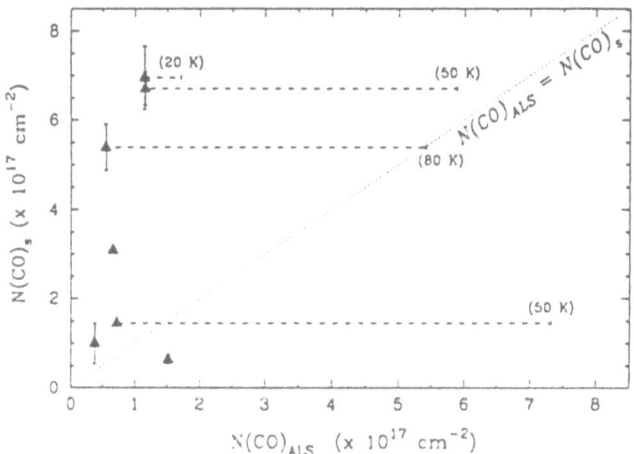

Fig. 2. Observed solid-CO column density $(N(CO)_s)$ *vs.* total (gas+dust) CO column density predicted using the ALS model for each protostar. The triangles correspond to integrating the model for temperatures below 17 K. The integration was also allowed to proceed up to the temperature indicated by the polar component of the mixture, for the sources for which laboratory fits were possible. The result of that integration is given by the right hand end of the dashed lines.

4   Embedding an Outburst

During the pre-main sequence evolution, a solar-type star is thought to undergo 10 - 100 FUOri-type of outbursts (Bell et al. 1995). Typically, during such an event the luminosity of the object increases by a factor of 100 - 600. As that released energy diffuses outwards in the envelope, the grains will experience a temperature increase with decreasing amplitude with the distance from the star. This is shown in Figure 3a for outburst luminosities $L_{outburst} = f \times L$, for $f = 10,100,250$ and 630, L being the luminosity of the source before the outburst. The pre-outburst state is taken to be the ALS model for TMC1A.

Figure 3b illustrates the effect of a steepening of the temperature gradient caused by an outburst, on the column density of material as a function of temperature. Clearly, the grain mantle distribution will be dramatically affected by

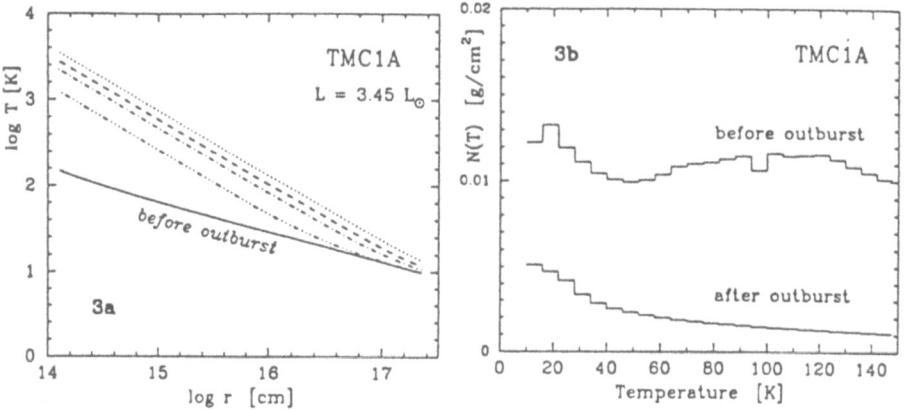

Fig. 3. The effects of an outburst on the a)temperature profile: maximum temperature reached by the grains along the radius, for outburst luminosities given in the main text; b)column density distribution: column density of material with a temperature within the bins shown by the extent of the horizontal lines: the distribution after the outburst results from taking the steeper temperature profile appropriate for the maximum temperature during the outburst.

such an energetic event: we expect to observe a significant decrease in the column densities of solid CO and H_2O, and the annealing of the ices, caused by the global warm-up of the envelope.

5   Acknowledgements

The United Kingdom Infrared Telescope is operated by the Observatories on behalf of the UK Particle Physics & Astronomy Research Council. TCT supported by grant BD/2095/92-RM from JNICT–Programa CIENCIA, Portugal.

References

Adams, F.C., Lada, C.J., Shu, F.H. (1987): Ap. J. **312**, 788
Bell, K.R., Lin, D.N.C., Hartmann, L.W., Kenyon, S.J. (1995): Ap. J. **444**, 376
Frerking, M.A., Langer, W.D., Wilson, R.W. (1982): Ap. J. **262**, 590
Sandford, S.A., Allamandola, L.J., Tielens, A.G.G.M., Valero, L.J. (1988): Ap. J. **329**, 498
Schmitt, B., Greenberg, J.M., Grim, J.A. (1989): Ap. J. **340**, L33
Tielens, A.G.G.M., Tokunaga, A.T., Geballe, T.R., Baas, F. (1991): Ap. J. **381**, 181
van Dishoeck, E.F., Blake, G.A., Draine, B.T., Lunine, J.I. (1993): in *Protostars and Planets III*, ed.s E.H.Levy and J.I.Lunine (Tucson,London,The University of Arizona Press), p.163
Whittet, D.C.B., Duley, W.W. (1991) A.& A. Rev. **2**, 167

Optical Constants of Amorphous Carbon Extracted from Recent Laboratory Extinction Measurements

Victor G. Zubko[1,2], Vito Mennella[3], Luigi Colangeli[3], and Ezio Bussoletti[4]

[1] Institute of Astronomy, N. Copernicus University, Chopina 12/18, 87–100 Toruń, Poland
[2] Main Astronomical Observatory, National Academy of Sciences, Golosiiv, Kyiv–650, 252127, Ukraine
[3] Osservatorio Astronomico di Capodimonte, via Moiariello 16, I–80131 Napoli, Italy
[4] Istituto di Fisica Sperimentale, Istituto Universitario Navale, via A. De Gasperi 5, I–80125 Napoli, Italy

Abstract. Optical constants of various amorphous carbon samples, possible analogues of interstellar and circumstellar dust grains are presented. They have been deduced from homogeneous laboratory data sets ranging from the extreme UV (0.04–0.05 μm) up to the far IR (2 mm), making use of the Kramers–Kronig technique. The calculations have been carried out for various grain extinction models: single Rayleigh and Mie spheres, continuous distribution of randomly oriented ellipsoids (CDE), homogeneous aggregates (HA) and fractal clusters (FC).

1 Carbonaceous Dust in Astrophysics

Numerous evidences have been accumulated about carbon as an important constituent of cosmic grains. There is an estimation of 60–70 % depletion of carbon from the gas phase (Cardelli et al. 1993). Graphite, SiC and diamond grains of probable interstellar origin have been found in meteorites (Draine 1994). Modelling of interstellar extinction curves requires graphite to be the principal carbon component with possible small amount of amorphous carbon (Aanestad 1995). On the contrary, the circumstellar extinction curves for a number of carbon-rich H-deficient stars (mainly RCB stars) show a wider UV peak shifted to 240 – 260 nm (Jeffery 1995) which may be modelled by both amorphous carbon (Muci et al. 1994; Jeffery 1995; Blanco et al. 1995) and graphite grains (Zubko 1995). Evidently, more refined extinction models need reliable optical constants of various carbonaceous materials. Recently, new measurements of the extinction coefficient of amorphous carbon grains ranging from 0.04 – 0.05 μm to 2 mm became available (Colangeli et al. 1993, 1995; Mennella et al. 1995). The purpose of this contribution is to present the optical constants of some amorphous carbon samples extracted from these new data by using the Kramers–Kronig technique. The full description of the theoretical approach, more extended results and their detailed discussion will be reported in a forthcoming paper.

2   Amorphous Carbon Samples

The amorphous carbon grains considered in this work have been produced under the following conditions:

1. arc discharge between amorphous carbon electrodes in Ar atmosphere at 10 mbar (ACAR sample);
2. arc discharge between the same type of electrodes in H_2 atmosphere at 10 mbar (ACH2 sample);
3. burning of benzene in air at normal conditions (BE sample);

Details on experimental set-up and sample preparation may be found in, e. g., Colangeli et al. (1993, 1995) and Mennella et al. (1995).

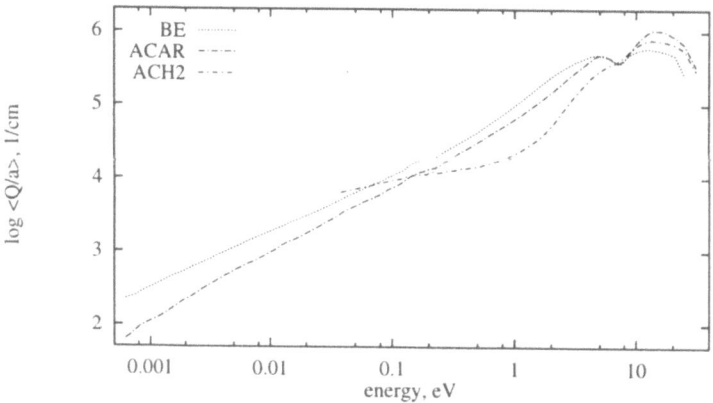

Fig. 1. Extinction coefficients of the analyzed samples

Amorphous carbon grains condense in chain-like structures containing single grains with typical spheroidal shape. In addition, grains are mostly aggregated in clusters containing some three to five individual grains. In turn, such clusters are linked together in a necklace-fractal structure. The average radius of the grains is 5–6 nm for ACAR and ACH2 samples and around 15 nm for BE sample. The extinction coefficients used for our analysis (Fig.1) have been obtained on grains simply sitting onto a substrate and are corrected to take into account the matrix effect.

3   Computation of Optical Constants

Our approach to compute the optical constants from the extinction coefficients, $< Q/a >$, consists of two main steps.

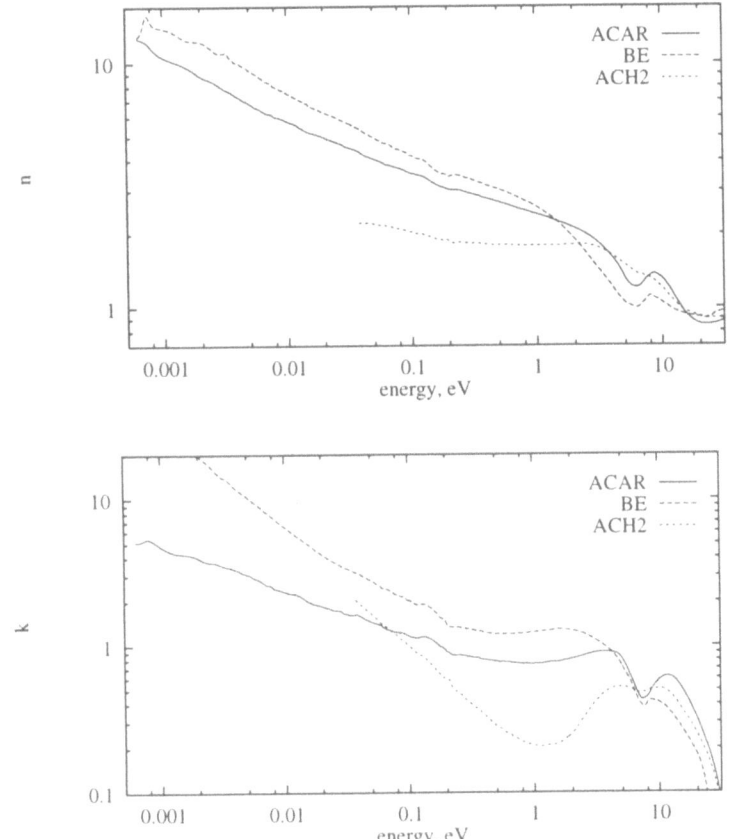

Fig. 2. Optical constants of amorphous carbon samples

As a first step, the polarizability per unit volume of a grain $A = A_r + iA_i$ (Ku and Felske 1986) is calculated. At wavelengths where data are available A_i is deduced by exploiting the suitable relation between A_i and $< Q/a >$ (Ku and Felske 1986). To evaluate A_r by means of the Kramers–Kronig approach, the high-energy tail of A_i beyond the range of measurements is needed. Here A_i is calculated by using the Mie theory and an estimation of the dielectric function for a carbon grain reported by Laor and Draine (1993). The low-energy tail of A_i is estimated by extrapolation on the basis of the spectral index of extinction coefficients in the far IR region: 0.9 ± 0.1 for ACAR, 0.8 ± 0.1 for BE and 0.4 ± 0.1 for ACH2.

Once A_i and A_r are known, to derive the optical constants a grain model has to be adopted. We have performed calculations for both single grains (Rayleigh sphere, Mie sphere, CDE) and aggregates (homogeneous aggregation and fractal clusters) mainly likewise to Rouleau and Martin (1991).

Both steps of calculations have been repeated in an iterative process to reach the convergency of dielectric functions, since to derive the high-energy tail of A_i the imaginary part of the dielectric function in the whole energy range $[0, \infty]$ has to be specified. In addition, the resulting optical constants are constrained to satisfy the sum rule for the effective number of electrons per atom which should go to six for carbon materials at sufficiently high energy.

As an example, the optical constants n and k for CDE single grains are reported in Fig.2. In agreement with Rouleau and Martin (1991), we find that the results for aggregated grain models are qualitatively mimiced by the simpler CDE model. In addition, the optical constants for energies > 1 eV become very close quantitatively for all the considered models.

References

Aanestad, P. (1995): Ultraviolet extinction to 10.8 inverse microns. ApJ **443**, 653–663

Blanco, A., Fonti, S., Orofino, V. (1995): A model for the amorphous carbon grains around C-rich objects. ApJ **448**, 339–345

Cardelli, J.A., Mathis, J.S., Ebbets, D.C., Savage, B.D. (1993): . ApJ **402**, L17

Colangeli, L., Mennella, V., Blanco, A., Fonti, S., Bussoletti, E., Gumlich, H.E., Mertins, H.C., Jung, Ch. (1993): Extreme-ultroviolet extinction measurements on hydrogenated and dehydrogenated amorphous carbon grains. ApJ **418**, 435–439

Colangeli, L., Mennella, V., Blanco, Palumbo, P., Rotundi, A. Bussoletti, E. (1995): Mass extinction coefficients of various submicron amorphous carbon grains: tabulated values from 40 nm to 2 mm. A&AS, in press

Draine, B.T. (1994): "Dust in diffuse interstellar clouds". in Proc. the First Symposium on the Infrared Cirrus and Diffuse Interstellar Clouds, ed. by R.M. Cutri and W.B. Latter (Astron. Society of the Pacific, San-Francisco)

Jeffery, C.S. (1995): The ultraviolet properties of cool material ejected by hydrogen-deficient stars. A&A **299**, 135–143

Ku, J.C., Felske, J.D. (1986): Determination of refractive indices of Mie scatterers from Kramers–Kronig analysis of spectral extinction data. J. Opt. Soc. Am. A **3**, 617–623

Laor, A., Draine, B.T. (1993): Spectroscopic constraints on the properties of dust in active galactic nuclei. ApJ **402**, 441–468

Mennella, V., Colangeli, L., Bussoletti, E. (1995): The absorption coefficient of cosmic carbon analogue grains in the wavelength range 20–2000 μm. A&A **295**, 165–170

Muci, A.M., Blanco, A., Fonti, S., Orofino, V. (1994): Ultraviolet spectra of amorphous carbon grains: comparison with the circumstellar extinction around C-rich objects. ApJ **436**, 831–836

Rouleau, F., Martin, P.G. (1991): Shape and clustering effects on the optical properties of amorphous carbon. ApJ **377**, 526–540

Zubko, V.G. (1995): On the interpretation of the extinction curves of RCB stars. MNRAS (submitted)

Part VII

Radiative Transfer

Radiative Transfer Models of Far-IR from W3 IRS 4 and IRS 5

Murray F. Campbell[1], Harold M. Butner[2], Paul M. Harvey[3], Neal J. Evans, II[3], Matthew B. Campbell[1], and Christopher N. Sabbey[1]

[1] Department of Physics and Astronomy, Colby College, Waterville, ME 04901 USA
[2] Department of Terrestrial Magnetism, Carnegie Institute of Washington, 5241 Broad Branch Road, NW, Washington, DC 20015-1305 USA
[3] Department of Astronomy, University of Texas, Austin TX 78712 USA

Abstract. Spherically symmetric, centrally heated radiative transfer cloud models were created to match KAO observations at 47 and 95 μm of dust continuum emission from the cores of W3 IRS 4 and IRS 5. After removal of an extended emission component from the source profiles, it was possible to fit the sources at both wavelengths. Depending on the dust grain properties assumed for the clouds, the models have either a uniform density or a power law radial density function with an exponent of -1. The models have inner dust-free cavities with radii of 0.06-0.08 pc, although these radii are not well determined. The clouds outer radii are 0.3-0.5 pc. Models for IRS 5 have central stellar luminosities of $2.5 - 3.5 \times 10^5 L_\odot$ depending on the dust type and cavity size, and models for IRS 4 have $9.3 \times 10^4 L_\odot$ for both dust types considered.

1 Introduction

W3 is a region of radio continuum emission with multiple HII regions and infrared sources which has been extensively observed at many wavelengths. Maps at 20 μm, 6 cm, 450 μm and 800 μm which are very useful for obtaining an overview of the region are presented by Ladd et al. (1993). When W3 was mapped with 30" resolution in the far-infrared by Werner et al. (1980), its emission was found to be dominated by peaks at IRS 5 and IRS 4 (whose peak is near W3 C), with extended emission associated with W3 A (which is associated with IRS 1). We present new observations and models of IRS 4 and IRS 5 for which full discussion will be given by Campbell et al. (1995)

2 Observations and Maximum Entropy Deconvolutions

W3 was mapped by scanning an eight detector linear array on the KAO at 47 and 95 μm in 1987, and again at 95 μm in 1989. Our new maps indicate much more nearly circular symmetry for IRS 5 and IRS 4 than the maps of Werner et al. (1980), and our lowest contours have somewhat different shapes than theirs.

We obtained precise radial source profiles by averaging individual scans which passed directly through the source centers. In the scans across IRS 5, diffuse emission from W3 A causes greater extension on the north side of the scans than the south. Symmetrical scans were created by locating the centers of the cores of emission and then reflecting the half-scans from the south side onto the north. Symmetrical scans were created in the same way for IRS 4.

We applied maximum entropy method (MEM) deconvolution to the symmetrical profiles for IRS 5 and IRS 4. For each source the deconvolutions give two components, a narrow core and a pedestal.

3 Radiative Transfer Models

We have applied the radiative transfer code of Egan, Leung, and Spagna (1988) for the modeling. It was modified to include the convolution of the cloud's emission with the KAO point source profiles (PSPs) to create model scans for comparison with the KAO scan data.

Our models assume spherical clouds with density which depends on radius as a power law specified by its exponent, α. The dust cloud is assumed to have a central dust-free cavity of radius r_c around a central star or group of stars and an outer radius of r_o. We specify the optical properties of the dust and the optical depth τ at 95 μm from center to edge of the cloud. Most of our trial models were made using dust optical properties from Mathis, Mezger, and Panagia (1983, hereafter MMP), but we also investigated Draine and Lee (1984, DL) dust-based models. The radiative transfer code requires the temperature and luminosity of the central star or dense group of stars as input assumptions. We used 25 000 K for IRS 5 and 35 000 K for IRS 4. For deeply embedded sources, the far-infrared emission is only weakly dependent on the temperature of the embedded star(s). We assumed a distance of 2.4 kpc. The code calculated the total flux density integrated along each scan and the scan profile at each wavelength. We base our modeling primarily on the 1987 data because both wavelengths were taken on the same observing flight. For all models presented, the flux density integrated along the scans agreed within 10% with that of the data scans at both wavelengths.

Model 1 was created in an attempt to fit the symmetrical profiles of IRS5, which we refer to as total scans because the pedestals found in the MEM deconvolutions have not been removed (see Table 1). The model emission convolved with the 1987 95 μm PSP matches the data but at 47 μm the model does not match the data. We could not find a model which matched the total profiles at both wavelengths. We characterize the quality of the fits numerically by the reduced χ^2 within the data FWHM. A good fit would be characterized by χ^2 of approximately two because of uncertainty in the PSP. At 47 μm, χ^2 is large for model 1 in the core, and the model does not provide a good overall fit to the total scan.

The pedestals of emission at both wavelengths shown by the MEM deconvolutions and the poor fit of model 1 to the 47 μm total scan both suggest that the total scans contain extended emission which is not due to the core. Approximate removal of extended emission from the total scans was accomplished by treating the pedestals in the MEM scans as baseline offsets. The pedestals were subtracted from the MEM scans, and the MEM profiles of the cores were re-convolved with the PSPs. We refer to these new scans as "core" scans. Model 2 is a good fit to the core scans of IRS 5 at both wavelengths. We searched for the minimum outer radius r_o which would provide a reasonable fit for a density function with $\alpha \leq 0$. The code also calculates the spectral energy distribution (SED) to compare to single beam photometric data. The model SED and photometry agree to about 0.5 in the log of the flux density, except in the near- and mid-infrared where one might expect the source of radiation to be hot dust close to the star which we have ignored in our models of far-infrared emission.

Model 3 also fits the IRS5 core scans but utilizes DL dust properties. Its SED is much lower than model 2's SED in the near- and mid-infrared, as would be expected from the difference in dust properties.

Because IRS 5 has only a small, weak radio continuum source, we might expect the far-infrared emitting cloud to have a small cavity. Model 4 has $r_c = 0.005$ pc. DL dust was chosen because its properties give broader profiles than those of MMP dust. It does not fit the data quite as well as models 2 or 3 for which r_c was treated as a free parameter. A variation of model 4 with MMP dust has a luminosity of $3.5 \times 10^5 L_\odot$ with all other parameters the same as those of model 4. We favor models 2 and 3 over model 4, but r_c is not precisely determined.

A parallel treatment of the scans for IRS 4 results in the models shown in the right hand portion of Table 1. Models 2 and 3 fit the core scans well.

Table 1. Radiative Transfer Models

Source: Model	IRS5: 1	IRS5: 2	IRS5: 3	IRS5: 4	IRS4: 1	IRS4: 2	IRS4: 3
$L/L_\odot$	5.0×10^5	3.1×10^5	2.5×10^5	2.8×10^5	3.0×10^5	9.3×10^4	9.3×10^4
$r_c(pc)$	0.08	0.08	0.08	0.005	0.06	0.07	0.07
α	-1	0	-1	0	0	0	-1
$r_o(pc)$	1.0	0.31	0.35	0.31	1.0	0.35	0.50
$\tau(95\mu m)$	0.30	0.30	0.30	0.32	0.30	0.24	0.24
Dust	MMP	MMP	DL	DL	MMP	MMP	DL
Scans	Total	Core	Core	Core	Total	Core	Core
$\chi^2(47\mu m)$	72.8	2.24	0.37	5.30	515.1	0.91	1.36
$\chi^2(95\mu m)$	2.23	0.26	0.63	0.31	9.93	1.68	2.67

4   Discussion

For each source, models 2 and 3 fit well. We might have expected smaller r_c because the dust temperature at r_c is typically only 200 K there, well below sublimation temperature. Furthermore, for IRS 4, r_c is twice the radius of the HII region W3 C. However, most sources which have been modeled require large dust-free or low density cavities. We might also have expected to find $-2 \leq \alpha \leq -1.5$ from star formation theory. For other objects, $-3 \leq \alpha \leq 0$. The scan profiles constrain r_o to be small; our values of r_o lies on the low end of the range of 0.3 - 2 pc reported for other objects.

Our IRS 5 models predict densities at r_c of $1.1 \times 10^5 cm^{-3}$ for model 2 (MMP dust) and $1.4 \times 10^6 cm^{-3}$ for model 3 (DL dust), and cloud masses of 440 and 2700 $M_\odot$ respectively. Small cavity models do not have substantially different total masses. Our models also predict broad emission profiles at 450 μm. For IRS 5, model 2 has a broad peak with a FWHM of 42" at 450 μm when convolved with an 8" beam. Ladd et al. (1993) and Oldham et al. (1994) observed 13-14" 450 μm peaks at IRS 5 and IRS 4 with 8" beams, and Oldham et al. (1994) estimated the mass of the 13" submillimeter core to be only about 200 $M_\odot$ using Hildebrand (1983) dust properties. For IRS4, models 2 and 3 give densities at r_c of $7.3 \times 10^4 cm^{-3}$ (MMP dust) and $9.8 \times 10^5 cm^{-3}$ (DL dust), and masses of 420 and 1600 $M_\odot$ respectively. They are considerably larger than the value of 83 $M_\odot$ given by Oldham et al. (1994) for a 13" diameter submillimeter emitting cloud. Finding both size and masses from the submillimeter emission different from those based on the far-infrared emission implies that different dust cloud components are emitting in each wavelength region. A compact disk located inside r_c could be responsible for the submillimeter peak while having very little effect on far-infrared emission of the cloud.

References

Campbell, M. F., Butner, H. M., Harvey, P. M., Evans, N. J., II, Campbell, M. B., and Sabbey, C. N.: (1995) High Resolution Far-infrared Observations and Radiative Transfer Models of W3 IRS 4 and IRS 5. ApJ, **454**, in press

Draine, B. T. and Lee, H. M.: (1984) Optical Properties of Interstellar Graphite and Silicate Grains. ApJ, **285**, 89–108

Egan, M. P., Leung, C. M., and Spagna, G. F. (1988): CSDUST3: A Radiation Transport Code for a Dusty Medium with 1-D Planar, Spherical or Cylindrical Geometry. Comp. Phys. Comm. **48**, 271–292

Hildebrand, R. H. (1983): The Determination of Cloud Masses and Dust Characteristics from Submillimetre Thermal Emission. QJRAS, **24**, 267-282

Ladd, E. F., Deane, J. R., Sanders, D. B., and Wynn-Williams, C. G. (1993): Luminous Radio-quiet Sources in W3(Main). ApJ, **419**, 186–189

Mathis, J. S., Mezger, P. G., and Panagia, N. (1983): Interstellar Radiation Field and Dust Temperatures in the Diffuse Interstellar Matter and in Giant Molecular Clouds. A&A, **128**, 212-219

Oldham, P. G., Griffin, M. J., Richardson, K. J., and Sandell, G. (1994): W3 – A Study of a Site of Massive Star Formation. A&A, **284**, 559–572

Werner, M. W., et al. (1980): High Angular Resolution Far-infrared Observations of the W3 Region. ApJ, **242**, 601–608

Models of Dusty Disks Including Transiently Heated Particles

Andreas Efstathiou[1] and Ralf Siebenmorgen[2]

[1] Division of Physical Sciences, University of Hertfordshire, Hatfield, Herts AL10 9AB
[2] ESA – ISO Science Operations, Vilafranca , P.O. Box 50727, E–28080 MADRID

Abstract. We present a method of incorporating small transiently heated particles, small graphites and polycyclic aromatic hydrocarbons (PAH) into axially symmetric radiative transfer calculations. The iterative method of solution of the radiative transfer problem in dusty disks accurately takes into account a distribution of large (classical) grains and multiple scattering from them. The method self-consistently accounts for the destruction of the small particles by a high density of energetic photons and sublimation of the large particles at a range of distances from the central source of radiation, depending on their size and composition. A number of models exploring the parameter space appropriate for regions of star formation is presented. We pay particular attention to regions of the parameter space that produce weak PAH emission features in accordance with observations. The geometry we have assumed is that of a *flared* disk.

1 Introduction

The formation and early evolution of stars occurs within dense regions of molecular clouds. In this environment, young stellar objects are associated with significant amounts of interstellar gas and dust. Observations indicate the presence of circumstellar disks. (Beckwith et al., 1990; Chini et al., 1991; Lada & Adams, 1992). However, the nature of the circumstellar dust is subject to a large debate. Transiently heated particles, which are seen in many astrophysical environments, have been observed in reflection nebulae, HII regions, Herbig Ae/Be stars and in the deeply embedded sources, such as WL16 in the Ophiucus cloud (Buss et al. 1990, Brooke et al. 1993, Hanner et al. 1992).

In this paper we report results from radiative transfer calculations that combine a detailed treatment of the emission of transiently heated dust particles with an accurate method of solving the radiative transfer problem in the case of axial symmetry. These models therefore represent the most complete treatment of the reprocessing of stellar radiation by dust in the disks surrounding Young Stellar Objects.

2 Radiative Transfer Model

The method of obtaining the intensity distribution at any point in the cloud and hence iterating for the temperature of each of the large grains is that of Efstathiou & Rowan-Robinson (1995, MNRAS 273, 649) The emission of the transiently heated particles is calculated according to the method of Siebenmorgen et al. (1992, AA 266, 501.). Inside the cloud we assume radiative equilibrium, which is equivalent to formulating the radiative balance condition for every grain species l at every point (r, θ) in the cloud:

$$\int C^l_{\nu,abs} J_\nu(r, \Theta) d\nu = \int \int C^l_{\nu,abs} B_\nu(T^l(r, \Theta)) P(T) \, dT \, d\nu \tag{1}$$

To obtain the intensity distribution at (r, Θ) we adopt the ray tracing technique of Efstathiou & Rowan-Robinson (1990, MNRAS 245, 275). In addition to their three components of the intensity, describing attenuated emission from the central source $I^{(1)}_\nu$, scattered light $I^{(2)}_\nu$ and thermal emission from big grains $I^{(3)}_\nu$, we introduce a fourth component $I^{(4)}_\nu$ for all transiently heated particles given by

$$I^{(4)}_\nu(r, \Theta, \theta, \phi) = \sum_{l=1}^{L_{gr}+L_{PAH}} \int_0^S C^l_{\nu,abs} \int_0^\infty B_\nu(T^l P(T) \, dT \, exp(-\int_0^{S'} a^{ext}_\nu ds') \, d\Sigma \tag{2}$$

where

$$a^{ext}_\nu = \sum_{l=1}^{L_{tot}} C^l_{\nu,ext} \quad n^l(r', \Theta) \tag{3}$$

S' is the distance of an integration point $P'(r', \Theta')$ from the radiation source point $P(r, \Theta)$, and S is the distance from $P(r, \Theta)$ to the edge of the cloud for the ray defined by θ and ϕ.

The abundances of the three grain components (classical grains, small graphites and PAHs) are as those postulated by Siebenmorgen & Krügel (1992) to explain the average interstellar extinction data. The continuous grain size distribution of the dust model is represented by adopting a discretisation with $L_{tot} = 8$ different sizes for the large particles, $L_{gr} = 5$ radii of small graphites and $L_{PAH} = 2$ kinds of PAHs: a molecule component consisting of N_C atoms (which is treated as a free parameter) and a cluster component made of 10 small molecules. The temperature distribution function P(T) dT describes the probability for finding a grain in the temperature interval T ... T + dT.

We define the grain number density distribution by means of a power-law (index β) in radius but assume that the density is constant with Θ. We explore models with $\beta = 0$ and 3/2. We also introduce Θ_1 as the opening angle of the *flared* disk. The thickness of a *flared* disk increases e.g. linearly with distance from the central source (Figure 1). The sensitivity of the results on Θ_1 is much smaller than the other parameters so we have decided to keep that constant at $30°$. All the models (unless otherwise stated) also assume that the central source of radiation is a blackbody at a temperature of $T_s = 25,000$K and that the ratio of inner to outer cloud radii r_1/r_2 is 0.001. The three main parameters we vary are the equatorial total A_V to the star, the sublimation temperature of the large grains T_1 and the number of carbon atoms in a PAH molecule, N_C.

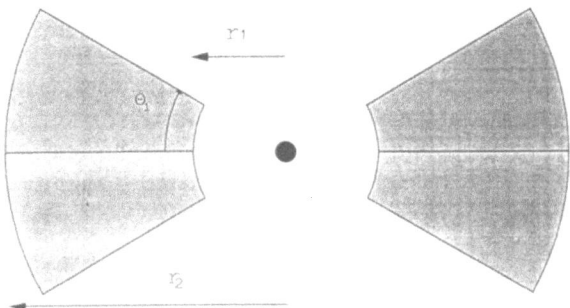

Fig. 1. Schematic of the assumed *flared* disk geometry.

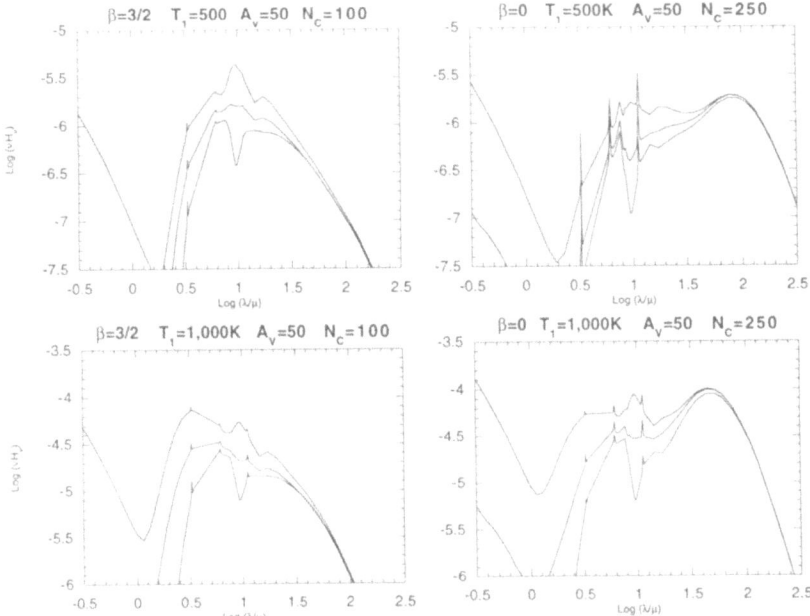

Fig. 2. Spectral energy distributions (SED) predicted for *flared* disk models that explore the basic parameters that affect the general shape of the continuum and prominence of the emission features. All models assume $\Theta_1 = 30°$ and $r_1/r_2 = 0.001$. Each panel shows the SED predicted for three different viewing angles θ_v of a model assuming the parameters shown on the top of the panel: $\theta_v = 0°$ or edge-on (bottom), $\theta_v = 28°$ (middle) and $\theta_v = 90°$ or face-on (top). All angles are measured from the equatorial plane.

3 Conclusion

We find that the strength of the features relative to the adjacent continuum decreases with increasing β (where β is the index of the power-law density distribution $n \propto r^{-\beta}$), dust sublimation temperature T_1, optical thickness of the disk in the equatorial direction, number of carbon atoms in the PAHs. A comparison of spherically symmetric and axi-symmetric models shows that the emission in the near and mid-infrared depends strongly on the viewing angle. The spherically symmetric models also generally show weaker PAH features (Figure 2). We present intensity maps at different wavelengths and inclinations. We show that the emission in the PAH bands is much more extended than the emission in the adjacent continuum (Figure 3).

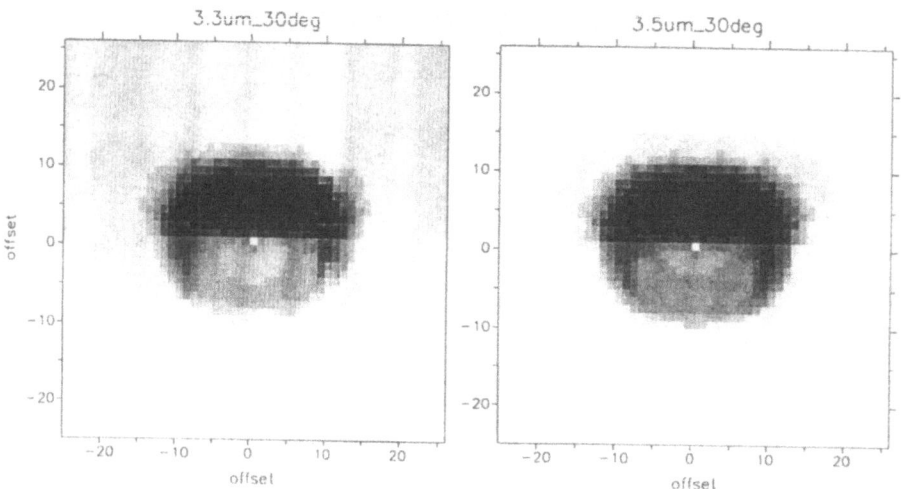

Fig. 3. Intensity distributions of the central region of a model with T_1 =1,000K, A_V =50, $\beta = 3/2$ for a range of wavelengths and viewing angles. Note how the intensity distribution at 3.3 and 11.3μm is invariably more extended than that at the neighbouring continuum. Each pixel is about one tenth of the inner radius of the disk which, of course, scales with the luminosity of the central source.

References

Beckwith, S.V.W., Sargent, A.I., Chini, R. and Güsten, R. (1990): AJ 99, 924
Buss, R.H. (Jr.), Cohen M., Tielens, A.G.G.M., et al. (1990): ApJ 365, L23
Brooke, T.Y., Tokunaga, A.T., Strom, S.E. (1993): AJ 106, 656
Chini, R., Krügel, E., Shustov, B., Tutukov, A. and Kreysa, E. (1991): AA 252, 220
Efstathiou, A., Rowan-Robinson, M. (1990): MNRAS 245, 275
Efstathiou, A. Rowan-Robinson, M. (1995): MNRAS 273, 649
Lada, C.J., Adams, F.C. (1992): ApJ 393, 278
Hanner, M.S., Tokunaga, A.T., Geballe, T.R. (1992): ApJ 395, L111
Siebenmorgen, R., & Krügel, E. (1992): AA 259, 614
Siebenmorgen, R., Krügel, E., & Mathis, J.S. (1992): AA 266, 501

Infrared Classification of Young Stellar Objects

Željko Ivezić and Moshe Elitzur

Department of Physics & Astronomy, University of Kentucky, Lexington, KY 40506-0055, USA

Abstract. The radiative transfer equation for a dusty envelope as close as possible to an embedded central source possesses scaling properties. For a given dust chemical composition, the solution depends only on overall optical depth and the functional form of the radial dust distribution. All other physical parameters (luminosity, overall density, etc.) do not affect the solution independently, only through their effect on the overall optical depth.

We model infrared emission for dust density distributions ranging from a stationary outflow ($1/r^2$) to a constant density, and dust grains composed of astronomical silicate and amorphous carbon. Preliminary results demonstrate that IRAS color–color diagrams can be parametrized in terms of scaling analysis. We find that the dust in envelopes around young stars resembles a $1/r$ distribution with the 100 μm optical depth ranging from 0.001 to 1. The evolution of these objects from the proposed class 0 to class III appears to reflect a decrease in overall optical depth.

1   Introduction

Infrared spectra of young stellar objects (YSO) form distinct groups. This observation initiated several proposals for detailed classification schemes. The most widespread classification involves α, the slope of the log-log plot of the spectral energy distribution between a near-infrared and an IRAS wavelength (Adams et al. 1987). YSOs are classified as protostars if $\alpha < 0$ (class I), pre-main-sequence stars for $0 \leq \alpha < 2$ (class II) and reddened main-sequence stars when $\alpha > 2$ (class III). Recently, a new class (class 0) was proposed for sources presumed to be even younger than protostars and surrounded by significantly larger amounts of circumstellar material (André et al. 1993). All other classification schemes were shown by Myers & Ladd (1993) to be mutually equivalent.

The separation of all Galactic infrared spectra into distinct classes can be seen directly in the IRAS color-color diagrams. The IRAS Point Source Catalogue lists 5687 objects with good quality fluxes in all four bands (12 μm, 25 μm, 60 μm and 100 μm). However, most listed fluxes are still contaminated by background cirrus emission at 100 μm, and to a lesser extent also at 60 μm (Ivezić & Elitzur, 1995; hereafter IE). The 821 uncontaminated sources are plotted in Figure 1. *These are the only Galactic point sources with reliable fluxes in all four wavelength bands.* The source distribution in the IRAS color-color diagrams displays a clear structure rather than random scatter. Note that IRAS colors for all black bodies with temperature $\gtrsim$ 2000 K are the same (Rayleigh-Jeans point). Distinct spectral classes are also reflected in various phenomenological associations of objects, delineated as boxes. It is obvious that infrared spectra of young and late-type stars are very different. Indeed, both associations and their differences are expected from general scaling properties of the radiative transfer problem.

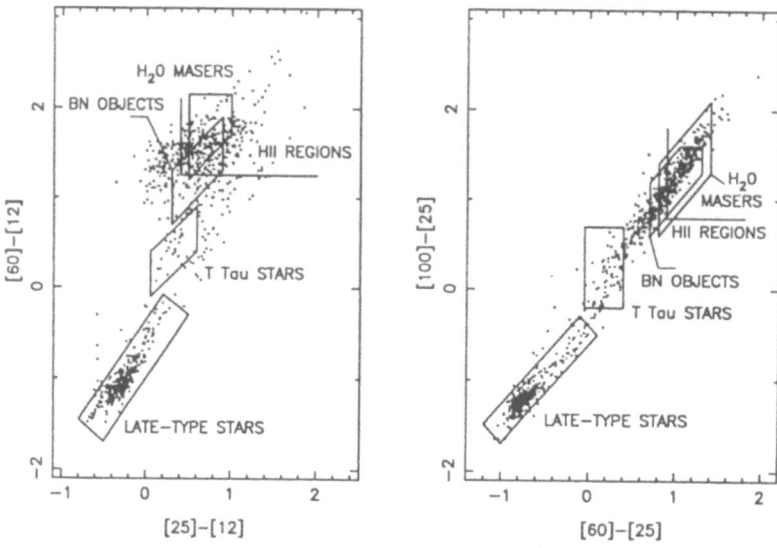

Fig. 1. Color-color diagrams of all IRAS sources, displayed as dots, with good quality fluxes at all four wavelength bands and uncontaminated by cirrus emission. Boxes delineate phenomenological associations: late-type stars (van der Veen & Habing, 1988), T Tau stars (Harris, Clegg & Hughes, 1989), Becklin-Neugebauer (BN) objects (Henning, Pfau & Altenhoff, 1990), HII regions (Wood & Churchwell, 1989) and H_2O masers detected in YSOs (Wouterloot & Walmsley, 1986). *1a* [60]-[12] vs. [25]-[12] diagram. *1b* [100]-[25] vs. [60]-[25] diagram.

2   Scaling Properties of the Radiative Transfer Problem

How do different spectral classes form and why are YSOs so different from late-type stars? We have shown (IE) that the radiative transfer problem in an isotropic medium surrounding a central source of radiation posses general scaling properties. When the dust is as close as possible to the central source, the solution of the radiative transfer equation is fully determined by two quantities: the normalized density profile $\eta = n(r)/\int n(r)dr$ and the overall optical depth τ at some fiducial wavelength. The radiative characteristics of the central source are largely irrelevant for the spectral properties of the emergent IR radiation. As a result, for a given dust chemical composition, each density profile η produces a family of solutions corresponding to a track in the color-color diagrams. All tracks start from the common origin, the Rayleigh-Jeans point, and position along the track is uniquely determined by τ. Other physical parameters (luminosity, overall density, size of the system etc.) do not affect the solution independently, only through their effect on the overall optical depth. Dust chemical composition does not vary significantly among infrared objects and overall optical depths span about the same range. Therefore, *young and late-type stars have such uniquely different infrared spectra because the functional forms of their dust density distributions are different.*

3   Results and Discussion

As a first approximation we have calculated emerging spectra for distributions described by power-laws[1] r^{-p}. The dust grains are composed of amorphous carbon and astronomical silicate with evaporation temperature of 800 K. Optical depths at 100 μm vary from 0 to 10. The central object is taken as a black body of 4000 K since the actual temperature is irrelevant at IRAS wavelengths as long as it is $\gtrsim$ 2000 K. The emerging spectra are convolved with the IRAS instrumental band profiles to produce the appropriate fluxes for the evaluation of IRAS colors. Preliminary color-color tracks are presented in figure 2 which shows that the model tracks properly delineate the region populated by IRAS sources. For a given dust composition, different families of objects can be associated with different p and τ. For example, late-type stars are only found in the region corresponding to $p \sim 2$ (we have verified this association from detailed modeling; IE). YSOs are distributed in the regions corresponding to $\frac{1}{2} \leq p \leq \frac{3}{2}$, in agreement with detailed modeling of some individual sources (e.g. Barsony & Chandler, 1993). Since position in a color-color diagram depends also on grain composition, the value of p for a particular source can not be uniquely determined without additional spectral data to provide information about the chemical composition (e.g. IRAS LRS spectra). However, most YSOs appear to congregate around the tracks for $p \sim 1$ irrespective of chemical composition. If correct, this result would provide a strong constraint on current theories of proto-stellar collapse. In free-fall one expects $p = \frac{3}{2}$ (Shu, Adams & Lizano, 1987), but this can change by effects of magnetic fields, outflows, rotation etc. For example, models for collapsing clouds with ambipolar diffusion produce $p \sim 0.9$ (Lizano & Shu, 1989).

The evolution of YSOs from class 0 to class III seems to correspond to an overall decrease in optical depth irrespective of the grains chemical composition or the precise value of p. Deeply embedded sources enter a diagram from the upper right corner and move toward the lower left as τ decreases, i.e. as their envelope mass decreases. It is noteworthy that IRAS 16293-2442, the best candidate so far for a protostellar collapse, is the reddest source in the diagrams. After all of the surrounding dust is cleared, a YSO finishes its journey through the infrared color-color diagrams as a naked star at the Rayleigh-Jeans point.

---

[1] Similar calculations were performed for BN objects by Gürtler et al. (1991). Here we consider all IRAS sources.

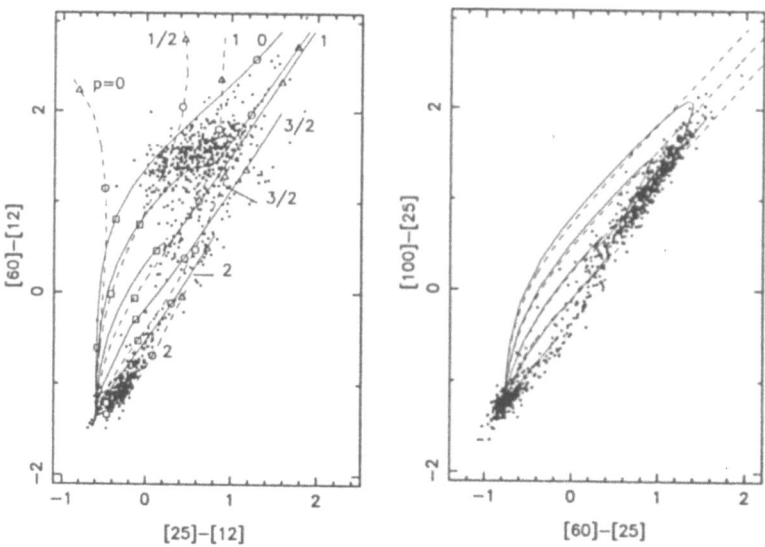

Fig. 2. IRAS color-color diagrams same as figure 1. Lines are preliminary model results for dust density distributions described by power-laws r^{-p} with $p = 0$, $\frac{1}{2}$, 1, $\frac{3}{2}$ and 2 (as marked) and grains composed of amorphous carbon (solid lines) and astronomical silicate (dashed lines). The common origin of all tracks is the Rayleigh-Jeans point. Distance from that common origin along each track increases with optical depth. Symbols indicate $\tau(100\mu m) = 0.001$ (squares), 0.1 (circles) and 0.3 (crosses).

References

Adams, F.C., Lada, C.J. & Shu, F.H. (1987): ApJ **312**, 788
André, P., Ward-Thompson, D. & Barsony, M. (1993): ApJ **406**, 122
Barsony, M. & Chandler, C.J. (1993): ApJ **406**, L71
Gürtler, J., Henning, Th., Krügel, E. & Chini, R. (1991): A&A **252**, 801
Harris, S., Clegg, P. & Hughes, J. (1989): MNRAS **235**, 441
Henning, Th., Pfau, W. & Altenhoff, W.J. (1990): A&A **227**, 542
Ivezić, Ž. & Elitzur, M. (1995): ApJ **445**, 415 (IE)
Lizano, S. & Shu, F.H. (1989): ApJ **342**, 834
Myers, P.C. & Ladd, E.F. (1993): ApJ **413**, L47
Shu, F.H., Adams, F.C. & Lizano, S. (1987): ARA&A **25**, 23
van der Veen, W.E.C.J. & Habing, H.J. (1988): A&A **194**, 125
Wood, D.O.S. & Churchwell, E. (1989): ApJ **340**, 265
Wouterloot, J.G.A. & Walmsley, C.M. (1986): A&A **168**, 237

2D Radiative Transfer Models of the Embedded YSOs HL Tau and L1551 IRS 5: What is Inside?

Alexander Men'shchikov[1,2] and Thomas Henning[1]

[1] Max Planck Society, Research Unit "Dust in Star-Forming Regions",
  Schillergäßchen 3, D-07745 Jena, Germany
[2] Astrophysical Institute Potsdam, Telegrafenberg A 27, D-14473 Potsdam, Germany

Abstract. We present 2-D radiative transfer models for two deeply embedded young stellar objects (YSOs), HL Tau and L1551 IRS 5. Both are surrounded by large non-spherical circumstellar envelopes (thousands AU in size) and drive bipolar outflows. Dense dusty environments of high optical depth hide the central sources from an observer. However, the optical thickness is much smaller in the polar (outflow) regions. In the latter, a high degree of linear polarization ($\sim$ 40%) is observed in the near infrared. Both objects are believed to be young stars surrounded by flat compact accretion disks (size $\sim$ 100 AU). The inferred geometry and parameters of the unseen sources, based mainly on indirect arguments and simple models which fit the observed fluxes, have to be verified by a detailed modeling. Attempting to recover structure and parameters of the hidden sources, we utilized our fast multi-component 2-D radiative transfer code [6] to explore a huge parameter space and to find self-consistent models which explain all available infrared continuum observations. Much simpler 1-D (spherically-symmetric) models are clearly inappropriate for this purpose: one must take into account anisotropic distributions of matter around the objects.

To emphasize a danger of using only spectral energy distributions (SEDs) for the derivation of source parameters, we present several *different* models of HL Tau and L1551 IRS 5 (Fig. 1). Our results demonstrate that the widely adopted method of fitting only SEDs is *absolutely insufficient* in reconstructing geometry and parameters of embedded objects. Depending on the *unknown* geometry, density structure, dust properties, optical depths, and viewing angles, derived luminosities of the sources can be in error *by a factor $\sim$ 30* or even more. The total envelope masses in the models of IRS 5 presented here differ *by the factor 54* (0.15, 0.47, and 8.1 $M_\odot$). There is an *intrinsic ambiguity* of the fitting of only a featureless continuum (cf. [7]) which makes this standard method for the derivation of source parameters useless or at least implies *huge error bars* in derived parameters. The only way to derive reliable parameters of embedded objects is to use *all of the spatial information* coded in observations *and to fit as many different data sets as possible*, in the frame of a self-consistent model. We emphasize that the photometry made with different beams is a readily available (but often ignored) source of spatial information which can help to test model predictions and constrain source parameters, and which is especially important for a large number of objects with no high-resolution observations.

We present a first detailed radiative transfer model of L1551 IRS 5 which explains very well most of the available infrared and submm/mm continuum observations (Figs. 2 and 3). The model perfectly fits the rich broad-band photometry in the whole range from visual to millimeter wavelengths. Intensity maps are in a very good agreement with available linear scans and maps at $50\,\mu m$ [1], $100\,\mu m$ [1], $1.25\,mm$ [2], and $1.3\,mm$ [8]. Model visibilities fit perfectly the interferometer measure-

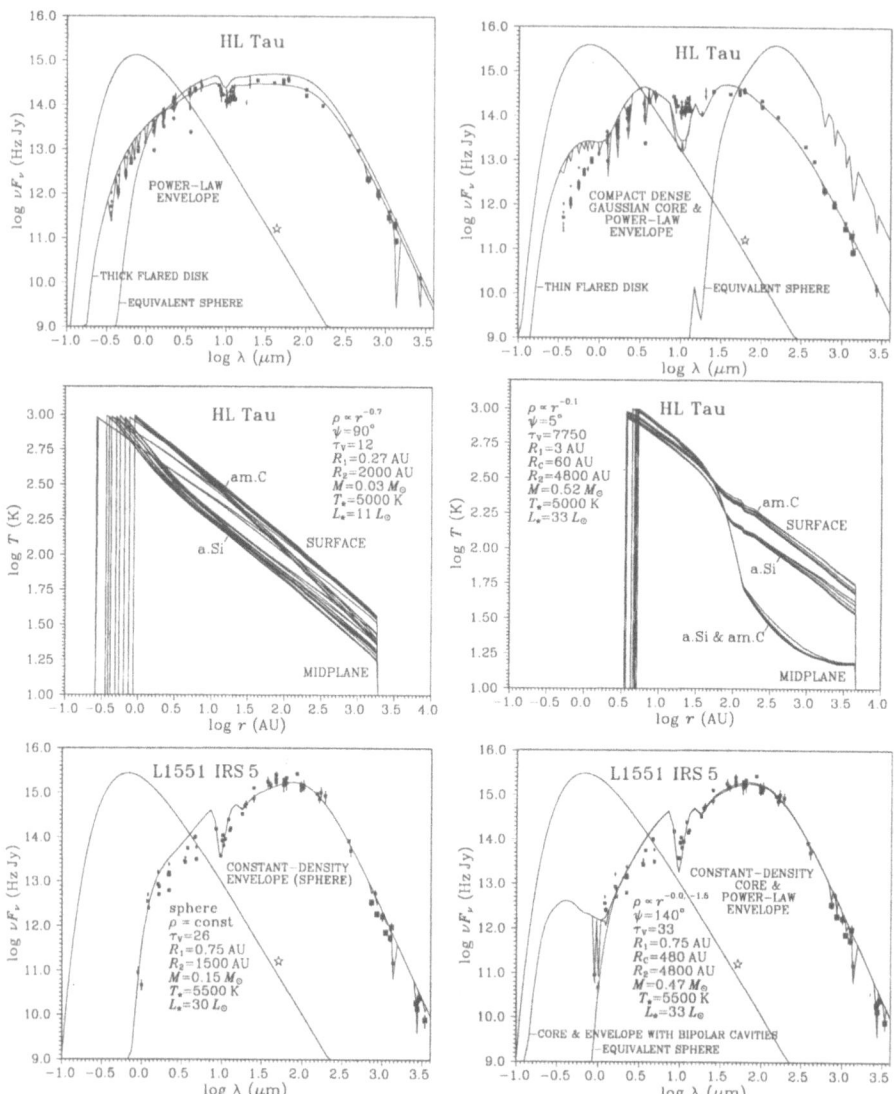

Fig. 1. (*upper left*) Thick flared disk model (opening angle $\psi = 90°$; $\rho \propto r^{-0.7}$) which fits well the photometry of HL Tau. $L = 11\ L_\odot$, $M = 0.03\ M_\odot$, midplane optical depth $\tau_V = 12$. Sharp "features" of the SED (except for $10\ \mu m$ feature) show effect of different beam sizes on the total model flux. Temperature profiles (*middle left*) are shown for each of the 8 grain radii ($0.005\ \mu m$ to $1\ \mu m$, $n(a) \propto a^{-4.2}$), for both astronomical silicate and amorphous carbon. (*upper right*) Thin flared disk model ($\psi = 5°$; dense gaussian core of $R_c = 50$ AU, $\rho \propto r^{-0.1}$ in the outer parts) which fits most of the fluxes of HL Tau. $L = 33\ L_\odot$, $M = 0.52\ M_\odot$, $\tau_V = 7750$. Dust temperatures (*middle right*), depend on the grain properties *and* location. (*lower left*) The simplest model (sphere of $R_2 = 1500$ AU, $\rho = $ const) explaining photometry of L1551 IRS 5. $M = 0.15\ M_\odot$, $\tau_V = 26$. (*lower right*) A more complex model (core of $R_c = 480$ AU, $\rho = $ const in the envelope of $R_2 = 4800$ AU, $\rho \propto r^{-1.5}$; $\psi = 140°$). $M = 0.47\ M_\odot$, $\tau_V = 33$.

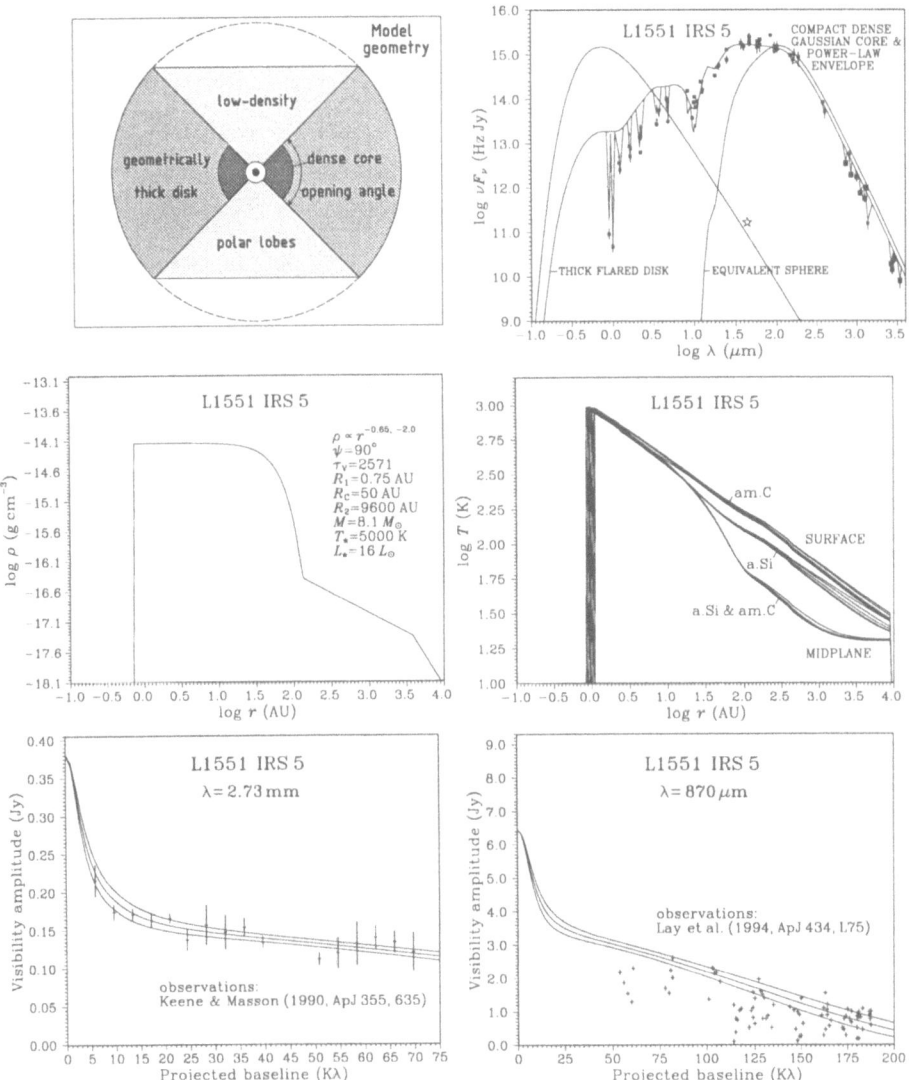

Fig. 2. Geometry, SED, density profile, dust temperatures, and visibilities, for our model explaining the whole data set for L1551 IRS 5. The thick flared disk has the opening angle $\psi = 90°$. A compact dense gaussian core (radius $R_c = 50$ AU) is inside the envelope ($R_2 = 9600$ AU) with two power-law density distributions ($\rho \propto r^{-0.65}, \propto r^{-2}$). The core with $n_{H_2} \approx 2 \cdot 10^9$ cm^{-3} contains only a small fraction (~ 0.1 %) of the total mass $M = 8.1$ $M_\odot$. Due to the non-spherical geometry, the stellar luminosity in our model is 16 $L_\odot$, *one half* of the commonly adopted value ~ 30 $L_\odot$. The temperature profiles are shown for each of the 8 grain radii ($0.005\,\mu m$ to $1\,\mu m$, $n(a) \propto a^{-4.2}$), for both astronomical silicate and amorphous carbon. Due to a high optical depth (midplane $\tau_V = 2570$), grains acquire similar temperatures in the midplane, independently of their composition or size. The far-infrared opacity $\kappa \propto \lambda^{-1}$ is essential for the good fit of the infrared continuum together with *both* visibilities at $870\,\mu m$ and 2.73 mm.

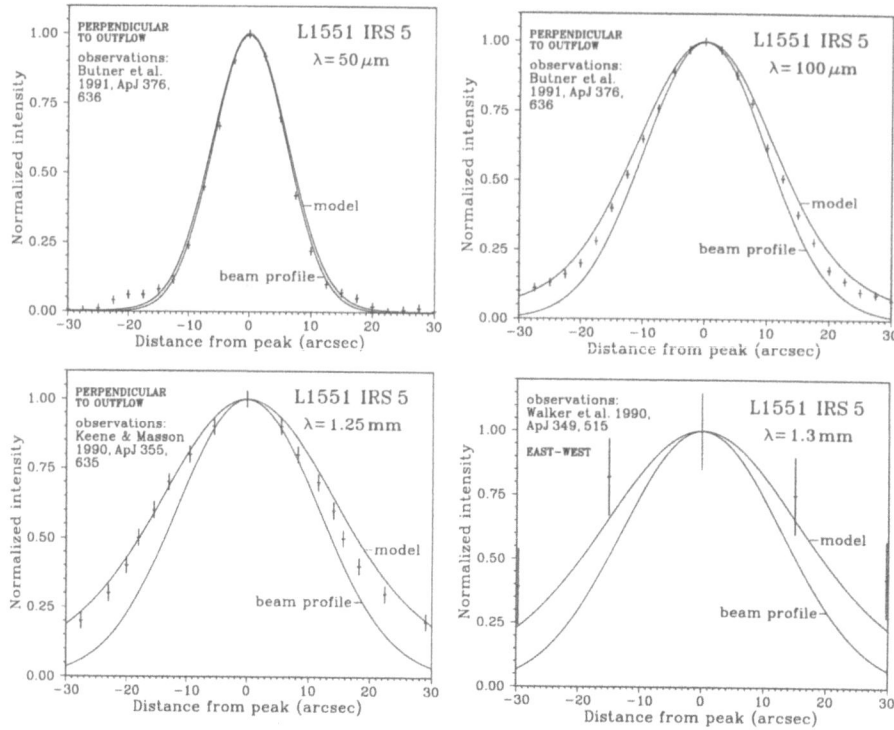

Fig. 3. Model intensity maps at $50\,\mu m$, $100\,\mu m$, $1.25\,mm$, and $1.3\,mm$ compared with observations of L1551 IRS 5. Intensity distributions in the direction perpendicular to the outflow are shown, together with the beam profiles used in the observations.

ments at submm/mm wavelengths ($870\,\mu m$ [4]; $2.73\,mm$ [2]) and confirm the presence of the compact, very dense core (radius $\approx$ 50 AU, $n_{H_2} \approx 2 \cdot 10^9\,cm^{-3}$) at the center of IRS 5. Model polarization map at $1\,\mu m$ (not shown) predicts both the polarization degree and pattern in agreement with the observed ones [5]. The thick flared disk model of IRS 5 with the opening angle $90°$ between the upper and lower conical surfaces can naturally account for the cross-shaped pattern observed at $730\,\mu m$ [3]. While our model of L1551 IRS 5 agrees well with all the observations, it implies a massive envelope ($8.1\,M_\odot$) and a low luminosity ($16\,L_\odot$), in contrast with previous models.

References

[1] Butner, H. M., Evans, II N. J., Lester, D. F., et al. (1991): ApJ **376**, 636
[2] Keene, J., Masson, C. R. (1990): ApJ **355**, 635
[3] Ladd, E. F., Fuller, G. A., Padman, R., et al. (1995): ApJ **439**, 771
[4] Lay, O. P., Carlstrom, J. E., Hills, R. E., Phillips, T. G. (1994): ApJ **434**, L75
[5] Lenzen, R. (1987): A&A **173**, 124
[6] Men'shchikov, A. B., Henning, Th. (1995): A&A, submitted
[7] Thamm, E., Steinacker, J., Henning, Th. (1994): A&A **287**, 493
[8] Walker, C. K., Adams, F. C., Lada, C. J. (1990): ApJ **349**, 515

3D Continuum Radiative Transfer

Jürgen Steinacker and Thomas Henning

MPG Research Unit *Dust in Star–Forming Regions*,
Schillergäßchen 2–3, 07745 Jena, Germany

Abstract. We present first results of a new 3D continuum radiative transfer code to determine the appearance of complicated dust configurations in spatially resolving images and spectral energy distributions. Contrary to former treatments, the full frequency-dependent problem is solved without any flux approximation using the linear equation solver BiCGSTABℓ. Given any dust density distribution, derived e.g. from a gas distribution produced by a hydrodynamical code, the program calculates self-consistently the radiation field and the temperature distribution of the dust particles. We discuss the results of simple test cases and the ambiguity problem with applying the 3D code to complex dust configurations.

1 Introduction

Many observations indicate that the distribution of accreting matter around young stellar objects deviates significantly from spherical symmetry (for a review see Sargent (1995)). In view of the high percentage of binary systems among the YSOs and the various internal or external forces that can distort an accretion disk, the surrounding distribution of gas and dust might not even show rotational symmetry for a lot of sources.

With the recent developments of observational techniques, especially in mm-wavelength interferometry, the spatial resolution exceeds a limit, where it is possible to obtain radial profiles (Ladd et al. (1995)), making the consideration of 3D effects necessary and crucial for the interpretation of the obtained data.

Continuum radiative transfer in axially symmetric dust configurations has been discussed in a number of recent papers. Efstathiou & Rowan-Robinson (1990) presented results from a raytracer for flared disks and discussed the dependence of the emergent spectrum on model parameters in Efstathiou & Rowan-Robinson (1991). Two different density distributions around a single star were investigated under the assumption of isotropic scattering by Collison & Fix (1991) using an iterative scheme. Sonnhalter et al. (1995) applied a flux–limited approximation to derive spectra, temperatures, and intensities of several combinations of disk–halo distributions. An approximate 2D method was proposed by Men'shchikov & Henning (1995), who caluclated the radiative transfer under the assumption that the density distribution throughout the disk depends only on the radial distance from the central star.

There are only a few papers about 3D continuum radiative transfer. Yorke (1986) solved the frequency–dependent problem, but used a flux–limited approximation. Stenholm et al. (1991) treated the frequency–averaged problem without any further approximation. In both papers, iterative methods are applied to handle the large number of unknown intensities arising from the fact that stationary 3D continuum radiative transfer incorporates 3 varibales in space, 2 in direction, and the frequency variable.

In this paper, we present a new efficient method to solve the 3D radiative transfer problem by applying a direct linear equation solver to the discretized transfer equation. The self-consistent temperature of the dust is obtained by applying the Newton method for nonlinear equations to the local energy density balance equation.

2   Method of Solution

To determine the total specific intensity $I_\nu^{tot}(\boldsymbol{x}, \boldsymbol{n})$ of the unpolarized radiation field in the dust configuration at the point $\boldsymbol{x}$ in the direction $\boldsymbol{n}$, we have to solve the stationary transfer equation

$$
\begin{aligned}
n\nabla_{\boldsymbol{x}}I_\nu^{tot}(\boldsymbol{x}, \boldsymbol{n}) = & -\sigma_\nu^{abs}(\boldsymbol{x})\ I_\nu^{tot}(\boldsymbol{x}, \boldsymbol{n}) - \sigma_\nu^{sca}(\boldsymbol{x})\ I_\nu^{tot}(\boldsymbol{x}, \boldsymbol{n}) \\
& +\frac{1}{4\pi}\sigma_\nu^{sca}(\boldsymbol{x})\int_\Omega d\Omega' p_\nu(\boldsymbol{n}, \boldsymbol{n}')\ I_\nu^{tot}(\boldsymbol{x}, \boldsymbol{n}') \\
& +\sigma_\nu^{abs}(\boldsymbol{x})\ B_\nu[T(\boldsymbol{x})]\ ,
\end{aligned}
\tag{1}
$$

where $\sigma_\nu^{abs}(\boldsymbol{x})$ and $\sigma_\nu^{sca}(\boldsymbol{x})$ are the absorption and scattering coefficients of the dust, respectively. $p_\nu(\boldsymbol{n}, \boldsymbol{n}')$ denotes the probability that radiation is scattered from the direction $\boldsymbol{n}'$ into $\boldsymbol{n}$, Ω is the solid angle, and B_ν is the Planck function. Altough it is difficult to solve this 6D partial integro–differential equation even for a given dust temperature $T(\boldsymbol{x})$, the nonlinear coupling between the radiation field and the dust temperature requires the simultanous consideration of the balance equation for the local energy density

$$
\int_0^\infty d\nu\ Q_\nu^{abs} B_\nu[T(\boldsymbol{x})] = \int_0^\infty d\nu\ Q_\nu^{abs}\int_\Omega d\Omega'\ I_\nu^{tot}(\boldsymbol{x}, \boldsymbol{n}')
\tag{2}
$$

to calculate intensity and temperature self–consistently. Here, Q_ν^{abs} is the absoption efficiency factor.

In order to allow for more complex inner boundary conditions like multiple stars in the dust configuration, we follow Efstathiou & Rowan-Robinson (1990) in splitting the specific intensity into an unprocessed passing stellar component and a processed component

$$
I_\nu^{tot} = I_\nu^* + I_\nu
\tag{3}
$$

leading to three equations

$$
n\nabla_{\boldsymbol{x}}I_\nu^*(\boldsymbol{x}, \boldsymbol{n}) = -\sigma_\nu^{ext}(\boldsymbol{x})\ I_\nu^*(\boldsymbol{x}, \boldsymbol{n})
\tag{4}
$$

$$n\nabla_x I_\nu(x,n) = -\sigma_\nu^{ext}(x)\ I_\nu(x,n) + \sigma_\nu^{abs}(x)\ B_\nu[T(x)]$$

$$+\frac{1}{4\pi}\sigma_\nu^{sca}(x)\int_\Omega d\Omega'\ p_\nu(n,n')\ I_\nu(x,n') + C_\nu^*(x) \tag{5}$$

$$\int_0^\infty d\nu\ Q_\nu^{abs}(x)\ B_\nu[T(x)] = \int_0^\infty d\nu\ Q_\nu^{abs}(x)\int_\Omega d\Omega'\ I_\nu(x,n') + D^*(x) \tag{6}$$

with the known contributions

$$C_\nu^*(x) = \frac{1}{4\pi}\int_\Omega d\Omega'\ p_\nu(n,n')\ I_\nu^*(x,n') \tag{7}$$

$$D^*(x) = \int_0^\infty d\nu\ Q_\nu^{abs}(x)\int_\Omega d\Omega'\ I_\nu^*(x,n')\ . \tag{8}$$

The first transfer equation (4) can be transformed to a path integral and thus easily be precalculated for all frequencies e.g. using fifth-order Runge-Kutta with adaptive stepsize control (Press et al. (1992)). To solve the second transfer equation (5), we discretize the spatial coordinates by down-winding finite differencing in a cartesian coordinate system x_j, y_k, z_l, where the indices j, k, l run from 1 to the number of grid points N_x, N_y, and N_z, respectively. Using the trapezoidal rule, the integral operator is discretized in spherical coordinates ϑ^g, φ^h, with the number of grid points N_ϑ and N_φ, while we introduce a logarithmic grid ν^f for the frequency.

With the abbreviation $I_{j,k,l}^{f,g,h} = I(x_j, y_k, z_l, \vartheta^g, \varphi^h, \nu^f)$ we end up with a large linear equation system

$$-\alpha_1^{g,h} I_{j-1,k,l}^{f,g,h} - \alpha_2^{g,h} I_{j,k-1,l}^{f,g,h} - \alpha_3^g I_{j,k,l-1}^{f,g,h}$$

$$+\eta_{j,k,l}^{f,g,h} I_{j,k,l}^{f,g,h} - \frac{1}{4\pi}\sigma_{j,k,l}^{sca;f}\sum_{m,n=1}^{N_\vartheta,N_\varphi} w^{m,n} p^{g,h,m,n} I_{j,k,l}^{f,m,n}$$

$$= \sigma_{j,k,l}^{abs;f}\ B^f[T_{j,k,l}] + C_{j,k,l}^{*;f} \tag{9}$$

where ω^m are the trapezoidal integration weights, $w^{m,n} = \sin\vartheta^m\ \omega^m\omega^n$, and where we have introduced the known constants

$$\alpha_1^{g,h} = \frac{\sin\vartheta^g\cos\varphi^h}{\Delta x}\quad \alpha_2^{g,h} = \frac{\sin\vartheta^g\sin\varphi^h}{\Delta y}\quad \alpha_3^g = \frac{\cos\vartheta^g}{\Delta z} \tag{10}$$

$$\eta_{j,k,l}^{f,g,h} \equiv \sigma_{j,k,l}^{ext;f} + \alpha_1^{g,h} + \alpha_2^{g,h} + \alpha_3^g\ . \tag{11}$$

Appropriate renumbering of the indices yields, for each frequency, to a sparse system of the form $AI = b$, where A is a lower block-triagonal matrix of size $N \times N$, with $N = N_x N_y N_z N_\vartheta N_\varphi$, I is the unknown intensity vector, and b contains the boundary values. The block system can be solved by simple back-substitution. Each dense submatrix is solved by the fast and stable BiCGSTABℓ

algorithm for non-symmetric matrices (Sleijpen & van der Vorst (1995)). Altough the total number of unknown intensities is huge for resonable resolutions, the numerical solution requires only storage of the order of $4N_\vartheta N_\varphi$ due to the used back-substitution scheme.

In order to obtain a self-consistent temperature field, we use the resulting intensity to calculate a temperature correction by applying the Newton method for nonlinear systems to the discretized form of (6) (Efstathiou & Rowan-Robinson (1991)). After i iterations, the correction is

$$
\Delta T^i_{j,k,l} \approx \left[\sum_{f=1}^{N_\nu} \omega^f Q^{abs;f} \left(\sum_{m,n}^{N_\vartheta, N_\varphi} \omega^{m,n} \, I^{i;f,m,n}_{j,k,l} - B^f[T^i_{j,k,l}] \right) + D^*_{j,k,l} \right]
$$
$$
\times \left[\sum_{f=1}^{N_\nu} \omega^f Q^{abs;f} \left. \frac{dB}{dT} \right|_{T^i_{j,k,l}} \right]^{-1} . \tag{12}
$$

To control the numerical error, we use the integral form of the flux conservation on a closed surface S

$$
\int_S d\boldsymbol{S}\,\boldsymbol{H} = 0 \tag{13}
$$

with the first moment of the radiation field $\boldsymbol{H}$ (Efstathiou & Rowan-Robinson (1990)), where its absolute value has in our notation the discretized form

$$
H_{j,k,l} = \sum_{f=1}^{N_\nu} \omega^f \frac{1}{2\pi} \sum_{m,n=1}^{N_\vartheta, N_\varphi} \omega^{m,n} I^{tot;f,m,n}_{j,k,l} \cos \vartheta^m \sin \vartheta^m . \tag{14}
$$

We are still in the process of implementing a full multigrid cycle with finer spatial grids which will yield to a much finer resolution without enhancing the runtime of the program considerably, contrary to common nested grid techniques.

3   Simple Test Cases

An extensive comparison with the published results of 2D/3D codes will be presented in a forthcoming paper (Steinacker (1995)). Here, we will just illustrate the ability of the code to handle 3D structures by a simple test case, which is a flat accretion disk with a close binary system in the center.

Figure 1 shows the image of the inner part of the circumbinary disk at $\lambda = 2\mu$m. To keep the results clear, we avoided to apply the code to a complicated structure or a sophisticated dust model. Hence, we assumed a simple power law for the dust density with a sublimation range close to the stars, were the radiation temperature was of the order of the sublimation temperature, and monosized dust with powerlaw opacity. As can be seen from the image, the program is able to resolve the inner edge of the disk and the binary, but the resolution is still limited without applying a multigrid technique to the interesting regions.

Fig. 1. Inner part of circumbinary disk and the binary as a simple 3D test case at $\lambda = 2\mu$m with an inclination to the plane of the disk of $20°$

In Figure 2, we have plotted the spectrum of the object using a logarithmic frequency grid of 16 grid points. Since the luminosity of the second star was 1/3 of the first star and the surface temperatures of the stars have been chosen comparable, the binary is not resolved in the spectrum. The source shows the typical flat spectrum in the frequency range, where the cold dust reemits the absorbed stellar radiation.

It should be emphasized that neither the inclusion of a more realistic dust model nor the consideration of more sophisticated density structures derived e.g. from (magneto) hydrodynamical calculations, but with the same resolution, will slow down the code.

4   Ambiguity in Multiparameter Applications

To deal with 3D dust distributions means to enter a wide range of free parameters, additionally to the parameters of the dust model and the properties of the stars inside the considered dust configuration. Trying to find constraints just by fitting a model e.g. with 6 free parameters to the spectral energy distribution is hard to achieve due to a large ambiguity in acceptable fits as it was shown by Stenholm et al. (1994).

The simultaneous consideration of spatial resolved images of the source at several wavelengths, however, will vastly reduce this ambiguity. But contrary to the case of a simple model for the accretion disk, no automated fitting procedure

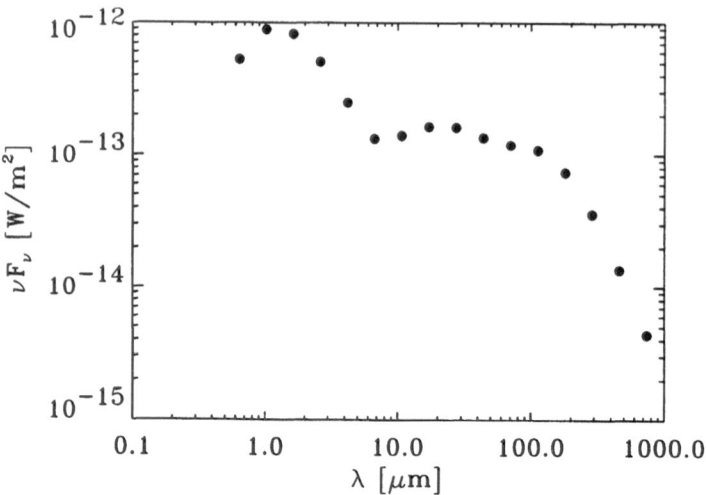

Fig. 2. Spectrum for the circumbinary disk.

like the Metropolis algorithm proposed in Stenholm et al. (1994) can be used to determine constraints for the free parameters.

On the other hand, the code provides a powerfull tool to test existing radiative transfer codes and to calculate the spectral and the spatially resolved appearance of complex dust configurations. Hence, it will be possible, for the first time, to confirm or rule out proposed more realistic distributions of the dust density using our code.

References

Collison, A.F., Fix, J.D. (1991): ApJ 368, 545
Efstathiou, A., Rowan-Robinson, M. (1990): MNRAS 245, 275
Efstathiou, A., Rowan-Robinson, M. (1991): MNRAS 252, 528
Ladd, E.F., Fuller, G.A., Padman, R., Myers, P.C., Adams, F.C. (1995) ApJ 439, 771
Men'shchikov, A.B., Henning, Th. (1995): A&A, submitted
Press, W.H., Teukolsky, S.A., Vetterling, W.T., Flannery, B.P. (1992): *Numerical Recipes* (Cambridge University Press, Cambridge)
Sonnhalter, C., Preibisch, Th., Yorke, H.W. (1995): A&A 299, 545
Sargent, A.I. (1995): in: *Disks and Outflows around Young Stars*, S.V.W. Beckwith, A. Natta, J. Staude (Eds.) (Springer, Berlin, Heidelberg), in press
Sleijpen, G.L.G., van der Vorst, H.A. (1995): Appl. Numer. Math., submitted
Steinacker, J. (1995): ApJ, in prep.
Stenholm, L.G., Störzer, H., Wehrse, R. (1991): JQSRT 45, 1, 47
Thamm, E., Steinacker, J., Henning, Th. (1994): A&A 287, 493
Yorke, H., W. (1986): in: *Astrophysical Radiation Hydrodynamics*, K.-H. Winkler, M.L. Norman (Eds.) (Reidel, Dordrecht), 141

Part VIII

Dust as a Catalytic Agent for Star Formation

Evolution of the Molecular Abundance in Protoplanetary Disks: Depletion of CO Molecules

Yuri Aikawa[1], Shoken M. Miyama[1], Takenori Nakano[2], and Toyoharu Umebayashi[3]

[1] National Astronomical Observatory, Mitaka, Tokyo 181, Japan
[2] Nobeyama Radio Observatory, National Astronomical Observatory ,
   Nobeyama, Minamisaku, Nagano 384-13, Japan
[3] Data Processing Center, Yamagata University, Yamagata 990, Japan

Abstract. We investigate the evolution of the molecular abundance in protoplanetary disks paying attention to the abundance of CO molecules in the gas phase. We take into account the adsorption of molecules onto grains as well as the reactions in the gas phase. We follow the molecular evolution for some model disks solving numerically the reaction equations. There is a critical distance from the star, R_{crit}, at which the temperature is equal to the critical temperature $\approx$ 20 K for the adsorption of CO molecules. At $R > R_{crit}$, CO molecules are depleted rather rapidly from the gas phase mainly due to the adsorption. For the minimum-mass solar nebula extended to the region of radius $R \approx 800$ AU, for example, molecules in the gas phase at $R > R_{crit} \approx$ 200 AU are depleted by a few orders of magnitude in 10^5 to 10^6 yr, while at $R < R_{crit}$ the depletion of CO is not significant in these time scales. This is consistent with the recent observations of the gaseous disks around some T Tauri stars.

1 Introduction

Since the exploration by the IRAS satellite, many protoplanetary disks have been observed with dust continuum. Recently, emission lines of gas-phase molecules are also observed with large radio telescopes. It would help us to understand the structure and physical and chemical evolution of the protoplanetary disks. However, it should be reminded that we can not observe the most dominant component, hydrogen molecules. We observe instead molecules such as CO, CS and their isotopes. So it is important to investigate the molecular abundance in protoplanetary disks theoretically.

Up to now, emission lines of CO molecules have been detected for GG Tau (age $\sim 3 \times 10^5$ yr) and DM Tau ($\sim 1 \times 10^6$ yr). The line profiles are double-peaked, strongly suggesting the existence of rotating circumstellar disks of several hundred AU radius. However, it is found that the amount of gas-phase CO molecules in these disks are one or two orders of magnitude smaller than that estimated from the observed intensity of dust continuum and the assumed CO/H_2 ratio ($\sim 10^{-4}$) (Dutrey, Guilloteau, & Simon 1994, Guilloteau & Dutrey 1995, Handa et al. 1995, Saito et al. 1995). Motivated by these observations, we investigate the abundance of gas-phase CO molecules in protoplanetary disks theoretically.

2   The Disk Model and the Reaction Network

As a disk model, we adopt the minimum-mass solar nebula (Hayashi 1981). In the cylindrical coordinates (R, Φ, Z) with the midplane of the disk at the $Z=0$ plane, the distribution of the temperature and the density are given by

$$T(R) = 28 \ (R/100AU)^{-1/2} \ K \tag{1}$$

$$n_H(R, Z) = 2.0 \times 10^9 \ (R/100AU)^{-11/4} \exp\left(-\frac{GM_*\mu m_H}{RkT} + \frac{GM_*\mu m_H}{kT\sqrt{R^2 + Z^2}}\right) \ cm^{-3}, \tag{2}$$

where G is the gravitational constant, M_* is the mass of the central star, μ is the mean molecular weight of the gas, m_H is the mass of a hydrogen atom, and k is the Boltzmann constant. We take $M_* = 1M_\odot$.

We solve the reaction equations for many kinds of species simultaneously, which include gas-phase reactions, adsorption onto grains, and thermal desorption from grains, and can be written as

$$\frac{dn(i)}{dt} = \sum_{j,k} \alpha_{ijk} n(j) n(k) + \sum_j \beta_{ij} n(j) \tag{3}$$

$$- \pi a^2 n(grain) n(i) S \left(\frac{8kT}{\pi m(i)}\right)^{1/2} + n(i \ desorbable) \nu_{osc}(i) \exp\left(-\frac{E_{ads}(i)}{kT}\right),$$

where $n(i)$ is the number density of species i in the gas phase, t is the time, α and β are the reaction rate coefficient, a is the grain radius (10^{-5} cm), $n(grain)$ is the number density of grains, $m(i)$ is the mass of the particle i, and S is the sticking probability of a particle when it collides with a grain, $E_{ads}(i)$ and $\nu_{osc}(i)$ is the binding energy and the oscillation frequency of the adsorbed particle i at the grain surface. We adopt the network of gas-phase reactions described in Prasad & Huntress (1980), and take $E_{ads}(CO) = 960K$ (Sandford & Allamandola 1990). We determine the initial abundance referring to the observed molecular abundance in the TMC1 cloud.

3   Numerical Results

3.1   Evolution of the CO abundance under fixed physical conditions

Fig. 1a shows the time variation of the abundance of CO in the gas phase for some values of the temperature. In the case of $T < 20K$, the adsorption becomes efficient and the abundance decreases exponentially at $\sim 10^4$ yr. The critical temperature for adsorption T_{crit} is determined by the balance between the adsorption term and the desorption term in (3), and is about 20K for the parameters we adopt. Fig. 1b and c show the results for some values of the density n_H and the sticking probability S. We can see that the time scale for adsorption τ_{ads} is inversely proportional to these parameters. This dependence can also be obtained by analyzing (3). If we neglect all the terms other than the adsorption term on the right-hand side, we find that $n(i)$ decreases exponentially in a time scale

$$\tau_{ads} = \frac{1}{\pi a^2 n(grain)S}\left(\frac{\pi m(CO)}{8kT}\right)^{1/2} = 4\times10^3\left(\frac{10^7 cm^{-3}}{n_H}\right)\left(\frac{0.3}{S}\right)\left(\frac{15K}{T}\right)^{1/2} yr.$$
(4)

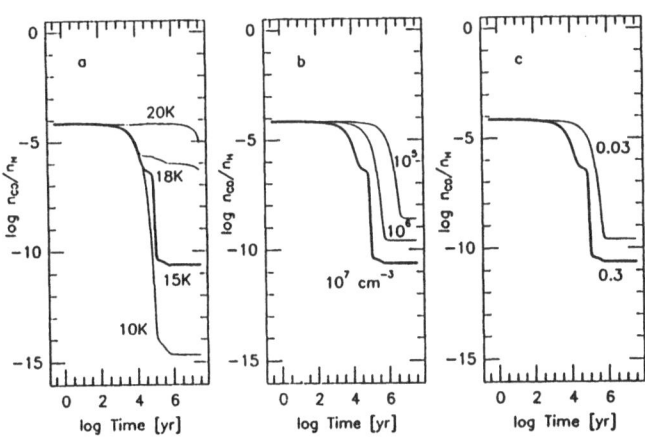

Fig. 1. Evolution of the abundance of CO in the gas phase in various situations. The thick line is for the case of $T = 15K$, $n_H = 10^7$ cm^{-3}, and $S = 0.3$.

3.2 Distribution of the CO abundance in protoplanetary disks

We now investigate the distribution of gas-phase CO molecules in a protoplanetary disk. By solving numerically the reaction equations (3) we follow the time variation of the abundances of various molecules at each point (R, Z) of the disk. For simplicity we neglect the mixing of matter in the disk. Fig. 2 show the distribution of the abundance of gas-phase CO molecules on the (R, Z) plane of the disk at 10^5 yr and 10^6 yr. The abundance is very low near the midplane because the time scale for adsorption, τ_{ads}, is much shorter than 10^5 yr owing to the high density. We can also see that the region of low abundance spreads as the time goes on. In the region of $R \sim 500$ AU, the CO abundance is significantly higher than in the vicinity. At the temperature of this region and at these epochs, the production of CO by the gas-phase reaction $C + O_2 \rightarrow CO + O$ counterbalances the adsorption temporarily. Fig. 3 shows the distribution of the column density of the gas-phase CO molecules. The dashed line shows the column density at the initial state with $n(CO)/n_H \sim 10^{-4}$. A notable feature is that the column density has a very steep slope at $R \sim 200$ AU. This is because the temperature in this region is close to the critical temperature for adsorption, $T_{crit} \sim 20$ K. We can also see that the column density of CO in the gas phase decreases with time. Comparing the thick line and the dashed lines in Fig. 3, we can see that

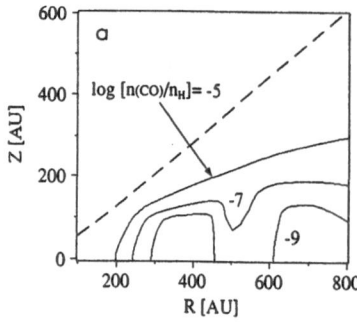 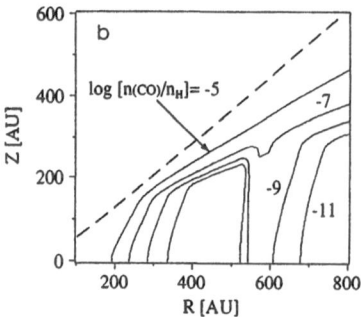

Fig. 2. The distribution of the abundance of CO molecules in the gas phase at (a)10^5 yr and (b)10^6 yr. The solid lines are contours of constant CO abundance. The dashed line represents the surface of the disk, which is taken at $n_H = 10^4$ cm^{-3}.

Fig. 3. The distribution of the column density of CO molecules in the gas phase at 10^5 yr and 10^6 yr.

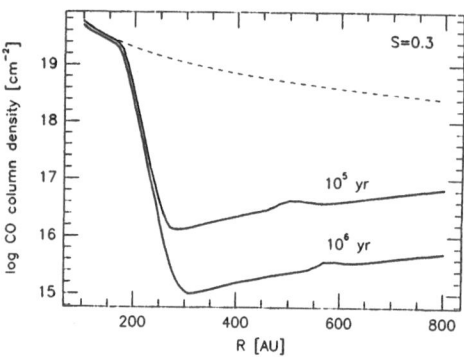

the CO molecules are depleted from the gas-phase by a few orders of magnitude at $R > 200$ AU. This is consistent with the observation of GG Tau and DM Tau. In addition there is an implication to the observations from our results. An observation with higher resolution would find a steep decrease of column density as shown in Fig. 3. Since the temperature in the region of this decrease is $T_{crit} \sim 20$K, it would tell us the distribution of temperature in the disk.

References

Dutrey, A., Guilloteau, S., & Simon, M. (1994): A&A **286**,149

Guilloteau, S. & Dutrey, A. (1995): A& A, in press

Handa , T., Miyama, S. M., Yamashita, T., Omodaka, T., Kitamura, Y., Hayashi, M., Onishi, T., Snell, R. L., Strom, S., Strom, K., Skrutskie, M. F., Edwards, S., Ohashi, N., Sunada, K., Saito, M., Fukui, Y., Mizuno, A., Watanabe, J., & Kataza, H. (1995): ApJ, in press

Hayashi, C. (1981): Prog. Theor. Physics. Suppl., **70**, 35

Prasad, S. S., & Huntress, W. T. Jr. (1980): ApJS, **43**, 1

Satio, M., Kawabe, R., Ishiguru, M., Miyama, S. M., Hayashi, M., Handa, T., Kitamura, Y., & Omodaka, T. (1995): ApJ, in press

Sandford, S. A. & Allamandola, L. J. (1990): Icarus, **87**, 188

Magnetic Fields, Interstellar Dust, UV Radiation, and Star Formation

Glenn E. Ciolek

Department of Astronomy and Astrophysics, University of Chicago,
5640 S. Ellis Ave., Chicago, IL 60637, U.S.A.

Abstract. We present calculations which demonstrate the formation (due to ambipolar diffusion) of collapsing, magnetically and thermally supercritical protostellar cores in otherwise magnetically supported model molecular clouds, accounting for the effects of interstellar grains (charged and neutral) and UV radiation. Charged grains couple to magnetic field lines either by direct attachment or by electrostatic attraction to electron-shielded ions ("quasiparticles"), which are themselves attached to the field. How grain-neutral drag increases the core formation timescale is discussed. We also show how ambipolar diffusion affects the grain abundance and the amount of mass and magnetic flux within in a protostellar core. Ionization by the interstellar UV radiation field effectively cuts off the (ambipolar-diffusion controlled) mass infall rate in model cloud envelopes.

1 Introduction

It was proposed nearly twenty years ago that star formation can be and is initiated by ambipolar diffusion in the deep interiors of massive, dense clouds, while the envelopes remain magnetically supported (Mouschovias 1976, 1977, 1978, 1979). The results of recent, increasingly more sophisticated numerical simulations support those early conclusions (e.g., see review by Mouschovias 1995; also Ward-Thompson in this volume). In fact, detailed comparison of the evolutionary models of Ciolek & Mouschovias (1994; hereafter CM94) with observations of the Barnard 1 molecular cloud was carried out by Crutcher et al. (1994). Overall, their model predictions were in excellent agreement (typically, to within a few percent) with actual observed quantities in the B1 cloud (such as core mass, radius, mean density, and mean magnetic field strength).

The theoretical models also include the effects of interstellar grains. In the cold ($T \simeq 10\ K$) interiors of molecular clouds, grains typically have a charge of 0 or $-e$, where e is the electronic charge (Elmegreen 1979; see also the review by LaFon in this volume). Thus, charged grains can couple to the magnetic field, and thereby slow the rate of ambipolar diffusion within a cloud. The timescale for the formation of a supercritical core in a magnetically subcritical cloud is the ambipolar diffusion timescale, which can be written as $\tau_{AD} \simeq \nu_{ff,tot}\tau_{ff}$ (Mouschovias 1987; see also Mouschovias 1982), where τ_{ff} is the free-fall timescale, and

$$\nu_{ff,tot} \equiv \frac{\tau_{ff}}{\tau_{coll}} = \nu_{ff}\left\{1 + \frac{\tau_{ni}}{\tau_{go,n}}\Delta_{g-}\right\} \tag{1}$$

is the *(total) collapse retardation factor* (Ciolek & Mouschovias 1993; hereafter CM93). In equation (1) the quantity τ_{coll} is the mean collision time of a neutral particle with all other species of particles attached to the magnetic field (such as ions, charged grains, or neutral grains). The quantities τ_{ni} and $\tau_{g_o n}$ are the neutral-ion and neutral-grain collision times, respectively; ν_{ff} ($\equiv \tau_{ff}/\tau_{ni}$) is the *(ion) collapse retardation factor*. Δ_{g_-} is the *charged-grain magnetic attachment parameter*. In the limit that charged grains are perfectly coupled to the magnetic field, $\Delta_{g_-} = 1$; in the opposite limit of grains that are completely decoupled from field lines, $\Delta_{g_-} = 0$. The term in braces in the last equality of equation (1) is ≥ 1, hence, grain-neutral drag can increase the length of time needed to form a protostellar core.

Charged grains will be at least partially attached to magnetic field lines if the condition $\left(\omega_{g_-} \tau_{g_- n}\right)^2 + \left[\left(n_i - n_e\right)/n_i\right]\left(\omega_i \tau_{in}\right)\left(\omega_{g_-} \tau_{g_- n}\right) \gtrsim 1$ is satisfied (CM93), where $\omega_{g_-} = eB/m_{g_-}c$ and $\omega_i = eB/m_i c$ are the gyrofrequencies of charged grains and ions, respectively, and $\tau_{g_- n}$ and τ_{in} are the mean collision times for charged grains and ions with neutral particles; n_i and n_e are the number densities of ions and electrons. The first term of this relation represents the direct attachment of charged grains to field lines. The second term represents indirect attachment due to electrostatic attraction between charged grains and electron-shielded ion "quasiparticles" (which are themselves attached to the field), each with an *effective charge* $Z_{eff} = e(n_i - n_e)/n_i$. *Neutral grains* are effectively coupled to charged grains if the timescale for the inelastic capture of electrons on neutral grains is much smaller than the neutral-grain—neutral collision time; otherwise, they move essentially with the neutrals (see CM93, § 3.1.2).

Below we discuss simulations which show the effects of grains on ambipolar diffusion in magnetically supported clouds. Models presented here could represent axisymmetric, isothermal H_2 clouds with a cosmic abundance of He, temperature $T = 10\,K$, mass $M = 100 M_\odot$, initial central density $n_{n,c0} = 2.6 \times 10^3\,cm^{-3}$, and initial central magnetic field strength $B_{z,c0} = 35\,\mu G$. The initial central mass-to-flux ratio is a factor $\simeq 4$ smaller than the critical value for gravitational collapse; hence, these clouds would remain magnetically supported in the absence of ambipolar diffusion. The grains have radius $a = 3.75 \times 10^{-6}\,cm$. The initial dust-to-gas mass ratio $\chi_{g,0} = 0.01$.

2   Evolution of Model Clouds

2.1   Core Formation Timescale and Reduction of Grain Abundance

Figure 1a displays the density (normalized to $n_{n,c0}$) in the central flux tubes of three model clouds (presented in CM94), as a function of time, in units of the initial central dynamical timescale $\tau_{dyn,c0}$ [$\equiv (r/|g_r|)^{1/2} = 1.1 \times 10^6\,yr$, where $g_r(r)$ is the radial gravitational acceleration]. A *star* on each curve indicates when the central mass-to-flux ratio becomes equal to the critical value for collapse. Model A neglects the collisional effects of grains; model B includes the collisional drag of ions, charged grains, and neutral grains. The drag from neutral grains is neglected in model C. The evolution of all three models is qualitatively the same: initially, there is a slow increase in the central density as the neutrals diffuse

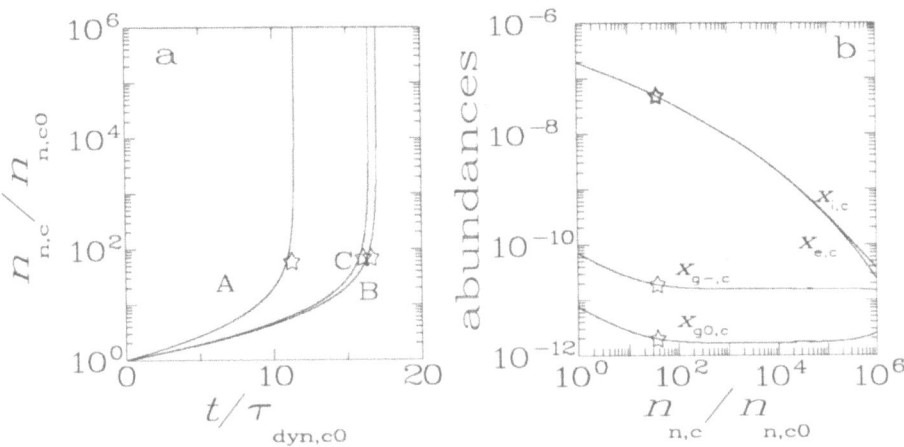

Fig. 1. Physical quantities in the central flux tubes of model clouds. The *star* on each curve indicates the time and density at which the central mass-to-flux ratio becomes equal to the critical value for gravitational collapse. (*a*) Central density $n_{n,c}$ of models A, B, and C, normalized to its initial value $n_{n,c0} = 2.6 \times 10^3$ cm^{-3}, as a function of time t, normalized to the initial central dynamical timescale $\tau_{dyn,c0} = 1.1 \times 10^6$ yr. (*b*) Fractional abundances (relative to the neutrals) of ions ($x_{i,c}$), electrons ($x_{e,c}$), charged grains ($x_{g_-,c}$), and neutral grains ($x_{g0,c}$) in model B, as functions of $n_{n,c}/n_{n,c0}$.

inward quasistatically (i.e., with negligible acceleration) through magnetic field lines that are "held in place" (Mouschovias 1978). This continues until the central mass-to-flux ratio becomes equal to the critical value for collapse. Beyond this, the neutrals begin to contract dynamically, and the density increases more rapidly with time. Magnetic field lines (and the plasma) are dragged in with the neutrals during this phase of the evolution. The evolutionary timescale is essentially the initial central flux-loss timescale $\tau_{\Phi,c0,B}$ [$\equiv \Phi_{B,c0}/(d\Phi_B/dt)_{c0}$, where d/dt is the (Lagrangian) time derivative comoving with the neutrals]; $\tau_{\Phi,c0,B}$ is equal to 11.8, 17.4, and $16.9\tau_{dyn,c0}$, respectively, for models A, B, and C.

From Figure 1a it is apparent that grain-neutral drag has retarded the evolution of model B with respect to model A by a factor $\simeq 1.5$. This is equal to the ratio $\nu_{ff,tot,B}/\nu_{ff,tot,A}$ (see eq. [1]; also, see CM94, § 3.3); hence, the increase in the evolutionary timescale of model B with respect to that of model A is exactly that which is predicted by the theoretical expression for τ_{AD} (see § 1). By the end of the run, grain-neutral drag exceeds ion-neutral drag in the central flux tubes of models B and C by more than an order of magnitude (see Fig. 5c of CM94). For $n_{n,c}/n_{n,c0} \gtrsim 100$, electrostatic attraction between charged grains and quasiparticles (see § 1), and not direct attachment, is the means by which charged grains remain partially attached to magnetic field lines. Finally, we note that model B evolves slightly slower than model C. Inelastic charge capture of ions and electrons on grains (see § 1) keeps the neutral grains well coupled to the charged grains (and, hence, the magnetic field) during the ambipolar-diffusion phase of contraction in model B. This increases the total grain drag acting on the neutrals (though, only by a small amount), and retards the evolution of model B with respect to model C.

Attachment of grains to magnetic field lines decreases the relative abundance of grains during the ambipolar diffusion phase of core formation. This can be seen in Figure 1b, which shows the fractional abundances of electrons ($x_{e,c} \equiv n_{e,c}/n_{n,c}$), ions ($x_{i,c} = n_{i,c}/n_{n,c}$), charged grains ($x_{g_-,c} \equiv n_{g_-,c}/n_{n,c}$, where $n_{g_-,c}$ is the density of charged grains), and neutral grains ($x_{g_0,c} \equiv n_{g_0,c}/n_{n,c}$, where $n_{g_0,c}$ is the density of neutral grains), in the central flux tubes of model B, as functions of $n_{n,c}/n_{n,c0}$. During the initial (quasistatic) contraction phase the grains and magnetic field lines are "left behind" by the inwardly diffusing neutrals, and this phase of the evolution. However, during the supercritical phase (i.e., for $n_{n,c}/n_{n,c0} \gtrsim 40$), when the core is dynamically collapsing, magnetic field lines and grains are dragged in with the neutrals; a neutral fluid element evolves with nearly constant fractional abundances of grains during this phase of the evolution. By the end of the run, $\langle \chi_g \rangle_{core}/\langle \chi_g \rangle_{env} = 0.256 = (1/3.9) = \mu_{d,c0}$, where $\langle \chi_g \rangle_{core}$ is the mean dust-to-gas mass ratio in the core, $\langle \chi_g \rangle_{env}$ is the mean dust-to-gas mass ratio in the subcritical (i.e., magnetically supported, *nonevolving*) envelope, and $\mu_{d,c0}$ [$\equiv (dM/d\Phi_B)_{c0}/(dM/d\Phi_B)_{crit}$] is the initial central mass-to-flux ratio in units of the critical value for gravitational collapse. In general, the dust-to-gas mass ratio in a core relative to that of the envelope in a cloud obeys the relation

$$\mu_{d,c0} \leq \frac{\langle \chi_g \rangle_{core}}{\langle \chi_g \rangle_{env}} \leq 1 \qquad (2)$$

(Ciolek 1993). The lower limit in equation (2) occurs if the grains are well attached to field lines during the subcritical phase of contraction, as in Figure 1b. If grains are not well attached to field lines (such as in the case of larger grains; see § 4.2 of Ciolek 1993), they move with the neutrals and the upper limit of equation (2) is attained instead.

If there is a distribution in the sizes of grains in a cloud, where $\eta(a, r)$ is the number of grains (per neutral particle) with radius a at position r [hence, $x_g(r) = n_g(r)/n_n(r) = \int_{a_{min}}^{a_{max}} \eta(a, r)da$], it follows that $\langle \chi_g \rangle_{core}/\langle \chi_g \rangle_{env}$ can be replaced by $\langle \eta(a) \rangle_{core}/\langle \eta(a) \rangle_{env}$ in equation (2). The ratio $\langle \eta(a) \rangle_{core}/\langle \eta(a) \rangle_{env}$ for small grains (which are well coupled to magnetic field lines, and, thus, left behind by the neutrals —see Fig. 1b) will be $\simeq \mu_{d,c0}$. On the other hand, $\langle \eta(a) \rangle_{core}/\langle \eta(a) \rangle_{env} \simeq 1$ for larger grains, because they are more easily detached from field lines by collisions with neutrals, and thus move with the neutral fluid. This may help to explain the observed increase in the size of grains at large optical depth (e.g., see, Vrba, Coyne, & Tapia 1993). That is, this phenomenon may be due to the fact that the abundance of smaller grains is actually *decreased* in a core forming by ambipolar diffusion. Of course, other effects, such as accretion onto grains, which are not accounted for in these models, may also contribute to the observed growth of grain sizes at high column densities.

2.2   Collapse of Cores, and Magnetic Support of Envelopes

Figure 2a shows the density profile (normalized to $n_{n,c0}$), as a function of r/R_0 (where $R_0 = 4.3 \, pc$ is the cloud radius) for a typical model cloud. The curves are shown at seven different times t_j ($j = 0, 1, 2, ..., 6$), chosen so as to have a central density enhancement $n_{n,c}(t_j)/n_{n,c0} = 10^j$. The times t_j are 0, 10.348, 12.643, 12.973, 13.027, 13.038, and $13.041\tau_{dyn,c0}$, respectively. This model is model B_{uv} of Ciolek & Mouschovias (1995; hereafter CM95), which is the same as model B discussed in § 2.1, except that it also accounts for the effect of an external (interstellar) UV radiation field acting on a cloud. A *star* on each curve locates the instantaneous radius of the supercritical core, inside which the total mass-to-flux ratio is equal to the critical value for collapse. The center of the *open circle* on each curve locates at each instant the *critical thermal length scale* $\lambda_{T,cr}(t_j)$ [$= C^2/2G\sigma_{n,c}(t_j)$, where C is the isothermal speed of sound, and $\sigma_{n,c}$ is the column density on the axis of symmetry], which is the scale over which thermal pressure forces in the core become important (Mouschovias 1991). For $t_j > t_1$, the cloud has a dynamically collapsing supercritical core, with mass $M_{core} = 6.1\ M_\odot$, radius $r_{core} = 0.12\ pc$, and mean core density $\langle n_n \rangle_{core} = 15.7 n_{n,c0} = 4.1 \times 10^4\ cm^{-3}$; by contrast, a model which did not include the effects of grains has a core that has 20% less mass (see § 3.3.1 of CM95). By time t_6, a nearly power-law density profile is established inside the core, with $n_n \propto r^s$, $-1.85 \lesssim s \lesssim -1.5$. The magnetically subcritical envelope, which contains about 90% of the mass of the cloud, remains effectively supported by the magnetic field, and does not evolve during the entire run (see Fig. 2a).

The mass infall rate for this model, in units of $M_\odot\ Myr^{-1}$, is displayed in Figure 2b at the same times t_j ($j > 0$). The mass infall rate becomes appreciable only inside the supercritical core, with $(\partial M/\partial t)_{max} \simeq 4.3\ M_\odot\ Myr^{-1}$ by the end of the run. However, in the subcritical envelope, UV radiation increases the degree of ionization to a level sufficient to keep the magnetic field essentially frozen in the matter. As a result, $\partial M/\partial t$ decreases precipitously (by more than four orders of magnitude) beyond the radius of the supercritical core.

3 Summary

We have discussed theoretical models which explicitly demonstrate the self-initiated (due to ambipolar diffusion) formation of protostellar cores in magnetically supported molecular clouds. These models also account for the effects of grains and the interstellar UV radiation field. Cores that formed in model clouds had physical properties that are in excellent agreement with observations of cores in star-forming molecular clouds. The mass infall rate is significant only in the supercritical core [$(\partial M/\partial t)_{max} \simeq 4\ M_\odot\ Myr^{-1}$], and is essentially cut off in the (UV-ionized, ambipolar-diffusion controlled) subcritical envelope.

When grains are attached to magnetic field lines, they are left behind by the neutrals during the ambipolar-diffusion phase of contraction, which decreases the abundance of grains in a protostellar core. In this case, the relative grain abundance in the core, compared to that in the envelope, is equal to the inverse of the factor by which a cloud's central flux tubes are initially magnetically subcritical. The abundance of small grains is reduced more than that of large grains, because smaller grains are better coupled to the magnetic field. This may explain observations which show an increase in grain size at large optical depths.

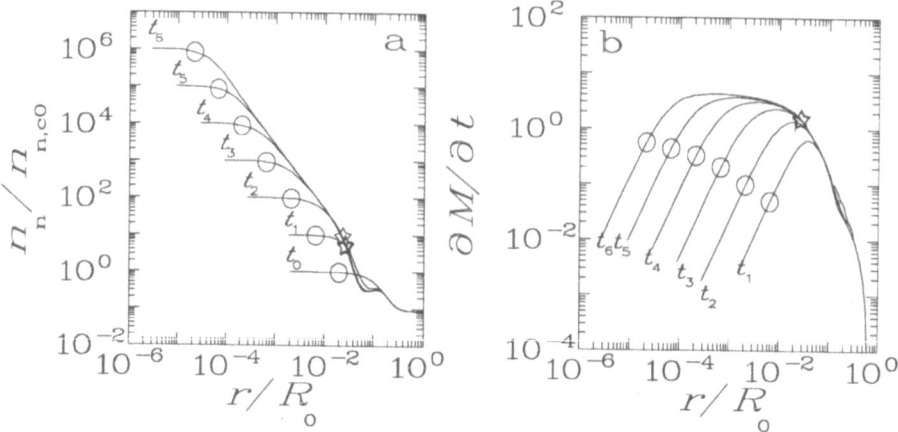

Fig. 2. Physical quantities in a typical model cloud, as functions of r/R_0 ($R_0 = 4.3\ pc$), at seven different times t_j (see text). A *star*, present only after a magnetically super-critical core forms, locates the instantaneous radius of the critical flux tube (see text). An *open circle* marks the instantaneous position of the critical thermal length scale (see text). (a) Density, normalized to $n_{n,c0}$. (b) Mass infall rate, in units of $M_\odot\ Myr^{-1}$.

Acknowledgement. G.C.'s attendance at this workshop was made possible by an international travel grant from the American Astronomical Society, which was funded by the National Science Foundation.

References

Ciolek, G. E. 1993, PhD Thesis, University of Illinois, Urbana-Champaign

Ciolek, G. E., & Mouschovias, T. Ch. 1993, ApJ, 418, 774 (CM93)

Ciolek, G. E., & Mouschovias, T. Ch. 1993, ApJ, 425, 142 (CM94)

Ciolek, G. E., & Mouschovias, T. Ch. 1995, ApJ, 454, 000 (CM95)

Crutcher, R. M., Mouschovias, T. Ch., Troland, T. H., & Ciolek, G. E. 1994, ApJ, 427, 839

Elmegreen, B. G. 1979, ApJ, 232, 729

Mouschovias, T. Ch. 1976, ApJ, 207, 141

Mouschovias, T. Ch. 1977, ApJ, 211, 147

Mouschovias, T. Ch. 1978, in Protostars and Planets, ed. T. Gehrels (Tucson: Univ. Arizona), 209

Mouschovias, T. Ch. 1979, ApJ, 228, 475

Mouschovias, T. Ch. 1982, ApJ, 252, 193

Mouschovias, T. Ch. 1987, in Physical Processes in Interstellar Clouds, ed. G. Morfill & M. Scholer (Dordrecht: Reidel), 491

Mouschovias, T. Ch. 1991, ApJ, 373, 16

Mouschovias, T. Ch. 1995, in The Physics of the Interstellar Medium and Intergalactic Medium, ed. A. Ferrara, C. F. McKee, C. Heiles, and P. R. Shapiro (San Francisco: ASP), vol. 80, 184

Vrba, F. J., Coyne, G. V., & Tapia, S. 1993, AJ, 105, 1010

Ambipolar Diffusion and Interstellar Chemistry

Michael P. Egan[1] and Steven B. Charnley[2,3]

[1] Phillips Laboratory, Geophysics Directorate, Hanscom AFB, MA 01731
[2] Astronomy Department, University of California, Berkeley, CA 94720,
[3] Space Science Division, NASA Ames Research Center, Moffett Field, CA 94035.

Abstract. Collapse of a magnetized self-gravitating layer of isothermal gas is examined using various models for the ionization state of the gas. We find that including PAHs in the cloud chemistry results in a higher ion mass density, which slows cloud collapse.

1 Introduction

The process of ambipolar diffusion is widely accepted as the mechanism which regulates star formation in molecular clouds. Models invoking this process have been able to explain collapse timescales and the magnetic field strength in molecular cores.

The degree to which ambipolar diffusion is effective depends on the ionization state of the cloud. Models of cloud ionization have progressed from simple equilibrium estimates (Elmegreen 1979) to ion-molecule chemical calculations (Umebayashi & Nakano 1990; Ciolek & Mouschovias 1993). Typical models include a few simple molecules, electrons, and dust grains. Ciolek (1995) has shown that charged grains dominate the ambipolar diffusion process. The presence of charged grains slows the rate of cloud collapse over what is expected if only neutral-ion collisions are modeled.

2 The Role of PAHs

The effect of polycyclic aromatic hydrocarbons (PAHs) on cloud collapse has not been examined. These large molecules, which are thought to be ubiquitous in the interstellar medium, may contain up to 10% of total carbon (Allamandola, Tielens & Barker 1985). Lepp & Dalgarno (1988) have shown that the presence of PAHs modifies the ionization structure of a cloud. Free electrons readily attach to PAHs and so, for PAH abundances (relative to H) greater than 10^{-8}, PAH$^-$ will be the dominant charge carrier in dark clouds. Lepp & Dalgarno predicted that a PAH-dominated chemistry leads to a higher fractional ionization than one where electrons and molecular ions like HCO$^+$ carry most of the charge. Hence, since they are much more massive than electrons, the presence of PAHs greatly increases the ionized mass fraction of the gas. We expect the effect of PAHs to be similar to that of grains, enhancing the coupling of the neutrals to the magnetic field. While PAHs are more abundant than grains by a factor of

10^5, dust grains are about a factor of 10^6 more massive than PAHs. Therefore, in order to have an appreciable effect on cloud collapse, nearly all of the PAHs would have to be ionized.

3   Cloud Model

We adopted the model of Shu (1983) for the MHD evolution of a self-gravitating isothermal slab. Although the 1-D problem is unrealistic, by directly coupling the detailed microphysics/chemistry to the MHD, this model allows us to quickly and self-consistently explore the effects of interstellar chemistry on gross dynamics. Our cloud model incorporates the solution of the hydrodynamics equations for an isothermal cloud with an equilibrium chemistry model. The chemistry equations are split from the hydrodynamics calculations. At each time step, the chemistry is solved on all spatial grid points, providing an updated ion mass density.

3.1   Hydrodynamics equations

Following Shu (1983), we solve the dynamics, magnetic field, and gravitational field equations for a self-gravitating layer of isothermal gas. Defining B as the magnetic field strength, ρ and ρ_i as the neutral gas mass density and the ion mass density respectively, surface density σ, and vertical coordinate z, the dynamical equations are

$$\frac{\partial}{\partial t}\left(\frac{B}{\rho}\right) = \frac{1}{\gamma}\frac{\partial}{\partial \sigma}\left[\frac{B^2}{4\pi\rho_i}\frac{\partial B}{\partial \sigma}\right]$$

$$\frac{B^2}{8\pi} + a^2\rho = 2\pi G\left(\sigma_\infty^2 - \sigma^2\right)$$

$$\frac{\partial z}{\partial \sigma} = \frac{1}{\rho}.$$

Here G is the gravitational constant, a the isothermal sound speed in the gas, and γ the drag coefficient for ion-neutral collisions.

3.2   Chemistry

For this preliminary study, we have followed previous work (Umebayashi & Nakano 1990; Ciolek & Mouschovias 1993) and assumed that the ionization state of the plasma over the time-scale of the dynamical evolution can be treated in a steady-state approximation. In contrast to previous studies we account for the exchange of volatile material between the gas and dust by accretion and evaporation (Charnley 1995). We assume that the gas at the start of the dynamical calculation is chemically well-evolved and so that the heavy elements are contained mostly in CO, N_2 and O_2. On grains, these molecules are desorbed into the gas at the rate required to reproduce the observed CO abundance in dense cores (i.e. $n_H \sim 10^4 - 10^5 cm^{-3}$) We assume spherical 'classical' grains of radius $0.1\mu m$, $T_{gas} = T_{dust} = 10K$, a cosmic ray ionization rate of $1.3 \times 10^{-17}s^{-1}$, and the dust abundance described below. We have included the chemistry of PAH molecules using, with extensions, reactions from Lepp & Dalgarno (1988) and Pineau Des Fôrets et al. (1988). It is unlikely that, once accreted, PAHs are efficiently returned to the gas in dark clouds. Hence, a steady-state solution containing PAHs necessarily requires that, in contrast to our treatment of volatile molecules, we ignore PAH-grain sticking. For a prescribed neutral gas density, the distribution of charge between the gas and dust and PAH species is easily computed and provides the value of ρ_i required by the MHD calculation.

4   Model Results & Discussion

To demonstrate the effect of the ionization of PAHs on cloud collapse, we present three models. Model 1, shown in Figure 1, is from Shu 1983. Here the ionized gas density is given by $\rho_i = C\sqrt{\rho}$. The plot shows the density profile of the cloud at $t = 0$, $t = 5$ Myr, and $t = 20$ Myr. Figure 2 contains the results of two models

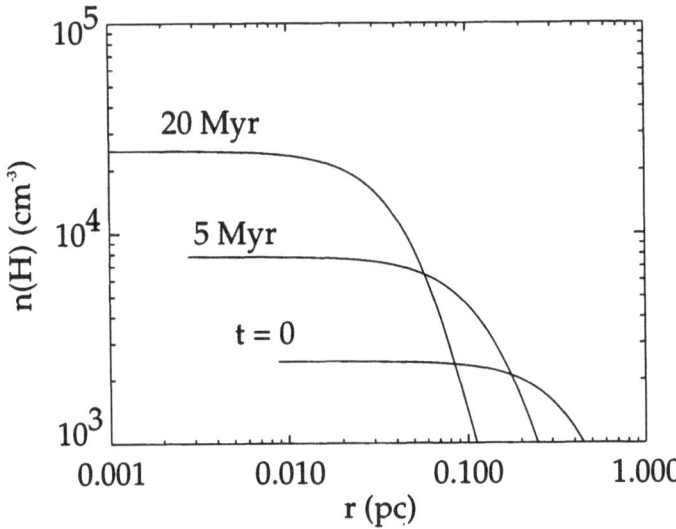

Fig. 1. Model from Shu (1983)

for which ρ_i has been calculated using the chemical model described above. Both models have the same initial conditions as Model 1. In both Model 2 and 3, the fractional abundance of dust (relative to H) is 10^{-13}. In Model 2, no PAHs are included in the calculation, while in Model 3, the fractional abundance of PAHs is 10^{-7}. The plot shows the density profile of the cloud at $t = 0$, $t = 20$ Myr, and $t = 50$ Myr. From these plots one can see that including dust slows the

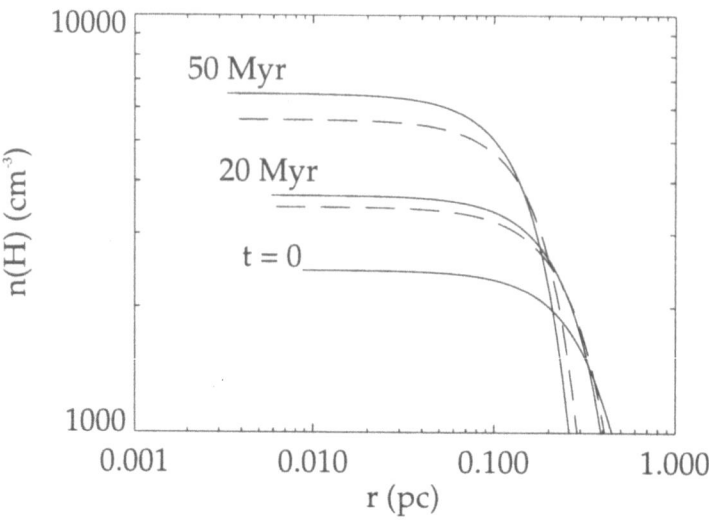

Fig. 2. Model 2, solid lines; Model 3 (PAH model), dashed lines

collapse timescale by approximately a factor of ten over the simple equilibrium model. This is expected from earlier results. The inclusion of PAHs tend to slow the collapse even further, the effect becoming more important as the collapse proceeds and the PAHs collect almost all the free electrons. In our model this process ionizes about half of the PAHs.

To fully quantify the effect of PAHs on the initial stages of star formation will require a fully time-dependent treatment of the chemistry in which PAHs are allowed to accrete unabated on to dust grains, as a dense core forms within a more diffuse interclump medium. These calculations are currently underway.

References

Allamandola, L.J., Tielens,A.G.G.M and Barker, J.R. 1985, ApJ, 290, L25
Ciolek, G. E. and Mouschovias, T. Ch. 1993, ApJ, 418, 774
Ciolek, G. E. 1995, Interstellar Polarization Conference, Troy, NY (poster paper)
Charnley, S.B. 1995, Astrophys. Space. Sci., 224, 441
Elmegreen, B.G. 1979, ApJ, 232, 729
Lepp, S. & Dalgarno, A. 1988, ApJ, 324, 553
Pineau Des Fôrets, G., Flower, D. and Dalgarno, A. 1988, MNRAS, 235 621
Shu, F. 1983, ApJ, 273, 202
Umebayashi, T. and Nakano, T. 1990, MNRAS, 243, 103

Simulating Dusty Gas Using SPH

S.T. Maddison and J.J. Monaghan

Mathematics Department, Monash University, Clayton, Victoria, 3168, Australia

Abstract. We briefly look at dust in astrophysical environments and its importance, with particular attention given to star forming regions and protostellar disks. Smoothed Particle Hydrodynamics (SPH) is used to simulate a two–phase flow comprising of dust and gas, coupled by gravity and friction. The SPH implementation is discussed with a brief look at the drag term. To test the code, we simulate of a sound wave in a periodic box, a collapsing stationary and rotating sphere. Our results compare well with known analytic solutions and expected results.

1   Introduction

Dust is found in practically all astrophysical environments, including the ISM, the arms of spiral galaxies, elliptical galaxies, the disks of young main sequence and pre-main sequence stars and the envelopes of AGB stars and planetary nebulae. Although dust makes up only 1% of ISM, it is observationally and dynamically important through its absorption, re–radiation and scattering of light. It is, for example, thought to be the dynamical driving mechanism of AGB star mass loss, with stellar radiation forcing the dust outwards with then drags the gas with it (see e.g. Netzer & Elitzur 1993).

Generally protostellar disks are modelled with gas alone although it is well known that dust is present (Smith & Terrile 1983, Beckwith et al. 1990, Simon et al. 1992). We want to include the dust in order to investigate its dynamical importance in protostellar environments. Noh, Vishniac & Cochran (1991) showed that dust acts to *increase* the growth rate of instabilities in protostellar disks for various azimuthal wave numbers m. We will be able to test this with our model and follow the nonlinear evolution.

2   SPH Implementation of Two–Phase Flow

Following the work of Harlow & Amsden (1975) and the SPH implementation by Monaghan & Kocharyan (1995), we have setup a two–phase dusty gas code. The applications for such a two–phase code in astrophysics extend beyond dust and gas to stars and gas (for galactic simulations) and luminous and dark matter (for cosmological simulations) by changing source terms in the momentum equation. The program could easily be extended to deal with three phases, to simulation stars, gas and dust or gas and two distinct dust populations.

In two phase flows the relative motion between the two fluids involves a momentum interchange, a heat interchange and possibly mass exchange (via a phase change). For simplicity in this test study we assume that the dust does not evaporate or coagulate and that the gas does not condense. In our SPH model the forces are due to pressure, viscosity, drag and gravity.

In this paper, the subscripts g and d will represent gas and dust particles respectively, and quantity $Q_{gd} = Q_g - Q_d$. The momentum equation of the two phase flow is given by

$$\frac{dv}{dt} = -\frac{1}{\rho}\nabla P \pm \frac{K}{\rho}(v_d - v_g) - \nabla\Phi , \tag{1}$$

where K is the drag term (see below for more details).

We define the local SPH density, $\hat{\rho}$, as a product of the global (average) density, ρ, and the void fraction, Θ, by

$$\hat{\rho} = \Theta\rho , \tag{2}$$

where $\Theta_d + \Theta_g = 1$. In astrophysical applications, $\Theta_g \sim 1$ and $\Theta_d \ll 1$. We also assume that the dust fluid exerts negligible pressure. The SPH particles are tagged as either *dust* or *gas* and experience the appropriate particle–particle interactions. The local densities and voids are assigned initially and updated as follows; at time t, average densities are calculated by the standard SPH summation, $\hat{\rho}_i = \sum_j m_j W_{ij}$; the dust void fractions are calculated as $\Theta_d = \hat{\rho}_d/\rho_d$; and the gas void fractions as

$$\Theta_g = 1 - \Theta_d = 1 - \sum_d \frac{m_d}{\hat{\rho}_d}\Theta_d W_{gd} = 1 - \frac{1}{\rho_d}\sum_d m_d W_{gd} . \tag{3}$$

The time stepping routine used is a predictor–corrector method, constrained by the force, Courant condition and viscosity, plus a drag constraint such that $\delta t_{drag} < \min(\hat{\rho}/K)$

3   Drag Force

From the SPH momentum equation, the drag force (per unit mass) is represented by

$$\frac{K}{\hat{\rho}}(v_d - v_g) , \tag{4}$$

and thus we need a representation for K. The drag force is dependent on the relative velocity, fluid viscosity and particle size. Valentine & Wohletz (1989) use

$$K = \frac{\rho_g\Theta_d c_D}{r}|v_{dg}|$$

for their pyroclastic applications, assuming constant density spherical grains of radius r. The drag coefficient, c_D, which depends on the Reynolds number of the flow and the shape of the body, is uncertain in astrophysical situations. In terrestrial applications it is generally determined experimentally. For our test cases, we use K = constant, to compare our results with analytic work and other simulations.

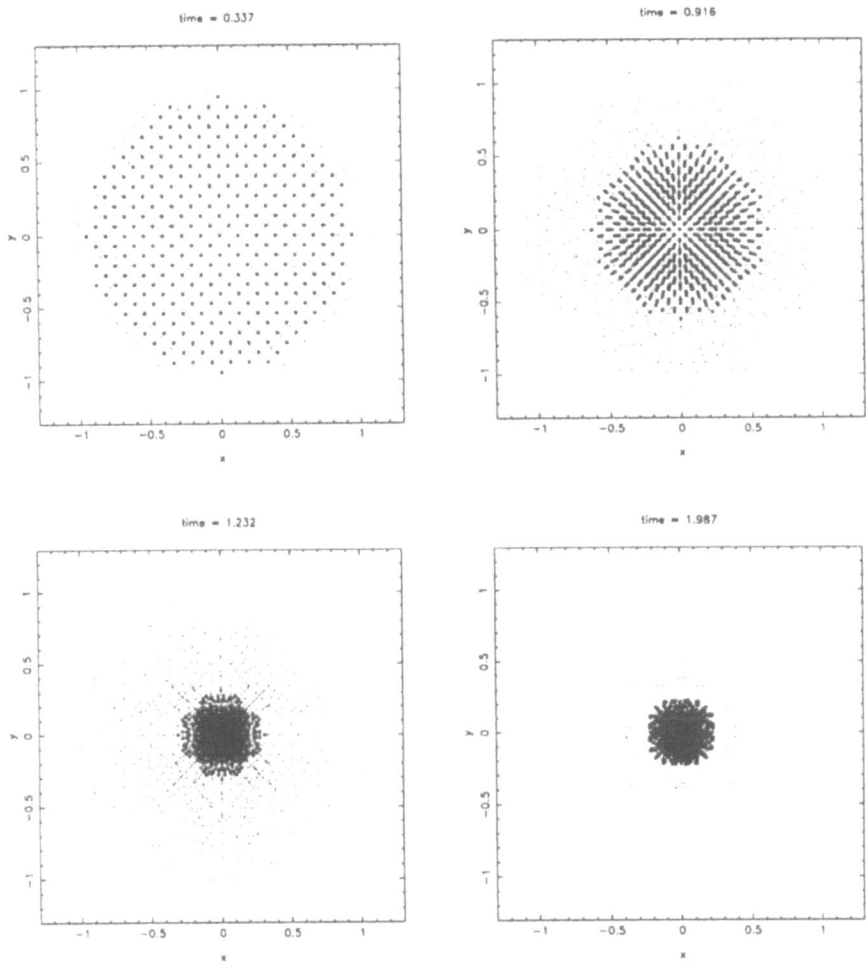

Fig. 1. Collapsing dusty-gas sphere, $K = 0$. The dust and gas are essentially decoupled and so the dust (open circles) collapses more rapidly than the gas (dots). The gravity solver is a multigrid, and thus some artificial symmetries are seen.

4 Test Cases

We set up an isothermal periodic box to model an infinite dusty-gas medium to test the code for stability and sound speed accuracy. For stability, we set the particles at rest and follow the evolution for 15 sound crossing times. The linear and angular momentum are conserved and the kinetic energy remains negligible. The choice of smoothing length can have an effect however. We found $h = 1.5$

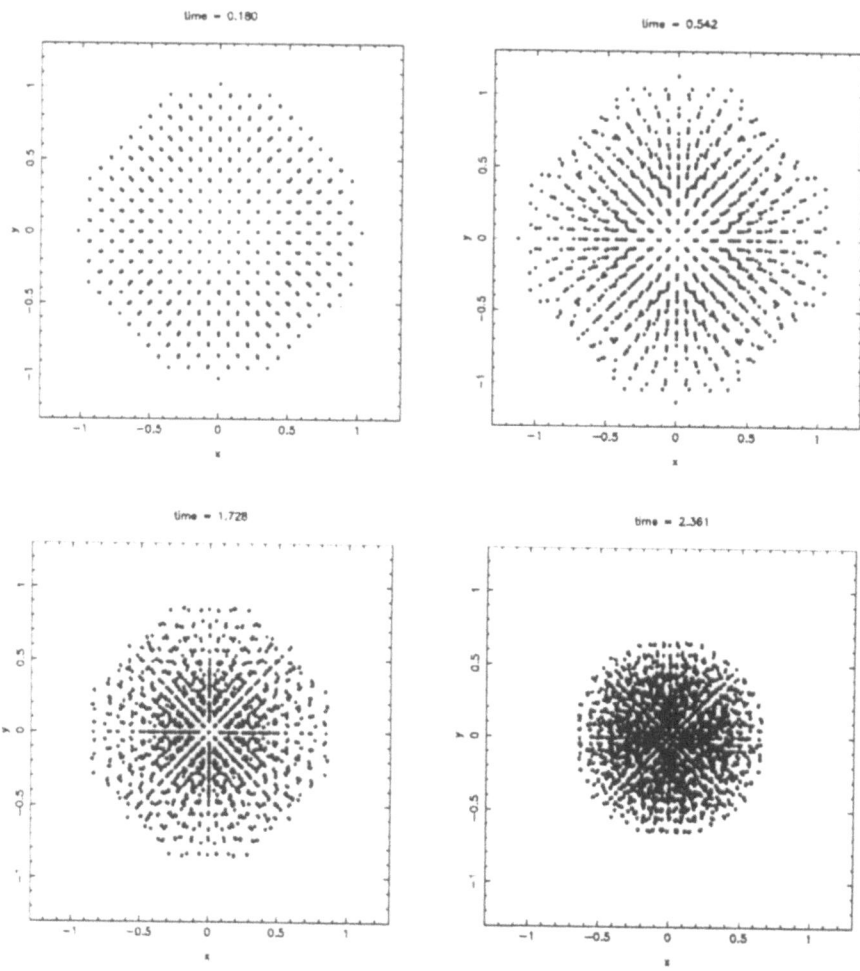

Fig. 2. Collapsing dusty–gas sphere with $K = 4$. Here the dust (open circles) and gas (dots) are tightly coupled and the dust is carried outwards initially with the gas and then they collapse inwards together. Both re-expand, the dust first as it is unaffected by viscosity. The systems eventually settles to an equilibrium configuration.

gave the worst results. To check the accuracy of the sound speed, we set up a sine wave in the box, with an amplitude of $\sim 10\%$ of the theoretical sound speed and test with $K = 0$ and compare with the limiting case solutions to the dispersion relation of Harlow & Amsden (1975), and also a non–zero value of K to compare with the solution of the full dispersion relation. The non–zero K case gave a sound speed accurate to within 5%, and for the $k = 0$ case, to within better than 2%. The error decreases as the resolution of the wave across the box in increased.

A collapsing initially isothermal sphere was also modelled and compared with known analytic and expected results. When $K = 0$, the gas and dust are decoupled and should collapse independently, whereas for $K \neq 0$, the fluids are coupled. We expect the gas to shock and dissipate as it is feels the effects of viscosity, which the dust does not. For a collapsing non–rotating sphere with $K = 0$ we expect the dust to fall first, followed by the gas which, unlike the dust, feels gas pressure. When $K \neq 0$, the dust will follow the gas outwards initially, then both fluids will collapse inwards. These effects were indeed observed (see Figs. 1 & 2).

We also modelled a rotating sphere, which evolves to a disk. Again, the gas shocks and dissipates energy because of the viscosity and remains in the central region whereas the dust passes through the newly formed disk and bounces around until an equilibrium is found.

5 Conclusion

We have implemented the two–phase SPH equations in our code, which has been tested for stability, sound speed accuracy and dynamical evolution. The code is now being used to study dusty–gas protostellar disks and the dynamical effects of dust on star formation.

6 References

Beckwith,S.V..W., Sargent,A.I., Chini,R.S. & Güsten,R. (1990), AJ, **99**, 924
Harlow,F.H. & Amsden,A.A. (1975), J.Comp.Phys, **17**, 19
Monaghan,J.J. & Kocharyan,A. (1995) Comput.Phys.Comm, **87**, 225
Noh,H., Vishniac,E.T. & Cochran,W.D. (1991) Ap.J, **383**, 372
Netzer,N. & Elitzur,M. (1993), Ap. J **410**, 701
Smith & Terrile, (1984), Science, **226**, 1421
Simon,M., Chen,W., Howell,R., Benson,J. & Slowik,D. (1992), ApJ,**384**, 212
Valentine,G.H. & Wohletz,K.H. (1989), J.Geophys.Res, **94**, 1867

Role of Dust in Protostar Formation

Telemachos Ch. Mouschovias

Departments of Physics and Astronomy, University of Illinois,
1002 West Green street, Urbana, IL 61801, U. S. A.

Abstract. We review the role of interstellar grains in the early, isothermal stages of *self-initiated* (due to ambipolar diffusion) star formation in massive, magnetically supported molecular clouds (up to a central density of about $10^{10}\ cm^{-3}$). Certain features of the formation and contraction of protostellar cores are virtually independent of the specific values of the (dimensionless) free parameters of the problem, provided that these values are kept within the observationally allowed range. (1) Core formation is a quasistatic process (i.e., one with negligible acceleration), occurring on the initial central flux-loss timescale. The neutrals contract through the essentially stationary magnetic field lines and the plasma (including grains). (2) Eventually a core becomes both thermally and magnetically supercritical and begins to contract dynamically. (3) Collapse, however, does not evolve into free fall at densities below $10^{10}\ cm^{-3}$. (4) The magnetically subcritical (but massive) cloud envelope evolves only on the much longer, local ambipolar-diffusion timescale. Beyond the well-known role of grains in the chemistry that determines the degree of ionization in *static* clouds, we quantify both their effect on ambipolar diffusion (hence, star formation) and the effect of the dynamics on grain abundance (hence, chemistry) in *evolving* dense cores. A number of commonly used assumptions are found to be incorrect.

1 Introduction

The observed spatial coincidence of young stellar clusters and dense, massive interstellar clouds (Baade 1944) led to the hypothesis that star formation takes place through the collapse and fragmentation of these clouds. Yet, both theory and observations have been showing for two decades now that the collapse of a cloud as a whole is neither necessary for nor commonplace during star formation, despite the fact that a typical cloud mass exceeds the thermal Jeans mass by a factor $\sim 10^2$ - 10^4 (see review by Mouschovias 1995). Most of the mass of a cloud remains magnetically supported while ambipolar diffusion initiates the formation and contraction of protostellar fragments (or cores) in the deep interior. Typically, only 10% of the mass is converted into stars.

Traditionally, interstellar grains have been thought as potentially affecting the evolution of magnetic clouds because of the key role they play in the chemical processes that determine the degree of ionization x_i (for a quasistatically contracting cloud, the ambipolar-diffusion timescale is $\tau_{AD} \propto x_i$; Spitzer 1968). Baker (1979) and Elmegreen (1979) also suggested that (charged) grains may couple to the magnetic field and thus play a role in ambipolar diffusion and star formation (see also Nakano & Umebayashi 1980). Gail & Sedlmayr (1975) had extended Spitzer's (1941) work and had shown that, for HI-cloud conditions ($T \simeq 100\ K$), the charge of a typical dust particle fluctuates between 0 and

$\pm e$. For dark molecular-cloud conditions ($T \simeq 10 \; K$), a grain is singly negatively charged approximately 90% of the time and neutral during the remaining 10% (Elmegreen 1979; Havnes, Hartquist, & Pilipp 1987; Draine & Sutin 1987). Elmegreen considered a spherical, uniform cloud contracting quasistatically and estimated the timescale for loss of magnetic flux *by the cloud as a whole*. It exceeded the free-fall timescale τ_{ff} by more than a factor of 10. Nakano & Umebayashi (1980) considered a similar model cloud contracting quasistatically and obtained a similar estimate for loss of flux by a cloud as a whole. They therefore concluded that ambipolar diffusion is ineffective at densities smaller than $5 \times 10^9 \; cm^{-3}$, for only above such densities does its timescale become comparable to τ_{ff}. Nakano & Umebayashi (1986a, b) estimated the drift speed v_D (denoted by v_B in their paper) between the neutrals and the plasma, found it to become comparable to the free-fall speed v_{ff} only at densities greater than $10^{11} \; cm^{-3}$, and concluded that magnetic flux loss is ineffective below these densities. Umebayashi & Nakano (1990), and Nishi, Nakano, & Umebayashi (1991) arrived at a similar conclusion.

There is a fundamental error in the above papers. It is the same one committed earlier by Nakano & Tademaru (1972) and pointed out by Mouschovias (1977). The assumptions that (1) the contraction is quasistatic and (2) the contraction timescale is equal to τ_{ff} (or, equivalently, that the contraction speed is equal to v_{ff}) are mutually inconsistent. Free fall is, by its very nature, anything but quasistatic. Moreover, τ_{ff} is a strict lower limit on the actual contraction timescale. The claim that a process is relevant only if its timescale is smaller than τ_{ff} badly underestimates the importance of the process. A similar logic applied to the Sun would lead to the conclusion that, since $\tau_{ff,\odot} \simeq 0.5$ hr, no process taking longer than 0.5 hr is relevant to solar evolution! In the case of molecular clouds, we had already pointed out that they are magnetically supported (not collapsing as a whole) and that τ_{AD} in the deep interior *is* the evolutionary timescale (Mouschovias 1977, 1979b). In this picture, fragments (or cores) form because of ambipolar diffusion, while the massive envelopes remain magnetically supported. (This support against gravity can include a contribution from hydromagnetic waves; see Mouschovias 1987a.) More recently, we have followed numerically in axisymmetric geometry the isothermal phases of this, *self-initiated* process of star formation (typically, from a central density of 3×10^3 to $3 \times 10^9 \; cm^{-3}$), starting from states which would be in exact magnetohydrostatic (MHS) equilibrium if it were not for ambipolar diffusion and/or magnetic braking. In § 2 we summarize some results from these investigations, in the form of an overview of the new theory of star formation, and in § 3 we focus on the effects of grains.

2   New Concepts - Basic Theory of Protostar Formation

A conceptual re-examination of the process of ambipolar diffusion concluded that it does not lead to loss of flux by a cloud as a whole (Mouschovias 1976, 1977, 1979b), in sharp contrast to the commonly made *assumption* that it does (e.g., see Nakano & Tademaru 1972; Nakano 1976; Mestel & Paris 1979). Its essence is *a redistribution of mass in the central flux tubes* of a self-gravitating, magnetically supported cloud. The neutrals, under the action of their self-gravity, contract through the field lines against thermal-pressure forces, provided only that the contracting mass exceeds the critical thermal ($\simeq$ Jeans) mass. The presence of the magnetic field does not prevent this instability against fragmentation (or core formation). It simply regulates it. The charged particles attached to the field lines give rise to neutral-plasma frictional forces, thus effectively transmitting the magnetic forces to the neutral particles and thereby converting the otherwise violent star formation process into a relatively quiescent one for most of its duration. The evolution takes place on the central flux-loss timescale $\tau_{\Phi,c} = \Phi_{B,c}/|d\Phi_{B,c}/dt| = \tau_{AD,c}/2$, where Φ_B is the magnetic flux. Written in the form

$$\tau_{AD} \simeq \frac{\tau_{ff}^2}{\tau_{ni}} \simeq 2 \times 10^6 \left(\frac{n_i/n_{H_2}}{10^{-7}}\right) \quad yr, \tag{1}$$

τ_{AD} is essentially independent of geometry (Mouschovias 1987b). It also allows us to define the *collapse retardation factor* $\nu_{ff} \equiv \tau_{AD}/\tau_{ff} \simeq \tau_{ff}/\tau_{ni}$ (Mouschovias 1982, 1987a; Ciolek & Mouschovias 1993), which is essentially the factor by which ambipolar diffusion retards the contraction of a core relative to free fall, because of ion-neutral drag. [1]

The redistribution of mass in the central flux tubes of a cloud effected by ambipolar diffusion eventually leads to a central mass-to-flux ratio that exceeds the critical value for collapse,

$$\left(\frac{dm}{d\Phi_B}\right)_{c,crit} = 1.5 \left(\frac{1}{63G}\right)^{1/2}, \tag{2}$$

calculated from a sequence of exact equilibrium states (Mouschovias & Spitzer 1976), and dynamic contraction ensues. [2] The more subcritical the central flux tube of a cloud is initially, the longer it must wait for ambipolar diffusion to increase its mass-to-flux ratio to the critical value for collapse, and the higher the density enhancement at which this is achieved (see Mouschovias 1991c, § 2.4).

During the collapse phase of the core, force balance is maintained *along* field lines (see Fiedler & Mouschovias 1993), the magnetic flux is essentially trapped inside, and the magnetic field increases with gas density almost as $B_c \propto \sqrt{n_{n,c}}$, the flux-freezing limit found by Mouschovias (1976). Typically, even at a central density as high as 10^{10} cm^{-3}, the *maximum* infall acceleration does not exceed 30% that of gravity, and the magnetic force dominates the thermal-pressure force everywhere in the supercritical core, except in a small region of size $\simeq$ 30 AU, in which the two forces are comparable (see review by Mouschovias 1995).

---

[1] When grain-neutral drag is accounted for, the timescale τ_{ni} in the expressions for τ_{AD} and ν_{ff} is replaced by τ_{coll}, the harmonic mean of the collision times of a neutral particle with ions and grains, provided that the latter are attached to the magnetic field.

[2] For a uniform (hence, nonequilibrium) thin disk of infinite radius, Nakano & Nakamura (1978) find a smaller critical value by the factor 1.19.

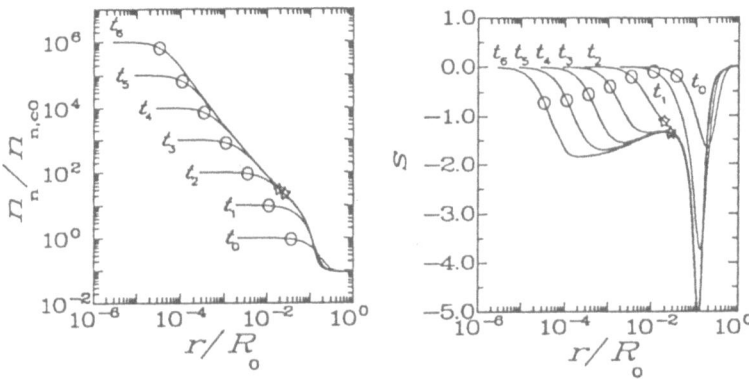

Fig. 1. Spatial profiles of physical quantities as functions of radius r, normalized to the initial cloud radius R_0 ($= 4.29$ pc), at seven different times $t_j (j = 0, 1, ..., 6)$ chosen such that the central density at t_j has increased by a factor 10^j with respect to its initial value. The times are 0, 10.667, 13.279, 13.614, 13.663, 13.673, and $13.675\tau_{ff,c0}$; $\tau_{ff,c0} = 1.3295$ Myr. A "star" or a curve, present only after a supercritical core forms, marks the instantaneous radius of this core. An open circle on every curve marks the instantaneous position of the critical thermal lengthscale $\lambda_{T,cr}$. (a, left) Neutral density, normalized to its initial central value $n_{n,c0} = 2.6 \times 10^3$ cm^{-3}. (b, right) The exponent $s \equiv \partial \ln n_n / \partial \ln r$.

Analytical calculations for both aligned and perpendicular rotators had shown that magnetic braking is very effective during the quasistatic phase of contraction, tending to keep the angular velocity near that of the background (Mouschovias & Paleologou 1979, 1980). Hence, centrifugal forces have a negligible effect on the evolution. Only after dynamical contraction sets in does magnetic braking become ineffective and angular momentum trapped inside the collapsing core (Mouschovias 1979a). Recent numerical simulations (Basu & Mouschovias 1994, 1995a, b) have confirmed this picture and produced detailed predictions, testable by observations (e.g., spatial profiles of the angular velocity at different times). During the collapse phase, the angular velocity inside the supercritical core varies almost as $\Omega_c(r) \propto n_n(r)^{1/2}$, the angular-momentum conservation law for a disk. As part of an extensive parameter study, the dependence of the physical properties of rotating protostellar cores on the initial mass-to-flux ratio μ_{c0} (in units of its thin-disk critical value) was determined (Basu & Mouschovias 1995b). The core mass is found to scale as $M_{core} \propto \mu_{c0}$, while its specific angular momentum scales as $(J/M)_{core} \propto \mu_{c0}^2$. Hence, $(J/M)_{core} \propto M_{core}^2$.

3 Effects of Grains

We use the proper creation/destruction rate equations to calculate in a self-consistent manner the abundances of charged species, i.e., electrons, molecular ions, atomic (metallic) ions, and charged grains, at each time step of the evolution; this includes the conversion of neutral grains to charged grains and vice versa. The high-energy cosmic-ray ionization rate is taken to be $\zeta_{CR} = 5 \times 10^{-17}\ s^{-1}$ (Spitzer & Tomasko 1968; Payne, Salpeter, & Terzian 1984). Grain-neutral frictional forces are also properly accounted for in the momentum equations. In the density range of interest ($\sim 10^3$ - $10^9\ cm^{-3}$), the dominant ions are HCO^+, Mg^+, and Na^+. As in the investigations ignoring grains, the initial central mass-to-flux ratio (μ_{c0}) is an important free parameter in the problem. The most significant new free parameter introduced by the physics of grains is most sensitive to the value of the grain radius a (see Ciolek & Mouschovias 1993, 1994). It affects both the rate of chemical reactions involving grains and the grain-neutral drag. In what follows we use the "standard" value $a = 3.75 \times 10^{-6}\ cm$. (Ciolek & Mouschovias 1996 account for a distribution of grain radii.)

The disklike model cloud has $\mu_{c0} = 0.256$ and is initially in an exact MHS equilibrium state, with thermal-pressure forces balancing gravity along field lines and magnetic plus thermal-pressure forces doing so perpendicular to the field lines. This state may be thought of as representing a cloud having the following physical parameters: central neutral density $n_{n,c0} = 2.60 \times 10^3\ cm^{-3}$, central magnetic field strength (along the axis of symmetry) $B_{c0} = 35.3\ \mu G$, temperature $T = 10\ K$ (hence, isothermal sound speed $C = 0.188\ km\ s^{-1}$), column-density lengthscale $l_{ref} = 0.858$ pc, radius $R_0 = 5l_{ref} = 4.29$ pc, mass $M_{cl} = 98.3\ M_\odot$, and a grain mass fraction $\chi_{g,0} = 0.01$.

Ambipolar diffusion is turned on at time $t = 0$ and the evolution is followed, as usual, to six orders of magnitude enhancement of the central neutral density $n_{n,c}$. It takes $13\tau_{ff,c0}$ to achieve supercritical conditions in the central flux tube (where $\tau_{ff,c0} = 1.3295$ Myr is the initial central free-fall time). Figure 1a shows the neutral density n_n as a function of radius r at seven different times t_j ($j = 0$, 1, ..., 6) chosen such that $n_{n,c}$ increases by a factor 10^j in the time t_j. A "star" on a curve, present only after a supercritical core forms, marks the instantaneous radius of the core. Once such a core forms, it is characterized by a compact, nearly uniform-density central region, whose size is determined by the instantaneous value of the critical thermal lengthscale $\lambda_{T,cr} = C^2/2G\sigma_{n,c}(t)$ ($\simeq$ Jeans length; see Mouschovias 1991a), marked by an *open circle* on each curve, and a "tail" of infalling matter left behind by the shrinking (both in size and in mass) central region. A near power-law density profile is established in the tail, $n_n \propto r^s$, with $-1.84 \lesssim s \lesssim -1.50$ in the region $1.0 \times 10^{-4} \lesssim r/R_0 \lesssim 8.0 \times 10^{-3}$ at time t_6 (see Fig. 1b). *A break in the slope of the density profile occurs almost exactly at the boundary of the supercritical core*, revealing the different physics governing the evolution of the collapsing core and that of the subcritical, magnetically well-supported envelope. At the end of the run, the mass and radius of the core are $M_{core} = 8.21 \times 10^{-2} M_{cl} \simeq 8.1 M_\odot$ (in excellent agreement with observations by Wood et al. 1994, showing that the most common core mass is somewhat smaller than $10\ M_\odot$), and $R_{core} = 2.67 \times 10^{-2} R_0 \simeq 0.11$ pc (which is also a typically observed core radius). The radius of the uniform-density central region is $\simeq \lambda_{T,cr}(t_6) = 3.38 \times 10^{-5} R_0 \simeq 30$ AU. Although the central density at this time is $2.6 \times 10^9\ cm^{-3}$, the *mean* density of the supercritical core is only $\langle n_n \rangle_{core} = 39.5 n_{n,c0} = 1.0 \times 10^5\ cm^{-3}$, comparable to the density of observed protostellar cores using NH_3 as a tracer molecule (e.g., see Bachiller et al. 1990; Crutcher et al. 1994). Similarly, the central field is 4 mG, but its mean value in the core is only $\simeq 65\ \mu G$.

Ion depletion onto grains would by itself tend to allow ambipolar diffusion to operate effectively even at high densities, during the collapse phase. However, grain-neutral drag almost compensates for the effect of ion depletion, and the exponent $\kappa \equiv d\ln B_c/d\ln n_{n,c}$ increases to 0.48 by the end of the run (see Fig. 2a), a value near the flux-freezing limit of 1/2. During the early, quasistatic phase of contraction, κ is very small, revealing that the neutrals are contracting through essentially motionless field lines (and plasma).

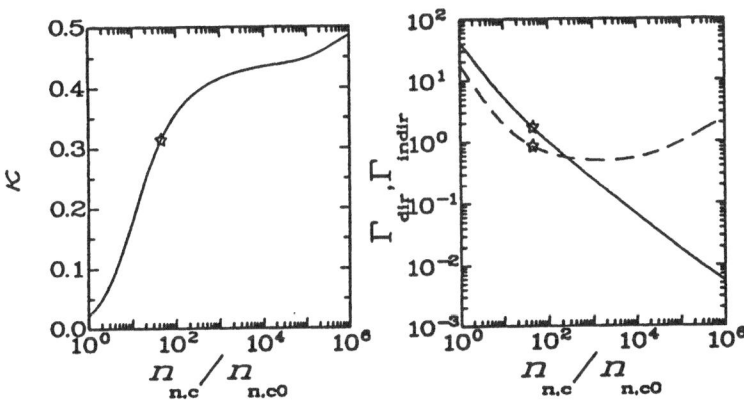

Fig. 2. (a, *left*) Exponent $\kappa \equiv d\ln B_c/d\ln n_{n,c}$ as a function of central density enhancement $n_{n,c}/n_{n,c0}$ for the model of Fig. 1. (b, *right*) Charged-grain direct (Γ_{dir}; *solid line*) and indirect (Γ_{indir}; *dashed line*) magnetic attachment parameters as functions of $n_{n,c}$.

The *direct attachment parameter*

$$\Gamma_{dir} \equiv \omega_{g_-}\tau_{g_-n},\tag{3}$$

where $\omega_{g_-} = eB/m_{g_-}c$ is the charged-grain gyrofrequency and τ_{g_-n} the grain-neutral collision time, reveals whether the charged grains are ($\Gamma_{dir} > 1$) or are not ($\Gamma_{dir} < 1$) directly attached to magnetic field lines. We find that, even at high densities, for which $\Gamma_{dir} \ll 1$ and one would expect complete decoupling of the grains from the field due to frequent collisions with neutrals, the grains remain at least partially attached to the field provided that

$$\Gamma_{indir} \equiv [(n_i - n_e)/n_i]\,\omega_i\tau_{in} = (n_{g_-}/n_i)\,\omega_i\tau_{in}\tag{4}$$

is $\gtrsim 1$. This is an *indirect attachment* mechanism of the grains to the field. It is due to electrostatic attraction between electron-shielded, magnetically attached ions ("*quasiparticles*"), each having an effective charge $e(n_i - n_e)/n_i$, and the detached grains; hence the appearance of the *ion* gyrofrequency $\omega_i = eB/m_ic$ and collision time with neutrals τ_{in} in the expression for Γ_{indir}. The quantities Γ_{dir} and Γ_{indir} are shown in Figure 2b as functions of central density. It is clear that, if it were not for the grain-quasiparticle attraction, the grains would detach from the field at $n_{n,c}/n_{n,c0} \gtrsim 10^2$. In reality, they remain partially coupled to the field even at the end of the run, at $n_{n,c} = 2.6 \times 10^9 \; cm^{-3}$.

Customarily, it is thought that the microscopic physical processes that determine the degree of ionization affect the rate at which ambipolar diffusion progresses and, therefore, the evolution of a cloud. Not much attention has been paid to the effect of the evolution on the microscopic physics. Figure 3a shows the central number densities, relative to that of the neutrals, of ions ($x_{i,c}$), electrons ($x_{e,c}$), charged grains ($x_{g_-,c}$), and neutral grains ($x_{g0,c}$) as functions of central density enhancement $n_{n,c}/n_{n,c0}$. It is clear that, during the quasistatic phase of contraction, the dust-to-gas ratio *decreases* in the central flux tubes of a cloud because the grains are well attached to the magnetic field and are thus left behind by the contracting neutrals. This reduces the rate of electron and ion capture by grains. If we represent the relation between the ion and neutral densities in the usual manner, $n_i \propto n_n^k$, *no constant k can accurately describe the ion density at all times*. Figure 3b exhibits the central value of the exponent k as a function of central density enhancement. It is seen (1) that $k_c > 1/2$ during the quasistatic phase (i.e., there are more ions than one would deduce from chemistry in static cloud models); and (2) that k_c decreases to Z 0.1 at the end of the run. *Hence, the magnetic field, through its effects on the evolution, affects significantly the chemistry in contracting cores.*

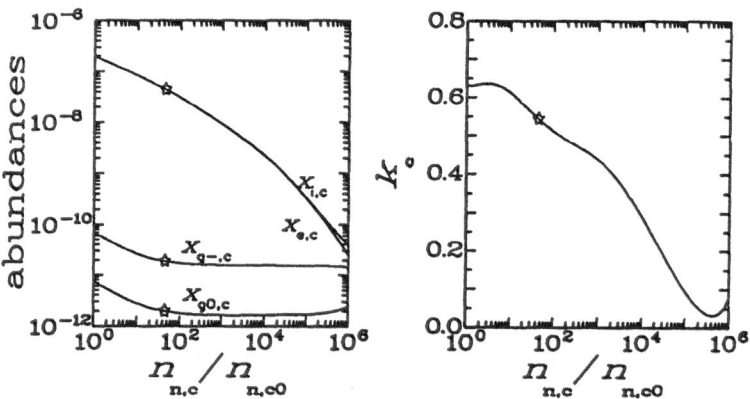

Fig. 3. (*a, left*) Central abundances of ions ($x_{i,c}$), electrons ($x_{e,c}$), charged grains ($x_{g_-,c}$), and neutral grains ($x_{g0,c}$), relative to neutral molecules, as functions of central density enhancement $n_{n,c}/n_{n,c0}$. (*b, right*) Exponent $k_c \equiv d\ln n_{i,c}/d\ln n_{n,c}$ as a function of $n_{n,c}/n_{n,c0}$.

The precise factor by which the dust abundance decreases in a supercritical core relative to its value in the envelope depends on the degree to which grains are attached to the field. For well attached grains (i.e., $a \lesssim 7 \times 10^{-6} \; cm$), this factor is equal to $1/\mu_{c0}$, the initial central mass-to-flux ratio in units of its critical value for collapse. In practice, the grain abundance in a core relative to that of the parent envelope may be easier to determine observationally than the *initial* mass-to-flux ratio of the core, which is essentially that of the envelope in the vicinity of the core. Hence, one may turn the problem around and, from the (observed) reduction of the grain abundance in a core, one may straightforwardly infer the initial mass-to-flux ratio relative to its critical value. Only after a core becomes magnetically and thermally supercritical, and dynamical contraction ensues, does its grain mass fraction become "frozen" in the matter. Since small grains are better attached to the magnetic field lines than more massive grains, the dynamical effect described above implies that the larger grains will tend to be more abundant in high density cores. To observations this would appear as grain growth with increasing density (see Vrba et al. 1981, 1993), when it may actually be a simple and direct consequence of the fact that the small grains, being better attached to the field lines, are preferentially left behind during the formation and contraction of protostellar cores.

These results have been used to construct a detailed evolutionary model for the Barnard 1 cloud (Crutcher, Mouschovias, Troland, and Ciolek 1994), with excellent agreement between the theoretical predictions and observations.

3.1   Effect of UV Ionization

In addition to cosmic-ray ionization and the microscopic and macroscopic effects of grains described above, Ciolek & Mouschovias (1995) have included the effect of UV ionization. We take $\zeta_{uv} = 1.2 \times 10^{-11} \; s^{-1}$ at the cloud surface, and we account for the attenuation due to grain absorption and carbon ionization. In addition to the ions HCO^+, Mg^+, and Na^+, the ions S^+, Si^+, and C^+ now play a significant role. Figure 4a shows the infall speed of the neutrals as a function of radius at six different times. The dashed line represents the velocity profile at the end of the run in the absence of UV ionization (the model cloud of Figs. 1 - 3). Clearly, although the infall speed within the core is virtually unaffected by UV ionization, infall is drastically reduced in the envelope. Figure 4b exhibits the mass infall rate dM/dt (in $M_\odot$/Myr) as a function of r/R_0 at six different times. The dashed line represents the infall rate at the end of the run in the absence of UV ionization, as in Figure 4a. Although the formation and evolution of the supercritical protostellar core is unaffected (except for the fact that it now takes 30% *less* time to form), *infall beyond the supercritical core is effectively cut off by the higher degree of ionization in the envelope.* The maximum mass that can be expected to accrete onto the forming star is therefore that of the supercritical core. Some of the accreting mass may be subsequently ejected in the form of a wind during the opaque (nonisothermal) phases of star formation.

4   Conclusion

Interstellar grains play a role in the microscopic processes that determine the degree of ionization and, therefore, the rate at which ambipolar diffusion leads to the formation and contraction of protostellar fragments (or cores) in molecular clouds. By being directly or indirectly attached to the magnetic field, they add grain-neutral drag to the already existing ion-neutral drag and lengthen

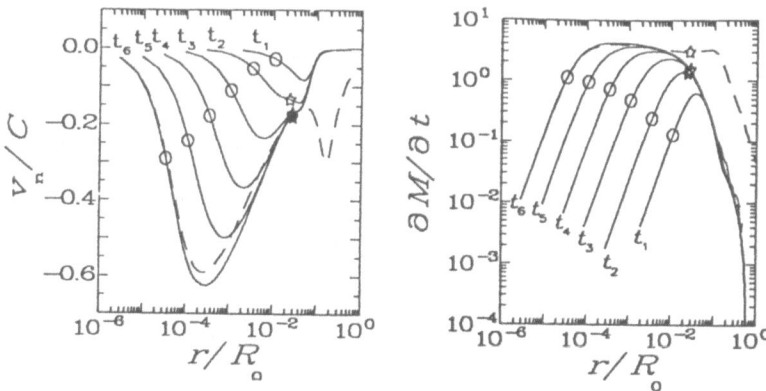

Fig. 4. Spatial profiles of (*a, left*) the neutral velocity v_n in the equatorial plane, normalized to the isothermal sound speed $C = 0.188$ km s^{-1}, and (*b, right*) the mass infall rate (in $M_\odot$/Myr), at six times t_1, t_2, ..., t_6 for a model identical to that of Figs. 1 - 3, except that UV ionization is now accounted for. The times are chosen such that the central density at time t_j exceeds its initial value by the factor 10^j; they are equal to 8.614, 10.507, 10.774, 10.818, 10.827, and $10.829\tau_{ff,c0}$; $\tau_{ff,c0} = 1.3295$ Myr. The "stars" and open circles have the same meaning as in Fig. 1.

the core formation time typically by 50%. Direct attachment occurs when the grain gyrofrequency exceeds the collision frequency with neutrals. However, even when grains would be detached from the field by collisions, they can remain at least partially attached because of electrostatic attraction with ion "quasiparticles". (These are electron-shielded ions, well attached to the field at densities $\lesssim 10^{10}$ cm^{-3}.) We find that the process of core formation itself affects the relative abundance of grains (hence, the chemistry) in dense cores in a profound way: grains attached to the field are left behind during the quasistatic phase of contraction, thus reducing the dust-to-gas relative abundance by the factor $1/\mu_{c0}$ (where μ_{c0} is the initial central mass-to-flux ratio in units of its critical value for collapse). Aside from its evident implications on the chemistry in dense cores, this result can be used to deduce the initial mass-to-flux ratio of a cloud if the relative abundances of grains in a core and the envelope are observed. Also, since small grains are better attached to the field than larger grains, we predict a stratification of grains according to size during core formation. This effect gives the *impression* of grain growth with optical depth. A number of commonly used assumptions (e.g., a fixed dust-to gas mass ratio as a function of density, and the parametrization of the ion density in terms of the neutral density as a power law with a single, constant exponent) are found to be incorrect. We have described a modern theory of star formation incorporating these new results.

Acknowledgement. This work was supported in part by the National Science Foundation under grant AST 93-20250.

References

Baade, W. 1944, ApJ, 100, 137

Bachiller, R., Menten, K. M., & del Rio-Alvarez, S. 1990, A&A, 236, 461

Baker, P. L. 1979, A&A, 75, 54

Basu, S., & Mouschovias, T. Ch. 1994, ApJ, 432, 720

Basu, S., & Mouschovias, T. Ch. 1995a, ApJ, 452, 386

Basu, S., & Mouschovias, T. Ch. 1995b, ApJ, 453, 268

Ciolek, G. E., & Mouschovias, T. Ch. 1993, ApJ, 418, 774

Ciolek, G. E., & Mouschovias, T. Ch. 1994, ApJ, 425, 142

Ciolek, G. E., & Mouschovias, T. Ch. 1995, ApJ, 454, 194

Ciolek, G. E., & Mouschovias, T. Ch. 1996, ApJ, to be submitted

Crutcher, R. M., Mouschovias, T. Ch., Troland, T. H., & Ciolek, G. E. 1994, ApJ, 427, 839

Draine, B. T., & Sutin, B. 1987, ApJ, 320, 803

Elmegreen, B. G. 1979, ApJ, 232, 729

Fiedler, R. A., & Mouschovias, T. Ch. 1992, ApJ, 391, 199

Fiedler, R. A., & Mouschovias, T. Ch. 1993, ApJ, 415, 680

Gail, H. P., & Sedlmayr, E. 1975, A&A, 41, 359

Havnes, O., Hartquist, T. W., & Pilipp, W. 1987, in Physical Processes in Interstellar Clouds, ed. G. E. Morfill & M. Scholer (Dordrecht: Reidel), 389

Mestel, L., & Paris, R. B. 1979, MNRAS, 187, 337

Mouschovias, T. Ch. 1976, ApJ, 207, 141

Mouschovias, T. Ch. 1977, ApJ, 211, 147

Mouschovias, T. Ch. 1979a, ApJ, 228, 159

Mouschovias, T. Ch. 1979b, ApJ, 228, 475

Mouschovias, T. Ch. 1982, ApJ, 252, 193

Mouschovias, T. Ch. 1987a, in Physical Processes in Interstellar Clouds, ed. G. E. Morfill and M. Scholer (Dordrecht: Reidel), 453

Mouschovias, T. Ch. 1987b, in Physical Processes in Interstellar Clouds, ed. G. E. Morfill & M. Scholer (Dordrecht: Reidel), 491

Mouschovias, T. Ch. 1991a, ApJ, 373, 169

Mouschovias, T. Ch. 1991b, in The Physics of Star Formation and Early Stellar Evolution, ed. C. J. Lada and N. D. Kylafis (Dordrecht: Kluwer), 61

Mouschovias, T. Ch. 1991c, in The Physics of Star Formation and Early Stellar Evolution, ed. C. J. Lada and N. D. Kylafis (Dordrecht: Kluwer), 449

Mouschovias, T. Ch. 1995, in The Physics of the Interstellar Medium and Intergalactic Medium, ed. A. Ferrara, C. F. McKee, C. Heiles, and P. R. Shapiro (San Francisco: ASP), vol. 80, 184

Mouschovias, T. Ch., & Paleologou, E. V. 1979, ApJ, 230, 204

Mouschovias, T. Ch., & Paleologou, E. V. 1980, ApJ, 237, 877

Mouschovias, T. Ch., & Spitzer, L., Jr. 1976, ApJ, 210, 326

Nakano, T. 1976, PASJ, 28, 355

Nakano, T., & Nakamura, T. 1978, PASJ, 30, 671

Nakano, T., & Tademaru, T. 1972, ApJ, 173, 87

Nakano, T., & Umebayashi, T. 1980, PASJ, 32, 613

Nakano, T., & Umebayashi, T. 1986a, MNRAS, 218, 663

Nakano, T., & Umebayashi, T. 1986b, MNRAS, 221, 319

Nishi, R., Nakano, T., & Umebayashi, T. 1991, ApJ, 368, 181

Payne, H. E., Salpeter, E. E., and Terzian, Y. 1984, AJ, 89, 668

Spitzer, L., Jr. 1941, ApJ, 93, 369

Spitzer, L., Jr. 1968, Diffuse Matter in Space (New York: Interscience)

Spitzer, L., Jr., & Tomasko, M. G. 1968, ApJ, 152, 971

Umebayashi, T., & Nakano, T. 1990, MNRAS, 243, 103

Vrba, F. J., Coyne, G. V., & Tapia, S. 1981, ApJ, 243, 489

Vrba, F. J., Coyne, G. V., & Tapia, S. 1993, AJ, 105, 1010

Wood, D. O. S., Myers, P. C., & Daugherty, D. A. 1994, ApJS, 95, 457

The Role of Dust in the Dissipation of Magnetic Fields in Molecular Clouds

Takenori Nakano[1], Ryoichi Nishi[2], and Toyoharu Umebayashi[3]

[1] Nobeyama Radio Observatory, National Astronomical Observatory, Nobeyama, Minamisaku, Nagano 384-13, Japan
[2] Department of Physics, Kyoto University, Sakyo-ku, Kyoto 606-01, Japan
[3] Data Processing Center, Yamagata University, Yamagata 990, Japan

Abstract. Dust grains play important roles in magnetic field dissipation through recombination of ions and electrons at grain surfaces and interaction of charged grains with magnetic fields. We show that not much dust is lost from the central part of the cloud even if much magnetic flux is lost.

1 Introduction

Dissipation of magnetic fields in molecular clouds is related to some important problems in star formation. The cloud with subcritical mass contracts quasistatically dissipating magnetic fields as investigated by Nakano (1979, 1982, 1983b) and followed by many authors (Lizano and Shu 1989; Tomisaka et al. 1990; Fiedler and Mouschovias 1993; Ciolek and Mouschovias 1994; Basu and Mouschovias 1994). The rate of the quasistatic contraction depends on the dissipation rate of magnetic fields, and characteristics of the contraction (e.g., nonhomology) depend on the density-dependence of the dissipation rate. The magnetic flux to mass ratio, M/Φ, of the cloud, in which the magnetic force nearly balances the self-gravity, is several hundred to 10^5 times greater than the ratio of the magnetic star (Nakano 1983a, 1984), and the magnetic flux of the cloud is not much less than this critical value. Therefore, most of the initial magnetic flux of at least the central part of the cloud must be lost at some stage of star formation. The phase of star formation at which extensive flux loss occurs depends on the density-dependence of the dissipation rate (Nakano 1984).

Dust grains play some important roles in magnetic field dissipation. First, ions and electrons recombine at grain surfaces. This lowers the densities of ions and electrons, and as a result enhances the dissipation rate. Second, a significant fraction of grains are charged and interact with magnetic fields suppressing the dissipation rate. Both have great effect on the dissipation of magnetic fields.

2 Dissipation Processes

As a dissipation process of magnetic fields one may think of, first of all, the Ohmic dissipation. However, this is quite inefficient in the ordinary molecular clouds. For instance, for a cloud of length scale $L \approx 0.1\,\mathrm{pc}$ and the ionization fraction $n_e/n_H \approx 2 \times 10^{-8}$ at a cloud density $n_H \approx 10^4\,cm^{-3}$ by hydrogen number, the time scale of the Ohmic dissipation takes a value $t_{od} \approx 4 \times 10^{17}\,yr$.

Mestel and Spitzer (1956) showed that there is a much more efficient dissipation process in a weakly ionized cloud, which is now called the ambipolar diffusion or the plasma drift: although ions (and electrons) are strongly frozen to magnetic fields, ions and magnetic fields can drift in the sea of neutral particles in a time scale much shorter than t_{od} because of low ionization fraction.

Because electrons and ions can stick to dust grains, they are charged even in molecular clouds and interact with magnetic fields. Because of large mass, grains are not strongly frozen to magnetic fields. Elmegreen (1979) investigated the field dissipation taking into account the effect of partial freezing to magnetic fields assuming that all grains have the same size.

The strength of coupling of a charged particle ν to magnetic fields is characterized by the quantity $\tau_\nu \omega_\nu$, where τ_ν is the viscous damping time of the motion of ν relative to the neutrals and ω_ν is the cyclotron frequency: when $\tau_\nu \omega_\nu \gg 1$, the particle is strongly frozen to magnetic fields. The value of $\tau\omega$ differs greatly among particles: e.g., $\tau\omega$ for a grain of charge $-e$, e being the unit electric charge, with radius 0.25 and 0.01μm is, respectively, 3×10^8 and 4×10^5 times smaller than that for an electron. Because $\tau\omega$ decreases as the density increases, even ions may not be frozen in very dense clouds. Thus we need a method of treating magnetic field dissipation by taking into account the effect of partial freezing for all kinds of charged particles. Nakano (1984) and Nakano and Umebayashi (1986a) made such a formulation which can be applied to the clouds containing arbitrary kinds of charged particles by extending Elmegreen's method.

We summarize some of their results. We restrict ourselves here to non-rotating axisymmetric clouds. In the poloidal plane on which the magnetic field B and the magnetic force $j \times B/c$ lie, the charged particle ν and the magnetic field drift relative to the neutrals parallel to the magnetic force with velocities

$$v_\nu = \frac{\omega_\nu}{\tau_\nu \Omega_\nu^2} \frac{A_1 \tau_\nu \omega_\nu + A_2}{A} \frac{1}{c}|j \times B|, \tag{1}$$

$$v_B = \frac{A_1}{A} \frac{1}{c}|j \times B|, \tag{2}$$

respectively, where j is the electric current density, $\omega_\nu = q_\nu e B/m_\nu c$ is the cyclotron frequency of particle ν with mass m_ν and electric charge $q_\nu e$, $\Omega_\nu^2 = \omega_\nu^2 + \tau_\nu^{-2}$, $A_1 = \sum_\nu \rho_\nu \omega_\nu^2/\tau_\nu \Omega_\nu^2$, $A_2 = \sum_\nu \rho_\nu \omega_\nu/\tau_\nu^2 \Omega_\nu^2$, $A = A_1^2 + A_2^2$, and ρ_ν is the mass density of particle ν: ω_ν is defined to have a negative value for a negatively charged particle. The dissipation time of magnetic fields can be given by $t_B \approx L/v_B$, where L is the length scale of magnetic fields.

When dominant charged particles are strongly coupled with magnetic fields, or $|\tau_\nu \omega_\nu| \gg 1$, we can easily confirm that the velocity v_B given by equation (2) reduces to the drift velocity of the ambipolar diffusion. Similarly, when dominant charged particles are not frozen, or $|\tau_\nu \omega_\nu| \ll 1$, the dissipation time t_B reduces to the Ohmic dissipation time $t_{od} \approx 4\pi\sigma_c L^2/c^2$, where $\sigma_c = \sum_\nu q_\nu^2 e^2 \tau_\nu n_\nu/m_\nu$ is the electrical conductivity, with n_ν the number density of particle ν. Thus our formalism includes the ambipolar diffusion and the Ohmic dissipation as limiting cases and can be applied to any situations of weakly ionized clouds.

3   Densities of Charged Particles

Densities of various charged particles are crucial in the dissipation of magnetic fields. We consider here the molecular clouds shielded from the ultraviolet radiation. Ions are first formed mainly through ionization of H_2 molecules and He atoms by cosmic rays and radioactive elements. The ionization rate of an H_2 molecule is given by $\zeta = \zeta_{CR} \exp(-\Sigma/\chi) + \zeta_R$, where $\zeta_{CR} \approx 1 \times 10^{-17} s^{-1}$ is the ionization rate by cosmic rays in the interstellar space, Σ is the depth from the cloud surface, $\chi \approx 96 \, g \, cm^{-2}$ is the attenuation length of ionization rate (Umebayashi and Nakano 1981), and $\zeta_R \approx 6.9 \times 10^{-23} s^{-1}$ is the ionization rate by radioactive elements. The ionization rate of an He atom is given by 0.84ζ. The effect of radioactive elements becomes dominant only at $\Sigma > 1 \times 10^3 \, g \, cm^{-2}$.

Ions undergo various gas-phase reactions such as formation of molecular ions from atomic ions, dissociative recombination of molecular ions with electrons, charge transfer of molecular ions to metallic atoms, and radiative recombination of atomic (mainly metallic) ions with electrons. We adopt here the reactions and their rate coefficients similar to those in Umebayashi and Nakano (1990).

In addition to the gas-phase reactions, ions and electrons recombine at grain surfaces. A significant fraction of grains is charged because ions and electrons stick to grains when they collide. Although the sticking probability of an ion is very high ($S_i \approx 1$), that of an electron, S_e, was rather uncertain. Umebayashi and Nakano (1980) showed that this is not very small (see also Nishi et al. 1991 for the effect of the curvature of grain surface). We adopt here $S_e = 0.6$. In addition to the Coulomb force we have to consider the polarization force in the collision of a grain with an ion or an electron (Draine and Sutin 1987). We consider seven charge states of grains: 0 (neutral), $\pm e$, $\pm 2e$, and $\pm 3e$. Recombination of ions and electrons occurs, e.g., in such a way: when an ion X^+ collides with a grain of charge $-e$, X^+ and the extra electron on the grain recombine and the atom X or the molecules formed by recombination leaves the grain surface using the energy released by the recombination.

The recombination rates of various ions at grain surfaces depend on the charge state distribution of grains, which is determined by the densities of ions and electrons. Inversely the ion and electron densities depend on the recombination rates at grain surfaces in addition to the gas-phase reactions. Therefore, the charge state of grains and the ion and electron densities have to be determined simultaneously. All the reactions can be regarded as in a steady state.

We adopt the population I chemical composition. A significant fraction of heavy elements is in grains. We introduce the depletion factors of elements in the gas phase: δ_1 is the fraction of C and O remaining in the gas phase, and δ_2 that of metallic elements such as Na, Mg, Al, Ca, and Fe. In diffuse clouds we have $\delta_1 \sim 0.2$ and $\delta_2 \sim 0.02$ (Morton 1974). We consider ice-mantled grains. We assume a power-law size distribution (Mathis et al. 1977) with respect to the

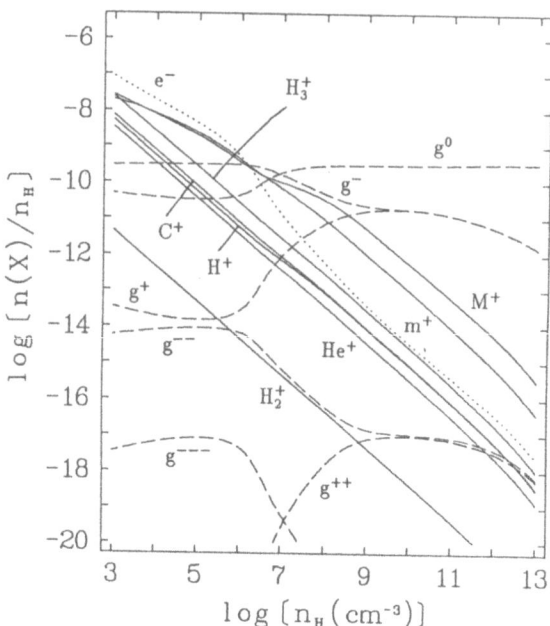

Fig. 1. The abundances of various particles as functions of the cloud density for the case of $\delta_1 = 0.2$, $\delta_2 = 0.02$, and $a_c^{(min)} = 50\,\text{Å}$. M^+ represents the metallic ions, m^+ the molecular ions other than H_3^+ and H_2^+, and g the grains.

radius of the grain core, a_c; $dn_{gr}/da_c = A n_H a_c^{-3.5}$, between $a_c^{(min)}$ and $a_c^{(max)}$, where $a_c^{(max)} \approx 2500\,\text{Å}$, $a_c^{(min)} \lesssim 100\,\text{Å}$, and $A_{gr} = 1.5 \times 10^{-25}\,cm^{-2.5}$. We assume that the thickness of the ice mantle, b_m, is independent of a_c: e.g., we take $b_m = 153\,\text{Å}$ for the case of $\delta_1 = 0.2$, $\delta_2 = 0.02$, and $a_c^{(min)} = 50\,\text{Å}$.

Figure 1 shows an example of the abundances of various particles as functions of the cloud density. As for the cloud configuration, see §4. At $n_H \lesssim 10^6\,cm^{-3}$ ions and electrons are the dominant charged particles, and at higher densities charged grains are dominant. Figure 2 shows the charge state distribution of grains as a function of a_c for the same case as in Fig. 1.

4   The Dissipation Time and Drift of Grains

We consider an oblate cloud which has contracted somewhat and is in quasi-equilibrium along field lines. The half-thickness of the cloud along field lines is given by $Z \approx 0.08(T/10\,K)^{1/2}(10^4\,cm^{-3}/n_H)^{1/2}$ pc, where T is the temperature of the cloud. This is nearly equal to the Jeans length. The mean magnetic force can be scaled by the mean field strength B as $|j \times B|/c = (B/B_{cr})^2|j \times B|_{cr}/c$, where B_{cr} is the critical field with which the magnetic force balances with the gravity, i.e., $|j \times B|_{cr}/c \approx GM\rho/R^2$; M, R, and ρ are the mass, radius across

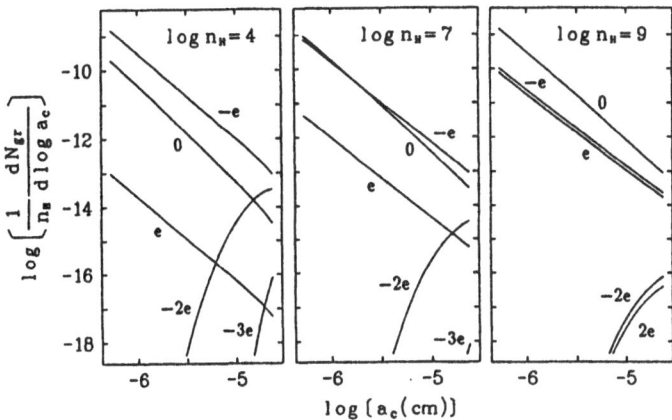

Fig. 2. The charge state distribution of grains as a function of the core radius a_c for the same case as in Fig. 1.

field lines, and mass density, respectively, of the cloud. We have $B_{cr} \approx 1.2 \times 10^{-5}(T/10\,K)^{1/2}(n_H/10^4\,cm^{-3})^{1/2}$ gauss.

The quasistatic contraction of a magnetized cloud induced by field dissipation is highly nonhomologous and the central part of the cloud with size nearly equal to the Jeans length contracts rather rapidly leaving the outer part almost unchanged (e.g., Nakano 1979, 1982, 1983b). The time scale of such contraction is given by the field dissipation time in the central part, $t_B \approx Z/v_B$.

Figure 3 shows the dissipation time t_B as a function of the cloud density for a cloud of $B = B_{cr}$ and $a_c^{(min)} = 50\,\text{Å}$ for three cases of the element depletion. The free-fall time t_f is shown by the dotted line. The dissipation time is rather insensitive to the element depletion. There is a critical density $n_H^{(cr)}$ only above which $t_B < t_f$ holds. At $n_H \lesssim 0.1n_H^{(cr)}$, t_B takes a value between 10 and 400 times t_f for clouds with $B = B_{cr}$. For $B < B_{cr}$ the time scale t_B is longer: $t_B \propto (B/B_{cr})^{-2}$ at $n_H \ll n_H^{(cr)}$ and t_B is rather insensitive to B at $n_H \gtrsim n_H^{(cr)}$. The critical density $n_H^{(cr)}$ takes a value between 3×10^9 and $6 \times 10^{10}\,cm^{-3}$ depending on the grain model and the element depletion. As mentioned in §1, most of the initial magnetic flux of the cloud core must be lost at some stage of star formation. When the magnetic flux is slightly smaller than the critical flux, or B is slightly weaker than B_{cr}, the cloud contracts dynamically (the gas pressure cannot keep the cloud in equilibrium for a long time because an isothermal cloud is gravitationally unstable). Because the dynamical contraction is not much slower than the free fall and $t_B \gg t_f$ at $n_H \ll n_H^{(cr)}$, extensive magnetic flux loss occurs only at $n_H \gtrsim n_H^{(cr)}$. At $n_H \ll n_H^{(cr)}$, the contraction of a cloud with $B \approx B_{cr}$ induced by field dissipation is quasistatic because $t_B \gg t_f$.

Figure 4 compares the drift velocity v_g of the grains of charge $-e$ with that of magnetic fields, v_B, at some cloud densities for the same case as the solid line in Fig. 3. At low densities small grains move almost with magnetic fields, but large

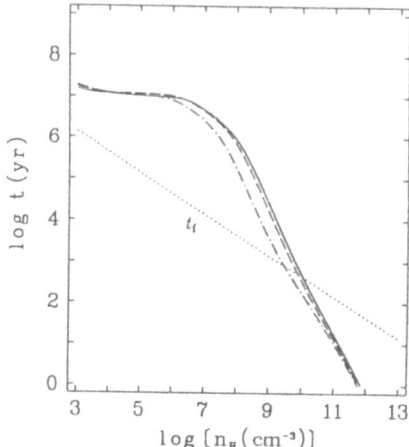

Fig. 3. The dissipation time t_B for a cloud of $B = B_{cr}$ and $a_c^{(min)} = 50$ Å for three cases of element depletion: the solid line is for the case of $\delta_1 = 0.2$ and $\delta_2 = 0.02$, the dashed line for $\delta_1 = 0.02$ and $\delta_2 = 0$, and the dot-dashed line for $\delta_1 = \delta_2 = 0$. For comparison the free-fall time t_f is shown by the dotted line.

grains do not. At high densities even small grains do not move with magnetic fields. For clouds with $B < B_{cr}$ the ratio v_g/v_B is smaller because the freezing to magnetic fields breaks down easier for grains than for ions and electrons at weaker magnetic fields.

We discuss here how much dust is lost from the cloud core by interaction with magnetic fields. A cloud with supercritical mass $[M/\Phi > (M/\Phi)_{cr}$, or $B < B_{cr}]$ contracts dynamically because the magnetic force is weaker than the gravity. Because t_B is longer in such a cloud than in the critical cloud with $B = B_{cr}$, the magnetic flux, and then grains, are hardly lost at $n_H \ll n_H^{(cr)}$. Significant flux loss occurs only at $n_H \gtrsim n_H^{(cr)}$, where even the smallest grains do not move with magnetic fields. Thus grains are hardly lost from such clouds.

A slightly subcritical cloud $[M/\Phi$ is only slightly smaller than $(M/\Phi)_{cr}$, or B is only slightly stronger than $B_{cr}]$ contracts quasistatically by losing magnetic flux from the central part. In this contraction a small decrease of magnetic flux brings about a large increase of the density (e.g., Nakano 1979, 1982, 1983b). As a result the cloud becomes very dense and even the smallest grains lose strong coupling with magnetic fields. Thus although only a small fraction of small grains may be lost, large grains are hardly lost.

A substantially subcritical cloud $[M/\Phi \lesssim 0.5(M/\Phi)_{cr}$, or $B \gtrsim 2B_{cr}]$ may contract quasistatically by losing the magnetic flux from the central part if the expansion of the cloud is inhibited by an external pressure or external magnetic fields. In this case the density increase induced by the flux loss is not so large as in the case of the slightly subcritical clouds, and a significant fraction of small grains may be lost from the central part, though large grains are not much

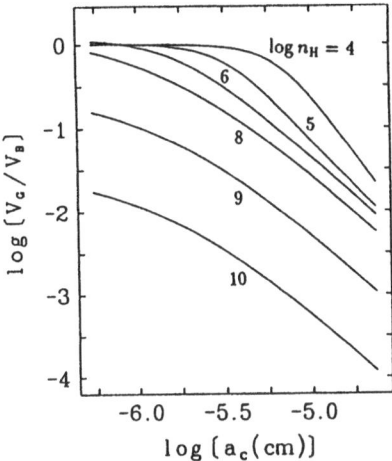

Fig. 4. The ratio of the drift velocity of grains of charge $-e$, v_g, to that of magnetic fields, v_B, at some cloud densities for the same case as the solid line in Fig. 3. Each curve is labeled with the value of $\log n_H$ in cm^{-3}.

lost. However, in such a cloud the magnetic force is more than about four times stronger than the self-gravity, and the cloud expands nearly freely unless it is inhibited by an external pressure or external magnetic fields. In the computer simulations of the quasistatic contraction of magnetized clouds, the magnetic field is usually assumed to approach a fixed uniform field far away from the cloud. In other words the field lines are pinned at the outer boundary of the computation region. With such a boundary condition the expansion of even a cloud with $B \gg B_{cr}$ is inhibited. There are no such pins in the universe, and substantially subcritical clouds are not realistic.

Thus the interaction of grains with magnetic fields cannot bring about significant loss of grains even from the central part of the cloud. This is consistent with the observational fact that the abundance of heavy elements in the interstellar medium is not significantly higher than that in population I stars.

5 Discussion

The charge of grains changes stochastically by sticking of ions and electrons. The relative abundance among charge states as indicated in Fig. 2 shows the relative stay time in each charge state. The change in the charge state has some effect on the grain motion, which Elmegreen (1979), Nakano (1984), and Nakano and Umebayashi (1986a) neglected. Ciolek and Mouschovias (1993) considered this effect for neutral and $-e$ grains of radius $a = 0.1\,\mu$m. At low densities this has significant effect on the largest grains though the effect on small grains is negligible compared with the frictional force of the neutrals. At high densities the effect is small even for the largest grains since the low ionization fraction makes the charge-change time much longer than the viscous damping time. Because the coupling with magnetic fields is mainly contributed by small grains, whose motion is hardly affected by charge change, this has little effect on t_B. Because this has an effect of decreasing the drift velocity v_g of $-e$ grains, v_g cannot be greater than that shown in Fig. 4. Therefore, the conclusion on the loss of grains in §4 holds even if this effect is taken into account.

Ciolek and Mouschovias (1993) refer our previous papers (Nakano and Umebayashi 1980, 1986a, b; Umebayashi and Nakano 1990; Nishi et al. 1991) as if we concluded that the ambipolar diffusion is inefficient at $n_H < n_H^{(cr)}$ (the value of $n_H^{(cr)}$ differs somewhat among these papers because of different grain models, etc.). What we have concluded in these papers is that the extensive magnetic flux loss from the critical flux occurs only at $n_H \gtrsim n_H^{(cr)}$. We have shown in these papers the figures of the time scale t_B as a function of the cloud density similar to Fig. 3 in this article. One can find out the efficiency of the ambipolar diffusion (more generally the field dissipation) as a function of n_H from these figures. Ciolek and Mouschovias also assert that the comparison of t_B with t_f and the comparison of v_B with the free-fall velocity in our papers is meaningless because the subcritical clouds (with $B > B_{cr}$) cannot contract freely. However, if we find $t_B \gg t_f$, we can conclude that the contraction of subcritical clouds induced by field dissipation is quasistatic, and from the ratio t_B/t_f we can find out how slow the contraction is compared with the dynamical contraction. Thus Ciolek and Mouschovias' criticisms on our previous papers are quite irrelevant.

References

Basu, S., Mouschovias, T. Ch. (1994) Ap. J. **432**, 720
Ciolek, G. E., Mouschovias, T. Ch. (1993) Ap. J. **418**, 774
Ciolek, G. E., Mouschovias, T. Ch. (1994) Ap. J. **425**, 142
Draine, B. T., Sutin, B. (1987): Ap. J. **320**, 803
Elmegreen, B. G. (1979): Ap. J. **232**, 729
Fiedler, R. A., Mouschovias, T. Ch. (1993) Ap. J. **415**, 680
Lizano, S., Shu, F. H. (1989) Ap. J. **342**, 834
Mathis, J. S. Rumpl, W. Nordsieck, K. H. (1977): Ap. J. **217**, 425
Mestel, L. Spitzer, L., Jr. (1956): Mon. Not. Roy. Astr. Soc. **116**, 503
Morton, D. C. (1974): Ap. J. **193**, L35
Nakano, T. (1979): Pub. Astr. Soc. Japan **31**, 697
Nakano, T. (1982): Pub. Astr. Soc. Japan **34**, 337
Nakano, T. (1983a): Pub. Astr. Soc. Japan **35**, 87
Nakano, T. (1983b): Pub. Astr. Soc. Japan **35**, 209
Nakano, T. (1984): Fund. Cosmic Phys. **9**, 139
Nakano, T., Umebayashi, T. (1980): Pub. Astr. Soc. Japan **32**, 613
Nakano, T., Umebayashi, T. (1986a): Mon. Not. Roy. Astr. Soc. **218**, 663
Nakano, T., Umebayashi, T. (1986b): Mon. Not. Roy. Astr. Soc. **221**, 319
Nishi, R., Nakano, T,. Umebayashi, T. (1991): Ap. J. **368**, 181
Tomisaka, K., Ikeuchi, S., Nakamura, T. (1990): Ap. J. **362**, 202
Umebayashi, T., Nakano, T. (1980): Pub. Astr. Soc. Japan **32**, 405
Umebayashi, T., Nakano, T. (1981): Pub. Astr. Soc. Japan **33**, 617
Umebayashi, T., Nakano, T. (1990): Mon. Not. Roy. Astr. Soc. **243**, 103

The Effects of Gas-Grain Interactions on Protostellar Line Profiles

J.M.C. Rawlings[1]

Department of Physics and Astronomy, University College London, Gower Street, London WC1E 6BT, Great Britain

Abstract. The chemical role of dust in infalling star-forming clouds is discussed. The depletion of gas-phase material as a result of freeze-out onto the surface of dust grains has important implications. Whilst the high velocity wings of some molecular species (such as NH_3) are suppressed, non-linear chemical effects result in chemical enhancements of others (eg. HCO^+). Preliminary results from a self-consistent chemical/hydrodynamical model of the protostellar core B335 are presented.

1 Introduction

The importance of the role of dust in protostellar collapse dynamics is well established. However, the effects of gas-grain interactions have previously only been considered in a very rudimentary fashion. This interaction may have important controlling effects, particularly in the determination of the ionization level within partially magnetically supported clouds. Since the efficiency of that support is critically dependent on the ionization level, the significance of gas-grain interactions is obvious (Hartquist et al. 1993). In addition, depletion of molecules from the gas phase into icy grain mantles has important implications with respect to the observability of protostellar inflows. We describe here the results that are beginning to emerge from models whose aim is to describe self-consistently the chemistry and hydrodynamics of star-forming cores. These models should be able to identify, unequivocally, the spectral signatures of collapse, specify the collapse parameters (such as the mass of the protostellar core and the collapse age) and provide some sort or arbitration in the much debated arguments concerning the dynamical nature of the collapse process.

2 Spectral Line Identifications of Protostellar Inflow

The cloud cores associated with star formation are cold ($\leq 20K$) and nearly isothermal, implying that low lying molecular rotational levels are optically thick. In addition, the velocities and velocity dispersions associated with infall are not expected to be large ($\lesssim 1$–2 kms^{-1}) so that most of the material in a core is radiatively coupled and the line profiles are hard to deconvolve. Moreover, the higher velocity material may be depleted from the gas phase into icy mantles. At later stages of evolution contamination of the inflow features by the ever-present (bipolar) outflows becomes a serious problem. Never the less, these problems are minimised in the lowest mass protostellar cores. Studies of the morphologies of molecular maps and of spectral energy distributions have their use in the identification of possible collapse candidates but often have ambiguous, model-dependent interpretations. Unambiguous identification of infall can only come from multiple spectral line observations of several molecular species. The lower energy transitions are often optically thick in which case they may exhibit the double peaked (self-reversed) profile with the blue wing stronger than the red wing that is characteristic of infall (see Zhou et al. 1993 for a simple explanation). However, amongst other criteria, this requires the presence of positive temperature and velocity gradients.

There is a very narrow 'window' in the protostellar evolutionary track where these conditions may be met and only a handful of candidate objects (including B335, IRAS 16293 and L1527) have been identified. Class 0 objects show signs of early dynamical activity, but they do not form a physically homogeneous group. The youngest sources are in the earliest stages of isothermal collapse (and hence do not exhibit asymmetric line profiles), whilst the later Class 0 sources have a central source and deviate significantly from isothermality. Non-polar ices (such as N_2, O_2, CO and CH_4) start to evaporate from grain mantles at temperatures as low as 20–30K (Mumma et al. 1993) with the result that the infall features in the line profiles become highly confused with hot material desorbed from grain mantles (eg. as seen in IRAS 16293–2422, Blake et al. 1994).

3   Chemical Models

Further complications arise when we relax the (invalid) assumption that the chemical fractional abundances are temporally and spatially invariant. In cold, dark cores, the dominant chemical process is the depletion, or freeze-out, of molecular material onto the surface of dust grains. High resolution VLA observations are now showing direct evidence for NH_3 depletion by 2–3 orders of magnitude in the innermost regions of protostellar cores (Wootten 1995). This may seem very discouraging; if the innermost, high velocity material is depleted then infall line broadening may be very difficult to detect.

However, in a model of the L1498 core, Rawlings et al. (1992) showed that, owing to non-linearities in the chemistry, some molecular species (eg. HCO^+, CH, OH, N_2H^+) are temporarily enhanced as a result of the depletion process and peak at non-zero infall velocities. The intrinsic line profiles of these species are expected to be significantly broader (with respect to species such as NH_3 and H_2O which simply decline monotonically as freeze-out ensues) and characteristic of the infall dynamics, being uniquely determined for each molecular transition. Thus the observation of a number of molecular species, coupled to descriptive chemical models may act as a very powerful diagnostic probe.

As a test case we have considered the well known infall candidate, B335. This object is both spatially and kinematically well resolved (Zhou et al. 1990,1993) and shows very clearly the spectral features described above (Choi et al. 1995).

It is also cold (13K) and is at least *consistent* with the Shu (1977) inside-out collapse model (with $r_{infall}=0.03$pc).

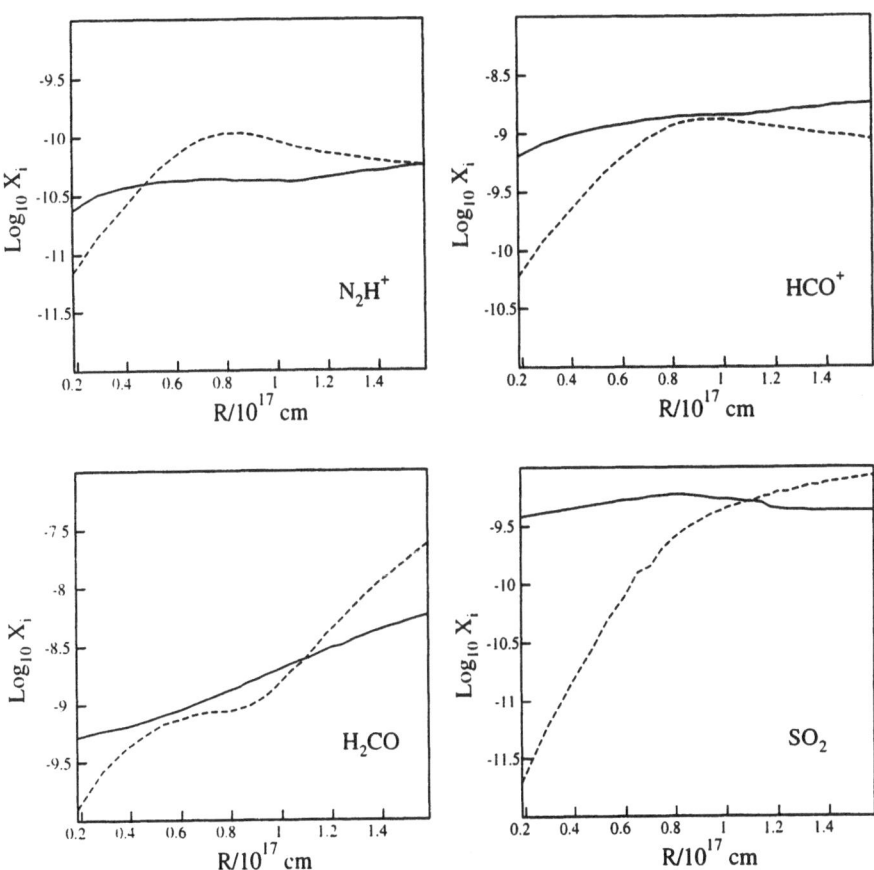

Fig. 1. Preliminary results from modelling of B335. Results are presented as the logarithm of the fractional abundances as functions of radius. Two possible solutions are shown, both of which are consistent with the chemical and dynamical constraints. The solid lines show the abundances obtained for $A_V=5.5$ and immediate collapse after the formation of the isothermal pressure balanced core. The dashed lines are for $A_V=7.0$ and allowing for a period of hydrostatic equilibrium (prior to the collapse) of some 5×10^5 years.

When studying the chemistry it is *essential* to realise that the chemical abundances, being time-dependent, are strongly dependent on the dynamical evolutionary history of the core. It is *not* satisfactory to simply 'add' chemical processes to one part of the dynamical evolution. Indeed, since the timescales for depletion and dynamical evolution are similar a critical sensitivity may be implied. In our (preliminary) models we have assumed the following dynamical model; isothermal, pressure balanced, protostellar cores form by simple collapse from interclump material that is chemically enriched as a result of continual cycling between clump and interclump media (driven by mass loaded, shocked stellar winds). Hydrodynamic collapse (Shu 1977) is then assumed to start after an unknown period of metastability. The modelling procedure is as follows; the input to the model consists of the density and temperature profiles, and the spectral line profiles. Using a (Monte-Carlo) radiative transfer model, average values of the molecular abundances are established which are then used to constrain the chemical model. The models were initially constrained by the infall radius (assuming the validity of the Shu model), and the abundances of H_2CO (3.6×10^{-9}), CS (3.2×10^{-9}), and NH_3 ($< 2 \times 10^{-9}$). Additional observations of other molecular species are now providing further chemical constraints. The output from the chemical model (in the form of the spatial variations of the molecular abundances) are then fed back into the radiative transfer model which is used to calculate the next generation of line profiles. This procedure is then repeated in an iterative fashion. Examples of the results obtained from the chemical model in the first cycle of such an iteration are given in Fig. 1. Recent modelling of the CS line profiles in B335 by Choi et al. (1995) has indicated that the average CS abundance is 9×10^{-9} in the outer regions and 5×10^{-9} in the core.

Obviously, it is very difficult to draw any conclusions at such an early stage but, for example, by looking at the different abundance profiles shown in Fig. 1 it is clear that we should soon be able to say whether or not, in the case of B335, the collapse is more hydrodynamic (as in Shu 1977) or quasi-static (as in Ciolek & Mouschovias 1994) in nature. Eventually, it should be possible to constrain the large number of free parameters in the model by the self-consistent modelling of many spectral lines.

References

Blake, G.A., Van Dishoeck, E.F., Jansen, D.J., Groesbeck, T., & Mundy, L.G. (1994): ApJ **428**, 680

Choi, M., Evans II, N.J., Gregersen, E.M., & Wang, Y. (1995): ApJ **448**, 742

Ciolek, E.E., & Mouschovias, T.Ch. (1994): ApJ **425**, 142

Hartquist, T.W., Rawlings, J.M.C., Williams, D.A., & Dalgarno, A. (1993): QJRAS **34**, 213

Mumma, M.J., Weissman, P.R., & Stern, S.A. (1993): *Protostars & Planets III* (University of Arizona, Texas), p.1177

Rawlings, J.M.C., Hartquist, T.W., Menten, K.M., & Williams, D.A. (1992): MNRAS **255**, 471

Shu, F.H. (1977): ApJ **214**, 488

Wootten, A. (1995): Ap&SS **224**, 43

Zhou, S., Evans II, N.J., Butner, H.M., Kutner, M.L., Leung, C.M., & Mundy, L.G. (1990): ApJ **363**, 168

Zhou, S., Evans II, N.J., Kömpe, C., & Walmsley, C.M. (1993): ApJ **404**, 232

Chemistry in Molecular Clouds Without and with Dust Coagulation

Ralf M. Sablotny and Thomas Henning

Max Planck Society, Research Unit "Dust in Star–forming Regions", Schillergäßchen 3, D – 07745 Jena, Germany

Abstract. Although calculations of molecular cloud chemistry now include gas–dust interactions in the form of freezeout and desorption processes, these models consider the dust grains to be spherical particles and the effects of grain coagulation have not been taken into account. But as there are strong hints for collisional growth of dust grains to occur in the environment of cold dark clouds, the gas phase abundances of molecules might depend on the morphology of these dust conglomerates. In computer simulations, aggregates are grown by the Ballistic Particle Cluster Aggregation (BPCA) process and used to calculate effective absorption and desorption properties of the dust. These values are then included in a chemical network model of cold, dark molecular clouds.

1 Introduction

In the last years the importance of dust as an active part in the chemistry of molecular clouds became more and more apparent (Millar and Williams 1993). The accretion of gas phase material onto dust is an obvious process to be included in simulations of the chemistry in cold, dense molecular clouds, while desorption seemed to be ineffective (see, *e.g.* Turner 1989). On the other hand, recent detailed investigations of desorption processes demonstrated their relevance to the chemistry of such regions (Bergin et al. 1995, Willacy and Williams 1993, Hasegawa and Herbst 1993). In these studies, however, dust grains have always been considered to be isolated spherical particles. But observational evidence as well as computer simulations (Ossenkopf 1993, Weidenschilling and Ruzmaikina 1994) show, that dust aggregates can grow in molecular cloud cores by collisions and subsequent sticking. This process is known as *coagulation*. The objective of the work presented here is to study the effect of coagulation upon the chemistry of molecular clouds by altering the total dust surface available for accretion and desorption.

2 Coagulation of Dust Grains

The physical properties of aggregates grown by coagulation processes considerably differ from the ones of the isolated particles (Blum et al. 1994, Smirnov 1990, Meakin 1988). For molecular clouds, the physical conditions ensure that ballistic collisions of the dust grains can always be assumed, as they are negligible in size compared to their mean free path. For mono–dispered spheres, two

growth scenarios will represent the extreme cases possible: In the Ballistic Particle Cluster Aggregation (BPCA) a single spherical particle collides with a cluster
much larger in size, while the collision of two equally–sized clusters is called Ballistic Cluster Cluster Aggregation (BCCA). This second mechanism results in
very open *fluffy* aggregates, while the first case yields very compact clusters. A
realistic growth in the ballistic regime will always be a mixture of these scenarios.
In the astrophysical literature, coagulation is studied in molecular clouds and
circumstellar disks (*e.g.* Henning et al. 1995, Weidenschilling and Ruzmaikina
1994, Sterzik and Morfill 1994, Ossenkopf 1993, Weidenschilling et al. 1984).

3   Influence on the Chemical Evolution of Molecular Clouds

For the two extreme growth scenarios of BPCA and BCCA, the influence of
morphology and sticking probability on the deposition of gas was studied by
Henning and Sablotny (1995). For the BCCA case, gas deposition is unaffected
by the growth process or the sticking probability. The BPCA case, however,
yields very compact clusters. For high sticking probabilities, the inner parts of
such aggregates are well shielded from the deposition of gas material. For low
sticking probabilities, no influence was found. Several processes are discussed in
the literature which result in material desorption from grain surfaces (Bergin
et al. 1995, Willacy and Williams 1993, Dzegilenko and Herbst 1995, Hasegawa
and Herbst 1993). In this work, we only included continuous processes. Discon-
tinuously working processes, like shock fronts or supernovae, are considered to
be very drastic in their effects: The energies involved are sufficiently high to
break up molecular bonds, hence "resetting" the chemistry, and to destroy not
only the aggregates but also the individual components, *e.g.* by sputtering. For
the simulation of the chemical evolution of a static molecular cloud (hydrogen
number density $n_H = 10^4 \, \mathrm{cm}^{-3}$, temperature $T = 10 \, \mathrm{K}$), we used the UMIST
database for chemical reactions, RATE94 (Farquhar and Millar 1993, Millar
et al. 1991), enlarged by reactions for accretion and desorption as described by
Willacy (1993). The species file contained 79 gas phase species. Out of this set
30 species could freeze out. The initial abundances were cosmic ones with all
elements in their atomic form, except C, which we took to be ionised (Willacy
1993). For the sake of simplicity, we have not yet considered any chemical reactions on the surfaces apart from the one for the formation of molecular hydrogen. Due to the high mobility of physisorbed H, however, we did not include
any morphology–induced changes to the reaction leading to H_2. The sensitivity
of the gas phase chemistry on the presence of dust grains is demonstrated in
Fig. 1: The solid curves show the case, in which dust grains have no influence
on the model except shielding the molecular cloud from the external radiation
flux and to serving as a platform for the formation of H_2. If accretion onto the
surfaces of isolated, spherical grains is included as the only additional form of
gas–dust interaction, total depletion of gas phase material occurs within a couple
of million years, while the lifetime of molecular clouds are expected to be in the

order of several 10^7 years. Due to desorption, a new steady state can occur as shown by the third class of curves in Fig. 1.

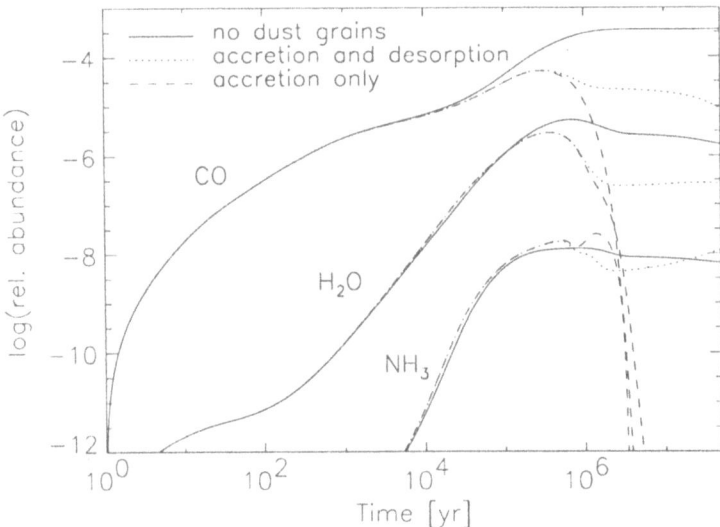

Fig. 1. Chemical evolution without and with dust–gas interactions. For physical conditions, refer to the text. The abundances are normalized relative to n_H.

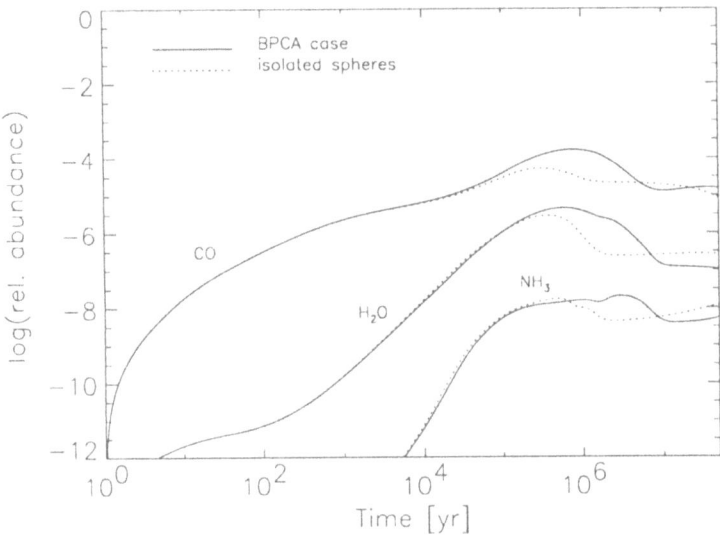

Fig. 2. Evolution of gas phase chemistry for three selected molecules. The influence of coagulation of the dust particles according to the BPCA growth process on accretion and desorption is shown. All abundances are normalized to n_H. For physical conditions, refer to the text.

For aggregates grown by the BPCA-process, the ratio of "visible" surface of the cluster to the total surface of all its components depends on the size and decreases as the clusters become larger. In a computer simulation using 40 BPCA aggregates with 2048 components each, we found a value of 0.2 for this ratio serving as an accretion efficiency by measuring the ratio of the aggregates' surface available for accretion of gas to the total surface area of the components. The diffusion through the aggregate of gas being desorbed from individual components was also simulated. By including the possibility of another accretion event, the effective desorption relative to desorption from isolated spheres was computed. For the same sample of 40 BPCA clusters, we found an desorption efficiency of 0.015. The effect of including these two efficiencies in the chemical network is shown in Fig. 2.

References

Bergin, E. A., Langer, W. D., and Goldsmith, P. F.: 1995, *Ap.J.* **441**, 222

Blum, J., Henning, Th., Ossenkopf, V., Sablotny, R., Stognienko, R., and Thamm, E.: 1994, in M. M. Novak (ed.), *Fractals in the Natural and Applied Sciences*, A–41, pp 47–59, Elsevier Science B. V. (North Holland)

Dzegilenko, F. and Herbst, E.: 1995, *ApJ.* **443**, L81

Farquhar, P. R. A. and Millar, T. J.: 1993, *Analysis of Astronomical Spectra* **18**, 6

Hasegawa, T. I. and Herbst, E.: 1993, *M.N.R.A.S.* **263**, 589

Henning, Th., Michel, B., and Stognienko, R.: 1995, *Planet. Space Sci.* **43**, 1333

Henning, Th. and Sablotny, R. M.: 1995, *Adv. Space Res.* **15(2)**, (2) 17

Meakin, P.: 1988, *Phase Transitions* **12**, 335

Millar, T. J., Rawlings, J. M. C., Bennett, A., Brown, P. D., and Charnley, S. B.: 1991, *Astron. and Astrophys. Suppl. Ser.* **87**, 585

Millar, T. J. and Williams, D. A. (eds.): 1993, *Dust and Chemistry in Astronomy*, The Graduate Series in Astronomy, Institute of Physics Publishing, Bristol and Philadelphia

Ossenkopf, V.: 1993, *Astron. and Astrophys.* **280**, 617

Smirnov, B. M.: 1990, *Physics Reports* **188(1)**, 1

Sterzik, M. F. and Morfill, G. E.: 1994, *Icarus* **111**, 536

Turner, B. E.: 1989, *Space Science Reviews* **51**, 235

Weidenschilling, S., Chapman, C., Davis, D., and Greenberg, R.: 1984, in R. Greenberg and A. Brahic (eds.), *Planetary rings*

Weidenschilling, S. W. and Ruzmaikina, T. V.: 1994, *Ap.J.* **430**, 713

Willacy, K.: 1993, *Ph.D. thesis*, Victoria University of Manchester

Willacy, K. and Williams, D. A.: 1993, *M.N.R.A.S.* **260**, 635

Part IX

Miscellaneous

Optically Thick Main Sequence Evolution for Still Accreting Massive Stars

Paolo A. Bernasconi

Geneva Observatory, Ch. des Maillettes 51, CH-1290 Sauverny, Switzerland

Abstract. We have extended the accretion paradigm to treat the formation of massive stars, modelling the accretion rate history of the observed dynamical state of molecular cloud cores. In this scenario massive stars can spend up to 2-2.5 Myr forming, sensibly contributing to the integrated far infrared (FIR) luminosity of the Galaxy. We find that many peculiar observational facts can be accounted for, if massive stars undergo a preliminary nuclear evolution while still accreting matter from the surrounding, optically thick placental medium.

The formation and early evolution of massive ($M \geq 10 \, M_\odot$) stars have challenged right from the start our theoretical schemes, since they totally lack of an optically recognizable pre main sequence (MS) stage. Indeed it is by now established that protostars do not appear high on their Hayashi adiabat, never exceeding in the Hertzprung-Russel diagram (HRD) an upper envelope broadly defined by strong CO outflows sources (Palla & et al. 1992). This very observational constraint has found a satisfactory interpretation in the framework of the accretion scenario, and it's the aim of the present contribution to show what its extension to higher final masses can tell us concerning the obscured evolutionary journey that bring massive stars into final view (Bernasconi et al. 1995).

Before doing this, let us just recall a few puzzling observational facts about the way massive stars are distributed at the luminous end of the HRD. With decreasing degrees of confidence we find that (Garmany et al. 1982, Massey et al. 1995):

- there is an apparent lack of the youngest (< 2 Myr) massive ($> 40 \, M_\odot$) stars close to the theoretical zero age MS (ZAMS);
- the upper mass limit for star formation does not seem to exceed $120 \, M_\odot$ in the Galaxy and Magellanic Clouds and is sensibly the same for all three galaxies, in spite of a difference of a factor of ten in their respective metallicity;
- comparison between the observed initial mass function (IMF) for massive field stars and associations show a systematic steeper IMF for the former.

Two simple equilibrium models of spherically symmetric clouds were constructed and solved numerically for their non homologous collapse to yield the accretion rate history. The isothermal (30 K) case gives a constant accretion rate of $10^{-5} \, M_\odot \, \mathrm{yr}^{-1}$. A modified density profile was also computed by interpreting the nonthermal contribution to the velocity field of giant molecular clouds and dense cores, expressed in the power law form $\sigma_{NT} \propto r^m$ ($0.3 < m < 0.7$), as arising from locally isotropic, turbulent motions (Myers & Fuller 1992). The supplemental pressure support within the cloud $p = \rho\sigma^2 = \rho(\frac{kT}{\mu m} + \sigma_{NT}^2)$ increases the available (now constantly rising) accretion rate, which can reach more than 10^{-4} M $_\odot$ yr^{-1} for the higher mass, still accreting stars.

The photospheric properties (or *birthlines*) of the accretion models are displayed in Fig.1 . For both cases the starting model consisted of a 0.8 M$_\odot$ core evolved canonically along its Hayashi track toward the deuterium MS. Until 2.6-2.75 M$_\odot$ the protostar stays fully convective in response of the thermostatic action of deuterium which burns at its equilibrium rate. The subsequent withdrawal of the convective envelope faster than the radiative diffusion timescale lets the core underluminous for its actual mass. When the luminosity front finally reaches the surface, the external shells expand and cause the models to follow a rather broad path in the HRD which defines an upper envelope for the optical appearence of low to intermediate pre MS stars.

The hydrogenic sequence is reached only at M=7.1-8.2 M$_\odot$ depending on the adopted accretion scenario, and from there on the models ascend the MS burning deuterium in a thin subphotospheric layer. The early departure of the birthline from the theoretical ZAMS toward lower effective temperatures is certainly the most prominent feature of the tracks. As seen from Fig. 1, the thermal scenario cannot account for masses higher than approximately 65 M$_\odot$, while stars as massive as 85-100 M$_\odot$ are inferred from HRD of open clusters such TR 14/16 or CygOB2 in our Galaxy. A far better agreement is however obtained by allowing turbulence as a supplemental pressure support in the parental cloud.

For conciseness we resume our present results in the following points:

1) in the accretion paradigm of star formation there is no place for an optical pre MS phase for masses M$\gtrsim$ 7-10 M$_\odot$, unless to accept accretion rates which appear inconsistent with the present physical state of molecular cores in the Galaxy. With the adopted parameters, the observed envelope for the stellar distribution is well matched, as are the positions of stars with outflows;

2) as noted elsewhere (Stahler 1985), low mass pre MS objects are to be regarded as younger than previously thought. In particular the viscous timescale associated with a depleting disk can become comparable to the quasi-static contraction phase leftover by the accretion interlude, and this can have important consequences on the possible residual accretion that young stellar objects (YSO) undergo in their elderly, visible stages;

3) an optically thick MS lasting 1-2 Myr is predicted for massive stars. All along this period their luminosity is effectively reprocessed into IR radiation by dust, sensibly contributing to the total IR power leaving the galactic disk. We may well ask which relationship link this compelling evolutionary stage to ultra compact HII regions;

4) accretion through a geometrically thin disk may effectively shield matter from the opposing action of radiation pressure. This circumstance could ensure the sole accretion timescale to be the dominant reason for a truncation in the IMF around 85-150 M$_\odot$, where hydrogen has been mostly burnt out in the stellar core. Moreover since the accretion rate has here been linked to the turbulent state

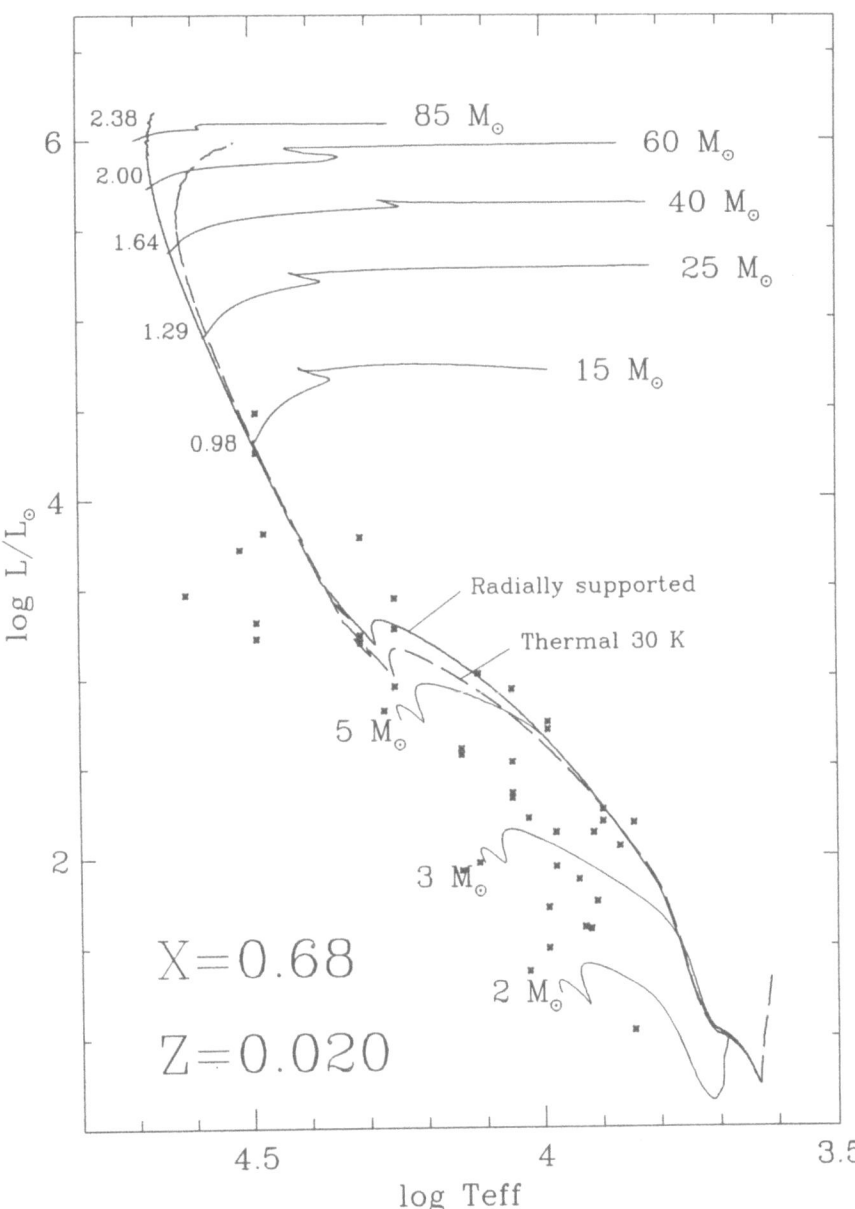

Fig. 1. Extended birthlines for two accretion rate histories (thick lines). The thermal scenario uses a constant $\dot{M} = 1 \times 10^{-5} \, M_{\odot} \mathrm{yr}^{-1}$ accretion rate. In the radially supported one, the accretion rate constantly increases and reaches $1.2 \times 10^{-4} \, M_{\odot} \mathrm{yr}^{-1}$ at 85 $M_{\odot}$. The accretion time for this latter case at masses corresponding to the displayed canonical post evolutionary tracks, are given in million years. Crosses trace the location of Herbig Ae/Be stars following Berrilli et al. (1992).

of giant molecular clouds, the universal character of a Kolmogorov type energy cascade gives us a universal, metallicity free mechanism able to account for the similarities of the upper HR diagrams in galaxies of various metallicities;

5) the observed IMF inferred by counting stars between theoretical mass bins do not reflect the true IMF, since as mass increases, stars appear in later stages of their nuclear evolution reducing their actual stay on the MS. In this picture then, the observed slope of the IMF is sensitive to the time history of the accretion rate, and consequently on the physical conditions prevailing in the molecular cloud out of which the stellar association has formed. If the possibly existing partition in the original cloud parameters are such as to produce higher accretion rates in OB associations than in massive field stars formed in relative isolation, then we would expect a steeper IMF for these latter;

6) as they first appear on the MS, massive stars are not chemically homogeneous objects. In particular the internal structure differs from that of a star with identical mass evolved canonically, since the size of the convective core is reduced by 5-10% as a consequence of prior nuclear evolution.

A number of cautionary notes must now be put forward if we want to correctly interpret the preceding results. The accretion tracks here obtained depend only on the time history of the accretion rate, and must really be viewed as a way to constrain $\dot{M}(t)$ in order to reproduce the distribution of stars in the HR diagram. Accordingly we need low (nearly thermal) accretion rates about 10^{-5} $M_\odot yr^{-1}$ to fit the position of YSOs, and constantly increasing ones somewhat in excess of 10^{-4} $M_\odot yr^{-1}$ to match the detached borderline of appeareance for early type stars. From this point of view, differing nonthermal supports in the cloud (other than turbulence, as magnetic fields) can possibly meet the numerical constraint on $\dot{M}(t)$. Finally let us stress that we are actually dealing with matter that succeeds in reaching the stellar surface, which can differ in rate by that taking part to the dynamical (flattened) collapse of the parental cloud and probably necessary to overcome the intense UV flux emerging from the central star.

Present efforts aim at coupling the mutually interacting dynamical evolution of the cloud-disc-star system. These simulations could give us a valuable insight into a still rather obscured period of the event-rich life of massive stars, and help in the interpretation of future high-quality observations of star forming regions in various perturbed environments.

References

Bernasconi, P. A., Maeder, A. 1995, A&A, in press

Berrilli, F., Corciulo, G., Ingrosso, G., Lorenzetti, D., Nisini, B., Strafella, F. 1992, ApJ, 398, 254

Garmany, C. D., Conti, P. S., Chiosi, C. 1982, ApJ, 263, 777

Massey, P., Lang, C. C., DeGioia-Eastwood, K., Garmany, C. D., ApJ, 438, 188

Myers, P. C., Fuller, G. A. 1992, ApJ, 396, 631

Palla, F., Stahler, S. W. 1992, ApJ, 392, 667

Stahler, S. W. 1985, ApJ, 293, 207

Dynamical Aspects of Dust Radiation Interaction Around Young Stars

Nicole Berruyer[1,2], Bruno Lopez[1,3], and Jean-Pierre J. Lafon[1,4]

[1] **Groupement de Recherches "Milieux Circumstellaires" du CNRS**
[2] Observatoire de la Côte d'Azur, Dpt Cassini/URA CNRS 1362, B.P. 229
06304 Nice Cedex, France
[3] Observatoire de la Côte d'Azur, Dpt Fresnel/URA CNRS 1361, B.P. 229
06304 Nice Cedex, France
[4] Observatoire de Paris-Meudon, DASGAL/URA CNRS D0335
92195 Meudon Cedex, France

Abstract. Young stars are surrounded by flat discs of matter including dust. Dust dynamics and so the disc structure are sensitive to several forces among which the radiation pressure plays a significant role. Conversely, the radiative transfer, which depends critically on the density distribution, is consistent with the dynamics. This is still an unresolved problem in which the present work tries to give a new insight. We successively check and discuss the effects of the structure of circumstellar discs, represented by the typical parameters appearing in the physical or empirical laws involved in the available models, onto the radiative force.
This is a preliminary qualitative and quantitative physical investigation necessary to allow dynamical modelization with good input physics.

1 Introduction

In spite of a lot of observations allowing semi-empirical descriptions of circumstellar dust envelopes of very young stellar objects, there is still no physically consistent model accounting for even the main characteristic features. Indeed, such models exist only in spherical or quasi-spherical symmetry, a situation irrelevant in the case of objects where matter is obviously concentrated in flat disc-shaped envelopes.

Dust provides an important contribution to dynamics and energy transfer in a strongly coupled self consistent system. A first step towards the understanding of the structure and the evolution of such discs is to analyse the relative strengths of the (numerous) forces at work, two of which certainly belong to the set of dominant forces: radiation pressure and gravity. The ratio β of the radiative force (F_R) to the gravity (F_G) was evaluated and discussed, showing that β is constant for an optically thin medium, so that it characterizes the whole disc and becomes a local parameter in an optically thick medium.

Using a numerical solution of radiative transfer throughout a flat disc, the exchange of momentum between the photons and the grains, through absorption or scattering, was calculated. The radiative force was obtained everywhere, taking into account the scattered photons and the contribution of the thermal radiation from the dust. The variation of the radiative force with the distance from star differs from the usual r^{-2} law, only valid for optically thin discs. The model is described in Sect. 2 together with the optical properties of the grains. Sect. 3 is devoted to the variations of the radiative force and of the gravity force. The main parameter is the total radial optical depth at $\lambda = 1\,\mu m$. In Sect. 4 the radiative force is compared to the gravity force in steady state. The ratio $\beta = F_R/F_G$ was computed for different disc structures. The conclusion (Sect. 5) is that the radiative force on dust competes with the gravity force.

2 Model

Observations suggest that the disc lies in the equatorial plane of the star with a physical thickness much smaller than the other typical sizes. Nevertheless, for simplicity of the model, constraints imposed by transfer and dynamics are not the same. Gravity is always radial so that, for this topic, an infinitely flat disc is the best and simplest description of the actual disc, but the radiative transfer is very difficult to solve under such conditions. A realistic compromise is a "quasi flat disc": we choose to solve the transfer in the framework of the so called "flare disc description".

The method of numerical simulation used for solving the problem of radiative transfer in a flat disc, is basically similar to those proposed by Lefèvre et al. (1982, 1983), then by Lopez (1994) for axisymmetric media. The stellar radiation and the thermal radiation from grains are described by individual photons. The star is characterized by an effective temperature $T_\star$. The dust disc has inside and outside boundaries at radial distances R_{in} and R_{out} respectively (in units of stellar photospheric radius). The grains are spherical with a unique radius. Three number density variations $n(r,\theta)$ were checked $(n(r,\theta) = n_0\, r^{-\alpha}$ with $\alpha = 1.5, 2.$ and $2.5\,; for\, n_0 = 0\; \theta > 8°)$.

The amount of dust is characterized by the extinction opacity defined along the equatorial radial direction. This is a kind of "flare disc":

$$\tau_{ext}(\lambda) = \int_{R_{in}}^{R_{out}} n(r,0)\pi a^2 Q_{ext}(\lambda)\, dr \ , \tag{1}$$

where a is the grain radius and $Q_{ext}(\lambda)$ the extinction efficiency obtained with Mie's theory. A Monte Carlo method is used to solve the problem of radiative transfer: "photons" trajectories are simulated, for each interaction with a grain a fraction of the energy leaves the disc, a fraction is absorbed and a fraction is scattered. The new direction of propagation is generated by a random process with a probability determined by the phase function $S(\theta)$. Since the momentum of the photon is known before and after the interaction, the momentum got by the grain is easily calculated. The disc is divided into concentric zones. Contributions from the direct stellar light, the scattered stellar light and the thermal radiation from grains are known everywhere in the disc and at all wavelengths. Integration over the spectrum gives the radiative force exerted on a grain.

Grains are assumed to be particles with the optical properties of Draine's (1985) "astronomical silicate". Spheres with radius $a = 0.1 \, \mu m$; $\varrho_S = 3.28 \, g \, cm^{-3}$ corresponds to the bulk density of a material which is probably rather fluffy in reality.

Star parameters: $T_{eff} = 5 \, 10^3 \, K$, $M_* = 1 \, M_\odot$, $R_* = 2 \, R_\odot$.

The internal radius of the disc is taken equal to $10 \, R_*$ as suggested by observations, with the assumption that the temperature does not exceed $1200 \, K$, $R_{out} = 10^4 \, R_*$. The total mass of the disc we used is $10^{-6} \, M_\odot$. Thus variations of α in the density distribution are correlated with variations of τ_{ext}.

3 Radiative Force and Gravity Across the Disc

When the optical depth in the disc is very low at all wavelengths, a grain at distance r from the star is illuminated by the unattenuated direct starlight and the radiative force exerted on it is:

$$F(r) = \frac{\pi^2 a^2 R_*^2}{r^2 c} \int Q_{pr}(\lambda) B_\lambda(T_*) \, d\lambda \; . \tag{2}$$

where Q_{pr} is the radiation pressure efficiency defined by:

$$Q_{pr}(\lambda) = Q_{abs}(\lambda) + (1- <\cos\theta>) Q_{sca}(\lambda) \tag{3}$$

$<\cos\theta>$ is the anisotropy factor for scattering. The respective contributions of absorption and scattering clearly appear: a photon which is absorbed gives a radial momentum $h\nu/c$ to the grain and a photon which is scattered in the direction θ gives $h\nu(1 - \cos\theta)/c$. Forward scattering provides no momentum while backward scattering contributes an amount $2h\nu c$. $F(r)$ varies as r^{-2} for optically thin media.

For higher optical thicknesses, $\pi B_\lambda(T_*) (R_*^2/r^2)$ must be replaced by the net outward flux $H_\lambda(r)$, which requires a detailed treatment of the radiative transfer throughout the disc.

The gravitational force is always radial. Its intensity is given by

$$F_G = m_{gr} GM_*/r^2 \quad with \quad m_{gr} = \frac{4}{3}\pi \varrho_S a^3 \tag{4}$$

where m_{gr} and ϱ_S are respectively the mass of a grain and the bulk density of the grain material.

4 Comparison Between F_F and F_G

Fig. 1 displays the variations of β versus the radial distance. Where the medium is optically thick, β is no longer a constant and is a function of the radial distance. This is due to the fact that, at each point, the net enlightening does not only result from the direct stellar light. The scattered stellar light and the thermal radiation from grains strongly contribute to the radiative force. Moreover, when the optical depth increases, the curves come very close to each other. Finally, β is rather underestimated because the bulk density of the material of the grains is probably closer to $2.2 \, gcm^{-3}$ than 3.2, due to porosity.

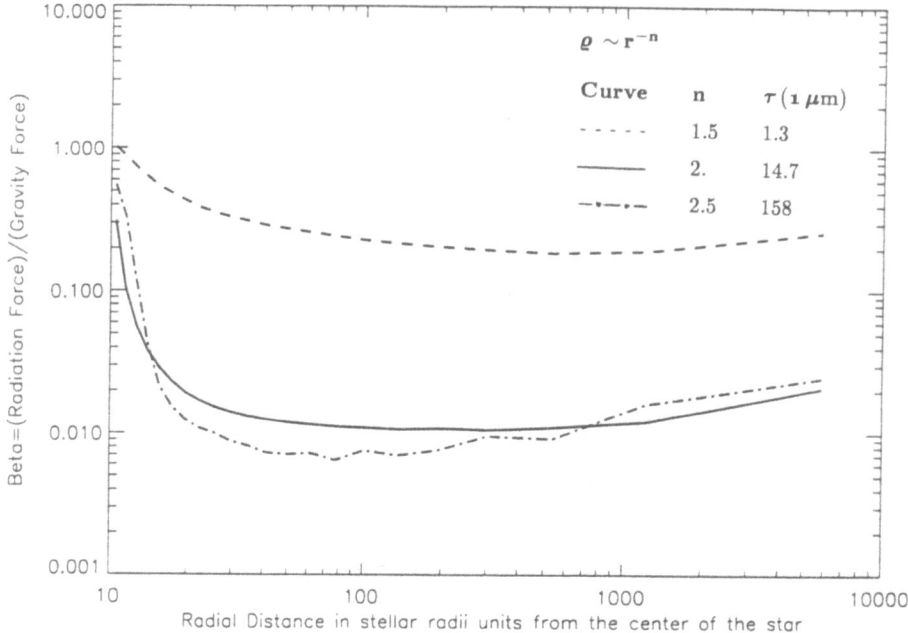

Fig. 1. Variations of $\beta = F_R/F_G$ versus the radial distance from the center of the star. The curves are given for three density profiles.

5   Conclusion

1. The radiation pressure decreases more rapidly than r^{-2} in thick discs. However the thermal infrared radiation weakens this effect.
2. The ratio β of the radiative force to gravity decreases with increasing optical depth.
3. The radiation force on dust competes with the gravity force in the inner regions of the disc.

Further improvements require careful and consistent modelling of the flow, including a description of other stellar phenomena responsible for the motion of matter, the evolution of grains etc.

References

Draine, B.T. (1985): *Astrophys. J. Suppl.* **57**, 587
Lefèvre, J., Bergeat, J., Daniel, J.Y. (1982): *Astron. Astrophys.* **114**, 341
Lefèvre, J., Bergeat, J., Daniel, J.Y. (1983): *Astron. Astrophys.* **121**, 51
Lopez, B. (1994): "Etude de la Perte de Masse des Etoiles Evoluées. Apport des Mesures à Haute Résolution Angulaire". Thesis, Université de Nice-Sophia-Antipolis

The Dust Size Distribution and Grain Shattering in Shock Waves

A.P. Jones[1,2], A.G.G.M. Tielens[1], D.J. Hollenbach[1], and C.F. McKee[2]

[1] NASA Ames Research Center, MS 245-3, Moffett Field, CA 94035-1000, USA
[2] Space Sciences Laboratory, University of California, Berkeley, CA 94720-7450, USA

Abstract. We consider the shattering of grains in grain-grain collisions, and follow the generation of small grains in supernova blast waves. We find that small grains are efficiently produced in shock waves, and discuss the implications of this for the interstellar size distribution.

1 The Interstellar Dust Size Distribution

The interstellar dust size distribution can be described in terms of power law size distributions (Mathis, Rumpl and Nordsieck 1977; MRN). These distributions give a good fit to the interstellar extinction curve and have the advantage that they are easy to deal with analytically. Recent work (Kim, Martin and Hendry 1994; KMH) fits the extinction curve, as a function of the ratio of total to selective extinction, using size distributions that are only roughly approximated by MRN power laws, and that extend to larger sizes with a possible exponential cutoff. The interstellar dust size distribution is thus reasonably well constrained by the extinction data, and may be the product of the balance between the formation and destruction processes that operate in circumstellar shells and in the interstellar medium (Bierman and Harwit 1980).

2 Grain Shattering Scheme

Based on models of crater formation and on experimental studies of micrometeorite impacts we have developed an algorithm for the shattering that occurs to grains in grain-grain collisions (Jones et al. 1995). In this scheme the shattered fragment mass distribution is a power law of the form $dn(m) = F_i m^{-\gamma} dm$, where F_i is determined by the total shattered mass, the index $\gamma = 1.77$ (slightly less steep than an MRN mass distribution power law index of 1.83), and the shattered fragment mass bounds are $m_{f-} \leq m \leq m_{f+}$. The lower fragment mass limit, m_{f-}, is taken to be that equivalent to 5Å grains, the lowest mass particles considered in our calculations, in order to ensure that the shattered mass is conserved. For non-catastrophic collisions (total mass fraction of target grain vaporised and shattered < 0.5) the upper fragment mass limit is $0.02 f_{sh} m_T$, and for catastrophic collisions it is $0.01 m_T (v_{cat}/v_r)^3$, where f_{sh} is the fraction of the target grain shattered, m_T is the target grain mass, v_{cat} is the catastrophic velocity for destruction of half of the target, and v_r is the relative projectile-target grain velocity. For any size distribution with an upper size limit of 2500Å the maximum fragment radius possible is 540Å. However, most of the fragment mass ends up in much smaller particles.

The threshold pressure for shattering is more than an order of magnitude smaller than that for vaporisation, and therefore grain shattering in grain-grain collisions will be the dominant grain mass redistribution process in shocks with velocities less than 200 km s^{-1}.

Using the above scheme to follow test particles through shocks of velocity $50 - 200$ km s^{-1} we find that secondary collisions, i.e., the collision of the shattered fragments with other grains, can be neglected. This is in agreement with the conclusions of Borkowski and Dwek (1995), and leads to a considerable simplification of the numerical problem. The grains are followed through a relatively simple mass-velocity space and we consider only the sputtering destruction of the fragments during their (post-shattering) deceleration with respect to the gas.

3 Small Grain Formation in Shock Waves

We have calculated the graphite and silicate grain destruction for MRN size distributions of homogeneous spherical particles (50Å $\leq$ radius $\leq$ 2500Å) for a range of shocks with initial velocities of 50, 100, 150, and 200 km s^{-1}, preshock densities of 0.25 cm^{-3}, and preshock magnetic fields of 3 μG (J. Raymond; private communication), typical of supernova shock waves in the warm intercloud medium. The results of these calculations are presented in Figure 1, where we show the percentage of the grain mass destroyed as a function of the shock velocity for the destructive processes of non-thermal and thermal sputtering, and vaporisation. Also shown in Figure 1 is the fraction of the initial grain mass that ends up in sub-50Å fragments.

In Figure 2 we show initial and final size distributions for grains that have been subjected to shocks with velocities of 50, 100, and 200 km s^{-1}. We note that, although the shattered fragment mass distribution is less steep than a MRN distribution, the final mass distribution is somewhat steeper than MRN for the 50 and 100 km s^{-1} shocks. The steepening of the final mass distribution is due to the cumulative effects of shattering over a wide range of grain-grain relative velocities. The most dramatic effects that we observe are the almost complete loss of grains larger than 1000Å from the size distribution accompanied by a shift of the size distribution to smaller grain sizes.

The small grain (radius < 50Å) formation timescale from the shattering of larger grains is essentially just the effective interval between low velocity shocks (i.e., 50 km s^{-1}) because of the low velocity thresholds for shattering in grain-grain collisions, and is of order ten million years.

4 Dust Destruction Timescales

Using the method of Jones et al. (1994) we have determined the lifetimes of dust against destuction in the interstellar medium, and find lifetimes of 6×10^8 yr and

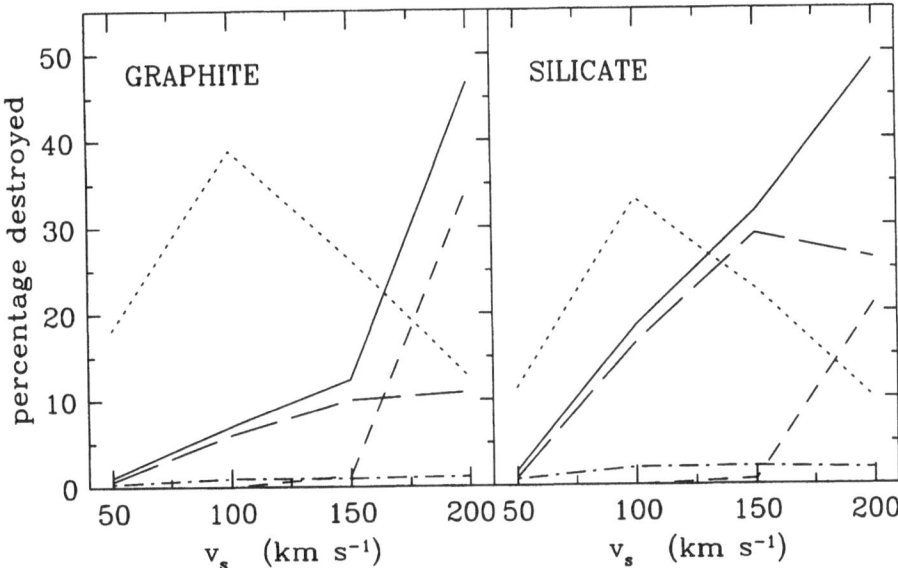

Fig. 1. Postshock graphite and silicate grain destruction (grain mass lost to the gas) as a function of the shock velocity, v_s, and destructive processes; total destruction (solid), non-thermal sputtering (long-dashed), thermal sputtering (short-dashed), and vaporisation (dash-dotted) destruction contributions, for fixed preshock density ($n_0 = 0.25$ cm^{-3}) and magnetic field ($B_0 = 3\,\mu$G). The dotted lines show the percentage of the initial grain mass that ends up in sub-50Å fragments.

4×10^8 yr for the MRN graphite and silicate dust, respectively. Additionally, the upper limit lifetimes for grains larger than 1000Å are 5×10^7 yr and 7×10^7 yr for graphite and silicate, respectively. These timescales are significantly shorter than the typical stardust injection timescales of 2×10^{10} yr.

5   Discussion and Conclusions

Our results indicate that the destruction timescale for dust in the interstellar medium is short compared to the injection timescale, for large grains (> 1000Å) the destruction timescales are even shorter due to shattering in grain-grain collisions. Thus, large grains must be efficiently re-formed in some phase of the interstellar medium and may be aggregates of smaller particles formed by grain-grain coagulation in dense clouds. If this is the case then the nature of the grains in shocks models must be re-visited using porous grains rather than solid homogeneous particles. We await physical models for the mechanical properties of these types of grains before attempting to include interactions between porous grains, and between solid grains and porous grains.

The final size distributions in our shattering calculations are steeper than those for MRN or KMH distributions, and extend to much smaller grain sizes.

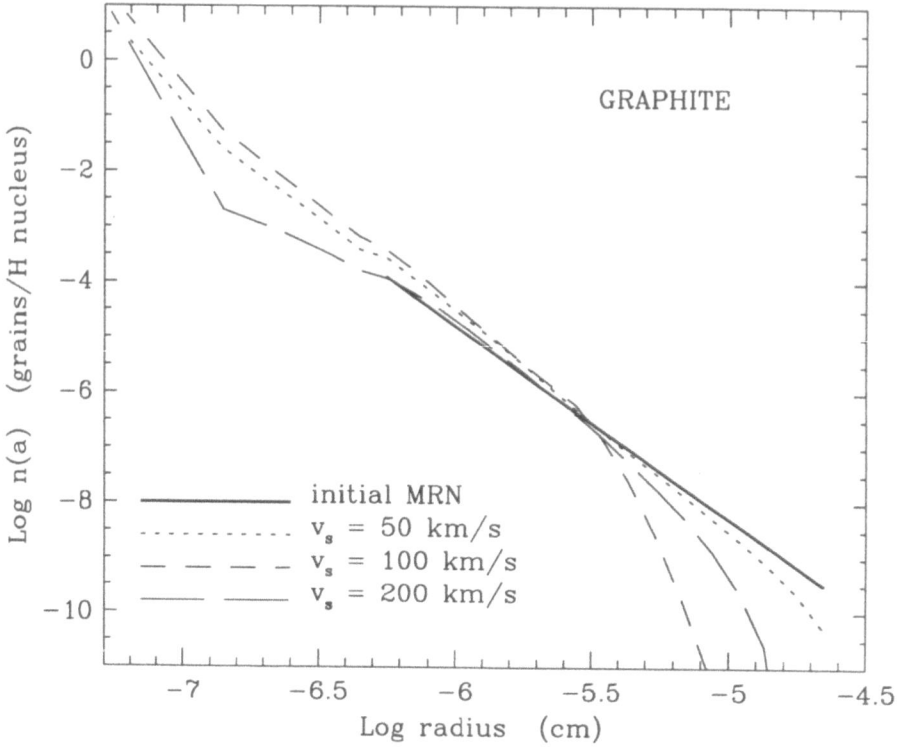

Fig. 2. Graphite grain initial MRN (solid), and postshock size distributions for shock velocities, v_s, of 50 km s^{-1} (dotted), 100 km s^{-1} (short-dashed), and 200 km s^{-1} (long-dashed).

Thus, the interstellar size distribution does not appear to be the result of shattering alone. A competing process that would flatten this shattered fragment distribution is coagulation in dense clouds. This would preferentially remove the smaller particles by accretion onto the larger particles and lead to a size distribution that may more closely approach the inferred interstellar size distribution, although this has yet to be demonstrated by viable grain coagulation models.

References

Bierman, P., Harwit, M. (1980): ApJ, 241, L105
Borkowski, K.J., Dwek, E. (1995): ApJ, submitted
Jones, A.P., Tielens, A.G.G.M., Hollenbach, D.J., McKee, C.F. (1994): ApJ, 433, 797
Jones, A.P., Tielens, A.G.G.M., Hollenbach, D.J., McKee, C.F. (1995): in preparation
Kim, S-H., Martin, P.G., Hendry, P.D. (1994): ApJ, 422, 164 (KMH)
Mathis, J.S., Rumpl, W., Nordsieck, K.H. (1977): ApJ, 217, 105 (MRN)
McKee, C.F., Hollenbach, D.J., Seab, C.G., Tielens, A.G.G.M. (1987): ApJ, 318, 674
Tielens, A.G.G.M., McKee, C.F., Seab, C.G., Hollenbach, D.J. (1994): ApJ, 431, 321

Dust Grain Processing:
A New Computational Method Applied
to a Protostellar Accretion Shock

D. Krüger, A.B.C. Patzer, and E. Sedlmayr

Institut für Astronomie und Astrophysik, Technische Universität Berlin,
Hardenbergstraße 36, D-10623 Berlin, Germany

Abstract. By applying a new computational multi–component method for the modelling of dust grain processing to a protostellar accretion shock we gain insight on the exact evolution of the grain size distribution function in the postshock region, where various processes act size–dependently on the grains.

1 Introduction

The supersonic infall of gas and *dust* during the collapse of a molecular cloud core to form a protostar is associated with a high–density accretion shock, which stands at the protostellar disk. The accreting dust particles are processed within such shocks by various physical and chemical processes, which generally act differently on grains of different composition, size, and velocity. Consequently, the grain size spectrum is dramatically changed as the dust grains pass through the shock front in entering the protostellar nebula. This development of the grain size distribution function *can generally not* be investigated simply by following separately the motion and size evolution of a few selected grain sizes.

Adopting the dissociative accretion shock model of Neufeld and Hollenbach (1994) we present an application of a *new computational multi–component method* for the modelling of dust grain processing (Krüger et al. 1995), which also accounts for drift velocities of the grains relative to the gas flow. Thereby, we gain insight of how exactly the size distribution function of silicate grains evolves as it is being affected by the evaporation of heated grains and by grain sputtering.

2 Basic Equations of the Multi–component Method

The fundamental differential balance or conservation equation for the grain size distribution function derived by Krüger et al. (1995) accounts for discontinuous processes (e.g. coagulation and shattering; the number and/or the size of the affected grains are abruptly changed due to grain–grain collisions) and continuous processes (e.g. sputtering; smooth grain size changes arising from interaction between gas particles and grains). The essential idea of the computational approach is to discretize the size distribution on a *Lagrange-like grid in grain size space* which moves according to the size evolution of selected sample grains. In this way, the grid is self-adapting according to the evolution of the size distribution, which seems much more appropriate than the use of a static grid.

Some simplifications arise from the application to the adopted shock model. In a first step we do not consider possible nucleation or grain growth behind the shock front. Instead, the processing of a *given* initial size spectrum of spherical grains is investigated (cf. Sect. 3). Thus, dust particles of radius a_* are injected at the base of the flow N_H^0, where the calculation starts. The sole independent variable of the *plan–parallel shock wave* is the hydrogen column density N_H behind the shock front. $v_d(a, N_H)$ is the velocity of a dust particle of radius a at position N_H; $a_{N_H}(a_*)$ denotes the radius of the grain (at position N_H) whose initial radius was a_*:

$$a_{N_H}(a_*) = a_* + \frac{a_0}{3} \int_{N_H^0}^{N_H} \left[n_{<H>} v_d(a_{N_H'}(a_*), N_H') \tau_{net}(a_{N_H'}(a_*), N_H') \right]^{-1} dN_H' \ .$$

τ_{net} is the net rate of change in grain size and a_0 is the hypothetical radius of a monomer. Details of the gas density structure $n_{<H>}$ are given in Sect. 3. For the considered case the general balance equation simplifies to the following special form:

$$-v_d(a_*, N_H^0) f(a_*, N_H^0) + \frac{da_{N_H}(a_*)}{da_*} v_d(a_{N_H}(a_*), N_H) f(a_{N_H}(a_*), N_H) = 0 \ .$$

In order to evaluate this equation we follow the velocity and size evolution of an initial grid of grain radii $a_{*,i}$. Details of the numerical scheme, grid generation, derivation of macroscopic quantities of the dust component etc. are outlined in Krüger et al. (1995).

3   Model of the Dissociative Accretion Shock

We adopted the protostellar accretion shock model of Neufeld and Hollenbach (1994) behind a dense, dissociative J–shock of shock velocity $v_s = 50$ km/s, and preshock density $n_{<H>}^0 = 10^9$ cm^{-3}. The shocked molecular gas is treated as a single fluid, in which the gas (neutrals, ions, electrons, but **not** the dust grains) has one common temperature and velocity. In single fluid J–shocks the gas is heated to high temperatures immediately behind the shock front, and then cools radiatively. The gas temperature profile $T_g(N_H)$ can be characterized by two temperature plateaus and by two drops of the temperature at column densities $\sim 10^{16}$cm^{-2} and $\sim 10^{21}$cm^{-2} behind the shock front (see Neufeld and Hollenbach 1994 for details). Since the thermal pressure in the postshock region is approximately constant, the gas compresses as it cools, and the gas density structure can be approximately calculated from the temperature profile with $n_{<H>}(N_H) = n_{<H>}^0 v_s^2 \mu m_H / (k T_g(N_H))$. As has been argued by Neufeld and Hollenbach (1994), betatron acceleration of the grains is unimportant at the high gas densities occuring in this model.

We investigate the evolution of an initial MNR power–law size distribution of silicate grains (Mathis et al. 1977) influenced by evaporation of heated grains and by grain sputtering. The evaporation of heated grains is treated according to Neufeld and Hollenbach (1994). The equilibrium grain temperature is determined by balancing the effective cooling via IR emission against the total heating rate, which comprises absorption of UV radiation, heating by impact energy of inelastically scattered gas particles, and energy input due to temporary accomodation of gas particles. These grain heating processes are treated using the methods presented by Hollenbach and McKee (1979) and Draine (1986). Grain sputtering is described by using semianalytical yields derived from experimental data (Tielens et al. 1994).

4 Results and Discussion

Absorption of UV-radiation is the most efficient heating process. Heating due to collisions with gas particles is of less significance, since most gas particles are scattered elastically and simply bounce off the grain surface without transfering much energy. Grain vaporization sets in, if the heating rate per unit grain area exceeds the maximum IR emissivity of the grain at the vaporization temperature, which is approximately proportional to the grain radius, i.e. small grains are heated to their vaporization temperature more easily. Vaporization of heated grains turns out to be the dominant mechanism of grain destruction. There is only little mechanical sputtering of the grains, amounting to a radius change of $\sim 0.5\%$. Due to the moderate gas temperatures thermal sputtering does *not* affect the grains.

The **evolution of the grain size distribution** $f(a, N_H)$ is presented in Fig. 1 by a sequence of snapshots taken at various column densities N_H behind the shock front. Two alterations of f have to be distinguished. The small grains are completely vaporized, which causes a drastic reduction of the value of the size distribution for small radii a. The second effect is the compression of the dust fluid which subsequently follows the compression of the radiatively cooling gas (at $N_H \sim 10^{16}$ cm^{-2} and 10^{21} cm^{-2}, cf. Sect. 3) with a specific time lag due to the size dependent distance required to decelerate a grain. This results in a successive upward shift of f, where sections of the size distribution correponding to smaller grain sizes rise first. The shape of the *inbetween* size distribution functions during grain deceleration and vaporization, which **cannot** be described by a power–law (cf. Fig. 1), is a *computational result* of the multi–component method that **cannot** be obtained by simply following separately the motion of test grains through the shock. Macroscopic dust quantities, as for example the absorption coefficient κ_d of the dust component and the optical depth τ_d corresponding to κ_d, can be easily derived from the obtained size distribution by integration. The optical depth of the dust component is responsible for the weakening of the UV radiation, which in turn determines the grain temperatures and thus the evaporation of grains. In our calculation this interaction between the dust grains and the UV radiation is consistently accounted for.

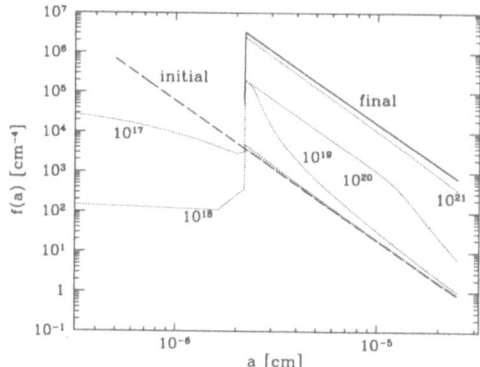

Fig. 1. Evolution of the grain size distribution as function of the particle radius a. $f(a)$ is depicted at various positions $N_H = 10^m$ cm^{-2} behind the shock front. The dashed line indicates the initial MNR distribution (Mathis et al. 1977). The solid line designates the final grain distribution.

5   Conclusions

In order to demonstrate the capabilities of the new computational multi–component method for the description of dust grain processing, we have investigated the evolution of the grain size distribution in a protostellar accretion shock. In this way, it has been possible not only to compute the final grain size distribution after processing by the shock but also to determine the shape of inbetween size distributions during grain processing, which cannot be described by a power–law. Essentially, we found that large grains above a certain limit size $a_{vap} = 2.2\,10^{-6}$cm, which depends on the grain material and is roughly proportional the UV-flux $F_s \approx 1/2\,n^0_{<H>}\,\mu\,m_H\,v_s^3$ emitted by the shock, are almost unharmed while those below a_{vap} completely vaporize. These results apply to silicate grains and also to other refractory materials such as iron. Ice mantles, on the other hand, will vaporize very quickly regardless of their thickness.

References

Draine, B. T. (1986): Multicomponent, reacting MHD flows. MNRAS **220**, 133–148

Hollenbach, D., McKee, F. (1979): Molecule formation and infrared emission in fast interstellar shocks. I. Physical processes. ApJS **41**, 555–592

Krüger, D., Woitke, P., Sedlmayr, E. (1995): A general multi-component method for the description of dust grain processing. A&AS **113**, 593–602

Mathis, J. S., Rumpl, W., Nordsieck, K. H. (1977): The size distribution of interstellar grains. ApJ **217**, 425–433

Neufeld, D. A., Hollenbach, D. J. (1994): Dense molecular shocks and accretion onto protostellar disks. ApJ **428**, 170–185

Tielens, A. G. G. M., McKee, C. F., Seab, C. G., Hollenbach, D. J. (1994): The physics of grain-grain collisions and gas-grain sputtering in interstellar shocks. ApJ **431**, 321–340

Electrical Charging of Dust: A Review

Jean-Pierre J. Lafon[1,2]

[1] Groupement de Recherches "Milieux Circumstellaires" du CNRS
[2] Observatoire de Paris-Meudon, DASGAL/URA CNRS D0335
   92195 Meudon Cedex, France

Abstract. Dust grains, as well as larger bodies embedded in plasmas and submitted to energetic photons, carry an electric charge depending on many physical processes: plasma particle collection, secondary emission, photoelectron emission, field emission, etc. The equilibrium charge, if any, strongly depends on the nature of the grain material, the state of the grain surface, the fluffyness of the grain, the photon spectrum, the plasma parameters etc. which can be characterized by unidimensioned typical numbers. All these effects, as well as the physics involved and the consequences for the dynamics, thermodynamics, chemistry etc. of the grains and the gas are reviewed and discussed in detail in relation with what could be observed with the VLT.

1 Introduction

Electrical charging of body surfaces is a very common problem in astrophysics (Spitzer, 1941). Indeed, the ambient medium is practically almost always more or less ionized and/or crossed by more or less energetic photons so that charge neutrality cannot be satisfied, due to a lot of processes contributing to its violation. Of course, the problem is similar for very small bodies like dust grains, whichever their nature and their composition.

However, in spite of many and many works devoted to it starting at the beginning of the century, this is still an unsolved problem: only the main physical processes involved and the orders of magnitude are fairly well known, while the details of the behaviours in various particular cases are still under discussion; besides, new processes at work are still discovered and investigated, with subsequent new unexpected behaviours of charged bodies.

In this paper, we shall concentrate on dust grain charging, notwithstanding referring to works concerned with charging of bodies of larger sizes under similar conditions.

We shall successively discuss three problems: the processes through which dust grains can acquire electric charge, the amount of charge that a grain can carry in various astrophysical media, and the effects produced by electrical charging on the behaviour of the grains and of the ambient media.

2 Electrical Grain Charging

Electric charge neutrality on the surface of dust grains can be broken by physical processes of three types: deposition of charges by electrons or ions coming from the ambient plasma, if any, and striking the grain surface, emission of charged particles, ions or electrons by the grain surface under bombardment by ambient particles, and emission of charged particles from the surface submitted to external processes (photons, heating etc.).

Particles falling onto the grains are of two kinds: first, there are those coming directly from the gas, and then those backscattered after a first contact but without enough energy to get free and leave definitively the grain surface. Subsequent backscattering can also produce new impinging particles, and so on.

A first kind of charged particles leaving the grain is made of those purely reflected by the grain surface (specular reflection), which occurs for particles with small energy. More energetic impinging charged particles produce backscattering, that is reflection with a different energy spectrum. For still higher energies, secondary (electron or ion) emission occurs with different energy ranges and spectra.

Finally, in the particular case of media with a porous structure, we recently discovered a new process able to strengthen the secondary electron emission, producing high emission yieds, and we called this process "Field Dependent Secondary Emission".

Other particles can leave the grain as consequences of other physical processes independent of the plasma particles surrounding the grain. Electrons can be emitted by heated grains as a result of an increase of the random thermal motion of the electrons in the solid (thermoemission). The grain is also often embedded in a photon soup (for instance in the interstellar medium UV photons from hot neighbouring hot stars); then photoelectric emission also contribute to grain charging. Finally highly charged grains generate close to their surfaces so high electric fields that electrons or even ions can be extracted by field emission.

The difficulties for computing the grain charge or potential come from the grain nature (chemical composition, optical properties), the grain structure (state of the surface, porosity etc.), its electrical properties (conductor or dielectric) and the geometry of the problem. Spherical symmetry is often assumed for grain models and also for particle collection.

This is compatible with an accurate description of some effects such as thermoelectron emission, which is related to the global temperature of the grain; such effects are important, for instance, close to the sun surface where they can change not only the intensity but also the sign of the charge. Grains can carry charges with opposite signs at the same place in the solar system; moreover grains which could have been destroyed by excessive internal tensile strengths due to strong charging may in fact be present close to the sun (Millet et al., 1980)

Of course, phenomena like illumination introduce asymmetries, even for spherical bodies; though the problem is not easy to solve, methods exist to analyse such cases and understand what happens, in particular concerning an accumulation of electrons on the enlightened side (Lafon, 1975a,b; 1976).

Now, spherical grains are only simple "theoretical" grains, i.e. models fairly far from reality, since actual grains have complicated shapes with "fluffy" surfaces and it can be shown that, while in some situations this does not invalidate the results obtained with spherical grains, it is not always the case, in particular for very small grains. Fluffy grains can be characterized by at least two parameters, the typical size of the protrusions from the surface in terms of the grain size, and a parameter characterizing both the degree of attachment of the protrusions to the grain surface and the enhancement of the electric field at the end of a protrusion (Lafon and Millet, 1984).

A general theory for deriving grain charge from ambient conditions has been developed by Lafon, Millet and Lamy in a series of papers (see references). At least concerning electrical charging, most of the grains have an electrostatic permittivity greater than 3, so that they usually behave as conductors, even if their optical properties, which depend on the dynamical permittivity, are those of dielectric bodies. It is only in the case porous grains for which FDSE is efficient, that dielectric electrostatic properties must be taken into account (Millet et al., 1989). In such cases, the yield of secondary electron emission (in emitted electrons per incident electron with a given energy) is strongly enhanced, which dramatically changes the surface charge; for instance, in media like Jupiter's magnetosphere, porous grains can be positively charged where bulk spherical grains would be very highly negatively charged without FDSE.

All these points, for a discussion of which room is missing here, are fully discussed in a forthcoming paper (Lafon, 1995).

3   Electric Charge on Grains in Astrophysical Conditions

Most of time, grain charging is analysed for spherical grains submitted to strong interactions with many particles or photons, so that the interactions are modelled on a statistical basis for large numbers of interactions: this leads to the assumption that charges carried by grains are more or less well described by a "continuous electric fluid" spread over a smooth conductor surface, so that an equivalent description can be developed in terms of a "surface potential". This is reasonable for a lot of cases, but is no longer meaningful for grain carrying small amounts of charges, since, physically, any charge is made of discrete particles with finite numbers of unit charges.

Continuous charging of continuous surfaces of bodies has been extensively discussed in the litterature. The general formalism has been given for bodies with any sizes by Lafon (1975a,b), and for small bodies (in terms of the smallest Debye length of the problem, and so for dust grains), by Lafon, Lamy and Millet (1981). Depending on the characteristics of the grains and also of the medium (e.g. density, ionization ratio, energy and intensity of the ambient radiation), which can vary by many orders of magnitude from the interstellar medium to the circumstellar, circumplanetary or interplanetary medium, the charge carried can vary from thousands of unit negative charges to thousands of positive unit charges, and the potential can reach hundredths of volts with both signs.

A broad oversight of the ranges of values reached by charges on dust grains in the interstellar medium is given by the analysis of Bel et al. (1989) for interstellar clouds. They give the first determination of the charge carried by grains consistent with the whole chemistry of the considered clouds. A weakness of the model lies in its simplified geometry: the clouds are represented by slabs of homogeneous distributions of matter (one or several of them) embedded in an isotropic flux of ambient photons. However, it seems that this is a small draw-back which does not alter significantly the main results (except, maybe, for very clumpy moderately diffuse clouds; see later), and this allows a complete treatment of the chemistry of the clouds based on about four hundredths of chemical reactions including all the components governing grain charging; among them the most important appear to be the electrons and the ions of H, He, C, Mg, Si, S, Fe. Other weaknesses of the model are that hydrostatic equilibrium is also assumed, which should be revised according to recent observations; moreover the grains are made of Draine's (1985) "astronomical silicate or graphite", which is not consistent with the fully discussed gas chemistry, but nothing better could be obtained at the present step of knowledge on processes of grain nucleation and growth. Finally, the gas temperature and the grain temperature are both estimated and not derived consistently with the chemistry; while this is probably not important concerning the gas, the results may be more sensitive to the assumed temperature of the grain that may depend significantly on the position of the grain inside the cloud through the way in which energetic photons can penetrate the cloud. This can be inferred from the results obtained with assumed grain temperatures.

In any case, one can basically distinguish three types of interstellar clouds classified using three main parameters to which the results are sensitive: the hot diffuse clouds, the moderately diffuse clouds and the dense dark clouds (or cloud cores). The parameters are displayed in order of decreasing sensitivity in table 1.

For clouds of the first class the electric charge carried by grains is practically constant throughout the (optically thin) cloud (except for a small decrease close to the edge). It is about 150 unit positive charges, weakly sensitive to the grain size for silicate grains with radius 0.1 μm in H1, and is roughly equal to half this value in H2. It is smaller by a factor larger than 2 for graphite. This illustrates the role of radiation in competition to particle plasma collection in grain charging.

The charge carried by grains decreases rapidly towards the center for denser clouds with higher optical thicknesses. The typical scale size for the decrease in moderately diffuse clouds as defined by the table is of the order of 1/4 of the thickness of the thickest cloud (M1), whichever the total optical thickness of the cloud (to which the actual thickness is closely related). This is still a problem of light penetration inside the cloud. This is confirmed by the increase by a factor 2 of the charge obtained when the ambient light flux is increased by a similar factor. All this may lead to some doubts concerning the results if the clouds are in fact clumpy, as suggested by some recent observations.

Dark clouds and cloud cores usually belong to the domain in which parameters are such that the charge is extremely small and the statistical "electric fluid" description used till now no longer holds and must be replaced by a new formalism.

Table 1. Dust charging in "typical interstellar clouds"

	Density ($H2/cm^{-3}$)	Grain temperature Kelvins	Optical depth at 547 nm
Hot diffuse cloud			
H1	10	100	0.05
H2	100	100	0.05
Moderately diffuse cloud			
M1	100	25	2
M2	100	25	1
M3	100	50	2
M4	100	50	0.05
Dense dark cloud			
D1	1000	25	1
D2	10000	10	5

It is out of the scope of this review to discuss the problems linked with computing charge in the case where it is of the order of a few units; some aspects of the problem have been recently discussed by Draine and Sutin (1987) and a full detailed analysis will be given elsewhere (Lafon, 1995). Let us only note that, in this case, the time variation of charge becomes more important, especially when the charge is close to the unit charge. Indeed, there are fundamental differences between the behaviours of a neutral grain and of a charged grain, and this qualitative and quantitative difference can be observed as soon as a unit positive or negative charge appears on the grain surface (see Sect. 4)

4   Examples of Strong Effects of Dust Charging

Grain charging may produce particular phenomena when magnetic field is present. For instance, if a grain carrying enough charge is not spherically symmetric, a rotation of its principal axis around some privileged direction aligned with the local magnetic field can be expected; this may explain the polarization observed in some magnetic interstellar clouds like those analysed in Sect. 3 (Bel et al., 1993).

Another example concerned with star forming regions is the coupling between neutral (or little ionized) gases and grains carrying no charge or a few unit charges, through ambipolar diffusion-like processes in dark cloud cores. In this case, charged grains with small Larmor radii are strongly coupled to the magnetic field lines and, since they are also strongly coupled to other lighter particles by collisions, the result is a (surprising) fairly strong coupling between a little ionized material and the magnetic field. Ciolek and Mouschovias (1993; see also this book) discuss models of cloud contraction under such conditions.

Grain charging can also significantly perturb the propagation of waves in circumstellar envelopes. It is expected that, at least for stars leaving the main sequence, and maybe for more evolved stars, mass loss can be initiated by Alfvén waves or acoustic waves, and also of course shock waves (Lafon and Berruyer, 1991).

Alfvén waves are transverse waves propagating in highly resistive ionized media where a magnetic field is present. Matter is coupled to magnetic field lines by ambipolar diffusion, so that magnetic field lines vibrate as material strings, and the higher the resistivity of the plasma, the better the coupling between the matter and the magnetic field, the smallest the dissipation rate of the waves: perfect Alfvén waves cannot be dissipated. Thus, it is obvious that the coupling between the matter and the magnetic field can be changed if it is due not only to ions and electrons, but also to charged grains. Indeed, the electric charge carried by grains is taken from the plasma (global charge neutrality is assumed): this alters the ionisation degree, while coupling is reinforced by charged grains which have collisionnal cross sections with the neutral particles higher than those of the ions and electrons. The effects are complicated if the charged grains have not reached an equilibrium charge, so that the electric charge they carry can fluctuate with time. Then, one can consider that the neutral grains and the charged grains are in fact different interacting fluids, also separately interacting with the ion fluids and the electrons.

An instance of the effects of grain charging on the propagation of circularly polarized Alfvén waves in dusty, weakly ionized media consisting of three gaseous fluids (electrons, ionized and neutral atoms or molecules) and of one size of grains that are either neutral or singly ionized, has been investigated by Pilipp et al. (1987). The results show that, of course, the effects are different for the two polarizations: the left hand polarized wave is less sensitive than the right hand polarized wave (this results from the fact that, in this example, the grains carry essentially no charge or an electron charge). Now, for the right hand polarization, there is some resonance frequency, at which the absorption rate is steeply increased, corresponding to a strong decrease of the transmission of energy (the intensity of the Poynting vector is decreased). Both the resonance frequency and the wave dissipation rate are altered by the presence of grains with fluctuating charges: in this case, the grains allow an easier dissipation of the Alven waves.

The propagation of acoustic waves in dusty plasmas has recently been also broadly investigated in relation with propagation of instabilities in planetary rings, mainly to find an explanation to the so called "spokes" (Melandso et al., 1993). When the plasma current is the dominant charging mechanism, the charge carried by the grain is strongly dependent on the dust and plasma conditions. In particular, if the surface charge is constant during a wave period, the dispersion relation is similar to the usual ion acoustic dispersion relation. Now, for low frequency waves, the wave frequency can be of the order of or less than the charging frequency of dust grains; then, a phase difference between the dust charge variation and the wave can lead to a strong damping of the wave. If, moreover, the ratio of the dust space charge to the electron space charge is large enough, the phase velocity of the wave can also be affected. Compressive dust acoustic waves can probably exist and propagate only locally within small regions of dust rings of the outer planets.

Let us at last look at the effects of the grain charge on grain dynamics. In the solar system, dust grains are always charged (Lafon et al., 1981). Thus, they interact with the magnetic field linked to the solar wind. Their trajectories in the interplanetary space results from four main forces: the gravity force, the radiative force (radiation pressure), the Poynting - Robertson force and the Lorentz force.

The Poynting - Robertson force, though of the order of a perturbation of the gravity and the radiative force, is of the order of the total force due to these competing effects. The Lorentz force comes from the coupling of the charged grain with the magnetic field carried with the solar wind. As a consequence of gravity and radiation pressure, the trajectory of an interplanetary grain is elliptical and can be characterized by its usual angle and size parameters (Kepler's dynamics). As a consequence of the Poynting - Robertson force, which is always dissipative, the elliptic trajectory is not stable: the grain always falls onto the sun, but very slowly in terms of the period of rotation along the ellipse. The tilt angle i of the plane of the ellipse with the ecliptic and the semi-major axis A of the ellipse also evolve with time, on time scales larger than the rotation period on the ellipse, under the influence of the Lorentz force.

Now, for any parameter, say z, one can define a characteristic evolution time (z / dz/dt). It appears that the various characteristic times of phenomena linked with the magnetic field are fairly well sensitive to the coupling between the grains and the magnetic field, which in turn is sensitive to the charge carried by the grain, and through it to the grain size. Finally, these characteristic times depend significantly on the grain size. The Poynting - Robertson force also depends on the grain size.

Mukai and Giese (1984) compared the characteristic evolution time of the ellipse under the Poynting - Robertson drag, to those characteristic of the evolution of the semi-major axis A and the tilt angle i, for various assumed (fixed) grain sizes, under the conditions of the solar wind at the distance of the earth from the sun. This allows to determine three size ranges in which the extrema time scales are not the same. For small grains (radius a smaller than 1 μm), the most and the less slowly efficient effects are respectively the Poynting - Robertson drag and the Lorentz variation of the tilt angle i. In the range of radii between 1 and 10 μm, the most (the less) slowly evolving parameter is the semi - major axis A (respectively the tilt angle i). Finally, for large enough grains, the most and the less slowly efficient effects are respectively the Lorentz variation of the semi - major axis A and the Poynting - Robertson drag! One can easily deduce from these results that the size distribution of the interplanetary dust grains is highly sensitive to variations of forces like the Poynting - Robertson force or the Lorentz forces, depending on the illumination and the magnetic field, and so depending on the position of the grains in the solar system.

5   Conclusion

Grain charging is an interesting but complicated field of physics with a lot of involved phenomena. Moreover, it has a lot of consequences for the behaviour of grains in circumstellar and interstellar media (dynamics, thermodynamics, chemistry etc.). In spite of many improvements of both theory and models obtained recently, better input physics is still necessary (physical processes), together with more accurate descriptions of the grains themselves (e.g. material, structure, properties) and also of the ambient plasmas (composition, densities, distribution functions).

The main difficulty to check the theories is that the grain charge cannot be directly observed or measured: it is only through observation (photometry, spectroscopy, imaging) of the most sensitive effects (spatial distribution, dynamics, thermodynamics) that it can be estimated: however, this implies using complex models. Now, laboratory experiments mimicking what occurs in astrophysical media but with media parameters that can be imposed now appear (this book). Modelling is also obviously in progress.

Together with observations at always higher spectral and angular resolution (VLT or VLTI, for instance are well fitted, especially for circumstellar envelopes) this provides a good hope for significant breakthroughs pretty soon.

References

Bel N., Lafon J.-P. J., Leroy J.L. (1993): Astron. Astrophys., **270**, 444
Bel N., Lafon J.-P. J., Viala Y.P., Loireleux E. (1989): Astron. Astrophys., **208**, 331
Ciolek G.E., Mouschovias T.C. (1993): Astrophys. J., **418**, 774
Draine B.T. (1985): Astrophys. J. Suppl. Ser., **57**, 587
Draine B.T., Sutin B. (1987): Astrophys. J., **320**, 803
Lafon J.-P. J. (1975a): Plasma Phys., **17**, 731 and 1175
Lafon J.-P. J. (1975b): Plasma Phys., **17**, 741
Lafon J.-P. J. (1976): Radio Science, **11**, 483
Lafon J.-P. J. and Berruyer N. (1991): Astron. Astrophys. Rev., **2**, 249
Lafon J.-P. J., Lamy P.L., Millet J.M. (1981): Astron. Astrophys., **95**, 295
Lafon J.-P. J., Millet J.M. (1984): Astron. Astrophys., **134**, 296
Melandso F., Alaksen T., Havnes O. (1993): Planet Space Sci., **41**, 321
Millet J.M., Lafon J.-P.J., Gonin J.C. (1989): Astron. Astrophys., **214**, 327
Millet J.M., Lafon J.-P.J., Lamy P.L. (1980): Astron. Astrophys., **92**, 6
Mukai T., Giese R.H. (1984): Astron. Astrophys., **131**, 355
Pilipp W., Hartquist T.W., Havnes O., Morfill G.E. (1987): Astrophys. J., **314**, 341
Spitzer L. Jr (1941): Astrophys. J., **93**, 369

Dust as a Tool to Study the Neutral Outflows from Luminous YSOs

H.J. Staude and T. Neckel

Max-Planck-Institut für Astronomie, Königstuhl 17, D-69117 Heidelberg, Germany

Abstract. The partially ionized jets often seen on the polar axis of bipolar nebulae associated with young stellar objects apparently carry only a minor fraction of the momentum flux required to accelerate their associated molecular outflows (See Staude and Elsässer (1993) for a review). This leads to the hypothesis that the jets are embedded within much broader primary bipolar winds leaving embedded YSOs and pervading their associated bipolar nebulae, which are are essentially neutral and therefore difficult to observe optically.

In search of a tool to study quantitatively the neutral outflows in the optical range, we looked for deeply embedded (IRAS Class I) YSOs associated with bright nebulosities showing strong $H\alpha$ emission. If $H\alpha$ is emitted by the unseen YSO and scattered by the dust in the associated nebular lobes, the variation of the line profile with position in the lobes reflects the velocity field of the scattering dust. If the dust is in motion, we can assume that it is carried along by a neutral flow and attribute to this flow the velocity field derived for the dust. The amount of scattering dust in the neutral flow can be estimated from brightness and color of the scattered stellar continuum. Using the standard dust-to gas ratio, this yields the amount of neutral gas in the flow and, together with its velocity field, the rate of neutral mass loss and momentum flux. The method works only with very luminous sources, whose mass loss is high enough for the associated dust to be observable.

The first object which we have studied successfully in this way is the bipolar nebula associated with the Class I source IRAS 08159-3543 ($L = 2 \times 10^4 L_\odot$). Analysis of our longslit spectroscopy and direct imaging shows: A neutral wind flows through both lobes at $v = 570$km/s. Within the entire volume Herbig-Haro emission results from internal shocks in the wind. From surface brightness and color of the scattered stellar continuum we derive the dust and neutral gas density within the outflow. The mass loss rate, $\dot{M} \approx 6 \times 10^{-5}...2 \times 10^{-4} M_\odot yr^{-1}$, fulfills the $\dot{M}/L$ relation derived by Levreault from CO observations of molecular outflows: on this basis we identify the neutral dust-carrying wind as the main accelerator of secondary molecular outflows. The fast dust in the lobes moves at the same velocity as the primary wind leaving the YSO: thus (momentum conservation!) it must have been accelerated near the YSO with the wind itself (rather than being entrained farther out). This result favours the model of a cold disk wind (as opposed to a hot stellar wind, in which the dust could not survive). A detailed account of this work is given by Neckel and Staude (1995).

References

Neckel, Th., Staude, H.J. (1995): IRAS 08159-3543: Optical Detection of the Dusty, Bipolar Wind of a Luminous Young Stellar Object. Astrophys. J. **448**, 832–847

Staude, H.J., Elsässer, H. (1993): Young Bipolar Nebulae. Astron. Astrophys. Review **5**, 165–238

The Distribution of CS and NH₃ in Star-Forming Regions

Stephen D. Taylor[1], Oscar Morata[2], and David A. Williams[1]

[1] Dept. of Physics and Astronomy, UCL, Gower St., London WC1E 6BT
[2] Dept. d'Astronomia i Meteorologia, Universitat de Barcelona,
   Av. Diagonal 647, E-08028 Barcelona, Spain

Abstract. The molecules CS and NH₃ are expected on theoretical grounds to trace high density interstellar gas in a similar fashion. However, observations show that CS is often widespread while NH₃ appears concentrated in dense cores. This result is interpreted here in terms of a clumpy model of molecular clouds, and a differential rate for chemistry of the two species. Crucial to this model is the accreting role of the dust particles. These CS cores are unresolved and numerous; most of them probably disperse rather than collapse to form a star.

1 Introduction

Since molecular cores were found to be associated with sites of low mass star-formation much effort has been concentrated on detecting such cores, particularly those in the process of collapse. These cores are dense, cold objects and hence tracers have been molecular species that are only excited at higher density, chiefly ammonia but also molecules such as HC_5N, H_2CO and CS . Zhou et al. (1989) looked for the CS J=2-1 and 3-2 transitions in a number of cores previously mapped in ammonia. They found the CS to be more widespread despite theoretically being excited at higher density n, and also that the CS linewidths were rather larger. In a systematic study of 16 cores Myers et al. (1991) found CS to be about twice as large in extent on average. Pastor et al. (1991) looked at a number of star-forming regions previously mapped in the NH₃ (1,1) inversion transition using the CS J=1-0 line. These transitions are expected to show emission at similar n - again they found the same lack of correlation. The effect seems to occur regardless of source distance, but of particular interest are their observations of the nearby (d ≈ 150pc) objects L1524 and L43, the maps of which clearly show that the CS emission is far more extended than NH₃ and that the intensity peaks of the two molecules are displaced from each other.

Separations in the emission peaks of molecules have been noted in the exceptionally chemically rich source TMC1, which appears to be a sequence of cores when mapped in CCS. Howe et al. (1995) presented chemical models of the collapse of such cores that confirmed how CS was likely to be associated with cores in earlier stages of collapse than those showing strong ammonia emission. That study has suggested the motivation for this work: can the different time-dependencies of the CS and NH₃ chemistries account for the different spatial distributions observed for these two molecules?

2   Gas Phase Chemistry of CS and NH₃

Since both molecules are used extensively to trace regions of possible protostar formation, it is important to realise that they are formed and destroyed in very different ways. Traditional chemical models of dark clouds (i.e opaque to visual and UV radiation) have started from the premise that all elements are in atomic form except for hydrogen which is fully molecular, a situation that could have a physical basis if most dark clouds have collapsed from more translucent origins (Howe et al. 1995). An appreciable fraction of the gas may then be ionised, in the form of C^+, S^+ and metal ions.

In most circumstances CS is formed from its ion,

$$CS^+ + H_2 \rightarrow HCS^+ + H \tag{1}$$

followed by recombination to CS. CS^+ can be formed from atomic and ionised sulphur,

$$H_3^+ + S \rightarrow HS^+ + H_2 \tag{2}$$

$$HS^+ + C \rightarrow CS^+ + H \tag{3}$$

$$S^+ + CH \rightarrow CS^+ + H \tag{4}$$

Large abundances of CS can be formed because it has no neutral reaction destruction pathways through O and C unlike other sulphur molecules. Many reactions with molecular ions simply recycle CS, and the primary destruction occurs through dissociative charge exchange with helium ions.

Ammonia is formed from ions undergoing successive hydrogen abstraction reactions followed by recombination,

$$N^+ \overset{H_2}{\rightarrow} NH^+ \overset{H_2}{\rightarrow} NH_2^+ \overset{H_2}{\rightarrow} NH_3^+ \overset{H_2}{\rightarrow} NH_4^+ \tag{5}$$

where the atomic nitrogen ion is created by cosmic-ray ionisation. Ammonia is prevented from containing a large fraction of the total nitrogen by charge exchange with ions such as H^+, C^+, He^+, O_2^+, and by cosmic ray induced photodissociation.

3   Models of CS Emission

Two possibilities seem to exist that may explain the observations; either that contained within the beam are a number of unresolved clumps, or that a low density envelope surrounds a dense core. Fuller (1988) has discussed the latter extensively, and here we concentrate on a model that represents the cloud as a conglomeration of clumps, which is favourable for a number of reasons. It can explain the lack of coincidence of the molecular emission peaks and also the larger CS linewidth. It can continue to explain the observations even when the

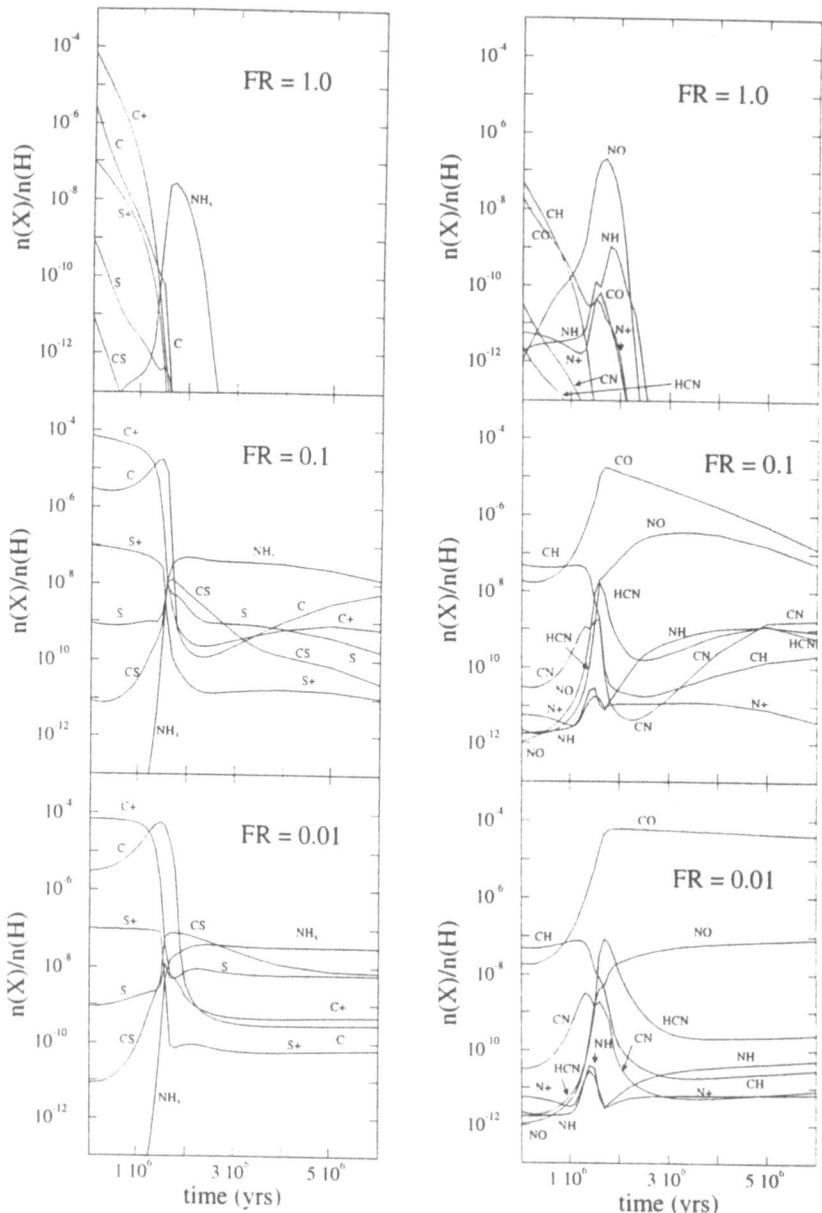

Fig. 1. Chemical fractional abundances (relative to H nuclei) as a function of time. (a) freeze-out parameter FR=1.0, corresponding to an average value of the product of the dust to gas number density ratio and square of the grain radius $<n_d a^2> = 2 \times 10^{-22}$ cm^2. (b) FR=0.1. (c) FR=0.01.

CS lines are optically thin, as appears to be very nearly the case in L43. Although a core-envelope chemical model shows CS to be much more abundant than NH$_3$ in the envelope much higher abundances could be attained in a clumpy medium.

We consider that the difference in emission between CS and NH$_3$ is due to the relative evolution of these clumps. This evolution is both chemical and physical due to differences in density or in collapse stage. The collapse may occur from diffuse conditions that would imply that only hydrogen is fully molecular. The gas phase chemistry of section 2 is supplemented by including the effects of freeze-out of molecules onto dust grains. The sticking efficiency and changes in the surface area and number density of grains from a canonical value are combined into a single parameter FR. No mantle desorption processes have been included. We model the collapse as free-fall, but halting at an arbitrary density that in practice depends on the balance of forces. Details of the chemistry and dynamics of the model can be found in Taylor,Morata & Williams (1995).

Fig.1 shows results for a collapse with initial density 1000 cm^{-3} and visual extinction 0.5 (edge to centre),and final density 5×10^4 cm^{-3}. C,Mg and S are taken to be ionised initially. Each graph shows chemical profiles (abundance per hydrogen nuclei) as a function of time for different values of FR. For FR=0.01 both NH$_3$ and CS are able to achieve high abundances, NH$_3$ doing so at later times than CS. At FR=0.1 however freezeout affects CS preferentially whereas NH$_3$ is still able to achieve a high abundance before falling off as all species eventually do. At high FR neither species achieves high abundance. From this picture it can be seen that for a maximum fraction of CS to exist in the gas it is necessary for the effective rate of freeze-out onto grains to be not too rapid, and this being the case NH$_3$ only achieves its maximum abundance at > 2 Myr. Since the peak abundances are affected detrimentally by values of FR as low as 0.1 it seems very likely that rather than the sticking efficiency being very low there are in fact desorption processes occuring that limit the mantle growth. If the abundance of NH$_3$ at the peak of the CS fractional abundance is such that its emission is not observable it is reasonable to associate the relatively large extent of the CS emission with unresolved cores that have existed on timescales of 2Myr or less, and which therefore have not had time to form large quantities of ammonia. Furthermore, rather than remain static it is likely that the density of the collapsed core will continue to slowly increase over time hence increasing the column of ammonia further and rendering it observable. This increase in density will only offset the decrease in the CS column due to chemistry over this time, hence the peak in CS emission need not coincide with that of NH$_3$.

References

Fuller, G.(1989): Ph.D. Thesis,University of California, Berkeley

Howe, D., Taylor, S., Williams, D. (1995): accepted by MNRAS

Myers, P., Fuller, G.,Goodman, A.,Benson, P. (1991): ApJ,**376**,561

Pastor, J. et al. (1991): A&A,**252**,320

Taylor, S.,Morata, O., Williams, D. (1995): submitted to A&A

Zhou, S., Wu, Y.,Evans, N.,Fuller, G.,Myers,P. (1991): ApJ,**376**,561

Low Mass Star Formation in Globular Filaments: Evidence from Dust and Molecular Line Emission

Helmut Wiesemeyer[1], Rolf Güsten[1], Robert Zylka[1],
Dirk Fiebig[2], and Melvin C.H. Wright[3]

[1] Max-Planck-Institut für Radioastronomie, Bonn, Germany
[2] Institut für Theoretische Astrophysik, Heidelberg
[3] Radio Astronomy Laboratory, University of California, Berkeley

Abstract. Despite the number of solar mass protostars estimated from the local star formation rate that should be detectable with current receiver technology, only little observational evidence for low-mass candidate protostars undergoing collapse is available. We present results of a search for evidence of protostellar collapse in globular filaments, which led to the identification of two sub–solar mass candidates. Dust observations and radiative transfer modeling confirms the interpretation of the observed CS(2–1) profiles.

1 Introduction

According to the selection criteria of the survey by Schneider & Elmegreen [1], globular filaments are characterized as condensations of markedly higher than average opacity. These fragments, appearing more or less periodically along an elongated dust cloud, are potential sites of low-mass star formation. As a first result of our systematic search for line profiles evidencing protostellar collapse, we present observations of GF9, a nearby ($\sim$ 150 pc distant) globular filament [2]. It extends over 5.9 pc (long axis, with an axial ratio as small as 1:30), has as total mass of $\sim$ 500 $M_\odot$, and contains 10 regularly spaced and dense ($\langle n_{H_2} \rangle \sim 10^4$ cm^{-3}) clumps of $\sim$ 10 $M_\odot$ each. They are best characerized by absorption measurements (HI, H_2CO), or by maps of their visual extinction, which are derived from star counts (Wolf diagram), using the Digitized Sky Survey [1]. The A_V maps will be used to continue our search for condensations in other globular filaments.

2 Observations

The single dish continuum data of GF9–3 were obtained with the 19–channel MPIfR bolometer and reduced with the *Mapping Software* package developed by R. Zylka [3]. For the interferometer data, an interactive, CLEAN based procedure

---

(Clark algorithm) was applied to each velocity channel separately, in order to deconvolve the maps from the interferometer response to a point source [2]. For the spectral line observations, the missing visibilities at short spacings were computed from the 30m CS(2-1) map and added to the interferometer data.

Table 1. Observational Parameters

transition	frequency (GHz)	telescope	Δv [1] (km s^{-1})	Beam (FWHM) (arc sec)
CS(2-1)	97.981	IRAM 30m	0.03	27
		IRAM PdB	0.24	5.2 × 3.8
CS(3-2)	146.97	IRAM 30m	0.04	18
C^{34}S(2-1)	96.413		0.06 [2]	27
C^{18}O(2-1)	219.56		0.03	12
continuum	230	IRAM 30m		11.1
	99.5 [3]	IRAM PdB		4.8 × 3.8

[1] not corrected for autocorrelator function, [2] smoothed to improve signal to noise, [3] double sideband (each 500 MHz wide) average

3   Line Radiative Transfer

The velocity structure of collapsing cloud cores makes the standard microturbulent or LVG approach useless, unless justified as a valid approximation. The LVG case fails, as even for a narrow linewidth, two distinct regions along the line of sight are interacting radiatively. Therefore, the level population is sensitive to the velocity structure (under sub-thermal conditions), which enters the rate equations by virtue of the solid angle averaged and frequency integrated radiation intensity, weighted by the absorption profile (assuming complete redistribution). To evaluate it, an adaptive integration algorithm is applied. The excitation modeling has been done under consideration of the collapse velocity field. The philosophy of a code developed by L.G. Mundy (private communication) is used: Physical quantities, such as density, abundance, velocity, kinetic and excitation temperature, are computed at discrete shell radii. The equation of radiative transfer is solved numerically by Gaussian quadrature along finer line-of-sight-elements. The quantities needed at the line of sight grid points are derived from the ones given as input, or resulting from the former iteration, by means of a spline interpolation. The residuals are minimized to ensure the

---

[2] using the IRAM *GILDAS Software Package*

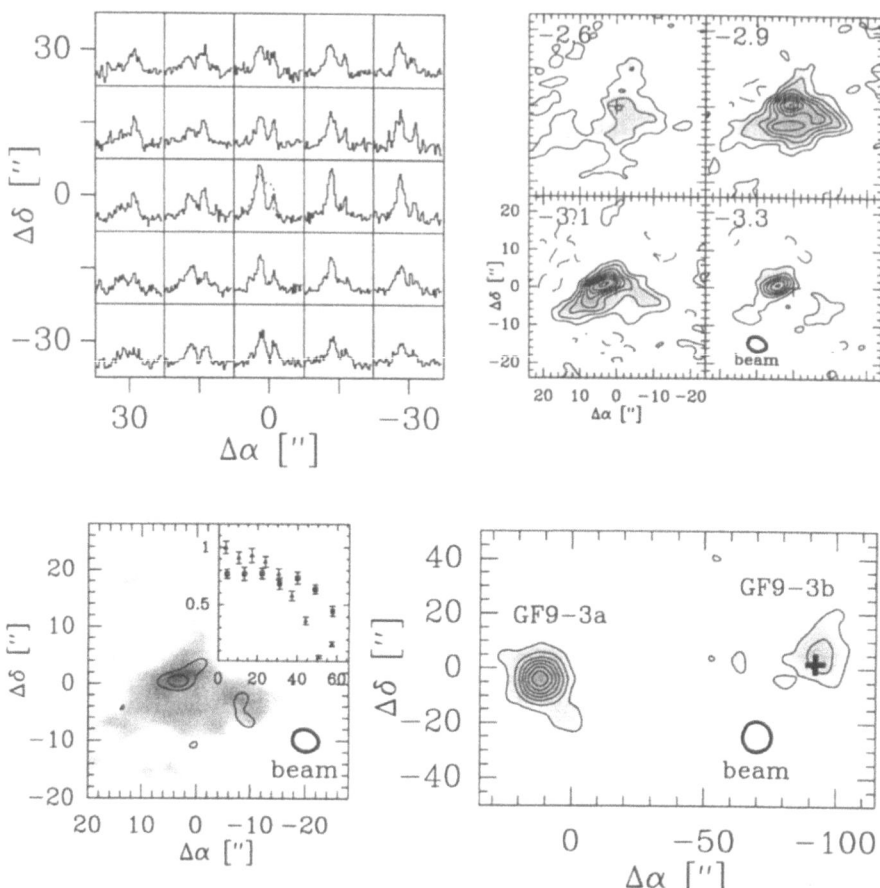

Fig. 1. Top left (a): 30m CS(2–1) spectra of GF9–2 ($v_{lsr}=$ −4 to −1 km s^{-1}, $T_{mb}=$ −0.5 to 2 K. Dotted spectrum at center position: C^{18}O (2–1) (×0.5). Top right (b): Interferometer maps, box marking: v_{lsr}. Contours: −50 to 400 by 50 mJy/beam. Bottom left (c): Velocity integrated interferometer CS(2–1) map. Contours: 50 and 75% level of λ 3mm dust emission. Insert: Visibility amplitude (rel. units), integrated parallel to the major (minor) source axis vs. uv distance (kλ) on minor (major) axis. d) 230 GHz map of GF9–3a/b. Cross: IRAS source position. Contours: 18 to 144 by 18 mJy/beam.

convergence of the Λ–iteration towards the desired solution [4]. The observed spectra (Fig. 1a) can be explained at least qualitatively in terms of an isothermal (20 K), collapsing envelope, given a central density as high as 10^6 cm^{-3}, a constant CS abundance of 6 · 10^{-9}, and a central infall velocity of 0.8 km s^{-1}. The squared linewidth varies as the inverse density. Within a 100 AU radius, all input quantities are kept constant. Within 1000 AU, the density is decreasing as $r^{-1.5}$, the radial velocity as $r^{-0.5}$. Outside this radius, a static envelope at modest (r^{-1}) density decline – to reproduce the spatial extent of the absorption feature – reaches out as far as 0.1 pc. Note that a large amount of foreground gas outside this radius is unlikely because of the small axial ratio of the filament.

4   Results

The search for evidence of protostellar collapse in GF9 successfully identified two candidates (GF9–2, GF9–4) [4][5] with the expected signature (Fig. 1a): Weaker emission at the peak that is red–shifted with respect to the source velocity, strongly asymmetric line profiles restricted to central lines of sight, and symmetric line profiles for optically thin emission. In the case of GF9–2 (a virial mass estimate yields ~ 0.5 $M_\odot$), the rotation of the core breaks the radial symmetry of the absorption feature with respect to the central region, allowing for an estimate of the projected rotational velocity. It is unlikely to exceed $\Omega \cos i \approx 10^{-12}$ s^{-1}, estimated from the linewidth at $\Delta\alpha = -30''$ (0.02 pc) from the center. The interferometric observations (Fig. 1b) evidence a rotation of the core at smaller radii ($\sim 2 \cdot 10^{-3}$ pc), which does not exceed $\Omega \cos i \sim 10^{-11}$ s^{-1} (estimated from the velocity dispersion). As expected, spherical symmetry is broken on these scales, evidenced by the dust continuum visibilities (Fig. 1c). The small–scale velocity integrated CS(2–1) emission coincides remarkably well with the dust emission, excluding an unusual anomaly of the CS abundance with respect to the dust mass. The results from the spectral line radiative transfer are only preliminary – it has to be modeled under cylindrical symmetry, in order to account for the flattened structure and rotation, as demanded for a crucial test of star formation theory.

The GF9–3 condensation (Fig. 1d) harbours a binary core, separated by 0.08 pc (projected distance), of total flux 380 and 220 mJy, respectively. GF9–3a appears as a point source, whereas GF9–3b is resolved (~ 1500 AU diameter). With a dust absorption cross section per H atom of 1.4×10^{-26} cm^{-2} at 230 GHz and a dust temperature of 20 K, the fluxes indicate a mass of 0.15 and 0.09 $M_\odot$, respectively. For the GF9–3a core, freeze–out onto dust grains, leading to a gas–phase underabundance of CS, becomes important, given the non–detection of CS emission, the low temperature (no IRAS source observed), and the high density. On the contrary, GF9–3b harbours a weak IRAS source, driving a diffuse outflow, and therefore is in a later evolutionary stage than GF9–3a.

References

1. Schneider, S., Elmegreen, B.G. 1979, ApJS **41**, 87
2. Fiebig, D., Güsten, R., Ungerechts, H. 1996, in preparation
3. Zylka, R. 1995, *"Pocket Cookbook" for the Mapping Software*
4. Auer, L. 1987, in *Numerical Radiative Transfer*, ed. W. Kalkofen, Cambridge University Press, p. 101
5. Güsten, R. 1994, in *The Cold Universe*, ed. Th. Montmerle et al., Editions Frontières, Gif-sur-Yvette, p. 169

Part X

Concluding Remarks

Conference Summary

Christoffel Waelkens

Instituut voor Sterrenkunde, Celestijnenlaan 200 B, B-3001 Leuven, Belgium

The title of this workshop, *The Role of Dust in the Formation of Stars*, is extremely ambitious. I realized that when I tried to anticipate beforehand what would be the main topics to address in this summary. Listening to the contributions and reading the posters has confirmed how difficult the subject really is. Maybe not more than 15% of the contributed papers directly addressed the question on how the presence of dust affects the star formation process. The majority of the contributions concerned dust as a *diagnostic* for understanding star formation or were devoted to the study of dust properties for its own sake.

But it was a good title and an excellent workshop. The subject is of obvious importance, and throughout the scientific debate was lively and of high quality. The difficulty just is that so far we lack the observational tools to address the question directly. Answers will come from ambitious instrumental developments, and from the VLT in particular.

The presence of dust probably is not an absolute necessity for forming stars: the first generation of stars must have formed in a dust-free environment. The question is whether the chemical enrichment of the universe and the increasing importance of the dust component in the interstellar medium has affected the star formation process significantly. Theory seems to tell us that it should have in many ways: the presence of dust must affect the cooling processes in contracting clouds and their fragmentation, the importance of ambipolar diffusion depends on the presence of a dust component, in a dusty environment radiation pressure assumes a much more prominent role.

It is then striking that the *observational* evidence is still so scarce. Present data do not suggest an initial-mass function that critically depends on metallicity. The rapid early chemical enrichment of our Galaxy proves the presence of an initial population of high-mass stars, but globular clusters of course do contain many low-mass stars. The observations of external galaxies do not point to IMFs which are very different from ours, although it may be doubted whether population synthesis models are a very critical discriminant in this respect. On one point, fortunately, theory and observations may readily agree: the role of dust is to form *planetary systems* (Artymovicz, this workshop).

The main conclusion of this workshop is then probably that an important task is ahead of us, a task which involves theory and modelling, laboratory astrophysics, and observations from the ground and from space.

The conclusion by Adams that the current star-formation paradigm by Shu, Adams and Lisano (ARAA **25**, 23-81 (1987)) is a *successful* paradigm, started a lively discussion which was not concluded today. Questions arose about the reality and nature of an inward-outward collapse and about the exact role of ambipolar diffusion, and about the precise relation between outflow and infall.

It appears that no unique scenario can explain both the formation of low-mass and of massive stars. The discussion about the interpretation of observations in terms of disks or envelopes is still ongoing, and it now appears that both probably coexist, at different distances from the star. Remarkable advances are being made in the development of radiative-transfer codes for circumstellar material. A particularly exciting field is the processing of dust in circumstellar envelopes and the interstellar medium, where a remarkable amount of formation and destruction processes play a role.

Personally I have been struck once more by the magic quantity which is eight solar masses. It sets the limit between high- and intermediate-mass stars, where high-mass stars are those that never develop a degenerate core during their main-sequence lifetime and hence go through the complete fusion chain. Eight-solar-mass stars are also about the most massive that *can* end up as white dwarfs, because the initial-final-mass relation for intermediate-stars yields a white-dwarf mass for them that is just below the Chandrasekhar limit. Third, the birth-line of observable pre-main-sequence stars intersects the ZAMS at eight solar masses, again. From Yorke's paper I have now learnt that it is also at eight solar masses that the effect of radiation pressure becomes important enough so as to counteract gravitational collapse and reduce the efficiency of star formation. If all this is just a coincidence, it is a striking one.

Several papers have underscored the importance of laboratory astrophysics for understanding the cosmic dust chemistry. It is well known how the rather tedious collection of basic atomical data is of fundamental importance for quantitative stellar spectroscopy. In a similar way, laboratory experiments are essential for our understanding of the formation and composition of cosmic dust. From the papers presented on this subject, I have understood that it is also a challenging field on its own, so that the continuation of these efforts is guaranteed, to the benefit of the whole community.

From the observational point of view, it appears rather clearly that the most important development in the near future is the launch of the *Infrared Space Observatory*, with which the ambitious Horizon 2000 program of the European Space Agency will yield its first scientific results. When these proceedings appear, the first ISO data may have become available, and there is every reason to expect that they will have an important impact on the subjects discussed in this meeting. However, the useful lifetime of ISO will be less than two years, and afterwards there will be a very heavy demand for follow-up observations. We should realize already now that no spaceborn successor of ISO has presently been commissioned: while NICMOS on HST will cover the near IR and ESA's FIRST the far IR, the thermal infrared will not be accessible anymore from space for quite some time. An important opening in this field would certainly be the airborne SOFIA project.

The good news is, however, that the possibilities of ground-based infrared astronomy are increasing rapidly. It is hardly necessary to recall how important this is for our field: the N-band is for the study of circumstellar and interstellar dust what the optical band is for the study of stars. In fact, as Ulli Käufl pointed out, more than 10% of the energy is radiated in this band by black bodies in the broad temperature range between 170 and 1000 K! There is a bright future for ground-based astronomy here, which deserves the development of the new instruments present-day technology makes possible. This workshop more than often has illustrated the crucial importance of the high spatial resolution which is achievable from the ground: without spatial resolution, an unambiguous interpretation of spectral energy distributions is not possible, without spatial resolution, young stars can hardly be distinguished from the companions they so often have. The workshop has also highlightened that we should be ambitious as far as spectral resolution is concerned if we want to understand the astrophysics of star formation and the chemistry of the dust.

In this context, our expectations for the VLT are especially high, for various reasons. Only a large-aperture telescope can provide the necessary angular resolution. This community is unanimous in pleading for the VLTI: many fairly bright objects are within reach of it, and spatial information is a crucial issue. A second major advance for our infrared needs is the prospect of having the VLT at the dry site of Paranal, which will offer conditions that can compete with those on Hawaii, and on the Canary Islands. Extremely important is also the southern location of the observatory: for our purposes, we are in a special location in the Universe, having the chance to be able to study star formation in two nearby galaxies with significantly different metallicities. Finally, the ambitious IR instrumentation for the VLT, that was discussed by Dr. Monnet, should enable us to carry out the research we want.

Dr. Monnet told us that he welcomed our suggestions, but that we would have most chances being heard if they were very negative or very positive. It is clear that our attitude is the second one. But with emphasis on the urgency that these nice prospects become reality as soon as possible. It is obvious that ISO will generate a heavy demand for ground-based infrared studies, and that this will happen before even the first VLT instrument is operational. Any development in infrared imaging and spectroscopy also on La Silla will turn out to be beneficial for an optimal exploitation of the unique opportunities ISO will soon offer to the European astronomical community.

It is appropriate to thank the two organisers of this most successful meeting, Ulli Käufl and Ralf Siebenmorgen. The attention of the participants during every session shows how right they were to choose this subject and how perfectly the workshop was organised. We also thank ESO for hosting this conference. Many colleagues will regret not having been here. In fact, the competition to get a place at the remarkable workshops ESO organises, is becoming as hard as that to get telescope time at La Silla. For the collection of more photons, ESO has been so wise as to undertake the VLT project. But I now wonder whether it will be possible to exploit the VLT without a Very Large Auditorium here in Garching.

Author Index

Subject Index

Source Index